TROPICAL
FORESTS
AND THEIR
CROPS

TROPICAL FORESTS
AND THEIR
CROPS

Nigel J. H. Smith, J. T. Williams,
Donald L. Plucknett,
and Jennifer P. Talbot

Comstock Publishing Associates

a division of *Cornell University Press*

Ithaca and London

First published 1992 by Cornell University Press.

Printed in the United States of America

⊗ The paper in this book meets the minimum requirements of the
American National Standard for Information Sciences–Permanence of
Paper for Printed Library Materials, ANSI Z39.48-1984.

Library of Congress Cataloging-in-Publication Data

Tropical forests and their crops / Nigel J.H. Smith . . . [et al.].
    p.   cm.
  Includes bibliographical references (p.   ) and index.
  ISBN 0-8014-2771-1 (cloth : alk. paper). — ISBN 0-8014-8058-2
(paper : alk. paper)
    1. Tropical crops.   2. Tree crops—Tropics.   3. Forest products—
Tropics.   4. Forest genetic resources conservation—Tropics.
I. Smith, Nigel J. H., 1949–   .
SB111.T76   1992
631.5′23—dc20   92-52772

Contents

Preface

The plight of tropical forests has galvanized world attention. Taxi drivers, porters, dentists, doctors, and politicians in many lands are increasingly aware that something is wrong with development trajectories in the tropics that are triggering so much forest destruction. Concern has focused mostly on the negative impact of tropical deforestation on global climate, soils, and the people who depend on forests for their livelihood. The loss of species, particularly wild plants and animals, has also received considerable attention. Philosophers deplore the loss on moral grounds, while scientists agonize over the disappearance of species that have not even been described, let alone studied. A major argument is that humankind is losing potential new crops and possible cures for cancers and other illnesses. But scant attention has been paid to one of the most important contributions of tropical forests to the future welfare of people living in many developing countries: genetic resources among wild populations of economically important crop plants and their near relatives.

Wild gene pools of many tropical crops are vital to raising and sustaining yields and nutritional quality—and thus to improving the diet and income of farmers and consumers. Crop breeders, who have traditionally worked with the better-known domesticated gene pool, are increasingly turning to wild material to locate desired genes, and advances in biotechnology have removed some of the barriers to widecrossing. Wild populations of a cultivated species sometimes harbor genes for resistance to diseases or pests that are apparently absent in the domesticated gene pool. Forest cousins of domesticated crops can be used in crossing programs to transfer valuable traits or for rooting stock or grafting material to bolster productivity on large plantations and small farms.

Our discussion of tropical crops centers on perennial, mostly woody species that were domesticated in forests often thousands of years ago. We focus on crops with wild populations still growing in jungles, sometimes virtually indistinguishable from cultivated forms, as well as related species, the products of which are occasionally gathered for food or other purposes. To illustrate the contribution of tropical forests to agriculture, we sample crops grown for a variety of purposes, including beverages, fruit, starch, oil, resins, fuelwood, fodder, timber, spices, and nuts. Our coverage is pantropical and spans many botanical families. In our discussion, Central America encompasses all countries from Panama to Mexico. Crops of widespread dietary or monetary value receive particular attention, but we also explore crops of local importance. Our sample of several dozen crops is only illustrative; many others could have been chosen from the extensive list of plants domesticated in tropical forests.

Before discussing the genetic resources of each crop, we outline the crop's socioeconomic importance. We examine its uses, dietary significance if any, and where it is grown. We also explore the cultural fabric surrounding each crop to demonstrate how important domesticated plants are to people throughout the world. Lore, in particular, is discussed in relation to conservation. In depicting areas of cultivation, we distinguish between large-scale plantations, which are generally backed by modern breeding programs, and dooryard or small field plantings maintained by traditional farmers. Background information on the economic, cultural, and dietary importance of the crop provides a rationale for scientists to investigate the genetic resources available for improving the crop's productivity and helps establish some priority for policymakers who decide which crops merit particular attention.

In assessing the genetic resources for each crop, we examine the known extent of the gene pools, and we describe species distributions on the basis of literature reviews, discussions with botanists and agricultural scientists working with each crop, searches of records in herbaria, and ecogeographic surveys. Although in some cases information is fragmentary at best, we attempt to assess the status of wild gene pools by outlining their distributions with maps that distinguish between hard data and conjecture and by identifying deforestation threats.

Local people are important sources of information on populations of wild crops and their near relatives—they have often had long and intimate contact with tropical forests—and their observations on natural history can help scientists and crop breeders identify areas with rich assemblages of germplasm. We explore land use plans for forested areas that contain crop gene pools, and we establish the degree to which wild gene pools are protected, at least on paper, within the boundaries of parks or reserves.

Specific legislation, if any, against the felling of tree species belonging to wild gene pools of crops is identified.

The status of the domesticated gene pool of tropical forest crops is explored along two fronts: gene pools in the hands of traditional farmers and collections maintained at agricultural research institutions and other scientific centers, such as universities and botanic gardens. Traditional forms of a crop are expected to be diverse in the area where the plant was domesticated, but secondary zones of genetic richness are also common. As a crop spreads from where it was first brought into cultivation, new forms arise, some of which are spotted and maintained. We assess the degree to which traditional selections are being replaced by modern cultivars. Although overall, high-yielding cultivars have not yet resulted in a "genetic wipeout" of traditional cultivated forms, more breeding programs with tropical fruits and nuts, for example, are coming on line, and the potential exists for imminent loss of valuable traditional selections in some tropical cash crops.

Our examination of genetic resources maintained at agricultural research institutes concentrates on working collections for use by breeders and gene banks. Working collections are typically geared to current breeders' interests and are regularly grown to maturity or propagated. Gene banks contain materials in the form of seed under reduced moisture and temperature conditions, plants in field gene banks, or tissue cultures in laboratories. For each crop, we survey field, seed, and in vitro germplasm collections with respect to the numbers and sources of accessions, degree of duplication, extent of evaluation, and how comprehensive they are. We judge comprehensiveness according to the coverage of traditional and improved varieties as well as the wild gene pool. In so doing, we also identify collecting gaps for traditional varieties and wild material in field gene banks.

Tropical forest crops and their wild relatives in botanic gardens are also identified. Although botanic gardens rarely maintain more than a few individuals of each species, collectively they often represent a valuable sample of the genetic variation of a crop and its wild relatives because they contain material from widely scattered locations. To help justify their existence, some botanic gardens in the tropics are complementing the work of agricultural research programs by setting up germplasm collections of useful tree crops.

Although we concentrate on tropical woody crops, we also examine some cultivated species that arose in open environments within forest ecosystems. Tropical forest environments range from tall, closed-canopy forests in areas with heavy rainfall to more open formations with shorter trees in drier regions. Even in closed-canopy forests, second-growth species take advantage of light wells created by tree falls and lightning strikes as well as meandering avenues created by rivers. Farmers have been clearing tropical forests for millennia, providing sunlit pockets and even extensive patches for rapid-growing, and quickly changing, plant communities. Successionary tropical forest species that have been selected by people in the distant past include widely cultivated bananas and plantains, the much-appreciated guava, succulent papaya, and tart passionfruit. To some degree, human disturbances of rain forests increase the area of second-growth habitats, but that does not necessarily lead to an expanded gene pool for crops with ancestral homes in second growth. Although second-growth areas are usually not on priority lists for conservation, successionary communities have a rightful place in forest preserves, and they can be a gold mine of genes for crop breeders. We explore the conservation implications of ecotone and second-growth plant species important for human subsistence and commerce.

For some crops, little information is available about genetic resources, so coverage is necessarily brief. For others, such as avocado, more research has been done on the genetic potential of cultivated forms and wild relatives, so our discussion is more extensive. Length of text devoted to each species is thus not meant to imply greater importance of the crop to local peoples and regional economies.

We are grateful to the Rockefeller Foundation for supporting this project and to Peter Greening, former Executive Director of the International Fund for Agricultural Research (IFAR), for encouraging us to write this book. IFAR administered the project and Peter Greening's unfailing enthusiasm helped sustain momentum in numerous ways. Christopher Miller and Linda Moore obtained library materials for us, and Liz Brady was helpful in background research on leucaena. The Public Awareness Association of the international agricultural research centers provided a grant to the senior author to collect data for the book in Brazil and Costa Rica in 1990.

One of us (J.T.W.) acknowledges the International Board for Plant Genetic Resources for permitting him a six-month sabbatical in 1989. During this sabbatical, Lawrence E. Skog kindly granted unlimited access to the Smithsonian Institution's Department of Botany in Washington, D.C., and Ruth Schallert was most helpful. Colleagues at the USDA-ARS, Beltsville, and the National Agricultural Library also gave generously of their time.

We are grateful to Martin McKellar, Mark Plotkin, Richard Evans Schultes, and several anonymous reviewers for critiquing the entire manuscript. Piers Pool made a substantial contribution to the section on clove. The following individuals provided valuable inputs on specific parts of the book: P. Alvim (cacao), J. C. Ascenso (cacao, cashew, oil palm), O. Atkins (avocado), E. Barcelos (oil palm), R. Barnes (tropical pines), B. G. Bartley (cacao), E. T. Beauchamp (cacao), B. O. Bergh (avocado), S. R. Bhat (bananas and plantains, citrus), H. C. Bittenbender (coffee), S. Brandt (coffee), J. Brewbaker (leucaena), W. S. Castle (citrus), K. L. Chadha (bananas and plantains), C. Clement (peach palm), M. Coffey (avocado), P. J. Genú (avocado), D. C. Giacometti (avocado, cacao), F. Gmitter (citrus), J. Grosser (citrus), J. Hardon (oil palm), C. J. Hearn (citrus), E. Holcomb (citrus), D. V. Johnson (cashew, sago palm), R. Litz (papaya), E. Lleras (avocado), J. D. Mitchell (cashew), J. Mora Urpí (peach palm), S. Mori (Brazil nut, cashew), J. L. Pereira (cacao), R. Ploetz (bananas and plantains), H. Popenoe (avocado), N. Rajanaidu (oil palm), P. Rowe (bananas and plantains), K. Ruddle (sago palm), C. A. Schroeder (avocado), R. K. Soost (citrus), C. T. Sorensson (leucaena), D. Ragone (breadfruit), B. H. Waite (bananas and plantains), G. A. Zentmyer (avocado), and F. W. Zettler (vanilla).

During various travels, all of us visited many of the plant collections mentioned in this book. Robert J. Knight and Ray Schnell graciously found time in their hard-pressed work schedules to show N.J.H.S. germplasm collections at the Subtropical Horticulture Research Station, Chapman Field, Miami. Drs. Knight and Schnell also provided valuable background information on the history of accessions, breeding, and relevant literature. Likewise, Carl Campbell provided a guided tour of germplasm collections and access to data on accessions at the Tropical Research and Education Center near Homestead, Florida. In California, Gray Martin kindly showed N.J.H.S. the germplasm collection of avocado maintained by the University of California at the South Coast Field Station, Tustin.

In Costa Rica, Jorge Mora Urpí graciously arranged for N.J.H.S. to visit a peach palm plantation geared to the production of heart-of-palm for export. Also in Costa Rica, Jorge Leon generously made some time available to discuss the status of perennial crops in his country. In Kerala and Karnataka, India, P. K. Das of the Central Plantation Crops Research Institute showed N.J.H.S. germplasm collections of oil palm, arecanut, cacao, and cashew, and took him to several farms to observe and discuss the use of genetic resources and opportunities encountered by growers.

We are grateful to Cornell University Press—especially Robb Reavill, science editor; Helene Maddux, senior manuscript editor; Cynthia Gration, former marketing manager—and to Margo Quinto, copyeditor, for their encouragement and painstaking care in helping us prepare the manuscript for publication. Maps were prepared by Jan Coyne, Marilyn Ruiz, Steve Rogers, and David Bell of the Department of Geography, University of Florida.

The views, opinions, and interpretations expressed in this book are ours and do not necessarily indicate approval of any reviewers, collaborators, or organizations.

NIGEL J. H. SMITH
J. T. WILLIAMS
DONALD L. PLUCKNETT
JENNIFER P. TALBOT

Gainesville, Florida
Washington, D.C.

A Note on Technical Terms

To reach nonspecialists with information about the plight of genetic resources of tropical forest crops and their potential for improving the diet and income of people around the world, we have tried to avoid using highly technical terms. Scientific names of crops, near relatives, and relevant pests and diseases are usually given the first time the species is mentioned, and the common name is used thereafter.

Sometimes we needed to use words that may not be familiar to all readers. Below are definitions of these technical terms:

Accession	A plant sample maintained as seed or in vegetative form in a gene bank
Allele	A gene that exists in more than one form
Backcross	Crossing hybrid forms with a desired parental genotype, often many times, in order to arrive at a suitable, improved type
Clone	Plant material that is propagated vegetatively
Cryopreservation	Storage of plant materials at $-196°C$
Cultivar	A crop variety; a cultivated plant can have many cultivars or varieties
Ex situ conservation	Conserving plant material away from its normal growing environment
Field gene bank	Maintenance of plant germplasm in outdoor plots
Gene flow	Exchange of genes between individuals of plant populations
Gene pool	Material that can be crossed with a species or form, thereby contributing genes to it

Genetic erosion	The loss of genes due to the disappearance of traditional varieties, wild populations, or near relatives
Genotype	A distinct and unique combination of genes in an organism
Heterozygous	A genotype with one each of two differing alleles
Inarching	A form of grafting in which stems are taped to encourage them to grow together; then one stem serves as root stock only
In situ conservation	Conserving plants in their natural setting
In vitro	Culturing of plant materials in aseptic or germ-free nutrient media under laboratory conditions
Marcotting	Inducing rooting along a branch so that the plant can be cloned
Orthodox seeds	Seeds that maintain viability when dried and stored at reduced temperatures
Outcrosser	A plant fertilized by pollen from another plant of the same species
Pathogen	An organism, such as a fungus, that causes a disease
Ratoon	Harvesting a crop several times before replanting
Recalcitrant seeds	Seeds that cannot maintain viability when dried and stored at reduced temperatures
Ruderal	A weedy plant that establishes itself in waste or disturbed sites
Second growth	Successionary vegetation that establishes itself once the original vegetation has been destroyed
Sport	A mutation on a vegetative portion of a plant, such as a bud
Swidden	Slash-and-burn farming in which fields are abandoned after several years of cultivation and are taken over by successionary vegetation, sometimes reverting to forest
Topworking	Cutting off an established stem to graft on new material
Widecrossing	Breeding a crop by crossing with other species, usually near relatives

Throughout the book, amounts of money are given in U.S. dollars.

CHAPTER 1

A Threatened Resource

Many cultivated plants important for food and income in developing countries arose in tropical forests. Some tree and shrub crops of the tropics figure prominently in industrial countries as well. A cup of coffee starts the day for many Europeans and Americans, macadamia and cashew nuts are popular snacks, and most consumers of chocolate reside in the First and Second worlds. Many drive to work, or take off and land, on tires containing some natural rubber. Jungles are also the ancestral home of highly nutritious Brazil nuts; of African oil palm, which is grown for cooking oil, soaps, and margarine; and of tropical fruits such as mango and avocado, whose markets are increasing in developing and industrial nations. The forests that gave birth to these widely cultivated, economically important plants, as well as many lesser known ones, are increasingly threatened by encroaching farmers, livestock grazers, loggers, fuelwood gatherers, miners, and reservoirs. Ironically, forest and second-growth communities that may contain wild populations or near relatives of crops are sometimes cleared to make way for the cultivated forms. Rich pools of potentially useful genes are thus shrinking throughout the tropics.

None of the tropical forest-born crops is a "staff of life" in the sense of cereals such as wheat, rice, or maize, or even in the sense of major root crops such as cassava, yams, taro, or sweet potato, but they nevertheless provide important dietary supplements in the form of vitamins, protein, fats, and carbohydrates. Some perennial crops, such as bananas, plantains, breadfruit, jackfruit, and sago palm, supply a substantial proportion of calorie intake in parts of Latin America, Africa, and the Pacific. Furthermore, perennial crops provide valuable cash income for small farmers to purchase basic necessities that they are unable to grow or produce themselves.

Perennial crops underpin the economies of many parts of the Third World. The ability of developing countries to repay their external debts will hinge, in part, on the productivity of export crops such as coffee, cacao, and rubber. Rubber and African oil palm dominate the economy of Malaysia, and these perennial cash crops provide significant income and employment in other parts of Southeast Asia. Coffee production is a major foreign exchange earner for dozens of countries, particularly in Latin America and central and eastern Africa. Yet the germplasm base of these tropical crops is poorly understood and may be shaky in some cases. Vast plantations of coffee, cacao, and rubber, for example, represent very limited parts of the gene pool and thus are vulnerable to widespread destruction by pests and diseases.

New challenges to perennial crop production in the tropics are constantly surfacing. Coffee rust arrived in Brazil from Africa in 1970 and reached northern South America by the early 1980s. Coffee rust depresses yields and may put a severe dent in coffee production in the Americas, just as it halted commercial coffee production in Sri Lanka in the 1870s. Since 1960, a new virulent form of another fungal disease, black Sigatoka, has threatened banana production in Latin America and Africa and is spurring a major turnover of varieties. Likewise Panama wilt disease forced the early retirement of the once widely planted 'Gros Michel' banana variety in the early part of this century. The ability of plant breeders to respond to such challenges rests in large part on the genetic resources they can fall back on to generate new, resistant varieties. Chemical treatments for disease and pest outbreaks are not always feasible because of cost, the rapid development of resistant pest and pathogen populations, and concern for the environmental impact of pesticides. Genetic resistance bred into varieties is often a more viable option than changes in management practices alone.

Tropical forests provide a wide variety of environmental services and extractive resources, including industrial products (Table 1.1), and their destruction has stirred concern on several fronts. Scientists, and increasingly the general public, are worried about the massive loss of species and the resultant curtailing of evolutionary paths, the disappearance of numerous species whose role in ecosystems is unknown, and the obliteration of species, many as yet undescribed, that could prove useful in medicine, industry, and agriculture. The scale of tropical deforestation has been compared to a mass extinction that rivals the die-off of dinosaurs and Pleistocene megafauna (Simberloff, 1986; N. Myers, 1988c).

Another concern is the possibility of global climatic change as a result of continued large-scale tropical deforestation, which accounts for close to one-fifth of carbon dioxide being released into the atmosphere (N. Myers, 1988a; Schneider, 1989). If the greenhouse effect takes hold, major clima-

TABLE 1.1

Environmental benefits and extractive resources of tropical forests

Environmental benefits
  Catchment protection: controlled runoff, water supplies, irrigation
  Conservation of plant and animal resources: recreation, tourism, national parks,
    game, gene pool preservation
  Soil eroision control: windbreaks, shelter belts, reclamation of eroded lands
Indigenous consumption
  Fuelwood and charcoal: cooking, heating
  Agricultural uses: shifting cultivation, forest grazing, medicines, mulches, fruits and
    nuts
  Building poles: housing, buildings, construction, fencing, furniture
  Pit sawing and sawmilling: joinery, furniture, construction, farm buildings
  Weaving materials: ropes and string, baskets, furniture, furnishings
  Sericulture, apiculture, and ericulture: silk, honey, wax, lac
  Special woods and ashes: carving, incense, chemicals, glassmaking
Industrial uses
  Gums, resins oils, and latex: rubber, tannin, turpentine, distillates, resin, essential
    oils
  Charcoal: dry cells, chemicals, polyvinyl chloride (PVC), reduction agent for
    steelmaking
  Poles: transmission poles, pitprops
  Sawlogs: lumber, joinery, furniture, packing, ship building, mining, construction,
    sleepers
  Veneer logs: plywood, veneer furniture, containers, construction
  Pulpwood: newsprint, paperboard, printing and writing paper, containers,
    packaging, dissolving pulp, distillates, textiles and clothing
  Residues: particle board, fiberboard

Source: Adapted from World Bank, 1978.

tic shifts are expected, including longer droughts in some areas, more severe hurricanes and typhoons in others, and rising sea levels.

The disruption of the lives of indigenous peoples who depend on tropical forests for their livelihoods as a result of development efforts has also
sparked protests. Indeed, the cultural integrity of traditional peoples and
the conservation of crop genetic resources are linked. Even predominantly
agricultural peoples in tropical forest environments still rely heavily on the
forest for a wide variety of products, including seasonal fruits and nuts,
firewood, game, and medicinal plants. Traditional societies typically maintain an impressive array of crop varieties, and their knowledge about wild
gene pools of crops and their near relatives is extensive.

Despite greater awareness of the ecological and cultural dangers of rapid
deforestation, the shrinking of crop gene pools arising from tropical deforestation has thus far received little attention. Practical reasons can be
offered for saving tropical forests, among them their value for agriculture

(Lande, 1988; Raeburn, 1990). The potential economic and cultural value of crop genetic resources in remaining tropical forests is a powerful additional rationale for forest conservation. But the paucity of careful analyses and in-depth surveys of wild and domestic gene pools for most tropical perennial crops hampers efforts to devise suitable conservation and management efforts.

DISTRIBUTION AND COMPOSITION OF TROPICAL FORESTS

Tropical forests girdle a substantial portion of the earth's surface along the equator and poleward until rainfall decreases to such an extent that forest gives way to savanna landscapes with widely scattered trees. In many cases, forests remain essentially continuous until they reach 15 to 20 degrees north or south of the equator, but in other areas a combination of cold offshore currents and rainshadow effects, as along the western seaboard of South America, permit deserts to penetrate close to the equator (Figure 1.1).

The most extensive area of tropical forest occurs in the Americas. The New World encompasses nearly half of the world's tropical forests. Starting from the Gulf coast of Mexico, a strip of forest of varying width occupies the eastern portion of Central America, and in the vicinity of El Salvador, both coasts of the isthmus. The forest continues into Colombia and along the Pacific coast into Ecuador. The towering Andes and the arid Caribbean coast of Colombia break the continuity of the forest, which picks up again in the Amazon drainage and in the upper Orinoco.

The Amazon and Orinoco basins contain the world's largest contiguous area of tropical forest. Brazil's coastal highlands create a rain shadow that prevents closed-canopy forest formations in much of the Brazilian Northeast and on the granitic shield. A relatively thin strip of tropical forest, containing many endemic species, hugs the Atlantic coast of Brazil, where moisture is intercepted by highland bluffs (Figure 1.1). Small outliers of tropical forest are also found in the western Caribbean where volcanic islands protrude high enough to force orographic uplift of trade winds.

The second largest extent of tropical forest, with close to a third of the world's total jungle area, occurs in South and Southeast Asia. Tropical forests in Asia are less contiguous than in Africa and Latin America, because no major land mass straddles the equator in that region (Figure 1.1). Asian tropical forests are separated into insular and peninsular patches by seas, oceans, gulfs, and straits, and this checkerboard pattern is one reason that they are so ecologically diverse. Asian forests are dominated by dipterocarps, comprising more than 500 species. Many of the important trop-

FIGURE 1.1 Approximate global distribution of tropical forests. Many parts of the tropical forest belt have been heavily disturbed by human activities. (Adapted from M. Jacobs, 1988:5; Keay and Aubréville, 1959; N. Myers 1985a; Whitmore, 1985.)

ical hardwoods used in furniture making and in construction come from dipterocarp forests. Southeast Asian forests have also provided many fruits, some of which are now cultivated extensively.

Tropical forests also extend into northeastern Australia, particularly Queensland. We include this relatively slender portion of the world's tropical forests because they contain some species of eucalyptus used for pulp and fuelwood in many tropical regions and because Queensland's forests are home to a tree of growing commercial importance: macadamia nut. Macadamia plantations are replacing numerous sugarcane fields in Hawaii, while in several parts of Central America, especially around Turrialba in Costa Rica, macadamia orchards are sprouting alongside plantings of coffee and other crops. Costa Rica now produces a high-quality macadamia nougat that sells briskly in stores and provides added employment opportunities in a country that constantly needs to create jobs for its youthful population. In the 1960s, there probably was not a single macadamia nut tree in all of Costa Rica, but the continued success of this industry hinges on a supply of high-yielding and resistant varieties to overcome ever-changing pest and disease conditions. Expansion of macadamia nut plantations will also hinge on producing varieties that are suited to different soil conditions.

Most of Africa's tropical forest occupies the Congo basin, but substantial stands are still found in West Africa (Figure 1.1). Because fewer geographic barriers separate portions of tropical forests in Africa, species diversity among plants is less marked than in tropical Asia or in the Neotropics (Whitmore, 1985:3). The main exceptions to the generally continuous belt of tropical forest in Africa are Madagascar, a large lenticular island off the continent's east coast, and outliers of forest in Uganda, Burundi, Rwanda, the Boma Plateau of Southeast Sudan and Southwest Ethiopia. The forests of West Africa are broken by a narrow wedge of tree-dotted savanna in Benin and Togo, caused mainly by a rainshadow effect created by the Guinea highlands, the high relief of Cape Three Points, the Mampong Scarp, and the Akwapim-Togoland Mountains (Harrison Church, 1960:23). Cold offshore upwelling currents during the summer also dampen convectional activity around the Dahomey savanna gap, and intensive slash-and-burn agriculture probably also contributed to reduction of tree cover along the coast between Ghana and Nigeria. Not surprisingly, many unique plants and animals are found in these isolated tracts of forest (N. Myers, 1985a:212).

Within each major tropical forest region, many forest types occur. A thorough review of the various forest types and systems for classifying tropical forests is beyond the scope of our focus on crop genetic resources, but some indication of the diversity of forest environments will provide a useful backdrop to the discussion of wild gene pools of crop plants. Tropical forests have been classified according to moisture regimes, landforms, and physiognomic characteristics. The fact that no standard or universal method of classifying tropical forests has yet emerged is probably an indication of the enormous lacunae in our knowledge of such environments. But two salient features stand out regarding tropical forests: they are ecologically diverse and species-rich.

Tropical forests range from mangroves along brackish coasts to fog-shrouded, relatively cool groves near treeline on mountain slopes. On tropical lowlands inland from the coast, forests can vary markedly in species composition over short distances as a result of differences in soils, drainage, and history of disturbances. Tropical forests can be punctuated by islands of stunted trees and thickets growing on nutrient-poor sands. In watersheds with substantial areas of heath forest, the streams and rivers tend to be heavily stained by tannins and other organic compounds, thereby producing dark waters such as the Rio Negro in Brazil. Seasonally flooded forests, containing a mixture of plants and animals radically different from those in upland forest, grow along many rivers with gentle gradients.

Tropical forests are the most biologically diverse environments in the world, and it is not surprising that they have been the source of so many crop plants. More than 200 perennial species, excluding medicinals and ornamentals, have been domesticated from tropical forest plants (Appendix 1). The forested areas of temperate and cold climates, particularly in northern Asia, are larger than the forest zones of the tropics, but they contain far fewer species, and only a handful of crops, such as pear, apple, cherry, mulberry, and some timber species, are derived from them. Peoples in all tropical forest areas have adopted plants to secure a reliable supply of a variety of products. Some of the better known crops include avocado from tropical America, coffee from Africa, and mango and citrus from Asia (Table 1.2).

CENTERS OF DIVERSITY

The greatest diversity in a crop's gene pool is usually in its area of domestication (De Candolle, 1855, 1902). But the place of domestication often is not the place where the crop is mostly grown. Most of the commercial plantations of tropical forest crops are far from where the plants were first brought into cultivation, partly as a result of efforts to escape major pests and diseases. Thus the biggest producers of arabica coffee are Brazil and Colombia, thousands of kilometers from the crop's ancestral home in Africa. Conversely, West Africa produces more cacao than Brazil, where the crop was first gathered in the wild in Amazonia and eventually cultivated in backyards. The largest producer of African oil palm is Malaysia, not West Africa, where the majestic palm grows in the rapidly shrinking forest.

TABLE 1.2

Some major crops originating in tropical forests of Latin America, Africa, and Asia, and their main uses

Crop	Scientific name	Major uses
LATIN AMERICA		
Allspice	*Pimenta dioica*	Spice
Avocado	*Persea americana*	Fruit
Cacao	*Theobroma cacao*	Drink, confectionery, suntan oil
Caribbean pine	*Pinus caribaea*	Pulp
Guava	*Psidium guajava*	Fruit, juice, ice cream, jam, and flavorings
Leucaena	*Leucaena leucocephala*	Firewood, fodder, shade crop
Papaya	*Carica papaya*	Fruit, juice, meat tenderizer
Passionfruit	*Passiflora edulis*	Fruit, juice, ice cream, sherbet
Peach palm	*Bactris gasipaes*	Fruit
Rubber	*Hevea brasiliensis*	Tires, caulking, weatherproofing, shock absorbers
Soursop	*Annona muricata*	Fruit, juice, ice cream
Sweetsop	*Annona squamosa*	Fruit
Vanilla	*Vanilla fragrans*	Spice, flavoring
AFRICA		
Arabica coffee	*Coffea arabica*	Beverage, flavoring
Oil palm	*Elaeis guineensis*	Oil for cooking, margarine, and soap
Robusta coffee	*Coffea canephora*	Beverage, flavoring
ASIA		
Breadfruit	*Artocarpus altilis*	Fruit, staple food
Clove	*Syzygium aromaticum*	Spice, flavoring for cigarettes
Durian	*Durio zibethinus*	Fruit
Jackfruit	*Artocarpus heterophyllus*	Fruit
Lemon	*Citrus limon*	Juice, flavoring, jelly
Lime	*Citrus aurantifolia*	Juice, flavoring, jelly
Mango	*Mangifera indica*	Fruit, juice, ice cream, chutney
Nutmeg	*Myristica fragrans*	Spice
Rambutan	*Nephelium lappaceum*	Fruit
Sour orange	*Citrus aurantium*	Marmalade
Sweet orange	*Citrus sinensis*	Fruit, flavoring
Tangerine	*Citrus reticulata*	Fruit, flavoring liqueur

The Amazon basin is the home of the rubber tree, but Brazil imports rubber from Southeast Asia, where most commercial rubber plantations are located.

The fact that many commercial plantations are so distant from the ancestral homes, the locus of primary diversity, of the tropical crops they cultivate underscores the interdependence of nations on genetic resources. The fate of wild populations of arabica coffee in dwindling tracts of forest in Southwest Ethiopia is of major concern to the economies of dozens of countries in Latin America and Asia. Wild populations of cacao in the forests of Amazonia may contain genes for resistance to witches'-broom and swollen shoot virus, two diseases that plague cacao plantations in Brazil and West Africa, respectively. Brazil also contains most of the germplasm of rubber, and the future of Amazonian forests is thus of more than passing interest to Malaysia and Indonesia.

Much of the debate about who owns and controls plant genetic resources has been cast as a north-south tug-of-war between the industrial nations (north), which often grow food crops on a vast commercial scale, and developing countries (south), where much of the genetic diversity of crops is found but which are often chronic food importers. In the case of tropical forest crops, however, the linkages are strongly south-south. Cooperation among developing countries in sharing and conserving crop genetic resources is vital, considering the importance of cash crops to Third World economies.

BIODIVERSITY, DEFORESTATION, AND POPULATION GROWTH

Tropical forests represent only 7 percent of the earth's surface, but they contain more than half the world's biota (E. O. Wilson, 1988). Tropical deforestation thus has far more repercussions than destruction of an equivalent area of temperate forest. Tropical forests contain vastly greater numbers of wild populations of existing crops and potential crops than any other biome.

All tropical countries with forests are losing these complex and valuable ecosystems. Our examination of the rates of deforestation underscores the fact that time is running out for forests in some areas, and along with their loss will go plant and animal genetic resources whose value we may never know. In other parts of the tropics, sizable forests still stand, but much remains to be learned about their genetic resources and potentially useful plants.

Deforestation has accelerated dramatically in the tropics since World War II (Matthews, 1983; Richards, 1977). At current rates, developing countries will have lost close to 40 percent of their forest cover between 1978 and the turn of the century, whereas deforestation in industrial countries, has essentially run its course (N. J. H. Smith, 1981b:8; Westoby, 1989:89). Forests are actually gaining ground in parts of the United States and Europe. In some cases, genetic erosion has taken its toll from earlier clearing and replanting with higher-yielding timber genotypes, but natural succession is also advanced in many formerly farmed areas. New England has more forests now than in the days of the colonies, and second-growth woods containing a mix of hardwoods and pine trees cover old cotton plantations in the South. Logging is an active industry in North America, but there is little net change in forest cover because of extensive replanting. In Scotland, many glens are now in young coniferous forest in an attempt to reestablish tree cover cut long ago for fuel and construction.

Deforestation slowed down and essentially halted in industrial countries as their economies switched to fossil fuels in the latter part of the seventeenth century and early 1700s. London, for example, ran out of accessible fuelwood in the late Middle Ages just as coal was becoming available at a reasonable price. With a switch to fossil fuels, pressure on woodlands for cooking and heating fuel was reduced, and emerging heavy industries created an urban-based society. With fewer people on the land, population growth leveling off, and agriculture intensifying, the cultivated area actually retreated in some parts of the developed world. In North America and Western Europe, fewer than 4 percent of the population now live on farms. With the spasm of forest removal in developed countries largely over, a major concern now is not deforestation, but the loss of good agricultural land to urban sprawl.

The picture is vastly different in much of the Third World. Although some developing countries, such as Brazil and Mexico, are well along the road to becoming industrial nations with highly diversified economies, the populations of most Third World countries are still heavily engaged in agriculture and extractive industries. In Latin America, more than half the people live in villages and on homesteads. In Africa, close to three-quarters of the people live in rural areas, and in tropical Asia, about two-thirds of the people are directly engaged in farming or other primary activities. Most farming systems require forest removal, or at least substantial modification of forest environments, in order to grow crops. Herein lies much of the conflict between conserving crop gene pools in tropical forests and meeting the immediate needs of families.

Not only is a large segment of the population in Third World countries living in rural areas, but the population of all Third World countries is growing, in some cases rapidly. Population growth rates vary among developing countries, but excluding China, which has an annual population growth rate of about 1 percent, the population growth rate in the Third World averages 2.5 percent a year (Table 1.3). When the world's population levels off at around 10 billion in the latter half of the next century, 87 percent of the people are expected to be living in the area now encompassed by developing countries (Demeny, 1986).

Thirteen countries with tropical forests in the Third World have population growth rates of 3 percent, and two, Kenya and Côte d'Ivoire, have population growth rates of 4 percent (World Bank, 1987:15). Most of the high-growth-rate countries that have tropical forests are in Africa. Some

TABLE 1.3

Population growth in the Third World (excluding China), 1986

Region	Population growth rate (millions)	(percent)	Annual increment (millions)
Southeast Asia	414	2.2	9.1
Latin America	419	2.3	9.6
Indian subcontinent	1,027	2.4	24.6
Southwest Asia	178	2.8	5.0
Africa	583	2.8	16.3
TOTAL	2,621	2.5	64.6

Source: L. R. Brown and Jacobson, 1986:8.

countries, such as Zaire and Cameroon, still have extensive forests in spite of rapid population growth. Others, such as Uganda and Rwanda, are losing their remaining forests at a rapid clip.

How much forest will remain in tropical countries by the time population growth levels off is debatable, but further shrinkage is assured. Without concerted action, some countries, such as Côte d'Ivoire and Nigeria, may lose virtually all of their forests within thirty years (Repetto, 1988). It is hoped that intensifying agricultural production on the better lands and rehabilitating regions degraded by unsustainable farming and grazing activities will absorb or feed much of the expected population growth. But population growth rates are particularly acute in countries with tropical forests, some of which will surely give way to new farms, ranches, reservoirs, mines, and settlements. Virtually all countries with tropical forests will experience at least a doubling of their human populations by 2050 (World Bank, 1984). In many cases, populations are expected to triple or even quadruple.

Nigeria's population, which stood at 105 million in 1986, is expected to nearly double by the end of this century and burgeon to 500 million before it stabilizes (L. R. Brown and Jacobson, 1986:10; World Bank, 1984:74). Nigerian forests contain gene pools of African oil palm and crops of local importance as well as timber species, all of which might be useful candidates for replanting provided that a sufficiently broad germplasm base survives for breeders to work with. Ethiopia has safeguarded some of its remaining forest, but other patches containing coffee germplasm are being lost and the population of this predominantly highland country is projected to quadruple eventually. Although it is difficult to envisage how Ethiopia can support four times as many people as it currently has, a marked increase in population seems assured, and pressures will continue to build on the last wild stands of arabica coffee.

Although population growth rates are a little lower in Latin America, population pressures are building on forest resources. Mexico's population is projected to soar to nearly 200 million, some two and a half times its 1986 level before it levels off. Perhaps only a few patches of tropical forest along the Caribbean seaboard and in Oaxaca are likely to survive by then. At least eighteen crops have all or part of their wild gene pools in the lowland tropical forests of Mexico (Appendix 1). Significant portions of wild avocado germplasm, among other crops, will be one casualty of continued deforestation in Mexico.

If Brazil's population continues to grow at the current rate of 2.5 percent a year, it will reach some 600 million by 2050, up from its current 125 million (*Veja*, 15 July 1987, p. 68). Historically, the Amazon basin, which contains most of Brazil's tropical forest, has been sparsely populated. But the traditional isolation of Amazonia is breaking down as pioneer roads breach the forest perimeter and create avenues of colonization (Fearnside, 1986; Moran, 1981; N. J. H. Smith, 1982). The human population of the Brazilian North, home to the wild gene pools of at least two dozen crops (N. J. H. Smith and Schultes, 1990), is growing at 4.7 percent a year, in large part because of a flood of settlers. In Rondônia, for example, the influx of migrants from other parts of Brazil in search of land and jobs led to an annual population growth rate in that forested state of 14.6 percent during the 1970s (Fearnside, 1983). Few if any surveys of botanical resources are conducted in advance of pioneer forest highways in Amazonia; the few, usually rapid, surveys that are conducted focus mainly on stands of desirable timber for logging income.

The extent of tropical deforestation is serious in most developing countries. Only 10 percent of the forests that once covered Haiti, El Salvador, and Madagascar still stand (Daugherty, 1972, 1973). Forests of Central America (including Mexico) are among the most endangered tropical forest ecosystems on earth; some 400,000 hectares are cleared annually, representing close to 4 percent of the region's remaining forest (Tangley, 1987). Wild gene pools of avocado and vanilla, to name just two crops, are being rapidly eliminated in Central America.

A few fragmented patches of tropical forest on the Malaysian peninsula are expected to survive this century, mostly in reserves. In Amazonia, precise figures on deforestation rates are not available, but an estimated 48,000 square kilometers of Amazonian forest south of the equator went up in smoke in 1988 (Bonalume, 1989a, b; Booth, 1989; Roberts, 1988a). Only western Amazonia and the Congo basin are expected to emerge with sizable tracts of tropical forest remaining in the next century.

Global rates of tropical deforestation vary widely according to such factors as categories of forest considered and whether the estimate takes into account the likelihood of cleared areas returning to forest at some point. At the high end, a loss or serious degradation of forests in the tropics of

245,000 square kilometers a year has been suggested (N. Myers, 1980a). Other high estimates for the late 1970s range from 200,000 to 210,000 square kilometers (N. Myers, 1980b:175, 1985b), an area a little larger than the United Kingdom (*Nature*, 12 September 1985, p. 111). A more conservative figure, also for the late 1970s, is a tropical forest loss of 76,000 square kilometers a year, an estimate mainly of timber resources and considering outright destruction only (FAO/UNEP, 1982).

Between 76,000 square kilometers and 110,000 square kilometers of tropical forests are eliminated outright each year (N. Myers, 1988b; WRI, 1985a:3). Approximately 1 percent of the world's tropical forests is permanently destroyed each year, and another 1 percent is significantly altered by a range of human activities. In view of the paucity of reliable data, however, estimates of tropical deforestation on a global scale are extremely rough. Improvements in remote-sensing techniques, particularly with satellites, are likely to facilitate future attempts to delineate the extent of deforestation.

Substantial portions of the closed-canopy tropical forests have been cleared, logged, or otherwise degraded within the last few decades. Even estimates in the low range are cause for alarm from the viewpoint of crop genetic resources. Although forest degradation is not as serious as outright removal, heavy disturbance of tropical forests is likely to affect some crop gene pools. Many tropical forest species depend on insects, birds, and bats for pollination and on mammals and birds for seed dispersal. Removal of some trees from the delicate ecological web characteristic of tropical forests may destroy elaborate mechanisms for fertilization and dispersal.

DRIVING FORCES

Two main forces are propelling forest destruction in the Third World: population growth and the concomitant inability of societies to absorb this growth by providing sufficient jobs and livelihoods, and the need to extract raw materials for export and for domestic industries. Reliable data on the amount of deforestation resulting from different human activities are difficult to come by and only rough estimates are possible.

Proportionally, however, there is little doubt that farmers are responsible for more deforestation in the tropics than all other factors combined (J. C. Allen and Barnes, 1985; N. Myers, 1986). Farmers account for roughly half of the forest felling in the Third World every year, whereas the lumber industry accounts for approximately one-quarter of forest destruction or degradation annually (N. Myers, 1985a). In Africa, shifting cultivators are responsible for at least 70 percent of tropical deforestation (Lanly, 1982). The timber industry is particularly intense in Southeast

Asia, and farmers often move in after loggers and permanently convert the forest to farmland and second growth. Fuelwood gatherers account for some 10 percent of forest removal in the Third World, and their impact is particularly noticeable in drier forest areas and in highlands. Pasture development, concentrated in Latin America, is responsible for about 10 percent of tropical deforestation. In the Brazilian Amazon alone, pastures sown with African grasses have displaced at least 10 million hectares of forest (Hecht, 1985). Mining and reservoirs are the principal remaining causes of tropical deforestation.

Tackling the forces behind forest clearing in the tropics is a long-term task of economic development, conservation, and improving access to family planning. Simply suggesting that Third World governments arrest or slow deforestation is not helpful. Growing populations need room to raise food and cash crops, and minerals and timber earn valuable foreign exchange to help pay for development. Reservoirs, which in some cases drown substantial tracts of forest that contain wild gene pools of crops, are needed to generate hydroelectric power for industry and rural development. The problem of trying to reconcile the needs of expanding human populations and a growing appetite for natural resources with preservation of the remaining tropical forests is faced by governments across the ideological spectrum.

Strong economic arguments need to be offered, in addition to scientific and esthetic reasons, for protecting and managing tropical forests. We make a case herein that tropical forests hold invaluable resources for improving and stabilizing the yields of numerous tropical crops. More forest areas need to be set aside in relatively undisturbed states for perpetuity, while other forest lands can be managed for multiple purposes and still safeguard genetic resources.

One way to meet the growing appetite for resources in tropical forests and to exploit them without undercutting their capacity to provide food and income for future generations is to implement a strategy for replanting logged areas with timber trees, perhaps using a mix of species or varieties to reduce disease and pest pressure. In the past, most of the fuel and timber from tropical regions came from mature forests; in the future, such products will have to come increasingly from reforested areas or managed second growth. Vast areas of the tropics will eventually be planted in timber, fuel, and multipurpose trees, and attention needs to focus on the genetic basis for such ambitious undertakings. The scale of such repair work is daunting considering that deforestation currently outstrips reforestation in the tropics by a margin of more than 10:1.

Large-scale monocultures have a checkered history in the tropics, mainly because of disease and pest attack, so deploying a consortium of species in reforestation projects may prove more viable than planting with a single species. If only one candidate is to be considered for reforestation, then backup varieties should be waiting in the wings in the event that the first choice fails. Also, a desirable timber species is not likely to thrive on all soil types and under every climate of tropical forest areas, so either varieties will have to be tailored to suit varying ecological conditions or alternate, equally valuable species will have to be used. In both cases, foresters and nursery workers will need a rich germplasm base from which to mold suitable candidates. Reforestation schemes to help meet fuelwood and fodder needs will also depend on a diverse array of species and varieties to suit different cultural and ecological conditions.

Agroforestry is part of the agriculture-forestry land use continuum with interfaces between annual crops, trees, and livestock. Agroforestry is thus another way to help make people and the forest more compatible and is now recognized as a high priority for agricultural research in tropical environments (ICRAF, 1988). Agroforestry has been practiced by indigenous groups since the dawn of agriculture and is still the preferred farming method of many people in tropical forests today. But simply transplanting culture-specific agroforestry configurations from one indigenous group to other groups or settlers is unlikely to work. New configurations employing various combinations of arboreal and herbaceous crops will have to be devised to satisfy the myriad cultural, ecological, and economic conditions in the tropics. Breeding programs for many tree crop candidates for agroforestry, ranging from fuelwood species to fruit and timber trees, are still in their infancy or are nonexistent in some cases. An assessment of genetic resources available in germplasm collections and in the wild will help policymakers establish research and funding priorities for the agroforestry and forestry continuum.

CROP GENE POOLS

Some of the crops considered here are little different from their wild progenitors. Such closeness to their wild kin can be attributed either to recent domestication or to limited selection or breeding. Domestication can be seen as a broad spectrum of plant management and development, from tending spontaneous seedlings of wild or semidomesticated species that have germinated around villages and campsites to high-tech breeding operations in which varieties in many cases depend entirely on people for their survival.

Little breeding has been undertaken for most of the tropical perennial crops. For the most part, improved cultivars have arisen from spontaneous mutations, such as bud sports with oranges, or by chance seedlings. The farmer's keen eye for selection has been the key to progress for many tropical perennial crops, such as avocado and mango. And farmers alone have done most of the selection for some crops, such as breadfruit and rambutan. Nevertheless, modern breeding methods employing controlled crosses and biotechnology are beginning to be used for some of the perennial crops and are sure to be more common in the years ahead.

Breeders, whether they are working with cereals or perennial crops, dip into five main germplasm categories for desirable materials: breeders' collections, landraces or traditional varieties, conspecific wild and weedy materials, near relatives, and unrelated species. These five germplasm categories fall into three main gene pools: primary, secondary, and tertiary (Figure 1.2). The relative contribution of each gene pool varies considerably between cultivated species (Plucknett and Smith, 1987).

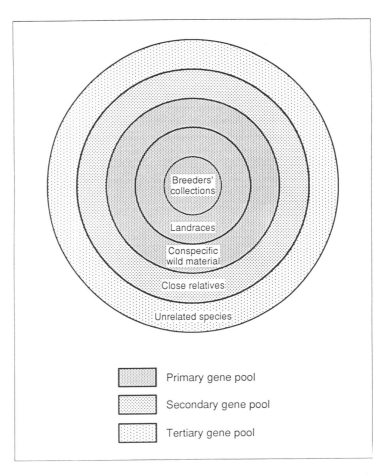

FIGURE 1.2 Germplasm categories and gene pools exploited by crop breeders.

The primary gene pool consists of all material belonging to the same species as the crop (Harlan and de Wet, 1971). This gene pool includes landraces and conspecific wild and weedy plants. Although materials in the primary gene pool are genetically diverse, they can be readily crossed because they belong to the same species. The secondary gene pool is composed of materials that are difficult to cross—closely related species and genera. The tertiary gene pool includes plants that can be crossed successfully with a crop only after considerable effort. Widecrossing with plants in the tertiary gene pool is possible only by employing specialized techniques such as chromosome doubling, embryo rescue, and the use of bridge species to transfer desirable genes.

Gene pools of crops sometimes overlap, or have the potential to do so if breeders find it necessary to tap wild germplasm. Breeders working with cultivated mangos can tap a common wild gene pool, and the genetic watersheds of breadfruit, jackfruit, and other cultivated *Artocarpus* species also partially overlap. Other cultivated tree species with gene pools in tropical forests that can be potentially shared include at least nine cultivated species of *Annona*, such as refreshing soursop and the much-appreciated sweetsop, and various species of citrus (Appendix 1). Gene flow between wild populations of cultivated species may be minimal under natural conditions, but breeders can marshal an array of techniques to loosen the boundaries of natural gene pools.

When confronted with a challenge, breeders always turn first to their working collections. In the case of many cultivated perennial plants in the tropics, little breeding has been attempted, so working collections are found only for some of the more economically important species such as rubber, cacao, and coffee. Within breeders' collections, we can distinguish between protovarieties and advanced breeding lines. Protovarieties are the first line of defense for breeders when a cultivar is experiencing problems in the field and needs to be replaced, because protovarieties are essentially ready for release.

If breeders come up empty in their own or other breeders' collections, they turn to landraces or check with scientists involved with germplasm enhancement and prebreeding pools. All the commercially important avocado and mango cultivars arose as spontaneous seedlings and were spotted in backyards or nurseries for their unusual properties. Such clonal varieties are then multiplied vegetatively through grafting, budding, or other techniques and are distributed widely to growers.

Traditional varieties or selections may have desirable genes, but often cannot be released directly because they typically also carry undesirable characteristics, such as low yield potential. Time-consuming backcrosses with elite material are normally required to create an appropriate product. In spite of such difficulties, landraces make valuable contributions to improving and stabilizing crop yields.

If landraces do not contain the trait in demand, then a breeder will usually turn to conspecific wild material, such as weedy forms. Wild species are being increasingly used in crop breeding, particularly among the cereal and pulse crops (Harlan, 1976; Plucknett et al., 1987). Although breeders working with tropical crops do not turn to wild material as much as cereal breeders, perennial crops have been improved in Third World countries as a result of crosses with wild populations or near relatives. Considerable untapped potential exists for broadening the genetic base of many tropical forest crops by dipping further into wild gene pools.

In the event that breeders cannot find what they are looking for in the primary gene pool, then they screen germplasm in the secondary gene pool. Close relatives are usually wild, but they can also be other cultivated species. The secondary gene pool has been tapped in several breeding programs involving perennial crops in the tropics. Short stature has been introduced to African oil palm, for example, by crossing it with an American cousin that grows wild along the banks of the Amazon.

Secondary gene pools hold considerable promise for further diversifying the germplasm base of many widely planted tropical tree crops and thereby providing a defense against pests and diseases. For example, a wild relative of leucaena (*Leucaena leucocephala*), a fuelwood and fodder crop in the humid tropics, may hold the answer to stopping damage caused by the jumping plant louse (*Heteropsylla cubana*). In 1986, this pysllid pest, which arose in tropical America or the Caribbean, made its way to the Pacific and Southeast Asia, where it began causing widespread damage (Vietmeyer, 1986). Plantings of leucaena in Africa, Asia, and Latin America are all based on a handful of selections, and this narrow genetic base increases the crop's vulnerability. Fortunately, resistance to the spreading insect pest has been located in *L. pallida*, and this leguminous highland shrub is being introduced into the leucaena breeding program at the University of Hawaii (J. L. Brewbaker, pers. comm.).

Unrelated species—the tertiary gene pool—rarely make a contribution to crop breeding programs because crossing barriers are so severe. With advances in biotechnology, however, tertiary gene pools are likely to become more important in the future. Since no one can predict how wide the net will be cast by breeders in the future, every plant extinction is a potential loss to agriculture.

CHAPTER 2

Beverage and Confectionery Crops

O f all the crops that have arisen in tropical forests, probably the ones most familiar to North Americans, Europeans, and the peoples of temperate Asia are coffee and cacao. When a Swedish businessman sips strong, black coffee in the morning before going to work or a Japanese schoolgirl nibbles on a bar of chocolate during a class break, neither probably realizes that those crops came from tropical forests. A Hershey bar is almost synonymous with American culture, but few U.S. citizens recognize its connection with the disappearing forests of Amazonia. Consumers in industrial nations are thus affected by the fate of tropical forests in a tangible way.

Cacao is native to the forests of the Amazon basin. Spontaneous populations of the understory tree also occur in parts of Central America, where deforestation is well advanced. A wide array of pests and diseases afflict cacao, and genes found in wild cacao may be crucial to the long-term health of cacao plantations in such widely separated countries as Brazil, Côte d'Ivoire, and Malaysia.

Arabica coffee, the most commonly consumed hot beverage, originated in a relatively small corner of Africa, covering parts of southeastern Sudan, southwestern Ethiopia, and northern Kenya. Thus, despite its name, arabica coffee is not native to the hot sands of the Arabian Peninsula, although Arabs were the first to roast and brew it in the form we appreciate today. When it comes to improving the yield, quality, and stability of coffee plantations in the humid tropics, breeders call upon the genetic resources found in the home of arabica coffee. The dwindling forests in East Africa may hold the key to the future of coffee.

A widespread drop in the yields of coffee or cacao plantations would disrupt local economies and would be more than an inconvenience for consumers in industrial nations. Coffee and cacao are major sources of foreign exchange for over fifty developing countries, many of them heavily indebted to banks in North America, Europe, and Japan. The ability of dozens of developing countries to service their external debts rests in part on their exports of coffee and cacao. Varieties of coffee and cacao that are genetically resistant to diseases and pests will not immunize exporting countries against the vagaries of international markets, but they will help lower costs of production and will reduce damage to the environment in coffee- and cacao-producing areas.

A little-known relative of cacao cultivated in parts of Amazonia, cupuaçu, may eventually be planted more widely. Cupuaçu is grown in backyards and in small groves for its football-sized fruits that contain a slippery, white pulp. The creamy pulp of cupuaçu is a regional delicacy and is used to flavor drinks, ice cream, puddings, and cakes. A sizable market for cupuaçu has developed in southern Brazil, and cupuaçu prices have been rising in real terms over the last decade or so. Large-scale plantings of cupuaçu are highly susceptible to witches'-broom, a disease that also afflicts cacao. Sources of resistance to this disease need to be located in cupuaçu populations in the forest, as well as in seedling genotypes raised in backyards, so production can be increased.

COFFEE

Coffee is one of the developing world's most important export commodities, surpassed only by petroleum and its derivatives (Kushalappa, 1989; Purseglove, 1976). Production and prices fluctuate, but coffee generates between $10 and $15 billion in export earnings for a total of some fifty Third World countries every year (*Economist*, 10 October 1987, p. 70; Kushalappa, 1989). Unlike petroleum, which is concentrated in a few fossil fuel–rich countries, coffee is grown in dozens of Third World countries.

Coffee is the single most important source of foreign exchange for Burundi, Colombia, El Salvador, Ethiopia, Rwanda, and Uganda. In Colombia, for example, coffee accounts for more than half of legal foreign exchange earnings (García and Montes, 1988). Brazil, which is grappling with over $110 billion in external debts, markets 30 percent of the coffee entering world trade. Most of Brazil's coffee is exported via Santos, hence the name of the popular variety. Latin America exports over two-thirds of the world's coffee and owes $420 billion to external creditors.

The yield stability of coffee in tropical countries thus means much more than just the price consumers pay for a cup of coffee in such cities as New York, Rome, and Stockholm. The ability of many developing countries to repay their external debt hinges in part on the vigor and productivity of their coffee plantings. A sharp downfall in coffee production because of inclement weather, an outbreak of disease, or the spread of insect pests could trip some developing countries into default and exacerbate the already strained ability of other developing countries to service their foreign debt. We all have a stake in the viability of coffee plantations in the tropics.

Except for often large plantations in southern Brazil, coffee is grown mostly by smallholders (Holm et al., 1979:488), particularly in mountainous areas with rich, volcanic soils (Figure 2.1). In the Kivu Province of Zaire, large families wedged on to 1-hectare plots depend on a dozen or so bushes of arabica coffee for the little cash income they can raise. Farmers of modest means account for most of the coffee grown on Madagascar, and coffee is one of the few sources of foreign exchange for the island (World Bank, 1980:8). Most of Mexico's 350,000 hectares in coffee is cultivated on small plots, averaging around 2 hectares (Alcorn, 1989). Coffee is also predominantly in the hands of smallholders in Colombia, Costa Rica, and the Dominican Republic (Hall, 1985:161; Lang, 1988:16, 100). Much of the Kona coffee in Hawaii, one of the world's most expen-

FIGURE 2.1 Farmer on Moorea, Society Islands, spraying herbicide around recently planted 'Agris' variety of arabica coffee (*Coffea arabica*). The coffee has been interplanted with taro (*Colocasia esculenta*). September 1989.

sive coffees, is tended by small-scale farmers. Thus the stability of coffee yields is a major social concern in the humid tropics.

How vulnerable are the coffee plantings in the tropics, and how intricately entwined is the crop in local and regional economies? The germ-plasm base of both arabica and robusta coffee plantings is exceedingly narrow (Charrier, 1980; Meyer, 1967). The narrow genetic limb supporting coffee production is especially worrisome in view of its vertical integration in local economies. Any large-scale collapse of coffee production would impair not only growers and seasonal pickers, but all those involved in the transportation and processing of the crop. Furthermore, coffee is avidly consumed in developing countries, particularly Latin America and Africa. To understand why coffee production rests on such a shaky genetic foundation and why the crop is so important to local cultures and economies, we briefly review its history.

DOMESTICATION AND SPREAD

Several species of coffee (*Coffea* spp.) have been taken into cultivation in Africa, the most important of which are arabica coffee (*C. arabica*) and robusta coffee (*C. canephora*). Arabica coffee is grown throughout tropical highlands of between 1,000 and 2,000 meters altitude, and because it produces beans with the best flavor and quality, it dominates global coffee production. Robusta coffee, native to the Congo basin, thrives in the lowland, humid tropics. Although its beans are of a lower quality than those of arabica coffee, it is hardy and better adapted to warmer locations. Uganda, Angola, and Brazil are major producers of robusta coffee.

Arabica coffee is indigenous to montane tropical forest in southwestern Ethiopia, the adjacent Boma Plateau of southeastern Sudan, and the Marsabit Forest of northern Kenya (Anthony et al. 1987; P. A. Jones, 1956; A. S. Thomas, 1942; Wrigley, 1988:1). People may have introduced arabica coffee plants to the Marsabit Forest and possibly the Boma Plateau long ago, perhaps because of their sought-after fruits (Charrier and Berthaud, 1985). Coffee berries could have been dispersed to outlying locations by birds or mammals. Alternatively, cooler temperatures may have prevailed at some point during the Pleistocene, as seems likely in Amazonia, thereby permitting arabica coffee to spread across low-lying valleys that are currently inhospitable to it.

Ethiopians and Sudanese never cultivated coffee on a large scale; rather, a few genetically heterogeneous bushes have traditionally been kept in the backyard, or the berries gathered in the wild. The earliest use of coffee berries may have been as a snack food. Flesh from ripe berries is still removed from the green beans to make coffee juice, some of which is fermented to make wine. Coffee juice is widely appreciated in Central Africa, such as in the Great Lakes Region of Zaire (Figure 2.2). Africans never developed the custom of roasting beans and brewing coffee as is the common practice today. The leaves and beans are sometimes chewed as a stimulant, similar to the habit of chewing coca leaves in the Andes to combat fatigue and hunger. Leaves of arabica coffee are also stewed to make "tea."

Arabs were the first to roast coffee beans and to cultivate the crop on a significant scale. Strong, black coffee was being brewed in Yemen by 1300, but how and when coffee arrived in Arabia is unclear. Migrants and settlers from the land of the Sabeans in Yemen repeatedly crossed the Red Sea to the highlands of Ethiopia during the first millennium B.C. (Wiedner, 1962:30). Some Sabeans also settled in northern Ethiopia (Davidson, 1969:13), and since they were active traders, it seems likely that coffee beans were taken to southwestern Arabia before the time of Christ. Alternatively, the crop may have crossed the Red Sea while the great Kingdom of Axum held power in Ethiopia from A.D. 100 to 700.

FIGURE 2.2 Farmer removing flesh from arabica coffee (*Coffea arabica*) berries in her backyard to make juice. Near Mulungu, Kivu Province, Zaire, March 1989.

Arabs esteemed coffee for its ability to stimulate mental powers. In the fourteenth or fifteenth century, some terraced plantations of arabica coffee were established in Yemen. Irrigated by water drawn from wells, the terraces were built primarily to facilitate irrigation, rather than to arrest soil erosion. The Mocha variety, with its subtle chocolate undertones, was selected from these early introductions to Southwest Asia.

From Arabia, coffee was introduced as a beverage to Egypt and Turkey. Venetians brought coffee to Europe. Romans first tasted coffee around 1600; previously their beverages were largely tea from herbs, fruit juices, and wine.

The café soon emerged as a gathering place for the intelligentsia, artists, and entrepreneurs. In England, a Lebanese merchant opened the first coffee house in 1650 at the Angel Inn in Oxford. Tillyard's coffee house, also in Oxford, was instrumental in the formation of the Royal Society. London's first coffee house was inaugurated in 1652. A couple of coffee houses in Change Alley, Cornhill, London, gave birth to the Stock Exchange. France's first coffee house was established at Marseille in 1671; this Mediterranean café received its coffee beans by boat from Alexandria. The first café opened in Paris a year later.

The emergence of coffee houses in the Mediterranean and Europe spurred the cultivation of arabica coffee. In response to the growing demand for coffee, European colonial powers began searching for places in their overseas possessions where the crop could be cultivated profitably on a larger scale.

Botanic gardens played a key role in spreading coffee to tropical lands in the eighteenth and nineteenth centuries. The Amsterdam Botanic Garden (now the Hortus Botanicus of the University of Amsterdam) received two arabica coffee plants from Java in 1706 (Wrigley, 1988:42). From Amsterdam, coffee seedlings were sent to Dutch Guiana around 1713 and to the Jardin des Plantes, Paris, in 1714 (Figure 2.3).

A coffee seedling was sent from the Jardin des Plantes in Paris to Martinique in 1721 (Aublet, 1775:50). That lone seedling started the coffee industry on Martinique. Progeny from that introduction to Martinique were sent to Jamaica in 1730, to Haiti (then Saint-Domingue), and to several locations in the American tropics. The selection of the seedling destined for Martinique was fortuitous because it ultimately led to development of the famous 'Blue Mountain' variety on Jamaica, one of the world's premier coffee varieties. Half a kilogram of savory 'Blue Mountain' sells for close to $40 in gourmet stores in North America and Europe.

The French made other introductions of arabica coffee to the Caribbean and tropical America from a staging point in the Indian Ocean: Réunion. French traders obtained coffee seedlings of the Mocha variety from Yemen and introduced them to Réunion in 1717 (Figure 2.3). The British tried

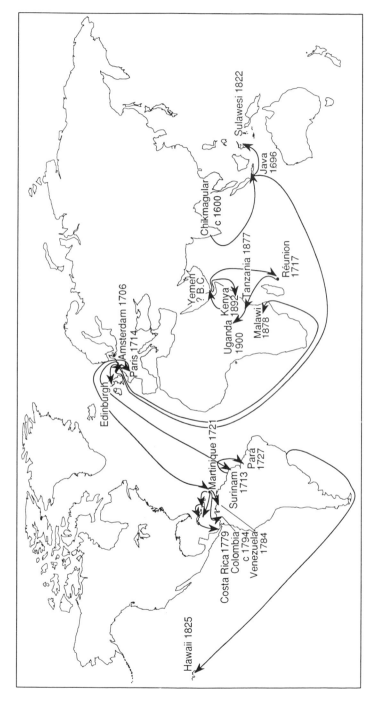

FIGURE 2.3 Introductions of arabica coffee (*Coffea arabica*) planting materials and routes taken. (Adapted from Aublet, 1775:51; P. A. Jones, 1956; B. Magalhães, 1980:38; N. J. H. Smith, 1986:24; Wallace, 1986:251; Wrigley, 1988:39.)

their hand at cultivating coffee when the Edinburgh Botanic Garden obtained coffee seedlings from Amsterdam and sent seedlings to Malawi (then Nyasaland) in 1878.

Missionaries also instigated coffee cultivation on new lands. In Hawaii, Don Francisco Martin is credited with the first planting of coffee near Honolulu in 1825 (Crawford, 1937). In Tanzania, Father Horner from Alsace brought arabica coffee plants to Bagamoyo from Réunion in 1877. The seeds were obtained from descendants of the introduction of the Mocha variety to Réunion in 1717 (Figure 2.3). Although the introduced coffee bushes fared poorly at Bagamoyo, their progeny were taken to other missionary outposts in Tanzania, such as the Kilema Mission on the slopes of snow-capped Kilimanjaro (O'Hare, 1986). The Kilema Mission supplied coffee seeds to the Bura Mission in the Taita Hills in 1892, which in turn provided seeds for St. Austin's Catholic Mission in Nairobi. The Fathers at St. Austin were probably not aware of the indigenous arabica coffee in northern Kenya.

Coffee quickly became the foundation for British settlers in Kenya at the beginning of the twentieth century. Missionaries needed a viable cash crop to help run their operations, and newcomers also needed to generate income to cover their start-up costs. The drama and hardships of this pioneer life based on coffee have been eloquently captured in Elspeth Huxley's *Flame Trees of Thika*. Thika is now a sprawling industrial town, but coffee remains an important cash crop for Kenya.

Coffee was introduced to Rwanda in 1905 by the Catholic mission of Mibirizi, which had obtained coffee seeds from Guatemala. Coffee is the major foreign exchange earner for Rwanda, and a coffee branch with berries is depicted on the country's ten-franc coin. It is ironic that coffee as a commercial crop arrived in East Africa and the Great Lakes Region by such circuitous routing, when to the north, people had been collecting and planting coffee on a small scale for millennia.

Spanish missionaries took coffee from Cuba to Guatemala and Puerto Rico. In 1784, José Antonio Mohedano, a Catholic priest, obtained seedlings from Martinique and launched Venezuela's coffee industry in the mild climate of the valley now largely occupied by ever-expanding Caracas. Today, coffee is Latin America's most important export commodity.

GENETIC RESOURCES AND CONSERVATION

The coffee gene pool can be divided into several sections, some of which overlap (Figure 2.4). *Paracoffea* is now recognized as a distinct genus, while most of the eleven species in the section *Argocoffea* have been transferred to the genus *Argocoffeopsis* (Carvalho, 1958; Wrigley, 1988:63). *Eucoffea* contains the arabica and robusta coffees. Although some sections

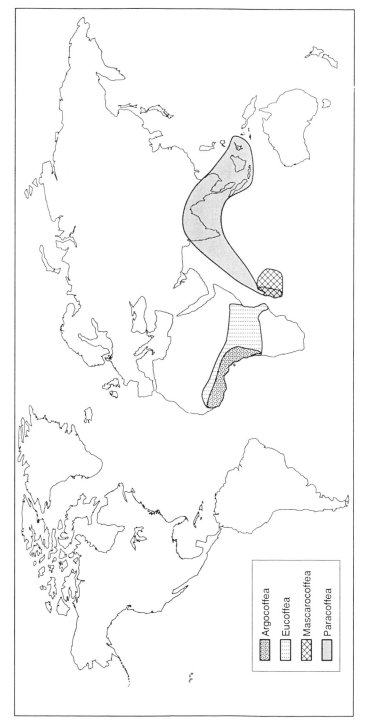

FIGURE 2.4 Sections within the coffee gene pool. (Adapted from Carvalho, 1958; Wrigley, 1988:63.)

Argocoffea
Eucoffea
Mascarocoffea
Paracoffea

have been renamed and the taxonomy remains confused (Charrier and Berthaud, 1985), the important point is that coffee's gene pool contains dozens of species that stretch from West Africa to Southeast Asia. Tracts of many different forest types will thus have to be preserved in order to safeguard representative portions of coffee's larger gene pool.

Wild coffees may prove useful to breeders in the future, if they survive the onslaught of forest clearing and development. Some species are confined to relatively small areas, such as *C. fadenii* and *C. mongensis* in the Taita Hills, Kenya, and the Usambara mountains in adjacent Tanzania (Berthaud, 1986:257). Cameroon and Gabon are a center of diversity for *Coffea*, since they contain heterogeneous populations of *C. brevipes*, *C. staudtii*, and water-loving *C. humilis*; undescribed species probably await collection.

Wild populations of arabica coffee are restricted to a relatively small area of between 1,370 and 1,830 meters altitude within its native range. Only about 400,000 hectares of forest containing wild coffee were left in Ethiopia by the mid-1980s (Wrigley, 1988:1). More worrisome still is the fact that the remaining forest is highly fragmented (Figure 2.5), thereby reducing the viability of coffee and other species in the forest patches. The Ethiopian government is taking measures to safeguard wild arabica coffee. With advice from J. G. Hawkes of the University of Birmingham, United Kingdom, the Ethiopian government has set up several in situ reserves for wild arabica coffee. In the Marsabit Forest in northern Kenya, apparently wild arabica coffee occurs in a narrow altitudinal range, from 1,500 to 1,550 meters, and is thus particularly vulnerable to land clearing (Anthony et al., 1987).

Little is known about the status of wild arabica coffee on the Boma Plateau, a remote area in a country mired in political turmoil. In the 1940s, members of the Kichepo tribe were cutting down still-abundant wild arabica trees on the Boma Plateau to harvest the berries (A. S. Thomas, 1942). In the Rume area of the Boma Plateau, increased forest clearing has been under way since the 1930s to make room for such crops as maize, tobacco, and sorghum. Farmers reportedly spare coffee bushes when clearing the forest so that they can harvest the berries in their fields (A. S. Thomas, 1942). Eventually, though, relict coffee bushes die or are killed by fires.

Wild arabica coffee is highly variable and a potentially valuable source of genes, and beans harvested from forests still account for approximately half of Ethiopia's coffee production (Ameha, 1991; Charrier, 1980; Meyer, 1967; Wondimu, 1987). Wild arabica coffee appears to be dispersed by monkeys and baboons (Meyer, 1967). Reserves set aside for wild coffee will therefore need to be designed to ensure that seed dispersal agents are not disturbed. And considering the high degree of variation in

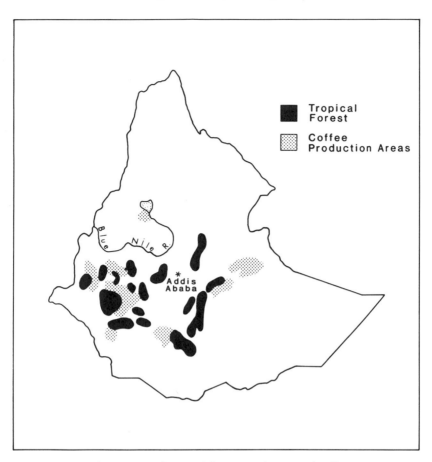

FIGURE 2.5 Remaining areas of tropical forest in Ethiopia and coffee-growing areas in 1976. (Central Intelligence Agency map #503188, 12-76.)

wild arabica coffee, several widely spaced reserves are needed to represent the species' natural heterogeneity.

Like many other perennial crops in the tropics, such as avocado, cacao, and rubber, some "wild" arabica coffee may be cultivated plants, or their descendants, that have been engulfed by returning forest. Coffee bushes can become feral after a homesite is abandoned, or they can sprout from seeds dropped along paths. How much selection was undertaken by the early cultivators of coffee is difficult to tell, but planted coffee probably differed little from wild populations. The line between truly wild arabica coffee and artificially enriched populations is thus hard to draw. Cultivated coffee is also genetically diverse in Ethiopia; farmers in the Harerge region recognize seventeen types of arabica coffee (Bellachew, 1987).

Despite the present genetic diversity, genetic erosion of both wild and cultivated coffee has reached alarming proportions in Ethiopia. A 1986 survey of 935 farms in the Habro, Webera, and Garamuleta Awraja areas revealed that 12 percent had abandoned coffee cultivation and 63 percent were on the verge of switching to other crops (Bellachew, 1987). Attacks by *Colletotrichum coffeanum*, which causes coffee berry disease, and the need to increase food production are the primary threats to coffee in its center of diversity. Relatively uniform varieties resistant to coffee berry disease have already replaced at least 10 percent of Ethiopia's heterogeneous arabica coffee (Ameha, 1991). Ethiopia's population is doubling about every twenty years, and plans have been discussed to settle as many as 300,000 people from the drought-plagued northern parts of Ethiopia to the south. Pressure on the remaining forests is thus likely to increase (Figure 2.5).

The wild gene pool of robusta coffee is in better shape than that of arabica coffee. Robusta coffee occurs in a much broader band of forest stretching from West Africa to Uganda, most of it still unpenetrated by roads. In the early part of this century, Ugandans were collecting apparently wild *C. canephora* in the wooded parts of Buganda province (Maclaren, 1924:215). As in the case of arabica coffee, robusta coffee was originally chewed rather than brewed. Robusta coffee has become increasingly important with the growing popularity of instant coffee over the last few decades.

Liberian coffee (*C. liberica*) is cultivated on a few farms in West Africa and accounts for less than 1 percent of world coffee production. Like oil palm, Liberian coffee was domesticated in the forest zone, together with vegetatively propagated root crops, such as yams. Liberian coffee is hardy and resists leaf curl, but it has progressively lost ground to robusta and arabica coffee because its tough, fibrous skin is difficult to remove from the berries (F. R. Irvine, 1930:173).

In parts of Sierra Leone, Guinea, and Côte d'Ivoire, upland or narrow-leaf coffee (*C. stenophylla*) is cultivated on a small scale, and its berries are also collected in the forest (Chandler, 1958:370; Wrigley, 1988:75). This minor cultivated species has fine-flavored beans that could be of use to breeders (Macmillan, 1935:342; R. O. Williams and Williams, 1951:130). Also, *C. stenophylla* may resist ubiquitous coffee leafminers (Wrigley, 1988:362). Other species of *Coffea* that are occasionally cultivated on a small scale include *C. bengalensis* in India; *C. zanguebariae* in Zanzibar, Tanzania, and Moçambique; and *C. eugenioides* in the Congo basin.

Little is known about the traits of uncultivated species of *Coffea*, some of which are harvested in the wild. In Moçambique, for example, rural peoples collected the berries of *C. racemosa* (Charrier, 1980). Much of the research on near relatives of coffee has focused on their levels of caffeine because of recent alarm about its effects on health. But as will be discussed later, the economic prospects for naturally caffeine-free coffees are not especially bright. Other potentially valuable characteristics of wild relatives of coffee, such as pest resistance, may ultimately prove more important. Also, near relatives of coffee are often used locally in medicines, and their strong wood is employed in construction (F. R. Irvine, 1961).

Despite the economic importance of coffee, only a few sizable gene banks have been established for *Coffea*. Coffee seeds remain viable for less than two years even under reduced temperature and moisture conditions (Berthaud, 1986:19), so germplasm is planted in field gene banks. The largest field gene bank, with 6,000 accessions, is maintained at Divo, Côte d'Ivoire. This gene bank is also the most comprehensive, since it contains a diverse array of wild species (Charrier, 1980). In Ethiopia, some 700 genotypes are kept at Chochie, Kefa (Worede, 1991). Smaller germplasm collections in Africa are maintained at Lyamungu, Tanzania, and Butare, Rwanda.

The largest coffee germplasm collection in Latin America is held at the Centro Agronómico Tropical de Investigación y Enseñanza (CATIE) on the outskirts of Turrialba, Costa Rica. As of June 1985, CATIE held 1,212 accessions of arabica coffee. In addition, CATIE has 49 accessions of robusta coffee and 26 accessions of Liberian coffee, as well as a few holdings of other *Coffea* species, such as *C. bengalensis* (4 accessions), *C. congensis* (7), *C. eugenioides* (6), *C. kivuensis* (1), *C. klainii* (1), *C. kopakata* (1), *C. mauritiana* (4), *C. racemosa* (27), *C. salvatrix* (2), *C. stenophylla* (2), and *C. travancorensis* (1). Germplasm from CATIE and other coffee research programs in Central America and Brazil is exchanged through a regional network, PROMECAFE (Programa Cooperativo para la Protección y Modernización de la Caficultura). PROMECAFE was established in 1978 and is supported by various donors, including the U.S. Agency for International Development and the European Community.

In the United States, the Subtropical Horticulture Research Station in Miami has a collection of 329 arabica coffee cultivars (Figure 2.6), but only 20 are accessions of landrace material from Ethiopia and Angola (Litz et al., 1983). In Asia, coffee gene banks are maintained in Karnataka, India, and the Institute for Industrial Crops in Bogor, Indonesia. For the most part, wild relatives of coffee are poorly represented in gene banks (Anthony et al., 1987).

Biotechnology promises to lower the cost of storing coffee germplasm. The Colombian Coffee Growers' Association (Federación Nacional de Cafeteros de Colombia) has raised over 200 coffee plants from tissue cultures (*Nature*, 19 June 1986, p. 721). Somatic embryogenesis in arabica coffee is not easily accomplished, however, nor is the task of maintaining the materials under slow growth conditions. More research is clearly required before coffee germplasm can be stored in vitro (Litz, 1987).

FIGURE 2.6 Field gene bank for arabica coffee (*Coffea arabica*). Subtropical Horticulture Research Station, Miami, 1987.

BREEDING CHALLENGES

A wider genetic base for coffee cultivation is essential to ensure the future health of plantations throughout the tropics. Only about thirty mutants of arabica coffee are grown worldwide (Charrier, 1980). Coffee plantations in Latin America and the Caribbean rest on a particularly shaky foundation. Only a handful of introductions are responsible for the multibillion dollar crop in tropical America and the West Indies. The Martinique introduction imposed a severe genetic bottleneck for much of the Neotropical plantings of arabica coffee. A coffee bush from Java is ultimately responsible for much of Latin America's coffee production (Wrigley, 1988:54).

The vulnerability of Latin America's coffee production was highlighted in 1970 with the arrival of coffee rust in Brazil. This fungal disease, caused by *Hemileia vastatrix*, wiped out commercial coffee production on Sri Lanka soon after it appeared there in 1869 (Morschel, 1971:9). Before coffee rust took its toll, over 100,000 hectares of coffee had been thriving on Sri Lanka (McCune, 1949). The fungal pathogen, which forms a deep orange covering over coffee leaves and significantly lowers yields, also devastated arabica coffee on Java and Sumatra in the latter part of the nineteenth century. When coffee rust outbreaks are severe, many growers switch to less-desirable robusta coffee or to other cash crops, such as cacao.

How coffee rust, which is native to Africa, reached Brazil remains a mystery. Spores may have been transported across the Atlantic on upper-level air currents (Bowden et al., 1971) or on board an airplane. The spread of coffee rust in Latin America has been relentless: by 1983 it had extended through Central America to Mexico (Saenz and Soleibe, 1987; A. Turrent, pers. comm.). Where prevalent, coffee rust has essentially driven out arabica coffee production below altitudes of 1,300 meters.

At the moment, copper-containing sprays are the main defense against coffee rust in Latin America, but the sprays have to be administered by hand, and the treatment is expensive (*Nature*, 19 June 1986, p. 721). Without the use of fungicides, coffee rust causes an average yield decline of 30 percent in Brazil (Eskes, 1989). Globally, chemical control of coffee rust costs between $1 and $3 billion.

The widespread economic significance of coffee rust has prompted a search for sources of resistance to the disease. Apparently, disease-free wild trees may not be a reliable indicator of resistance. The dynamics of disease and pest populations are different in natural ecosystems than in artificial settings. Wild trees may be relatively unscathed by disease because they are more dispersed than plantation trees or because their conditions may be less humid. Promising material needs to be tested under field conditions and exposed to an array of pathotypes of a disease before it can be considered reasonably resistant.

Some arabica coffees in Ethiopia reportedly resist coffee rust to varying degrees (Sylvain, 1955). Several germplasm collecting missions have been organized in arabica coffee's center of diversity since the 1920s (Berthaud and Charrier, 1988). A few, sporadic collecting trips were made mainly by the British until the 1950s and 1960s, when agricultural scientists from the United States, France, and several coffee-producing countries visited Ethiopia under the auspices of the Food and Agriculture Organization (FAO), the U.S. Agency for International Development (USAID), as well as the French Institute for Overseas Scientific Collaboration (ORSTOM, Institut Français de Recherche pour le Développement en Coopération) and the Coffee and Cacao Research Institute (IRCC, Institut de Recherche sur le Café et le Cacao) at Divo, Côte d'Ivoire. In 1964, for example, FAO sponsored a coffee germplasm expedition headed by F. G. Meyer of the U.S. National Arboretum in Washington, D.C., which collected some 500 different types of arabica coffee for distribution to leading coffee research centers around the world (Krug and De Poerck, 1968). The purpose of these visits to Ethiopia was to give technical advice on coffee growing and to obtain seeds, some of which were sent to CATIE near Turrialba, Costa

Rica (Meyer, 1967). In the 1980s, Ethiopia began restricting the export of native plant materials, thereby halting further shipments of arabica coffee germplasm from its native range (Doyle, 1985:201; Fowler and Mooney, 1990:104).

Spurred by concern over the arrival of coffee rust in Brazil, coffee breeders at the Agronomic Institute of Campinas (IAC, Instituto Agronômico de Campinas) obtained seeds in 1973 from 200 accessions of arabica coffee collected in Ethiopia and maintained at the CATIE gene bank (Carvalho et al., 1988). While progeny from these Ethiopian accessions did not match the yield levels of 'Mundo Novo', one of Brazil's leading varieties, several lines showed resistance to at least some races of coffee rust. Furthermore, some of the progeny exhibited other useful traits, such as precocious bearing and male sterility; the latter trait opens up the possibility of developing hybrids. Colombia is using wild arabica coffee from Ethiopia in its rust resistance breeding work (Prescott-Allen and Prescott-Allen, 1986:331).

A major obstacle confronting coffee breeders in their attempts to incorporate resistance to coffee rust is the highly variable nature of the pathogen. In the early 1970s, twenty-five races of coffee rust were known (Ferwerda, 1976), but by 1986, the number of identified races had climbed to thirty-three (*Nature*, 27 November 1986, p. 331). Tolerance to a broad range of rust races, rather than immunity to just a few pathotypes, will prove more durable and helpful to farmers. Some arabica coffee genotypes resist up to twenty-seven races of coffee rust. Ethiopian coffees may not contain genes for resistance to all races of coffee rust; resistance genes may have to be obtained from other sources. Ideally, various genes for resistance to coffee rust will be located in arabica germplasm to help confront the wide array of coffee rust races. Near relatives of arabica coffee offer some hope for locating resistance genes for coffee rust.

Diploid robusta coffee, which resists coffee rust, has been successfully hybridized with tetraploid arabica coffee by using colchicine to double the chromosome number. Interspecific crosses have been performed for some two decades to further breeding and to better understand species relationships (Berthaud et al., 1986; Carvalho and Monaco, 1968; Carvalho et al., 1989). Unfortunately, yields of the hybrids are often disappointing, and bean quality suffers (Wright, 1988:416). A protracted effort will be required to break linkages with undesirable genes during backcrossing work. In the distant future, sophisticated recombinant DNA techniques may permit the incorporation of the desired resistance genes into superior arabica coffee germplasm, thereby leaving behind unwanted traits such as poor bean quality and slow growth.

Coffee berry disease, caused by another fungus (*Colletotrichum cof-*

feanum), mummifies berries, thereby triggering a drastic decline in yield. The pathogen has caused serious losses among coffee growers in East Africa since the 1950s. Some wild arabica coffee collected on the Boma Plateau in 1942, dubbed 'Rume Sudan', has been used intensively in Kenya for resistance breeding to the disease since the 1970s; 'Rume Sudan' also exhibits partial resistance to coffee rust (Eskes, 1989). The loss of coffee germplasm to coffee berry disease in Ethiopia is particularly serious. The importance of coffee to the economies of eastern Africa and the irreplaceable genetic diversity of arabica coffee in Ethiopia warrant redoubled efforts to locate other sources of resistance to the disease.

Several insect pests attack coffee, some of which can cause serious damage. In the Cerrado region of Brazil, for example, outbreaks of coffee leafminer (*Leucoptera coffeella*), known locally as *bicho-mineiro*, depressed coffee yields in 1985 (J. C. A. Magalhães et al., 1987). This moth larva is also a pervasive pest in Africa. Many other pathogens, insect pests, and nematodes can cause extensive damage to coffee plantings (Owen, 1973).

Pests sometimes spread, posing new threats to coffee growers. Coffee berry borer (*Hypothenemus hampei*; Scolytidae) is indigenous to Central Africa and arrived in Jamaica in the late 1970s where it threatens famed 'Blue Mountain' coffee groves (Ford, 1981). Resistance to widespread insect pests, such as coffee leafminer, would reduce the use of pesticides on coffee plantings and would be particularly helpful to growers who cannot afford chemical control of insect pests.

The booming market for decaffeinated coffee has prompted a search for caffeine-free coffees. Decaffeinated coffees, first marketed in 1906 (Jacob, 1935:282), often lose some of their flavor because the water-leaching process or the use of chemical solvents robs some of the coffee's subtle flavors. Madagascar contains between forty and fifty wild species of *Coffea*, most of which are reputedly free of caffeine (Charrier, 1980; Jenkins, 1987:39). The only species of *Coffea* on the Comoro Islands, *C. humblotiana*, is also caffeine-free (N. Myers, 1983:41). Unfortunately, the habitat for *C. humblotiana* is rapidly disappearing as people harvest the trees for house construction and clear the land for farming (Charrier, 1980). Another coffee relative, *C. eugenioides*, which occurs wild in Zaire and is cultivated on a small scale in western Uganda, Kenya, and western Tanzania, has low levels of caffeine in the beans (Purseglove, 1974:458).

A program to widecross arabica coffee and robusta coffee with caffeine-free species of *Coffea* on Madagascar has been under way since the 1970s (Charrier, 1978). Attempts to produce fertile hybrids between robusta coffee and species of *Coffea* in the Malagasy Republic have proved largely fruitless since F_1 hybrids are virtually sterile. It has also proven difficult to cross arabica coffee with caffeine-free species of *Coffea*. Arabica coffee produces sterile triploid hybrids when pollinated by diploid species. Such crossing experiments have helped scientists understand the relationships between species of *Coffea*, but after nearly twenty years of effort, no marketable, caffeine-free arabica coffee has been produced.

Another impediment to efforts to develop a caffeine-free arabica or robusta coffee of high quality is the brisk demand for caffeine by the pharmaceutical and beverage industries. The extra cost of decaffeinating coffee is covered by the sale of caffeine for incorporation in diet pills, asthma drugs, headache remedies, stimulants, and colas. Medicinal uses account for about a third of caffeine consumption.

Wild species of coffee contain other traits that may eventually be transferred to arabica and robusta coffees. Several species on Madagascar tolerate drought (Wrigley, 1988:68). Drought-tolerant wild species of *Coffea* overcome lack of rainfall by dropping their leaves, among other physiologic responses. In areas subject to erratic rains, or in places with highly porous soils, increased resistance to low or highly uneven rainfall would be a desirable attribute. Unfortunately, about two-thirds of Madagascar's forests have already been felled (G. M. Green and Sussman, 1990), and much of whatever promise they offered the coffee industry might already have been destroyed.

PROSPECTS

Given the global economic importance of coffee, many research institutions are working on a host of issues related to the commodity. Progress is being made in understanding the complexity of some of the principal diseases, but more research is warranted to better understand variation in and some of the characteristics of wild coffees.

Research on genetic resources and crop management pays off. A major reason that Brazil has retained its position as the world leader in coffee production is the long and consistent investment in agricultural research. The Agronomic Institute of Campinas was founded in the State of São Paulo in 1877 to work on coffee and other crops; genetic research on arabica coffee began in 1933 (Carvalho, 1958; EMBRAPA, 1984:104; IAC, 1987). In Kenya, researchers continue to upgrade coffee varieties for farmers, thus helping ensure that Kenya remains an important exporter of quality coffees. Recently, for example, the Coffee Research Foundation in Kenya released 'Ruiri II', which resists coffee berry disease (Lele and Meyers, 1988:31).

In spite of impressive advances in understanding the genetics, chemistry, and agronomy of coffee, the safety of much of coffee's gene pool is endangered. Gene banks need to be upgraded with more wild material because they are convenient for breeders, but they can only complement in situ conservation. Many populations of coffee's near relatives on Madagascar and the Mascarene Islands have already disappeared.

More support is needed for intermediate quarantine services so that germplasm can be safely exchanged between countries. Because coffee materials are exchanged as seeds, seedlings, or rooted cuttings, vigilance is necessary to check the spread of pests and diseases. Seeds are the safest way to export coffee germplasm, since they do not carry the major fungal diseases. Fumigation is still necessary to help prevent pathogen and insect stowaways.

The Subtropical Horticulture Research Station in Miami, the USDA station in Glenn Dale, Maryland, and a facility in Montpellier, France, are the main locations offering intermediate quarantine for coffee. Such services need a further infusion of funds to facilitate the processing of their heavy work load.

High prices for coffee in 1976 and 1977 triggered a scramble to plant more coffee. As a result, a glut of coffee depressed prices for much of the 1980s (Akiyama and Duncan, 1982). In 1987 alone, coffee prices plunged 40 percent (*Economist*, 2 April 1988, p. 62; *Wall Street Journal*, 28 July 1989, p. C12; *Washington Post*, 26 July 1989, p. A26). Overproduction of coffee is not new; Brazil had to burn millions of sacks of coffee in the early part of this century to reduce stocks. Prices may recover somewhat as East European countries and the republics of the former Soviet Union open their economies and eventually create more discretionary income to purchase coffee and other "luxury" items. Price fluctuations for commodities are to be expected; farmers who can keep their costs lower—by planting varieties that require less pesticide, for example—have a comparative advantage during both upward and downward price swings.

Recent reports apparently contradicting the bad publicity about the effects of caffeine on health may help spur demand for coffee (*Time*, 22 October 1990, p. 59). Moderate coffee drinkers may not be any more at risk for stroke or heart complications than those who abstain from caffeine. Even a slight increase in the 350 million cups of coffee consumed in the United States every day would help coffee growers in many countries.

The piecemeal approach to evaluating and conserving coffee's genetic resources is surprising when the economic importance of the crop is considered. Many coffee improvement programs have not been as aggressive as they might in incorporating exotic germplasm in their breeding efforts, in part because of Ethiopia's ban on exports of arabica coffee materials. In spite of this embargo, other areas could be profitably canvassed for gene banks and to advise governments and private voluntary organizations on appropriate sites for conservation. Some accessions in gene banks need to be duplicated for safety, while others could be culled because they are redundant.

CACAO

Cacao (*Theobroma cacao*) is native to central and western Amazonia, where it occurs as a forest understory tree up to about 900 meters on the eastern flanks of the Andes (Figure 2.7). The sweet pulp surrounding cacao beans served mainly as a snack food for peoples of Amazonia in precontact times. Well before the dawn of agriculture, people on hunting and gathering trips into the forest broke open the elliptical pods from wild cacao to snack on the refreshing pulp.

Artificial concentrations of cacao may have aggregated on ancient camp and village sites, particularly along river banks, as has happened with a number of other useful plants in the tropics, such as certain palms. On Marajó Island in the Amazon delta, for example, cacao has been found growing on an archaeological mound. Cacao at this archaeological site is thought to have been introduced by Indians long ago; the last tribe inhabiting the area, the Aruã, were gone by 1800 (Meggers and Evans, 1957).

Cacao pods are likely to have been carried back to the village for later consumption, and discarded beans may have sprouted in rich kitchen middens around homes. In the Ecuadorian Amazon, concentrations of cacao are found on black alluvial soils (J. B. Allen and Lass, 1983:47), some of which are probably anthrosols laced with potsherds (N. J. H. Smith, 1980). Along the middle Tapajós, wild cacao is found only in the understory of forest thick with palms, especially babaçu (*Orbignya phalerata*), a valuable source of cooking oil (Ducke, 1925). Babaçu concentrations often develop after repeated slash-and-burn farming, suggesting that people may have had a hand in the density of wild cacao along the Tapajós and elsewhere in the Amazon basin.

In the upper Rio Negro and southern Orinoco, various indigenous groups cultivate cacao in their backyards. How long this small-scale cultivation has been practiced is not known, but it probably extends back millennia. The king of Portugal spurred wider growing of cacao in the lower Amazon after a 1679 edict encouraging its cultivation (Rangel, 1982:3). Cacao plantations were established around Alenquer, Pará, by the close of the seventeenth century (Bartley et al., 1987).

Cacao may not have been domesticated for chocolate in Amazonia, but the tree has long been cultivated on a limited scale for its sweet pulp and possibly other attributes. For example, rural folk in Amazonas State, Brazil, rub cocoa butter on bruises (Van den Berg and Silva, 1986). The shade-tolerant tree spread out early from Amazonia as a dooryard domesticate long before the arrival of Europeans and reached Central America as a domesticate centuries, if not millennia, before the Spanish arrived (Urquhart, 1961). Cuatrecasas (1964) stated that cacao spread in early times from South America through Central America and into Mexico, but did

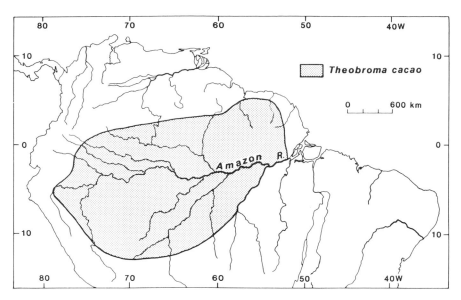

FIGURE 2.7 Distribution of wild cacao (*Theobroma cacao*). (Adapted from J. B. Allen, 1987; J. B. Allen and Lass, 1983:47; D. Clement and Lanaud, 1987; Coudreau, 1883:15; Huber, 1906; J. G. Myers, 1930; Stahel, 1920; Van Roosmalen, 1985:420; Vello and Silva, 1968.)

not specify whether such diffusion occurred naturally or with the aid of people.

Domesticated cacao may have left Amazonia along several routes. Schultes (1984a) suggested three possible ways cacao could have penetrated Central America from Amazonia. First, after reaching the mouth of the Amazon, domesticated cacao could have diffused up the coast and across northern South America among Arawak and Carib groups. Second, cultivated cacao could have spread from the Rio Negro into the Orinoco watershed via the Casiquiare Canal and then along the northern shores of South America, except perhaps the dry Guajira Peninsula. People may have moved into the Orinoco system from the Negro some 16,000 years ago (Lathrap, 1977). Charcoal samples from two preceramic sites along the middle Orinoco suggest the river was inhabited at least 9,000 years ago (Barse, 1990). Third, cacao could have been taken across the Andes through a low pass. Cobley (1956) also subscribed to the idea of cultivated cacao crossing the Andes in early times.

Why cacao should have been singled out for such early and long-range diffusion from its native hearth is somewhat of a mystery. Other equally appealing Amazonian fruit trees, such as various species in the Annonaceae, did not reach or at least take hold in Central America. Schultes (1984a) suggested that an early use of cacao might have been as a stimulant and the crop spread for that specific purpose rather than as a fruit or for chocolate.

Early dispersal routes of cacao are not just fertile ground for academic speculation: the forms in Central America diversified genetically. Remnants of early cultivated forms of cacao might well contain interesting patterns of variation or gene alleles. *Criollo* cacao in parts of Venezuela, for example, may be older than *criollo* cacao in Central America. In preconquest times, Indians near Mérida in the Venezuelan Andes cultivated a fine *criollo* cacao from which they prepared a chocolate drink called *chorote* (Pittier, 1935). The 'Porcelano' cacao of Lake Maracaibo's southwestern shores is indigenous to Venezuela, and the quality of this *criollo* cacao is classified as extra fine on the international market (Soria, 1970). Similarly, gene pools of cultivated cacao from antiquity may also be found along the humid coasts of Ecuador and Colombia.

Precisely how cacao arrived in Central America remains a mystery, but it has been cultivated there for at least 3,000 years (Cheesman, 1944). The Aztecs of Mexico and the Mayans of the Yucatán prepared a chocolate drink by pounding cacao beans with maize kernels then adding boiling water and fiery *Capsicum* peppers. In the 1920s, this traditional chocolate beverage was still being served in parts of Guatemala (Figure 2.8). Spaniards mellowed this concoction by substituting sugar and vanilla for the peppers. Cacao beans were also used for currency in precontact times in Mexico.

From the Yucatán Peninsula, cacao was taken to Trinidad and Venezuela in 1525 (Figure 2.9) and then, by Spanish galleons, to the Philippines from Mexico in 1670. Colonel José Ferreira Gomes took cacao as an ornamental from Brazil to the Portuguese colony of Principe in 1822 and thence to São Tomé in 1830, both in the Gulf of Guinea (J. Ascenso, pers. comm.). This introduction, now referred to as the West African *amelonado*, is arguably the single most important transfer of cacao germplasm considering its impact (B. Bartley, pers. comm.). Ghana obtained cacao from Fernando Po in 1879, and by 1951 West Africa accounted for 60 percent of global cacao production. The old British Imperial College of Tropical Agriculture in Trinidad was responsible for much of the transoceanic movement of cacao germplasm. Now part of the University of the West Indies, the cacao collection on Trinidad is still one of the best field gene banks for the crop. Currently, Côte d'Ivoire, Brazil, Malaysia, and Ghana are the main cacao exporters. Cacao is an important cash crop for smallholders as well as larger plantation owners in many parts of the lowland, humid tropics.

While Ghana's position has recently slipped as a major cacao producer, in part because of problems with disease, Malaysia is dramatically expanding its cacao production and promises to shortly become a major exporter.

FIGURE 2.8 Wilson Popenoe taking a break from collecting avocado (*Persea americana*) germplasm to drink chocolate mixed with pepper near San Cristobal, Verapaz, Guatemala, in 1924. The hot beverage is being served from a bowl made from calabash gourd (*Crescentia cujete*). (Courtesy of Hugh Popenoe.)

Cacao production in Malaysia, on sizable plantations as well as small farms, leaped from 2,300 metric tons in 1970, to 65,000 in 1983, to 125,000 in 1986 (Jain, 1988:10; G. A. R. Wood and Lass, 1987:544). On Kauai, Hawaii, the McBryde Sugar company has begun experimenting with cacao, perhaps as a hedge against any future reduction of the subsidy for sugar production in the United States.

With the emergence of Malaysia as a significant cacao producer, coupled with Brazil's ambitious plans to expand production of the commodity and surpass Côte d'Ivoire as the dominant world producer, overproduction threatens to undercut the profits of cacao growers worldwide. After rising in the 1970s, cacao prices dipped for most of the 1980s, thereby putting additional pressure on marginal producers (*Economist*, 19 December 1987, p. 61; World Bank, 1983).

Prospects for cacao demand, however, are bright. The 1980s witnessed a veritable chocolate binge in North America, particularly for luxury chocolates. Several up-market chocolatiers, particularly established Belgian com-

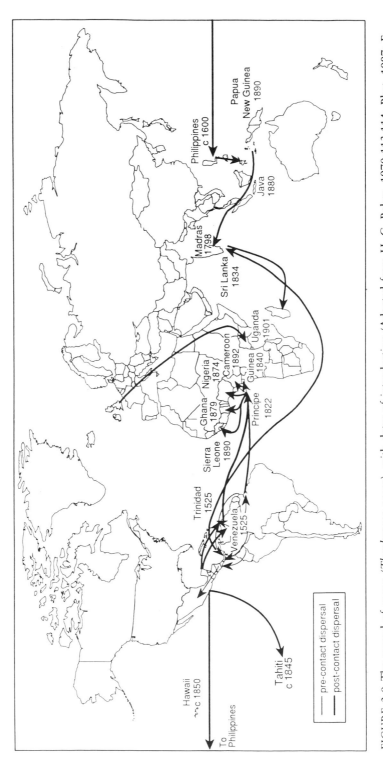

FIGURE 2.9 The spread of cacao (*Theobroma cacao*), with dates of introduction (Adapted from H. G. Baker, 1970:113,114; Bhat, 1987; E. Brown and Hunter, 1913; Cheesman, 1932; Cope, 1976; Crawford, 1937; Hardy, 1960:14; Hermann et al., 1989:22; Knapp, 1923; Laryea, 1981; Prior, 1984; Purseglove, 1974:573; Ratnam, 1961; Urquhart, 1961; G. A. R. Wood and Lass, 1987.)

Map labels:
- Philippines c 1600
- Papua New Guinea 1890
- Java 1880
- Madras 1798
- Sri Lanka 1834
- Uganda 1901
- Cameroon 1892
- Nigeria 1874
- Ghana 1879
- Guinea 1840
- Principe 1822
- Sierra Leone 1890
- Trinidad 1525
- Venezuela 1525
- Hawaii c 1850
- To Philippines
- Tahiti c 1845

Legend:
— pre-contact dispersal
— post-contact dispersal

panies such as Godiva, Neuhaus, Schoofs, and d'Orsay, have opened out-
lets in major metropolitan areas of the United States and Canada. The
boom in hand-crafted, superior chocolates, so typical of western Europe
and now increasingly common in North America, has created a growing
demand for high-quality cacao beans. The quality of chocolate is determined by numerous factors, such as the degree of fermentation and proper
drying techniques, but varietal selection is important. Also, concerns that
eating chocolate can raise blood cholesterol levels have been alleviated by
recent studies that demonstrate that the major component of cocoa butter—stearic acid—does not increase the fatty substance that can clog
arteries (Bonanome and Grundy, 1988; Roberts, 1988b). Reports widely
publicized on U.S. television that eating chocolate helps create an imper-
vious coating to the teeth, thereby providing a shield against decay-causing
bacteria, may further boost sales of chocolate confectionery.

Opportunities abound for genetically improving the quality of cacao
beans and other desirable characteristics. The genetic resources of cacao,
which occur as semiwild trees in Central America and Amazonia, have
hardly been tapped (Bartley, 1981). Near relatives of cacao have not been
used in cacao improvement at all. West Africa's cacao production rests on
a particularly narrow germplasm base, mostly simple selections rather than
highly bred materials (Cope, 1976; Juma, 1989:136; J. A. Williams,
1978). The potential for improving the yield, flavor, and hardiness of cacao can be realized only if the tremendous variation found in wild cacao
populations, traditional backyard plantings, and genebanks is screened
more fully and incorporated into breeding experiments.

DOMESTICATED RELATIVES OF CACAO

Several of cacao's relatives have been domesticated, but none has
achieved the prominence of cacao. Cupuaçu (*T. grandiflorum*) is native to
eastern and central Amazonia and is cultivated in parts of Amazonia for its
delicately flavored pulp, which is used locally to make aroma-rich drinks,
ice cream, and puddings. Cupuaçu's genetic resources and potential are
discussed later in this chapter.

Cacaui (*T. speciosum*), an understory tree in forests of northern South
America and southern Central America, is occasionally cultivated in the
Brazilian Amazon (Santa-Anna Nery, 1885:90). The Ka'apor tribe on the
eastern fringes of Amazonia, for example, sometimes grow cacaui (Balée,
1989b; Balée and Gély, 1989). In the Bolivian Amazon, the Chácabo also
plant *T. speciosum* where the species also occurs wild in the forest (Boom,
1989). More commonly, though, the tennis-ball-sized fruits (Figure 2.10)
are broken open by hunters and gatherers of forest products to snack on
the tart-sweet pulp.

FIGURE 2.10 Cacauí (*Theobroma speciosum*) with fruit in forest. Serra dos Carajás, Pará, Brazil, February 1990.

Along the Casiquiare Canal of southern Venezuela, *T. subincanum* and an unidentified species of *Herrania*, a genus closely related to *Theobroma*, are grown in some dooryard gardens for their juicy pulp (Sanchez et al., 1988). Another relative of cacao, *T. obovatum* (Figure 2.11), may be culti-vated on a minor scale along the Caquetá River in the Colombian Amazon (J. B. Allen, 1987).

In southern Mexico, *T. angustifolium* and *T. bicolor* are cultivated in Veracruz, Chiapas, and Tabasco to make chocolate. In the Ecuadorian Amazon, *T. bicolor* is also occasionally planted in dooryard gardens (J. B. Allen, 1987:15, 18); the pulp reputedly tastes like a blend of durian and papaya. The Runa who inhabit the headwaters of the Payamino in the Napo Province of the Ecuadorian Amazon consider the presence of *T. bicolor* as an indicator of former human occupation; the Runa cultivate *T. bicolor*, known locally as *patasi*, and protect *T. speciosum* and *T. subincanum* in the forest (D. Irvine, 1989). The Aguarana Jivaro cultivate *T. bicolor* along the Cenepa River, an affluent of the Marañon, in the Peru-

FIGURE 2.11 Fruits of *Theobroma obovatum*.

vian Amazon (Boster, 1983). At the turn of this century, Indians collected
fruits of *T. bicolor* along the Purus River in western Amazonia (Huber,
1906). Beans from *T. angustifolium* or *T. bicolor* seldom enter world trade
because they render an inferior chocolate.

EVOLUTION OF CULTIVATED CACAO

Three main types of cultivated cacao have evolved, often with considerable variation within each type (Soria, 1973). The current typology was
drawn up on the basis of geography and materials collected in the early
years of modern cacao improvement. *Criollo* cacaos originated in Central
America, *forastero* cacaos are from the Amazon basin, while *trinitario* cacaos arose on Trinidad. Classification of different morphological forms,
based principally on pod shape, has been confused and has no botanical
significance (León, 1960). Some cacao specialists prefer to subdivide *forastero* cacaos into subgroups, but until the genetics of cacao is better understood, only three main groupings are recognized here.

Criollo cacao developed in northern South America or Central America
and is characterized by thin-walled, red or yellow pods that are easily
opened. The beans are large, round, and contain white, or occasionally
pale violet, cotyledons when ripe. The plump beans are not astringent and
render the highest quality chocolate of all the cacao types (Ascenso, 1964;
Cheesman, 1932). Although they produce a superior chocolate, *criollo* ca-

caos are hardly cultivated anymore because they are low-yielding and highly susceptible to several diseases.

Criollo cacaos have been losing ground to improved materials throughout their former range in Central America and other regions since at least the 1930s. In Mexico, Tabasco and the Soconusco region of Chiapas have been traditional strongholds for *criollo* cacao, but practically all new plantings since the 1930s have been *trinitario* or *forastero* types (Cheesman, 1944). In 1943, G. E. Spencer of the former Imperial College of Tropical Agriculture sought *criollo* cacao in Guatemala, Honduras, and Nicaragua, and found the situation considerably changed since the visit of Paul Preuss to those countries in the late 1800s (Preuss, 1901). Spencer found only a few scattered *criollo* cacao trees in Guatemala, none in Honduras, and some *criollo* cacao in plantations of mixed cacao types around Rivas in Nicaragua (Cheesman, 1944). In Nicaragua, it had become hard to locate pure *criollo* cacaos by the early 1940s. By 1958, Nicaraguan *criollos*, also known as 'Cacao Real', had been largely replaced by *forastero* and *trinitario* types (Soria, 1966). In 1986, only five *criollo* cacao specimens were located in Guatemala after four collecting missions to Alta Verapaz, Izabel, Petén, San Marcos, and Malacatan that covered 4,383 kilometers by car and over 400 kilometers on foot (Rivera, 1986). In the Sinú Valley of Colombia, fungal disease and pests forced the replacement of high-quality *criollo* types with better-yielding *forastero* varieties in the 1920s (Parsons, 1952).

Many of the introductions of cacao in the sixteenth and seventeenth centuries (Figure 2.9) were of the *criollo* type, and vestiges of these white-beaned cacaos may be found on smallholdings. In 1798, for example, *criollo* cacao was taken from Ambon in the Moluccas to Madras. Isolated pockets of *criollo* cacao linger in parts of Colombia, Venezuela, Madagascar, Comoro Islands, and Samoa (Braudeau, 1974). *Criollo* cacao on Madagascar is derived from material sent from Trinidad to Sri Lanka in 1834, which in turn sent germplasm to Réunion. A few *criollo* cacao trees were noted on São Tomé in the early 1960s (Ascenso, 1964).

The near demise of *criollo* cacao is understandable in the face of higher-yielding and more vigorous cacao types. But the continued loss of *criollo* cacao is worrisome because of the high-quality chocolate derived from it. Many of the finer *forastero* cacaos contain *criollo* germplasm. Given the spectacular upsurge of interest in premium chocolates, it might pay to plant pure *criollo*. In spite of its lower yield, growers catering to the higher end of the market with *criollo* cacao may realize a profit. Some specialty growers on Jamaica have begun experimenting with *criollo* cacao (A. Todd-Bockarie, pers. comm.). The British East India Company introduced *criollo* cacao to Madras from Amboina in 1798; some of the progeny from this introduction under observation at the Kallar Fruit Station produce exceptionally fine chocolate (Ratnam, 1961). However, in Kerala, along India's Malabar coast, trials of *criollo* cacao have been disappointing because of disease problems and poor yields (Das, 1982).

Fortunately, *criollo* cacaos were introduced to Trinidad in 1893 from Nicaragua, and some of these selections proved useful in cacao improvement work at the former Imperial College of Tropical Agriculture. The International Board for Plant Genetic Resources (IBPGR) ranks *criollo* cacao high on its priority list for collecting and conserving germplasm. IBPGR has funded and organized trips to collect germplasm of wild and *criollo* cacao in Guatemala and Mexico and to look for it when collecting other cacao in Colombia.

Forastero cacaos are a diverse group, in both the wild and cultivated forms. The thick-walled pods of *forastero* cacaos are usually yellow when ripe and are characteristically smooth and rounded at the ends. The flattened beans usually contain violet cotyledons. Because of their high fruit production, *forastero* cacaos dominate world production. Traditional *forastero* cacaos, propagated locally by seedlings, can still be found in parts of Amazonia where their production was encouraged by the Portuguese and Spanish. The Madeira and Purús river systems, for example, are noted for their cacao; locals still fashion oblong blocks of ground cacao for making chocolate drinks and desserts. Cocoa butter is not expressed from these home preparations, and the resulting oily flavor of rustic chocolate is an acquired taste. In the late nineteenth century, the central Amazon between Santarém and Obidos had the densest plantations of cacao (Spruce, 1908a: 78).

Trinitario cacaos are highly variable because of their hybrid origin on Trinidad. After *criollo* plantations were severely damaged by a blast of unknown cause in the eighteenth century, *forastero* cacao was introduced to Trinidad and apparently cross-pollinated with some surviving *criollo* cacaos. *Trintario* cacaos are known in the trade as "fine cacao," and the excellence of their flavor may be partly due to *criollo* germplasm.

WILD GENEPOOLS AND IN SITU CONSERVATION

Wild cacao in the forest is highly variable, particularly in its core area, suggesting a potentially rich source of germplasm for future breeding efforts (Bartley, 1963; IBPGR, 1981). Populations of wild cacao are often quite uniform within a valley or restricted area but may vary markedly from one watershed to the next. Thus, although wild cacao has an ample distribution, destruction of forest in any part of its range can significantly reduce variation in the gene pool.

Wild cacao has apparently adapted to a variety of ecological conditions.

Along the Jari River separating Amapá and Pará in the Brazilian Amazon, wild cacao is found along floodplains of streams as well as on well-drained upland soils (Vello and Silva, 1968). In Amazonia, wild *T. cacao* is found in both flood-prone areas and on uplands, whereas in the Guianas wild cacao prefers low-lying, wet sites (J. B. Allen and Lass, 1987:48; Huber, 1906; J. G. Myers, 1930; Van Roosmalen, 1985:420). In Rondônia, Brazil, on the other hand, wild cacao favors deep red alfisols (*terra roxa*) derived from weathered basalt (Almeida and Almeida, 1987). In the vicinity of the President Medici settlement along the Cuiabá-Porto Velho highway in Rondônia, wild cacao reaches densities of 142 trees per hectare (Maciel and Lisboa, 1989). Such high concentrations of cacao would favor witches'-broom epidemics; the environs of President Medici might be a good place to look for resistance to the disease.

Populations of wild cacao in the "core" of the species' natural distribution, in the foothills of the eastern Andes in Ecuador, Peru, and Colombia, are self-incompatible. In contrast, populations closer to the periphery are self-fertile (Cope, 1976). Midges are regarded as the primary pollinators of cacao, but several species of ants (*Crematogaster* sp. and *Ectatomma tuberculatum*) and a picture-winged fly (*Euxesta* sp.) may be significant pollinators of forest cacao along the Napo River in the Ecuadorian Amazon (Wright, 1984). As an obligate outcrosser in its core area, cacao thus maintains its heterozygosity. The self-incompatibility of some Amazonian cacao populations has proved useful in developing "hybrid" varieties at the Cocoa Research Institute in Tafo, Ghana (Glendinning, 1967). Current cacao "hybrids" are really synthetic crosses between desirable parents achieved by hand-pollination.

Wild cacao is threatened in many parts of its extensive range (J. T. Williams, 1981). Government-sponsored schemes to siphon excess human populations from the Andean highlands to the Amazon lowlands in Peru and Ecuador are accelerating deforestation of wild cacao habitats. The dramatic increase in coca plantings in the Upper Amazon has probably also destroyed populations of wild cacao and other genetic resources. Indians in the Ecuadorian Amazon leave wild cacao alone when clearing forest for their fields, a practice not usually followed by colonists arriving from the Andes (J. B. Allen and Lass, 1983:16). Oil drilling in the Peruvian and Ecuadorian Amazon is disturbing some ecologically diverse forest habitats. The 1.6-million-hectare Jari concession in Pará State, now run by a consortium of Brazilian companies, has probably eliminated some cacao populations on well-drained uplands to make room for 80,000 hectares of pulp-producing eucalyptus, Caribbean pine, and Gmelina plantations. Brazil nut gatherers along the Jari clear wild cacao under giant *Bertholletia excelsa* to facilitate the gathering of the hard pods (*ouriços*) that contain the protein-rich nuts. The fact that wild cacao favors fertile soils in at least two parts of its range, black soils in the Ecuadorian Amazon and alfisols in Rondônia, renders them particularly vulnerable to clearing for farms or plantations. Ironically, the clay terra roxa soils in Rondônia are targeted for cacao plantations and other crops because they are the most fertile soils in the state.

Naturalized populations of cacao have also been found in widely scattered locations in tropical America and the Caribbean. Naturalized cacao was reported on Jamaica in the 1800s (Grisebach, 1864:91). Some "wild" cacao trees have been reported in the forests of the Cauca valley in Colombia and in western Ecuador (F. J. Pound, 1938:13). In Central America, presumably feral populations of *T. cacao* have been found on Barro Colorado Island in Panama (Croat, 1978:597; Standley, 1927:24), Costa Rica (Holdrige et al., 1971:155), the Caribbean slope of Nicaragua (Belt, 1888:345), the Lancetilla Valley in Honduras (Standley, 1931), the Caribbean coastal forest of Guatemala (Brigham, 1887:345), and in Mexico from Colima in Michoacan to Vera Cruz, Chiapas, and Tabasco (O. Lopez et al., n.d.; Price, 1917; Standley, 1924:805). "Wild" cacao in Mexico is highly variable.

These "wild" populations outside the native range of cacao are presumably derived from escapees from cultivation. The distinction between a truly wild cacao and a feral specimen can be hard to draw. Some naturalized cacao populations probably stem from plantations that were reclaimed by the forest centuries ago. From the perspective of plant breeders, both wild cacao and feral populations contain potentially valuable genes. Little is known about gene flow in the wild; such information would be invaluable for breeders. Feral populations are even more threatened than wild cacao in its native range. If cacao germplasm is conserved in situ in its core area, it could be argued that introduced Central American forms are of peripheral interest. Yet Mesoamerican cacao has apparently differentiated substantially from populations in Amazonia and therefore warrants conservation. The forests of Central America, the Cauca Valley of Colombia, and the Ecuadorian coast are falling especially fast. Deforestation in Nicaragua is threatening wild cacao (Sutton, 1989).

Sizable forest reserves are needed to secure wild cacao and its dispersal agents. Near water, cacao is dispersed by currents, but on land, monkeys, squirrels, and bats probably spread wild cacao (J. P. Allen and Lass, 1983: 48; P. Wilson, pers. comm.). In the Amazon, capuchin monkeys (*Cebus apella*), agoutis (*Dasyprocta* spp.), and macaws are occasional pests in cacao orchards and may be involved in dispersing the beans. Brown capuchins (*C. apella*) and white-fronted capuchins (*C. albifrons*) smash pods of wild cacao on branches to access the sweet pulp (Terborgh, 1983:76). Large tracts of forest are needed to sustain cacao's dispersal agents and pollinators.

Cacao has some twenty near relatives (Table 2.1). Cacao's near relatives have been largely ignored by breeding programs, in part because they have been perceived as impossible or very difficult to cross with *T. cacao*. Widecrossing experiments have been under way since the mid-1940s (Addison and Tavares, 1952). In Ghana, for example, cacao has been crossed with several near relatives, including cupuaçu, but all the interspecific crosses have been weak and have not borne fruit (Glendinning, 1967). Viable seeds have been produced after crossing *T. cacao* with *T. microcarpum* and *T. grandiflorum*, however. Cope (1976) suggests that cacao's near relatives do not have much to offer, but little is known about the potential of *Theobroma* or *Herrania*. No systematic search has been undertaken of potentially useful traits in the wild relatives of cacao.

Some of the near relatives surely contain genes for resistance to prevalent diseases and pests of cacao. As widecrossing techniques become more routine, their potential may be tapped in the future—provided that representative forest habitats where they occur can survive the onslaught of settlers and development.

TABLE 2.1

Cacao's near relatives (*Theobroma* spp.) and their approximate distributions, uses, and local names

Theobroma species	Distribution	Uses and local names
angustifolium[1]	Central America	Mixed with cacao to make cocoa in Mexico and Costa Rica; *cacao de mico, cacao silvestre* (Costa Rica), *cushta* (El Salvador)
asclepiadiflora	Panama	
bernoullii	Panama	
bicolor[2]	Veracruz, Chiapas, Tabasco (Mexico) to Pará (Brazil)	Pulp eaten and beans ground into cocoa in Mexico; *pataxte* (Mayan), *pataste* (Mexico to Costa Rica), *bacao* (Colombia), *cacau do Pará, cacau do Peru, cacauarana* (Brazil), *patasi* (Ecuador)
calodesmis	Amazon region of Peru; Colombia	
capillifera	Colombia	*cacao de monte*
chocoense	Colombia	
cirmolinae	Colombia (Pacific region)	
gileri	Ecuador; Colombia (Pacific region)	
glaucum[2]	Amazonia in Ecuador and Colombia	*challua cacao*

TABLE 2.1—*cont.*

Theobroma species	Distribution	Uses and local names
grandiflorum	Pará, Brazil	Cultivated in Amazonia for its pulp, which is eaten and used to make drinks, ice cream, and puddings; *cupuaçu* (Brazil), *copo azu* (Venezuela)
mammosum	Costa Rica	
microcarpum	Amazonia in Colombia and Brazil	*cacaurana, cabeça de urubu, cacau-jacaré* (Brazil)
nemorale	Colombia (Pacific region); Amazonia	*cacao de monte* (Colombia), *cacao montero* (Venezuela)
obovatum[2]	Amazonia, Brazil; Ecuadorian Amazon; Caquetá River, Colombia	*cacau cabeça-de-urubu* (Brazil)
simiarum	Costa Rica; Panama; Colombia	*cacao de mico* (Costa Rica)
speciosum[2]	Amazonia, northern South America to southern Mexico	*cacauí* (Brazil), *cacaurana*
spruceana	Pará, Brazil	*cacau azul*
stipulatum	Colombia (Pacific region)	
subincanum[2]	Venezuela; Amazonia in Peru, Ecuador; Pará, Rondônia, Amazonia (Brazil); French Guiana	*cupuí* (Brazil), *sacha cacao* (Peru)
sylvestris	French Guiana; Brazil; Colombia	*cacauí, cacau azul, cacau da mata* (Brazil)

Sources: J. B. Allen, 1987; J. B. Allen and Lass, 1983:16, 18, 35, 39; C. M. V. C. Almeida et al., 1987:14; J. O. P. Carvalho et al., 1986; Cavalcante, 1988:64; Hardy, 1960:308, 314; Holdridge et al., 1971:330; Le Cointe, 1947; Peters et al., 1989; F. J. Pound, 1938; Purseglove, 1974:571; Rodrigues, 1986; Sanchez et al., 1987, 1988; Silva et al., 1977; Spruce, 1908a:82; Standley, 1924; J. A. Williams, 1978.
[1]Wild species used to make chocolate in Central America.
[2]Cultivated on a limited scale.

Conservation areas for cacao's near relatives will have to be large to accommodate dispersal agents; monkeys, for example, open fruits of *T. microcarpum* along the Purus River to feed on the pulp (Huber, 1906). Western Amazonia has an especially large concentration of cacao relatives and therefore warrants particular attention for in situ conservation. In Costa Rica, Panama, and Colombia, cacao de mico (*T. simiarum*) is probably also dispersed by monkeys as its Spanish name suggests (Table 2.1).

COLLECTING GERMPLASM

Cacao seeds do not remain viable for long and tend to sprout if any delays are encountered in transporting them from the field to a gene bank or quarantine facility. Collectors have therefore experimented with various vegetative means of collecting germplasm. "Low-tech" approaches, such as budwood cuttings, have proved disappointing because the cuttings often die during transportation. The University of Nottingham, United Kingdom, has developed a useful method for collecting cacao germplasm in which the surfaces of small twigs are sterilized with bleach and inserted into tubes containing agar, fungicides, and antibiotics. This relatively simple method has dramatically improved the survival rate of vegetative germplasm samples during transportation (Withers, 1987).

Disinfecting twigs and sealing them in tubes has become the method of choice for collecting cacao germplasm. Already, this technique has been applied to some other woody crops, such as breadfruit. With certain modifications tailored to the needs of individual species, this technique is likely to be used in collecting trips for germplasm of perennial plants in the near future.

FIELD GENE BANKS

Cacao has unorthodox seeds that cannot be dried and frozen without killing the embryos. Because field gene banks are costly to maintain, only a small fraction of the genetic diversity of *Theobroma* is represented in ex situ collections. For example, Brazil's cacao research and development program (CEPLAC, Comissão Executiva do Plano da Lavoura Cacaueira) spends approximately $3,300 a year just on upkeep for 537 clones kept at a field gene bank near Itabuna, Brazil. Salaries of scientists and supporting staff increase dramatically the costs of maintaining and evaluating useful collections. Some 4,000 to 5,000 accessions of cacao are maintained in field gene banks worldwide. But there are many duplicates, and only a small portion of cacao's genetic resources are safeguarded in them (J. B. Allen, 1987:12; J. B. Allen and Lass, 1983:59; L. E. Lopez, 1981). In association with the London Cocoa Trade Board, IBPGR has initiated a project to document collections so that unique accessions can be identified and duplicated for safety and unnecessary duplication can be eliminated.

Habitat protection is the best way to conserve the range of diversity of wild cacao. For cultivated forms, targeted wild populations in imminent danger of extinction, and certain wild types of particular interest to breeders, field gene banks can serve both as a conservation measure and as a ready source of fresh germplasm for breeders. Gene banks could serve as holding areas for wild material to help restore some deforested areas.

The International Cocoa Genebank, Trinidad (ICGT), with 1,872 clones as of 1988, holds perhaps the most comprehensive collection of cacao germplasm (Table 2.2). This gene bank sends cacao material to interested parties throughout the world. In 1964 and 1965, for example, ICGT sent

TABLE 2.2

Major field gene banks for cacao and its near relatives

Institution	Location	No. of accessions
Cocoa Research Institute	Tafo, Ghana	6,000
International Cocoa Genebank	Trinidad	1,872
CEPLAC	Belém, Pará, Brazil	1,749
CEPLAC	Itabuna, Bahia, Brazil	537
CATIE	Turrialba, Costa Rica	518
INIAP	Pichilingue, Ecuador	500
Tropical Agriculture Research Station	Mayaguez, Puerto Rico	372
Subtropical Horticulture Research Station	Miami, Florida	320
IFCC	Abidjan, Côte d'Ivoire	300
INIAP	San Carlos, Ecuador	281
Cocoa Research Station	Tawau, Sabah, Malaysia	200

Source: J. T. Williams and Damania, 1981; N. J. H. Smith, field notes.
Note: See Appendix 2 for spelled-out names of institutions.

600 clones and 50,000 seeds to Nigeria (B. Bartley, pers. comm.). Significantly, ICGT is integrated with the Cocoa Research Unit of the University of the West Indies, which ensures that the collection is well tended and used. With support from the European Development Fund, ICGT is relocating its gene bank on a 35-hectare site with room to expand (Toxopeus and Kennedy, 1989).

The roots of the ICGT extend to F. J. Pound's collecting mission to the Amazon region of Ecuador, Colombia, and Peru in 1937–38 (Mabbet, 1988). More accessions have since been added, and the gene bank periodically receives duplicates of accessions collected in Ecuador by a recent London Cocoa Trade Amazon Project (J. B. Allen, 1987; J. B. Allen and Lass, 1983). Germplasm collected during this project is being held at a 4-hectare gene bank at San Carlos in Ecuador; the 281 accessions held there are being duplicated for distribution to the ICGT and other collections. ICGT will soon be adding cacao germplasm samples collected in Bolivia and Mexico. Nevertheless, some materials are being evaluated in areas where the environmental stresses are most severe, as in West Africa, where the cocoa swollen shoot virus occurs. Much work remains to be done since only a small portion of the upper Amazon germplasm maintained at ICGT has been evaluated and used (Toxopeus and Kennedy, 1989).

Another cacao germplasm collection designed to serve the needs of cacao research programs worldwide is held at the Subtropical Horticulture Research Station of the USDA Agricultural Research Service (ARS) at Chapman Field, Miami. Known as the Miami collection, this germplasm assemblage of 320 clones is kept in a glasshouse as a precaution against frost. A duplicate collection is held in the open at the Tropical Agriculture Research Station near Mayaguez, Puerto Rico. The value of duplicating germplasm collections was underscored in 1987 when a technician at the Mayaguez station inadvertently sprayed a powerful herbicide on the cacao collection thereby killing several accessions. The Miami collection has dispatched material to various agricultural research stations, such as in Papua New Guinea and Samoa, but both collections are vulnerable to hurricanes, and neither the Miami nor the Mayaguez gene bank is linked directly to any breeding program.

Other important cacao germplasm collections are maintained by CEPLAC near Itabuna, Bahia, and in the Amazon near Belém and Altamira. The Itabuna gene bank is well maintained on 6 undulating hectares. The 537 genotypes in the Itabuna collection are evaluated for yield and resistance to black pod, caused by various fungi (*Phytophthora* spp.). The cacao germplasm collection at CEPLAC's genetic research station for cacao (ERJOH, Estação de Recoursos Genéticos do Cacao José Heraldo) is 17 kilometers from Belém and contains 1,700 clones and over 23,000 seedlings. A major distinguishing feature of the Belém collection is that more than 90 percent of the material has been collected in the wild (C. M. V. C. Almeida et al., 1987:3).

Although the Belém gene bank was formally established only in 1979, CEPLAC has organized over fifty germplasm collecting trips since 1965 (Bartley et al., 1987). The Belém gene bank contains a small collection of nine of cacao's near relatives including *Herrania*. Access to the 269-hectare Belém gene bank is easy, but the site suffers from heavily leached soils that require fertilization, thereby adding to maintenance costs. Constant vigilance is also required to prevent squatter settlements from spilling across the rear boundary of the property. The open, flat site could use some nitrogen-fixing shade trees to reduce fertilizer costs and protect seedlings and recently grafted material. In spite of these difficulties, the well-tended collection is being evaluated systematically for fruiting characteristics and other traits. CEPLAC keeps a smaller cacao germplasm collection at kilometer 100 of the Altamira-Itaituba stretch of the Transamazon Highway, mainly to serve cacao growers along the rainforest road.

Colombia began assembling cacao germplasm in 1943 and now holds 641 accessions in field gene banks at Palmira, Caribia, and Tulenapa. These collections include local varieties, introductions, primitive cacao cultivars, and some near relatives of cacao. A total of 186 clones in the collections have been introduced from other countries. Colombia's cacao research program, part of the national agricultural research system (ICA, Instituto Colombiano Agropecuario) has wisely sponsored the collection of indigenous cacao germplasm with a view to identifying sources of disease resistance, among other traits (Camacho et al., 1991). Evaluation of the collections is under way.

Cacao gene banks need upgrading with more material and increased funding for maintenance and evaluation (Bartley, 1981). The cacao gene bank linked to the Cocoa Research Institute, Tafo, Ghana, is in one of the world's major cacao-producing regions, but the 6,000 accessions in the collection are progeny from no more than sixty trees (J. B. Allen, 1981). Close relatives of cacao are especially poorly represented in collections. The cacao germplasm collection maintained by CATIE (Centro Agronómico Tropical de Investigación y Enseñanza) near Turrialba, Costa Rica, is one of the more comprehensive collections but contains only twelve accessions of nine of cacao's near relatives. Nevertheless, a strength of the CATIE collection is the inclusion of *criollo* accessions gathered in Central America during the 1950s and 1960s (Toxopeus and Kennedy, 1989). The gene bank at the Subtropical Horticulture Research Station, Miami, contains over 300 accessions of cacao but only one accession of *T. bicolor* and three accessions of *T. grandiflorum*. Also, cooperative agreements need to be worked out so that materials can be safely exchanged as needs arise. No one country is likely to have all the germplasm for cacao improvement in the future.

BREEDING CHALLENGES

The gap between cacao yields achieved on experiment stations and productivity obtained by growers is larger proportionally than for most crops. Under ideal conditions, cacao can yield 3.7 metric tons of beans per hectare every year, but the productivity of cacao on smallholdings and plantations is often less than 10 percent of the potential (Table 2.3).

A crop's productivity is determined by a host of factors, such as management practices, planting material, and the availability of improved technologies. Market prices are also important, since they usually determine the level of inputs made in cacao groves. The steep decline of Ghana's cacao production in the 1970s and early 1980s, for example, is attributed to unfavorable incentives for growers, neglect of plantations, and insufficient support for the national cacao research program. The Cocoa Research Institute of Ghana (CRIG), which took over from the West African Cacao Research Institute (WACRI) in 1962 after Ghana's independence, has been passed from one ministry or government agency to another an average of once every three years from 1962 to 1988 (Eicher, 1989). Commodity research for cacao and other export crops has also deteriorated in Nigeria and Tanzania, because of unstable funding and lack of political support (Lele, 1988).

In Brazil and Côte d'Ivoire, in contrast, cacao yields have doubled on

TABLE 2.3

Productivity of cacao on some experiment stations, plantations, and smallholdings

Country	Management	Year	Beans (metric tons/ha/yr)
Ghana[1]	Experiment station	ca. 1980	3.7
Papua New Guinea[1]	Experiment station	ca. 1980	3.0
Bahia, Brazil[2]	Experiment station	1980	2.0
Bahia, Brazil[2]	Av. growers' yield	1980	0.7
Papua New Guinea[1]	Good plantation yield	ca. 1980	0.7
Côte d'Ivoire[2]	Av. growers' yield	1980	0.6
Papua New Guinea[1]	Av. plantation yield	ca. 1980	0.3
Papua New Guinea[1]	Av. smallholder yield	ca. 1980	0.2
Cameroon[2]	Av. growers' yield	1979	0.3
Ghana[2]	Av. growers' yield	1979	0.2
Trinidad[3]	Av. growers' yield	1980s	0.17

[1]Prior, 1984.
[2]Rangel, 1982:56.
[3]Hunter, 1990.

average since the 1960s. These productivity gains have been achieved because growers have been well served by their respective national cacao research and extension programs. Sustained support for cacao research in Côte d'Ivoire has been a major factor in the country's increasing share of the world trade in the commodity. In 1961–63, for example, Côte d'Ivoire accounted for only 9.2 percent of the global trade in cacao beans, but by 1983–85, its share had increased to 29 percent (Lele, 1988).

Even in countries with relatively well-supported cacao research programs, the gap between experimental yields and those obtained on plantations and smallholdings is still impressive. Although yields obtained under ideal conditions on experiment stations are usually not realizable under field conditions, much work can be done to close some of that gap. Diseases and pests and low soil fertility are major factors in depressing cacao productivity under field conditions. Between 20 and 30 percent of cacao production is lost to diseases and pests (Juma, 1989:136).

Cocoa swollen shoot virus (CSSV) is the most important pathogen of cacao in West Africa, particularly Ghana, and it is a significant quarantine concern. CSSV was first noted in Ghana in 1936, Nigeria in 1940, Côte d'Ivoire in 1945, Togo in 1955, and Sierra Leone in 1958 (H. G. Baker, 1970:11; Hardy, 1960:273; Lavabre, 1981). The viral disease prompted the establishment of WACRI in Ghana in 1944 (J. A. Williams, 1978). CSSV was partly responsible for Ghana's precipitous decline as a leading cacao exporter. Exports of cacao beans from Ghana declined from 571,700 metric tons in 1965 to an average of 200,000 metric tons a year by the 1980s (Ampofo and Osei-Bonsu, 1987). Attempts to control the disease have sapped extension efforts and have destroyed millions of cacao trees. By 1957, some 70 million cacao trees had been cut down in Ghana in an effort to control the disease, and by the end of 1986, 187 million cacao trees had been felled in an unsuccessful attempt to contain the disease (Ollennu and Owusu, 1987; Purseglove, 1974:592; Thresh et al., 1988).

Initially thought to be a soil nutrient deficiency, the virus has many strains, thus complicating the task of screening and breeding for disease resistance. Diversity in the pathogen also makes detection difficult when checking germplasm for international exchange. For example, only in 1987 was an ELISA (enzyme-linked immunosorbent assay) test developed for the pathogen, but this test may not detect all strains of the virus (Plucknett and Smith, 1989).

CSSV is confined to West Africa, where it is indigenous to forest trees belonging to the Sterculiaceae (to which cacao belongs), Bombacaceae, and Tiliaceae (H. G. Baker, 1970:117). The pathogen also infects a bombacaceous tree introduced from tropical America: the kapok tree (*Ceiba pentandra*). The kapok tree has been widely disseminated in tropical Africa and Asia for the soft fiber surrounding the tiny seeds, which is used to stuff mattresses and pillows. Both native and introduced trees, then, provide a reservoir for the pathogen, which readily infects cacao plantations.

Pruning infected branches and removing diseased trees have not eliminated the problem and have lead to lost production. Reinfection is common since the pathogen occurs in nearby forest. Insecticidal spraying of the pathogen's vectors has also proved ineffective. Mealybugs (*Planococcoides* spp.) transmit CSSV, and they are protected by a waxy coat supplied by ants (*Crematogaster striulata*). The ants feed on a sugary fluid exuded by the more than fourteen species of mealybugs involved in transmitting the disease, and in turn the ants protect the mealybugs (Adegbola, 1981). While insecticides temporarily reduce ant populations, mealybugs are hardly affected.

Large-scale and continued use of insecticides on cacao farms is impractical for several other reasons. The ants that protect mealybugs may serve to fend off other cacao pests. Cacao groves produce all year round in areas with no pronounced dry season, so insecticides can destroy midges and other insects that pollinate the crop. Continued use of insecticides is environmentally and financially costly. Furthermore, some insecticides impart off-flavors to cacao, and increasingly stringent food-quality standards in industrial nations will limit the range of insecticides available to cacao growers.

Genetic resistance to CSSV is the most efficient and environmentally benign approach to controlling the disease. Cacao research programs in

Côte d'Ivoire, Ghana, and Nigeria have been screening cacao germplasm for resistance to the disease for over four decades. Extensive screening of local materials in West Africa for resistance to CSSV proved fruitless because of the narrow genetic base of the crop (J. A. Williams, 1978). Consequently, the ICGT sent a broad selection of cacao materials from its collection to Ghana in 1944. A small part of this consignment, semiwild cacao from the upper Amazon, displayed vigor and precocious growth in the presence of the disease and was used in crosses to develop a commercial cultivar by the mid-1950s (Toxopeus and Kennedy, 1989).

Cacao breeders have tapped germplasm collections extensively in search of resistance to CSSV. Some of the wild and traditional cacao collected by F. J. Pound in the Peruvian Amazon in 1938 in his search for resistance to witches'-broom has proved especially useful in the quest for resistance to CSSV. Pound,[1] an agronomist employed by Trinidad's Department of Agriculture, obtained resistant material along the River Nanay (NA31-34), on Iquitos Island (IMC47, IMC60), at Parinari on the River Marañon (PA7, PA35), and Scavina 12 (Legg, 1981; Thresh et al., 1988). Such selections, deposited at the ICGT in Trinidad and subsequently forwarded to West Africa and other regions, have been used extensively in crosses to develop resistant and high-yielding material. The first Upper Amazon "hybrids" with reasonably good levels of resistance to CSSV were released in Ghana in 1986 (Thresh et al., 1988). Some *Trinitario* cacaos, such as T9/21 from Costa Rica, also contain genes resistant to CSSV.

Resistance to CSSV has been found in wild populations of cacao, even though the pathogen does not occur in the native range of the crop. Genes that code for some other function may confer resistance to the disease. Or, genetic resistance may be found in forest trees in West Africa known to harbor the pathogen. If resistance should be found in distant, or totally unrelated, species to cacao, it will be some time before the responsible gene, or genes, can be transferred to cacao.

CSSV is prompting a large-scale varietal turnover from the predominant 'Amelonado' cacao, a *forastero* type, to a more heterogeneous mix of clones and "hybrids." CSSV may be a blessing in disguise, since cacao orchards in West Africa are more genetically diverse than they were before CSSV reached epidemic proportions.

Another major disease of cacao, witches'-broom, caused by *Crinipellis perniciosa*, was first reported on cacao plantations in Surinam in 1895, Ecuador in 1918, Trinidad in 1928, and the Orinoco Delta of Venezuela in 1937 (Hardy, 1960:231; F. J. Pound, 1938; Purseglove, 1974:592). Endemic on wild cacao, its near relatives, *Herrania* species, and some wild

---

1. A pioneer collector of cacao germplasm, F. J. Pound died in a boating mishap off the coast of Liberia in 1945 (B. Bartley, pers. comm.).

species of *Solanum* in the Amazon basin (Wheeler and Mepsted, 1988), the disease occurs widely in northern South America, parts of the Caribbean, and is spreading upward through Central America. The pathogen was recorded in Panama in 1986 and threatens the cacao-growing areas of Costa Rica, Nicaragua, Guatemala, and Mexico. The symptoms of the yield-reducing disease include swelling and proliferation of shoots, which then atrophy (Figure 2.12).

Witches'-broom has essentially prevented Amazonia, the home of cacao, from becoming a major producer of the crop (Figure 2.13). In Brazil, the vast Amazon region produces less than 2 percent of national cacao production (Afonso, 1979), even though an estimated 120,000 hectares in Pará and 750,000 hectares in Rondônia are endowed with suitable soils and weather for cacao (Rangel, 1982:103). By encouraging orchard hygiene, however, Amazonia has recently emerged as a significant producer of cacao in Brazil. Between 1976 and 1988, some 90,000 hectares of cacao were planted in Amazonia, particularly in Rondônia (Alvim, 1989). The

FIGURE 2.12 Cacao attacked by witches'-broom, caused by *Crinipellis perniciosa*, near Belém, Pará, Brazil, 1988.

FIGURE 2.13 A recently established cacao grove not yet affected by witches'-broom near kilometer 80 of the Altamira-Itaituba stretch of the Transamazon Highway, Brazil, in September 1979. The child is the daughter of migrants of German extraction from Rio Grande do Sul.

Brazilian Amazon now produces more cacao than Mexico, Colombia, or Venezuela.

The dry cerrado separating Bahia from Amazonia, as well as care in exchanging germplasm, prevented the establishment of the disease in Bahia until the late 1980s. In 1989, two foci of the disease were confirmed in Bahia; one at Catolé, 40 kilometers northwest of Ilheus, and the other in the Municipality of Camacan, 100 kilometers south of Ilheus (P. Alvim, pers. comm.). Witches'-broom has now established a firm foothold in Bahia, and eradication efforts are no longer feasible. The recent arrival of the pathogen in Brazil's main cacao-producing region underscores the urgency of exploring for more sources of genetic resistance to the disease.

In Rondônia, control measures against witches'-broom are essentially restricted to removing infected parts of cacao trees (L. C. Almeida and Anderhan, 1987). If such measures are not followed carefully, losses can reach 90 percent (Rudgard, 1986). When cacao prices slump, as in the 1980s, growers tend to neglect their cacao groves, thereby increasing the amount of inoculum for infecting more cacao.

Genetic resistance would be a more practical method of controlling the disease. The hope of finding resistance to witches'-broom was the major rationale for F. J. Pound's collecting trip to the Upper Amazon in the late 1930s. Two of Pound's samples from the Scavina farm, SCA 6 and SCA 12, proved resistant to witches'-broom. Although SCA 6 and SCA 12 resisted the disease and were high-yielding, the beans were small.

Further crosses were necessary to develop productive clones with acceptable bean size as well as resistance to witches'-broom. A cross between SCA 6 and ICS 6, the latter a selection of the Imperial College of Tropical Agriculture (now the University of the West Indies) on Trinidad, has been widely planted in Latin America and the Caribbean (Mabbett, 1988; Purseglove, 1974:596). Clone IMC 67 from the vicinity of Iquitos, Peru, has some resistance to witches'-broom and is also vigorous, a precocious bearer, and well suited to a wide range of conditions.

Unfortunately, Scavina resistance to witches'-broom has broken down in some places, an indication that several races of *C. perniciosa* have evolved. In Ecuador, cacao with Scavina genes became highly susceptible to witches'-broom by 1950, only three years after resistant material was planted (B. Bartley, pers. comm.). By 1973, SCA 6 no longer withstood witches'-broom on Trinidad (Laker and Sreenivasan, 1987). Scavina germplasm was susceptible to the pathogen in the Brazilian Amazon by 1982.

Six geographically separated races of witches'-broom have been identified thus far (McGeary and Wheeler, 1988). Distinct isolates of *C. perniciosa* include: (1) Pichilingue and Rio Palenque in Ecuador and Chigorodo and Manizales in Colombia; (2) Sucua, Ecuador; (3) Manaus, Amazonas, Brazil; (4) Ouro Preto, Rondônia, Brazil; (4) Castanhal, Pará,

Brazil; (6) Trinidad and Tobago. Furthermore, the severity of disease symptoms on cacao varies considerably within some geographic groups of the pathogen (Wheeler and Mepsted, 1988).

Such heterogeneity in pathogen populations complicates the task of breeding for resistance since multiple-gene resistance will be needed. Perhaps the best that can be hoped for is reasonably high levels of tolerance to a broad array of *C. perniciosa* pathotypes rather than short-lived immunity or complete resistance to one or two forms of the disease (A. J. Kennedy, 1985).

New sources of resistance to witches'-broom will have to be found if there is any hope of controlling the disease in Tropical America and the Caribbean. In Venezuela, screening trials have indicated some resistance to witches'-broom among native *criollos* and some *trinitarios*. Attempts to infect *T. microcarpum* with the pathogen have failed, suggesting that resistance genes may be located in some of cacao's near relatives (G. A. R. Wood and Lass, 1987:283).

To accelerate the time-consuming process of screening and breeding for resistance to witches'-broom, researchers established an international collaborative research network in 1985. The International Witches' Broom Project (IWBP) links cacao pathologists and breeders in Brazil, Colombia, Ecuador, Trinidad, Venezuela, the United Kingdom, and United States in the search for effective control measures against the disease. Specific objectives of IWBP include studies on epidemiology, loss assessment, population biology of the pathogen, host resistance, and chemical control (Bonaparte, 1981). A commodity group, the International Office of Cocoa, Chocolate, and Confectionery Sugar (IOCCC) funds this exemplary collaborative research effort.

Black pod, caused by at least four species of pathogenic fungi (*Phytophthora palmivora*, *P. megakarya*, *P. capsici*, *P. citrophthora*), is the most widely distributed disease of cacao (Zentmyer, 1987). Black pod alone destroys between 10 and 30 percent of world cacao production (Juma, 1989:136; G. A. R. Wood and Lass, 1987:267). The fungi not only discolor the pods, but ruin the beans, kill seedlings, and damage young leaves. Several fungicides are effective against black pod, but they are expensive and often are not economically feasible (Persad, 1987).

Accelerated evaluation of cacao germplasm for resistance to this widespread disease is therefore warranted. Much of the early screening for black pod resistance was conducted before it was realized that several species of fungi could provoke the disease. Black pod disease is ripe for a research networking approach so that polygenic resistance can be developed more efficiently to the various pathogenic fungi. Some *trinitario* cacaos exhibit resistance to black pod, but more extensive screening is needed to locate multiple sources of resistance to the variable disease (Lockwood, 1985).

Watery pod rot, caused by *Moniliophthora roreri*, is a problem in Peru, Ecuador, Colombia, the Brazilian Amazon, Costa Rica, and Panama. The pathogen was first reported in Ecuador in 1914 and has been spreading ever since (G. A. R. Wood and Lass, 1987:293). It reached Panama by 1956 and southern Costa Rica in 1979 and has since devastated 80 percent of the cacao groves along Costa Rica's humid Caribbean coast (Hall, 1985:172). Watery pod rot vaulted the Andes by unknown means and penetrated the Amazon in the early 1980s.

Attacks of the fungus can be so severe that growers are sometimes forced to abandon their cacao plantings. Chemical control is largely ineffective and prohibitively expensive (Jimenez et al., 1987; G. A. R. Wood and Lass, 1987:299). Some *criollo* cacaos reputedly resist *M. roreri*, but breeding efforts in this direction have been discouraged because of *criollo* cacao's susceptibility to black pod and *Ceratocystis* wilt (Hardy, 1960: 241). Traditional 'Nacional', an indigenous *forastero* cacao from Ecuador, reputedly resists the disease, and EET 233, a cultivar resulting from the cross of 'Nacional' with an unknown "hybrid," shows promise in combating the mobile pathogen (G. A. R. Wood and Lass, 1987:299). Scientists in Costa Rica have identified resistant clones, including UF 273, CC 137, EET 67, EET 75, and EET 183, in the CATIE germplasm collection (Phillips and Galindo, 1987). The Ecuadorian cacao research program has identified EET 233 as resistant to both watery pod and witches'-broom (Aragundi et al., 1987).

Ceratocystis wilt, caused by *Ceratocystis fimbriata*, can rapidly kill cacao trees, particularly when they are pruned—to remove witches'-broom or large numbers of borers (*Xyleborus* sp.), for example. Such pruning facilitates penetration by the pathogen. First recorded in Ecuador in 1918, the virulent fungal pathogen may have been responsible for the "blast" disease that wiped out cacao plantations on Trinidad in 1727 (G. A. R. Wood and Lass, 1987:330; Wrigley, 1988:46). Ceratocystis wilt is now virtually pantropical and appears to be mutating. A more virulent race emerged in Ecuador in 1951. Curiously, the pathogen occurs in Africa but on hosts other than cacao (G. A. R. Wood and Lass, 1987:330). *Criollo* cacao is especially susceptible, but recently some wild cacao from the Peruvian Amazon has been found to resist the disease (Prescott-Allen and Prescott-Allen, 1986:285).

Another fungal disease on the move is vascular-streak dieback (VSD), caused by *Oncobasidium theobromae*. VSD reached epidemic proportions on Papua New Guinea in the early 1960s, but the responsible pathogen was described only in 1971 and its biology elucidated a year later (Prior, 1985). Originally confined to Papua New Guinea, VSD has spread to Sabah, Sumatra, Malaysia, and possibly India and the Philippines. A "hybrid" cacao recommended for planting in Papua New Guinea resists VSD but is highly susceptible to black pod and is thus not suitable for small farmers who cannot afford heavy use of fungicides (Prior, 1984). Multiple disease resistance, including resistance to as many pathogen races as possible, is needed in cacao-growing regions. Replacing cultivars of cacao made obsolete by the spread of a disease or other pathotypes is much more costly than with annual crops.

Numerous insects also attack cacao, none of which is yet adequately controlled by genetic management. Insect resistance has not been high on the agenda of cacao breeders, despite their importance as disease vectors and their direct damage to cacao (Figure 2.14). The cacao research program in Côte d'Ivoire is screening germplasm for resistance to several insects, including mirids such as *Sahlbergella singularis* and *Distantiella theobromae*, two major pests of cacao in Africa (Decazy and Coulibaly, 1981; Muller, 1981). CEPLAC scientists recently began screening germplasm in the Itabuna gene bank for resistance to various insects. Rather than proceed down the lengthy roll call of insects that damage cacao, we highlight here two relatively new pests.

FIGURE 2.14 Portable insecticide sprayer being refilled for use in a cacao plantation near Itabuna, Bahia, Brazil, in August 1988.

The cocoa pod borer (*Conopomorpha cramerella*) is potentially the most serious pest of cacao in Southeast Asia and the Pacific. Larvae of this small moth bore through cacao husks and disrupt bean development. First noted as a pest on Sulawesi (then Celebes) in the 1850s, the cocoa pod borer spread to Java in the 1890s, then reached Mindanao by 1930, Sabah by 1980, and Sarawak by 1983 (Mumford, 1985). The cocoa pod borer appeared on the Malayan Peninsula in the mid-1980s. The moth has a limited flight range, so movement of infested pods for planting stock is implicated.

Chemical control of the cocoa pod borer is difficult because the larvae are concealed inside the pods and because the pest lives on a range of tree hosts in the Leguminosae and Sapindaceae. These families are common in forests surrounding cacao plantations, contain many economically important species, and are often planted near cacao. Rambutan (*Nephelium lappaceum*) in the Sapindaceae, for example, is a popular fruit in Southeast Asia and serves as a host for the cocoa pod borer. Walnut-sized rambutan fruits are frequently transported over large distances to supply urban markets and may be involved in spreading the pest.

The best hope of containing the pest is a combined approach involving the search for host resistance and biocontrol. Biocontrol research is under way in Sabah and Mindanao. A cacao that bears extremely hard pods in the Belém gene bank could be a useful source of resistance. This durable-husked cacao came from a backyard of a smallholder near Benevides, virtually on the doorstep of the cacao gene bank.

A search is under way in Brazil for genetic resistance to another insect pest that bores into cacao pods, causing considerable damage. In 1981, larvae of a curculionid beetle, *Conotrachelus humeroptictus*, were first recorded attacking cacao plantations near Cacoal, Rondônia. The beetle has also infiltrated a new cacao germplasm collection at Ouro Preto d'Oeste in Rondônia (Mendes et al., 1987). This new and highly destructive pest threatens to undermine attempts to establish cacao as a viable cash crop for pioneer farmers in Amazonia. Resistance to the cocoa pod borer might also prove effective against curculionid beetles.

FUTURE NEEDS

Three main points are worth stressing to further cacao research and development. First, more and better quarantine facilities are needed for cacao in order to accommodate the stepped-up demand for shipments of germplasm. Second, tissue culture developments will bring mixed blessings for cacao. On the one hand, germplasm could be shipped more safely in vitro and genetic engineering might incorporate exotic genes, but micropropagation could increase the danger of genetic homogeneity in plantations. Third, more systematic efforts are needed to map and understand the variation of cacao in the wild, both for the benefit of breeding programs and to guide conservation efforts.

Quarantine is a severe bottleneck for several cacao breeding programs and gene banks (Imle, 1966). Cacao germplasm is usually shipped as budwood or seeds, both of which have a relatively short viability unless grafted or planted. Fewer cacao quarantine facilities are available today than in the 1960s. Several former colonial powers have phased out quarantine services for cacao and other tropical perennial crops since 1970 (Plucknett and Smith, 1989). The Royal Tropical Institute in Amsterdam, for example, terminated its quarantine service for cacao in the early 1970s, while the Royal Botanic Gardens, Kew, followed suit in 1984 (Plucknett and Smith, 1988:34).

Few quarantine operations remain to handle the growing shipments of cacao germplasm. The Subtropical Horticulture Research Station at Miami is one of the most heavily used quarantine services. This station indexes for CSSV by grafting susceptible material. In 1973, the U.S. Agency for International Development (USAID) withdrew support for quarantine work at Miami (Soderholm and Vasquez, 1985). Fortunately, the American Cocoa Research Institute and a consortium of European donors stepped in to continue support. Consequently, the Miami facility has been able to build two glasshouses and hire a technician to process quarantine shipments. This widely recognized quarantine service cannot use ELISA to screen for certain pathogens.

IRCC (Institut de Recherche du Café, du Cacao et des autres Plantes Stimulantes) and CIRAD (Centre de Coopération Internationale en Recherche Agronomique pour le Développement) jointly operate a quarantine service for cacao at Montpellier, France (J. B. Allen, 1987). Recently, the University of Reading in the United Kingdom has initiated a cacao quarantine service to help fill the increasingly global demand for this service.

The ICGT has used Barbados as a staging area for importing cacao germplasm for over fifty years. Barbados does not have a cacao industry and is 247 kilometers by sea to the nearest cacao-growing area on Grenada. The Barbados quarantine facility was recently upgraded to accommodate the growing volume of germplasm shipments. CEPLAC's cacao breeding program at Itabuna uses a quarantine facility in Salvador, the capital of Bahia, to check for the presence of witches'-broom and other diseases and pests. Cacao germplasm is quarantined in Salvador for one year before release. Salvador is well north of Bahia's cacao-growing region, but the quarantine facility has not always proved efficient. The screenhouses originally used to check germplasm shipments tore, so they were replaced with glass. Unfortunately, at 13° S, the glasshouses have proved too hot for some cacao germplasm shipments.

More and better equipped intermediate quarantine stations are needed to facilitate the safe exchange of cacao germplasm. Only when well-equipped "way stations" are in place to carefully screen cacao materials for pests and pathogens, can the full potential of cacao's genetic resources be realized for all cacao-growing regions. Continued testing and recruitment of fresh germplasm is vital for cacao breeding programs in tropical America, West Africa, and tropical Asia.

Research on tissue culture of cacao will surely pay dividends for breeding, propagation of desirable types, as well as conservation. In vitro techniques will also greatly facilitate the screening and shipping of cacao germplasm. Many hurdles will have to be surmounted, however, before biotechnologies can be applied regularly to cacao improvement and shipment. For example, excessive callus formation remains a problem in cacao tissue cultures (Adu-Ampomah et al., 1987).

But propagation by tissue culture could be a double-edged sword. Rapid multiplication in vitro could permit the mass production of genetically identical clones, unlike the current picture, where cacao plantations are established by seed with varying degrees of genetic heterogeneity. Widespread adoption of "super" clones would undoubtedly bring temporary benefits, but could set up growers' for catastrophic crop failures in the future.

Without a broad and secure germplasm base, fewer options will be open to improve cacao and meet future challenges to productivity. All the collections of cacao germplasm are in field gene banks, not a particularly secure way to safeguard genetic resources over the long term. Tissue cultures under slow growth could be a more efficient and safer way to maintain at least some of the genetic resources of cacao, in part because less space and labor are needed. Tissue culture techniques for cacao are still in their infancy, however. Excised immature embryos from seeds or small buds might be the explants of choice for cacao tissue culture.

Other priorities for cacao improvement include a more systematic screening of accessions in field gene banks, preparation of herbarium specimens of currently used cultivars to help sort out the confusion of names surrounding certain clones, development of purebred lines for the production of true hybrids, and long-term field trials (Hunter, 1990). Unfortunately, as the need for research on cacao genetics and inheritance patterns increased, several key cacao development institutions lost experienced scientists in the late 1980s and early 1990s. Progress can stall unless bright young scientists are recruited and given adequate working conditions to replace the departed experts.

CUPUAÇU

Cupuaçu (*Theobroma grandiflorum*), one of the cultivated relatives of cacao, is highly prized in the Amazon and Upper Orinoco for its creamy pulp, which surrounds beans encased in a large, rust-colored pod (Figure 2.15). Although the pulp, which occupies about a third of the fruit, has little nutritional value, its flavor is unique and immediately appealing to those who try it.

Cupuaçu pulp is used to make fresh juice, ice cream, jam, and tarts. The cupuaçu season falls in the rainy months from January to April and is eagerly awaited because insufficient quantities of pulp are available for freezing to meet demand during the remainder of the year. The delicate aroma of cupuaçu is a welcome component of the region's culinary delicacies.

Demand for cupuaçu sometimes outstrips the supply even in season, as evidenced by the high prices that the fruit commanded in 1990. In January 1990, roadside stands near Belém were selling cupuaçu for $1.50 each, well beyond the means of the majority of the poor. Even considering that 1990 was a poor harvest for cupuaçu, the high price of a popular fruit once affordable for all segments of society indicates great potential for increasing its supply.

Cupuaçu is native to the understory forest of eastern Pará State and western Maranhão in Brazil (Figure 2.16). Herbarium records[2] indicate that wild cupuaçu occurs from the Tapajós River to the eastern fringes of the Amazon rain forest; it is apparently absent in a native state north of the Amazon River. The broad Amazon River may have thus posed a for-

---

2. Twenty-two cupuaçu specimens were examined at two herbaria in the Brazilian Amazon: the Museu Goeldi (MPEG) and the Instituto Agronômico do Norte (IAN) herbarium at the EMBRAPA research station, both in Belém, Pará. Herbarium records were inspected to delimit the apparent natural range of cupuaçu as well as areas where it has been introduced. Locations where *T. grandiflorum* has been collected in the wild: (1) Rio Turiaçu, Mun. Monção, Ka'apor Indian Reserve (MPEG 118747); (2) Rio Gurupi, Mun. Viseu, Tembé Indian Reserve (MPEG 125864); (3) 65 km SSW of Tucuruí, Pará (MPEG 89448); (4) Zé Doca, km 180 BR 316, Maranhão (MPEG 67051); (5) 25 km S of Tucuruí Dam, Pará (MPEG 80738); (6) Km 16 BR 263, Tucuruí, Pará (MPEG 84326); (7) Colonia Tres Satubas, 3°5′ S, 45°45′ W (MPEG 71113); (8) Nr. Rio Itacaiúnas, 80 km SW Marabá, Pará (MPEG 89656); (9) Ilha das Onças, Mun. Bacarena, Pará, 1°25′ S, 48°27′ W (MPEG 111306); (10) Cachoeira Mangabal, R. Tapajós, Pará (MPEG); (11) Serra Buritirama, R. Itacaiúnas, Pará (IAN 128723, 128721; 128733); (12) EMBRAPA station nr. Belém, Pará (IAN 11838). Locations where cultivated cupuaçu has been collected: (1) Lake Tefé, Mun. Tefé, Amazonas, 3°20′ S, 64°50′ W (MPEG 11633); (2) Monte Dourado, Mun. Almerim, Pará (MPEG 125372); (3) Nr. Leticia, Colombia (IAN 20287); (4) São Gabriel, R. Negro, Amazonas, Brazil (IAN 16957); (5) R. Uaupés, Amazonas, Brazil (IAN 115064); (6) Tefé, Amazonas (IAN 29770); (7) Mouth of R. Xié, upper Rio Negro Basin (IAN 52724); (8) Leticia, Colombia (IAN 20332).

FIGURE 2.15 Cupuaçu (*Theobroma grandiflorum*) for sale in a street market in Manaus, Brazil, in 1980.

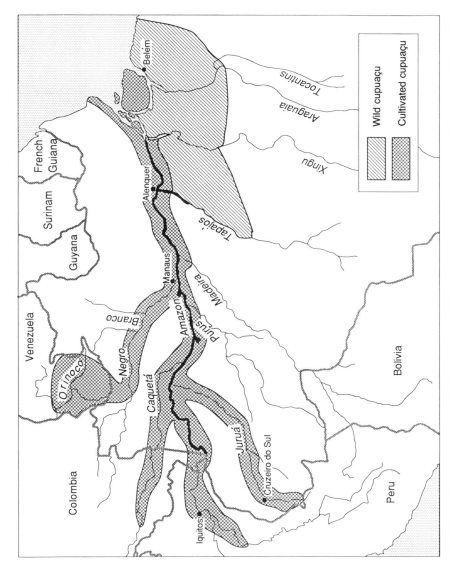

FIGURE 2.16 Approximate distribution of wild and cultivated cupuaçu (*Theobroma grandiflorum*). (Adapted from J. B. Allen, 1987:19; F. J. Pound, 1938; Sanchez et al., 1988; herbarium records [see n. 2].)

70

midable dispersal barrier for cupuaçu until humans penetrated the region in the Paleolithic.

According to rural informants, cupuaçu is dispersed in the wild by capuchin monkeys (*Cebus apella*), which break open the pods on a branch to access the pulp, and agoutis (*Dasyprocta* spp.), which bury some of the beans. Paca (*Agouti paca*) also feed on cupuaçu fruits, but they apparently do not store seed.

As they did with cacao, indigenous people may have enriched some areas with cupuaçu to provide refreshing snacks near villages and temporary camps. Several indigenous groups in eastern Amazonia, such as the Araweté, Ka'mapor, Tembé, and Wayãpi, have semidomesticated cupuaçu by planting and carefully protecting the trees, particularly in old second growth (Balée and Moore, 1991). Some of the denser stands of cupuaçu may have arisen from beans discarded by people after they removed the pulp. Cupuaçu beans germinate readily, particularly if they land in partial shade and on relatively fertile kitchen middens.

In precontact times, cupuaçu was taken up the Amazon and some of its tributaries (Figure 2.16). Cupuaçu is planted as far west as the lower Ucayali in Peru and the Caquetá River in the Colombian Amazon (J. B. Allen, 1987:19; F. J. Pound, 1938). Near Cruzeiro do Sul in southwestern Amazonia, cupuaçu has been incorporated into an agroforestry system that includes mango (*Mangifera indica*) and cacao (Vello and Rocha, 1967).

In remote times, cupuaçu seeds were evidently also taken up the Rio Negro, through the Casiquiare canal, and into the Upper Orinoco and its tributaries. The Desana of the Upper Rio Negro drink cupuaçu juice after it has been "blessed" by a shaman to facilitate difficult births (Buchillet, 1988). Such tight integration of a crop into an aboriginal culture suggests that cupuaçu has been grown along the Rio Negro and some of its tributaries for a long time. *Copo-azú*, as cupuaçu is known in Venezuela, is a backyard tree in San Carlos de Rio Negro and in San Pedro de Cataniapo on the Cataniapo River, an affluent of the Orinoco at 5°32' N (Sanchez et al., 1988). The Yekuana Indians along the Rio Ventauri, another tributary of the Upper Orinoco, refer to cupuaçu as *mamaku kawcuai*. The latter name is not derived from the Brazilian name, cupuaçu, suggesting that the tree has been in the Upper Orinoco for a long time.

Nowadays, cupuaçu is mostly produced in backyards from discarded or planted seeds. Few people bother to collect fruits from the widely dispersed trees in the forest. Also, cupuaçu in the forest tends to grow fairly tall, reaching as high as 20 meters, as the trees compete for light. Cupuaçu in dooryard gardens are easier to harvest because they tend to be bunchy and because they can be encouraged to put out more lateral branches if pruned.

Demand for cupuaçu is growing within Amazonia and in some extra-regional markets, such as Rio de Janeiro. Recently, small amounts of cu-puaçu pulp have been exported to an ice cream maker in the northeast United States and to France. Cultural Survival, based in Cambridge, Mas-sachusetts, is contracting with farmers in Tomé-Açu in the Brazilian Ama-zon to supply cupuaçu pulp to the United States; of the 60 metric tons requested in 1990, only 7.3 tons were available for export from the com-munity between April 1990 and March 1991 (A. Homma, pers. comm.).

Amazonia has always captured the imagination of Brazilians and many other nationalities, and consumers are interested in exotic jungle fruits. A rancher near Manacapuru, Amazonas, has interplanted 45 hectares of pas-ture with cupuaçu (P. Alvim, pers. comm.). Pulp is removed at the ranch and is frozen for sale, principally in Manaus, after the cupuaçu season is over when prices are higher.

Another modest cupuaçu plantation was established in the early 1960s by a rancher at Fazenda Itaqui, 14 kilometers east of Santa Isabel in the Bragantina zone of Pará. Cupuaçu trees at Fazenda Itaqui are in a sepa-rate, 25-hectare grove and are not intercropped. After the pulp is removed manually by laborers, the discarded cupuaçu beans are ground up and mixed with livestock feed. Cupuaçu groves, planted with a density of 170 trees per hectare, can produce 2,000 fruits per hectare after six or seven years.

AGENDA FOR IMPROVEMENT

Witches'-broom, the same disease that afflicts cacao, is the main reason that cupuaçu is not grown on a wider scale in Amazonia (Figure 2.17). All cupuaçu plantations are infected with witches'-broom, but disease symp-toms often vary markedly between trees because they are different ge-notypes. Plantations are thus convenient grounds for making selections for disease resistance and other traits. The owner of the cupuaçu plantation at Fazenda Itaqui has tagged several cupuaçu trees because of their apparent tolerance to witches'-broom and because they exhibit other characteristics, such as unusually large fruits or high yields.

Cupuaçu plantations are particularly useful sites for screening for resist-ance to witches'-broom because disease pressure is often severe when cu-puaçu or cacao are grown as monocultures. In the forest, cupuaçu trees are more dispersed and therefore less prone to infection with witches'-broom. An apparently healthy cupuaçu tree in the forest might be due to the ab-sence of the pathogen rather genetic resistance to the disease.

A more systematic search for resistance to witches'-broom on existing plantations, in backyards, and in the forest would be a high priority for any cupuaçu improvement program. Cacao growers may exaggerate the

FIGURE 2.17 Cupuaçu (*Theobroma grandiflorum*) infected with witches'-broom on a plantation intercropped with peach palm (*Bactris gasipaes*), near Manaus, Amazonas, August 1988.

dangers of witches'-broom in some cases in an attempt to stop the expansion of cupuaçu cultivation (FAO, 1986). If cacao and cupuaçu prices are high enough, however, the disease can be at least partially checked by periodically removing and burning or burying infected parts of the trees.

Cupuaçu plantations are also affected by several insect pests. At Fazenda Itaqui near Santa Isabel, some cupuaçu pods are infested with white mealybugs that were being tended by ants. Although the mealybug infestations are unsightly and would probably detract from the selling price, it is not clear whether the mealybugs affect productivity or the quality of the pulp. Also, the ants may be fending off other insects from the cupuaçu trees. At a field gene bank for cupuaçu near Manaus, an unidentified beetle larva, known locally as *broca de fruta*, is damaging some of the pods.

Cupuaçu has evolved a defense against herbivorous insects that needs to be safeguarded in any breeding or selection efforts. Fruits, young branches, the extremities of old branches, flower buds, and the calyx of opened flowers are covered by a rusty pubescence that attracts ants, particularly species of *Crematogaster* and *Camponotus* (Silva, 1976). Such protection is not entirely effective, since grasshoppers (Orthoptera) and caterpillars (lepidopteran larvae) cause some damage to leaves. Nevertheless, the velvety texture of fruits, buds, and young branches serves a useful purpose.

Cupuaçu selections can be readily grafted on rootstock of the same species or its near relatives (Venturieri et al., 1987). This technique permits the selection of clones with high-quality fruit, earlier flowering, and better shaped trees for harvesting. A major disadvantage of cloning is that orchards become genetically more homogeneous, particularly if only one clone is used, and thus more vulnerable to disease and pest epidemics.

A seedless cupuaçu, a deleterious mutation in nature, has elicited some excitement in the Brazilian Amazonia. This seedless cupuaçu was found in a backyard in Cametá on the lower reaches of the Tocantins and has the advantage of reducing labor costs because the pulp does not have to be removed from seeds. Unfortunately, it is highly susceptible to witches'-broom and is low-yielding.

Apart from higher yield and disease resistance, especially for witches'-broom, regular bearing is another desirable attribute. The 1989 cupuaçu crop was exceptionally good in the Belém area, whereas the 1990 harvest was disappointing. Consistent high yields are important if commercial plantings are to expand.

Near relatives of cupuaçu may eventually be used in efforts to improve disease and pest resistance in the crop, among other traits. Most wide-crosses with *Theobroma* produce no mature fruit, but viable hybrid seeds have been obtained by crossing cupuaçu with cacao, *T. obovatum*, and *T. subincanum* (Addison and Tavares, 1952). If cupuaçu's gene pool does not contain the desired characteristics for upgrading varieties, then breeders may tap closely related species.

GERMPLASM COLLECTIONS AND EVALUATION
EFFORTS

For the foreseeable future, gene banks for cupuaçu will be limited to trees since the seeds are short-lived and cannot be dried and stored under reduced temperatures. The Brazilian agricultural research system (EMBRAPA, Empresa Brasileira de Pesquisa Agropecuária) has identified cupuaçu as one of the minor fruits of Amazonia that deserves more research attention. At the Belém station of EMBRAPA, cupuaçu is a high priority for germplasm survey and collecting. Near Manaus, EMBRAPA's newly organized agroforestry research center (CPAA, Centro de Pesquisa Agroflorestal da Amazônia) has a collection of 100 cupuaçu seedlings (C. R. Clement, 1991). In 1988, this small field gene bank contained progeny from seventy-five mother trees growing in backyards in the Bragantina zone, Manaus, Benjamin Constant, and Manacapuru. Cupuaçu seedlings in the CPAA field gene bank are being evaluated for disease resistance and other qualities.

FUTURE TASKS

In the 1970s, several chocolate manufacturers, such as Nestlé in Switzerland, expressed interest in cupuaçu as a flavor for chocolate and also as a novel type of chocolate. Two immediate problems confront any moves to develop cupuaçu chocolate: first, insufficient beans are available for large-scale testing, and, second, although up to half the seeds contain a white fat similar to that of cacao, cupuaçu butter appears to have a lower melting point than cocoa butter. Cupuaçu chocolate would thus tend to melt at room temperature.

Although the chocolate market will continue to be dominated by cacao, some up-market chocolatiers might be interested in selling a specialty chocolate, such as from cupuaçu, even if it required constant refrigeration. Also, cupuaçu might enhance the flavor of some conventional chocolates. The fate of Amazonian forests has captured global attention, and marketing a natural product from Amazonia that provides local employment would be one way to help save the forest. Cupuaçu plantations could be established on degraded pastures and farmland, and conserving the wild gene pools of cupuaçu in the fast-disappearing forests of Amazonia would take on added importance.

CHAPTER 3

Major Fruits of the Forest

A part from bananas, which we discuss in Chapter 6, four tropical species or groups of related species are cultivated and traded extensively: mango, the citrus fruits, pineapple, and avocado. Guava, papaya, sapodilla, and passionfruit are also widely grown, but they travel less well or have not yet developed sizable export markets.

These categories are not hard and fast, and some of the less commercialized fruits may well assume greater importance in the future. Certain varieties of papaya, for example, are small with relatively tough skins that withstand transportation and handling and have therefore started to penetrate new markets in both developing countries and industrial nations.

In addition to supplying calories, these major eight tropical fruits are significant sources of minerals, vitamins, and, in the case of avocado, oil. They provide an important dietary supplement in tropical lands and are a significant source of cash income for many growers in both tropical and subtropical regions. These fruits also provide jobs at ports, packinghouses, and markets in urban centers in developing countries and industrial nations.

Tropical fruits come from a broad range of plant families, and the future of their gene pools depends largely on the fate of tropical forests and farming practices in and around forests. Genetic erosion of tropical fruits is linked to deforestation and to the widespread planting of outstanding clones. Cultivation of all the major tropical fruits has passed the stage of sowing seeds in favor of vegetative propagation of superior forms, particularly on commercially oriented farms and plantations. Although the major fruits are still propagated by seed in backyards and on small farms, a trend toward adoption of a few high-yielding cultivars with desirable flavor and travel characteristics is well under way.

Production trends of the major tropical and subtropical fruits are highly dynamic and vary markedly among species. Citrus production dominates the world trade in tropical fruits, with over four times the output of pineapple and seven times the production of mango. Papaya cultivation is largely confined to developing countries and is becoming a more important crop in humid regions. Commercial pineapple production is confined largely to tropical areas, although Hawaii dominates the export trade of fresh and canned pineapple. Both industrial and developing nations are involved in the export trade of mango and citrus fruits and juices. The tropical or subtropical portions of industrial countries account for less than 3 percent of papaya output. Avocado cultivation, on the other hand, has decreased steadily in developing countries but has become more of an export crop for the developed world. One-fifth of global avocado production now comes from industrial countries.

Pineapple, papaya, and bananas are backyard and plantation crops in the humid tropics, whereas citrus production is split roughly equally between First and Third World countries.

Such patterns raise important policy questions. To what extent do multinational corporations based in temperate countries bear some responsibility for the conservation of genetic resources of tropical perennial crops that enter world commerce? Business enterprises in temperate countries have a direct stake in the future health of citrus and avocado plantations. Private-sector producers and financiers in industrial countries have a vested interest in the collection, conservation, and evaluation of genetic resources of pineapple and papaya because of the strong and growing demand for those fruits in North America and Europe. At the moment, agribusiness investment in these crops is confined mainly to production agronomy, transportation, and marketing. Very few resources from the private sector are currently available to support research on expanding the genetic resources available for breeders of tropical perennial crops.

A second issue is the heavy emphasis on commercial aspects of tropical fruits. While export-oriented crop production does generate valuable foreign exchange and provides employment in tropical countries, the needs of small-scale growers also warrant consideration. Many of the major tropical fruits are cultivated extensively on small plots and in backyards, even in urban areas. Varieties developed for large-scale plantation conditions, where fertilizers and pesticides are employed regularly, may not be appropriate for smallholders. Also, tropical fruits developed for the export trade may not meet taste and texture criteria of local populations. How to address the needs of both constituencies requires input from farming system specialists, greater sensitivity to local conditions and market opportunities, and some reorientation of research efforts, particularly by the public sector.

Such points need to be aired and discussed because financial resources will have to be mobilized shortly if genetic resources are to be adequately conserved and evaluated. Such sharpening of priorities is always difficult when national needs and the dictates of the international marketplace differ. The International Board for Plant Genetic Resources (IBPGR) has established regional priorities for many cultivated plants, including perennial crops of the tropics, but a major stumbling block has been the lack of clear agreements among genetic resource programs to work together and share the limited external funding.

MANGO

Mango (*Mangifera indica*) is a well-known tropical crop familiar to many consumers in industrial countries. Mango fruits are generally eaten fresh and are prized by tropical peoples as well as consumers in temperate countries who can afford to buy them. The juicy and aromatic flesh has a hint of turpentine, but is usually sweet. The fleshy fruits are good sources of vitamin A and contain varying amounts of vitamins B and C. The generally high levels of vitamin A in mango are significant because diets in the humid tropics are frequently deficient in the vitamin. Some mango varieties are better sources of vitamin C than oranges (Chandler, 1958:269). Mango fruits vary greatly in size and shape, ranging from as little as 3 centimeters long to ten times that size, and from round to elongated (Figure 3.1). When ripe, the skin ranges from green to yellow, pink, red, or purple.

Mango fruits are also canned, and immature fruits are used to make pickles and chutneys. Dried mango fruits are ground with tumeric to make a powder to flavor dishes. Mango seeds and bark are employed in a variety of folk remedies. In the Society Islands, large mango trunks are occasionally dug out to make outriggers.

Given the esteem and range of uses of mango, it is not surprising that this ancient crop has been incorporated into mythology and religion. In India, numerous myths describe the origin of mango and attest to the sacredness of its wood and numerous white flowers. Mango is revered by Hindus and Buddhists. Buddha was once presented with a grove of mango trees in which he rested. Hindus consider the mango a wish-granting tree and a symbol of love and devotion (Gandhi and Singh, 1989:109). In India, villagers believe that mango trees sprout fresh leaves at the birth of a son, and they thus festoon doorways when a boy is born. Mango leaves are also used to decorate marriage ceremonies in the hope that the couple will bring forth a son. On holy days, Hindus use mango twigs to brush their teeth, and the leaves serve as spoons for pouring libations. Numerous icons depicting such themes adorn Hindu temples (S.M. Gupta, 1971).

FIGURE 3.1 Assorted mangos (*Mangifera indica*) in the germplasm collection held at the Subtropical Horticulture Research Station, USDA, Miami, Florida.

Mango has also been featured in architecture, art, and fabrics of India. Mango is depicted on the stupa of Barhut and Sarchi built in 150 B.C. (Gandhi and Singh, 1989:109). Yellow-fruited mango trees appear in the canvas *Pechhavai with Gupis, Cows, and Heavenly Beings*, painted in the late eighteenth century, suggesting mango's spiritual connotation. A watercolor, *Prince Resting after a Hunt*, painted in Rajasthan about 1740, portrays a young prince reposing in the welcome shade of a mango tree. A fruiting mango tree appears in a bucolic scene involving a courting couple in another canvas painted in Rajasthan in the eighteenth century, *Saranga Ragini* (Bhattacharya et al., 1973: plate 43). In *The Meeting of the Eyes* (Devgarh, ca. 1780), a mango tree with flowers symbolizes a couple's blossoming affection (Isacco et al., 1982:17). Mango fruits inspired the paisley fabric pattern (U. Lele, pers. comm.), now common in many parts of the world on ties, cravats, blouses, and women's dresses.

Mango adapts to a variety of soils as long as annual rainfall exceeds 750 millimeters and there is a dry season to promote flowering and pollination. It grows best when dry conditions prevail during flowering and fruiting and the rains coincide with vegetative growth. Thriving on infertile, marginal soils where many other fruit trees may encounter difficulty growing, mango is a towering evergreen tree that can live up to a century, and its broad canopy has made it a favorite shade tree in the humid tropics. The streets of Belém at the mouth of the Amazon are cooler than those of Manaus farther upstream because of the numerous, ancient mangos that line many of the city's narrow streets. In addition to providing welcome relief from the midday heat, mangos provide tasty treats for pedestrians during the early part of the rainy season.

India accounts for about two-thirds of the world's mango production, one of the few cases where the area of origin of a perennial crop is still the main region where it is produced. Pakistan, Mexico, the Philippines, Brazil, Indonesia, Haiti, Bangladesh, Kenya, Mali, and the United States are also significant mango producers. Mangos are frequently encountered in markets in much of the humid tropics (Figure 3.2).

ORIGIN AND SPREAD

Mango was domesticated from wild *M. indica* in northeastern India some 4,000 years ago (Figure 3.3). Cultivated forms gradually diffused eastward, particularly with the help of Buddhist monks starting around 500 B.C. Buddhist monks are not allowed to till the land for food, but must rely on providence and the generosity of the surrounding population for sustenance. Fruits falling to the ground are fair game, however, hence the large number of fruit trees around many Buddhist temples.

When mango reached Africa is still in dispute. Egyptians, Phoenicians,

FIGURE 3.2 Two varieties of mango, 'Leche' (foreground) and 'Matilla', Chichicastenango, Guatemala, 1988.

and Babylonians have traded at various ports along India's Malabar Coast as far back as the third century B.C. Persian traders have also long been plying the waters between Calicut, India, and the Red Sea (Rizvi, 1987). Equipped with seaworthy dhows and lanteen sails, Persians may have taken mango to East Africa by the tenth century, although the fruit may not have been planted in Africa until as late as the nineteenth century (Busson, 1965:344). Given the long history of trade between East Africa, Southwest Asia, and India, however, it seems likely that the highly esteemed mango would have been planted on Zanzibar, if not in Malindi or Mombasa in Kenya, before the arrival of Portuguese seafarers in the late fifteenth century.

The Portuguese, Spanish, British, and French took mango to the New World several times after 1500 (Harris, 1976; Mukherjee, 1972; Purseglove, 1976:25). The Portuguese took mango from India, probably Goa or

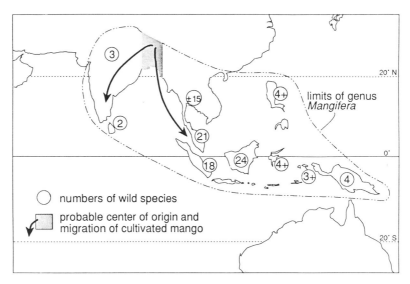

FIGURE 3.3 Area of domestication of mango (*Mangifera indica*) and distribution of wild *Mangifera* species. (From an unpublished IBPGR report by J. M. Bompard, © IBPGR.)

Calicut, to Brazil in the early eighteenth century (Figure 3.4). A British frigate purloined a French shipment of mangos from Mauritius on its way to Haiti and planted them on Jamaica in 1782 (Fawcett and Rendle, 1926:7).

The first mango introduction to Florida in 1833 failed, but the second attempt was successful in 1861. One of these grafted introductions, 'Mulgoa' (sometimes referred to as 'Mulgoba'), bore superb fruits on the property of Reverend Elbridge Gale at Mangonia on Lake Worth in 1898, thereby drawing attention to the exciting prospects for cultivating mango (Wolfe, 1937). Although 'Mulgoa' produces good-tasting fruit, it is highly susceptible to anthracnose, which disfigures the fruit, thereby reducing its market value. Also, 'Mulgoa' yields are modest.

Early Indian introductions, such as 'Mulgoa' from Poona and 'Sandersha' from Bangalore, produced some useful seedlings, however, thereby becoming the trunks of "family trees" that have lead to several famous varieties. 'Haden', a seedling of 'Mulgoa', sprouted in the backyard of Captain Haden in Coconut Grove and became the mother tree of one of Florida's most successful cultivars (Galloway, 1938).

'Haden', first propagated in 1910, is a fairly consistent bearer and has gained a place in commercial groves in parts of Latin America, especially Colombia, Peru, and Brazil. 'Haden' is esteemed enough in Brazil to be trucked 2,000 kilometers from the southern part of the country to Belém at the mouth of the Amazon (Cavalcante, 1988:155). Espírito Santo is the main producer of 'Haden' in Brazil. 'Haden' was also seen in the public market of Parauapebas, a pioneer town only a decade old at the foothills of the Serra dos Carajás in January 1990. In spite of the variety's susceptibility to anthracnose, 'Haden' fetches a higher price than other varieties grown in Brazil and is thus profitable to grow and transport over long distances. Peru exports 'Haden' mangos to Europe.

'Zill', a seedling of 'Haden', was planted by Carl King in Lake Worth, Florida, and first fruited in 1930. 'Zill' has been grown commercially since 1940. Salmon pink 'Keitt', selected on the property of Mrs. J. N. Keitt in Homestead around 1946, is also a 'Mulgoa' seedling (Lynch and Mustard, 1956; Ruehle and Ledin, 1956). Large-fruited 'Keitt' is another Florida selection that has proved popular in some developing countries. Some plantings of 'Keitt' have appeared recently near Bragantina, Pará, in the Brazilian Amazon. 'Tommy Atkins', another large, red Florida selection is grown in Colombia and exported to gourmet markets in London. Mangos sell for $3 each in London stores such as Harrods and Fortnum and Masons.

'Brooks', a seedling of 'Sandersha', first fruited in a Miami backyard in 1916 and soon became a popular variety in Florida (Lynch and Mustard, 1956). 'Kent', a 'Brooks' seedling, was planted in the backyard of Leith D. Kent in Coconut Grove in 1932. Selected in 1938, 'Kent' has unusually large fruits, yields well, and is still cultivated in Dade County, Pine Island, and overseas (Krome, 1956). Peru exports 'Kent' mangos to Europe.

Florida is not the only source of export-quality mango varieties adapted to the humid tropics. Costa Rican farmers clone a local selection, 'Mora', in Guanacaste and Nicoya for export to Germany and some other European countries. The alternating wet and dry seasons in the northern Pacific region of Costa Rica make it ideal mango country. In San José, the capital of Costa Rica, 'Mora' commands the same price as 'Haden'.

Mango reached Tahiti in 1848, and several local forms soon arose. A striking local red mango from Tahiti is depicted in Gauguin's *Vahine no te vi* (1892), and at least two mango types are depicted in his *Te arii vahine* (1896).

MANGO GENE POOLS

Mango is genetically very heterogeneous, and seedling progeny are extremely variable. Cultivated mango originated from wild *M. indica* in northeast India, where many wild mango trees still grow, particularly in the Assam-Chittagong Hills (IBPGR, 1986:10). Wild mango reportedly grows in the western Ghats (Hawkins, 1986:365). Mango also occurs wild in Bangladesh, Burma, Malaysia, Thailand, and Indochina. Herbarium

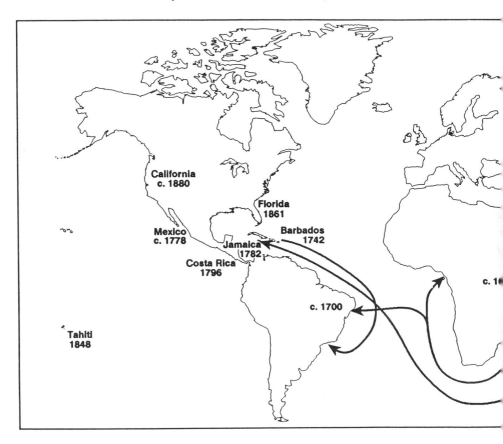

specimens of apparently wild *M. indica* have also been collected in China and Sri Lanka.

Wild mango is found as scattered individuals in lowland primary and secondary forests, where it is probably dispersed by bats (Whitmore, 1985: 44). More fieldwork is needed, particularly to inaccessible places, to decipher the true distribution of wild *M. indica*. Some herbarium records may be escapees from cultivation.

Feral mango may nevertheless be of interest to breeders, particularly as sources of disease and pest resistance. Islands are more susceptible to biotic invasions than continents, and particularly dense stands of feral mango are more common in insular environments. Spontaneous mango virtually dominates some old secondary forests on Bora Bora in the Society Islands, Polynesia. Spontaneous mango is also abundant on Tonga and Jamaica (R. O. Williams and Williams, 1951:215; Yuncker, 1959:170) and has naturalized on Samoa (Parham, 1972:70).

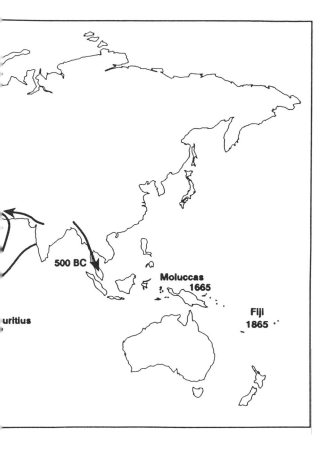

500 BC

Moluccas
1665

Fiji
1865

urltius

FIGURE 3.4 Successful introductions of mango (*Mangifera indica*) around the world. (Adapted from Fawcett and Rendle, 1926; Pétard, 1986:203; Purseglove, 1976:25; Valmayor, n.d.).

Mango is feral in parts of West Africa, Honduras, and Mexico (Owen, 1973:76; Standley, 1924:660, 1930a). In Central and South America, monkeys may disperse feral mangos since fruit bats in the New World are not as large as some of the species in tropical Asia. On Barro Colorado Island, for example, capuchin monkeys (*Cebus capucinus*) have been observed feeding on mango fruits (Oppenheimer, 1985:260).

The taxonomy of *Mangifera* is still in dispute, but between 40 and 60 species are found in South and Southeast Asia (Mukherji, 1949). People gather the fruits of over a dozen wild species of *Mangifera*, and several species are cultivated on a small scale (Mukherjee, 1985:8; Samson, 1986: 216). Cultivated mangos in Southeast Asia include the horse mango (*M. foetida*, Figure 3.5), kuini (*M. odorata*, Figure 3.6), binjai (*M. caesia*, Figure 3.7), lanjut or langoot (*M. lagenifera*), and *M. zeylanica*. Kuini has fibrous fruits with a strong resinous smell, whereas binjai has pale, rough-skinned fruit.

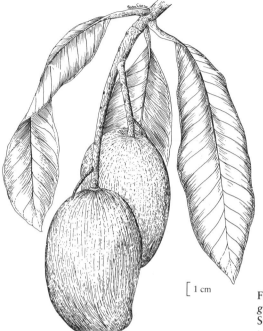

[1 cm

FIGURE 3.5 Horse mango (*Mangifera foetida*). (Illustration by R. Suraliso Nurbiantoro, in S. Sastrapradja et al., 1980.)

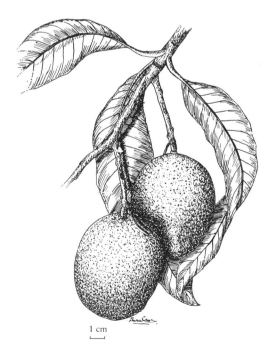

1 cm

FIGURE 3.6 Kuini (*Mangifera odorata*). (Illustration by R. Suraliso Nurbiantoro, in S. Sastrapradja et al., 1980.)

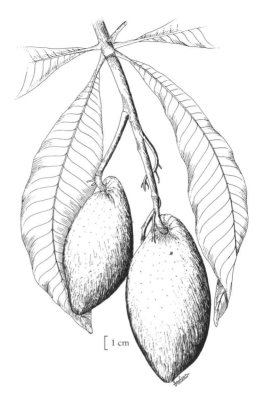

[1 cm

FIGURE 3.7 (*Mangifera caesia*). (Illustration by R. Suraliso Nurbiantoro, in S. Sastrapradja et al. 1980.)

Borneo and the Malayan Peninsula contain a particularly rich assortment of cultivated mangos (Table 3.1). At least sixteen of the twenty-four species of *Mangifera* are found in forests, riverbanks, and swamps of Kalimantan (B. Seibert, 1989). Borneo and Malaysia are being logged rapidly, mainly to supply the Japanese market.

Mango's cultivated relatives all have fleshy fruits and could become more widely cultivated if more selection of superior types were made available. Some of the mango relatives have very fibrous fruits, with flavors ranging from sour to bittersweet or sweet, and are often used for making curries, chutney, and pickles (Chin and Yong, 1985:13, 15). Dayak people of East Kalimantan province in Borneo establish temporary camps in the forest every year to gather wild mango and other fruits (Bompard and Kostermans, 1986).

Wild relatives of mango could be useful to breeders looking for several valuable traits. *M. decandra*, *M. inocarpoides*, and *M. gedebe*, for example, have been found growing in flooded conditions and could serve as rootstocks for waterlogged soils (Mukherjee, 1985). Towering *M. similis*, indigenous to Sumatra and Java (Mukherji, 1949), may be a good parent

TABLE 3.1

Numbers of species of *Mangifera* in Malaysia

Area	Wild	Semicultivated	Cultivated[1]
Borneo	19+	6	10
Java	4+	1	7
Lesser Sunda	5	1	4
Malayan Peninsula	16	6	8
Moluccas	5	1	4
New Guinea	3+	1	1
Philippines	2		3
Sulawesi	6		3
Sumatra	15+	2	7

Source: Adapted from Bompard and Kostermans, 1986.
[1]Includes *M. indica*, *M. caesia* complex, and *M. odorata*.

for breeding freestone mangos. Wild relatives of mango could be useful rootstocks to induce dwarfing in cultivars.

The rich assemblage of mango relatives in forests of the Malayan Peninsula, Sumatra, and Borneo are increasingly threatened by farmers and logging operations. Only relatively small pockets of forest are likely to survive into the next century in Malaysia. Borneo and Sumatra are still primarily forested, but settlers from overcrowded Java are clearing forests containing many species of *Mangifera*.

CONSERVING THE RESOURCE BASE

Only extensive tracts of forests in diverse environments are capable of capturing a significant portion of the wild gene pools of mango and its near relatives. Wild populations occur in primary forest, regrowth, swamp forests, and woods that are periodically flooded. Large and widely scattered forest preserves are needed throughout the ecologically variable range of *Mangifera* in South and Southeast Asia. In cooperation with the World Wildlife Fund, IBPGR has supported field surveys of wild mango and its near relatives, and specific reserves will be proposed when fieldwork is complete and data analyzed.

Fortunately, at least some reserves contain populations of wild mango and its near relatives. A biosphere reserve on the Mysore Plateau (Wynad-Nilgris), India, contains wild populations of *M. indica*. At least ten wild relatives of mango are found in Indonesia's extensive system of parks and reserves on Sulawesi, the Moluccas, Java, and Kalimantan (Mukherjee, 1985:11). In the Philippines, malapaho (*M. monandra*) occurs in the Puerta Galera Biosphere Reserve on Mindoro. Thailand has twelve near relatives of mango in parks or reserves, while Singapore has five near relatives of mango in reserves. Several reserves on Sri Lanka contain populations of the endemic *M. zeylanica*. At least twenty-seven wild relatives of mango occur in reserves throughout tropical Asia, but little information is available on their status and variability.

Major field gene banks for mango are held in a dozen countries (Table 3.2). India maintains over 1,000 accessions of mango distributed among 25 institutions. Approximately 400 cultivars are duplicated at several agricultural research centers in India. Several gene banks, such as in India, the Philippines, and Malaysia, are rich in local selections. Expansion of gene bank collections to incorporate more traditional selections, particularly from unusual ecological settings, would safeguard materials that might otherwise be lost.

TABLE 3.2.
Mango germplasm collections with at least 100 accessions

Institution	Location	No. of accessions	Remarks
FCAV	Jaboticabal, São Paulo, Brazil	495	450 seedlings from Brazil
IPB	Los Baños, Philippines	352	238 spontaneous plants from Philippines
DICOF	Havana, Cuba	350	Cultivars from Cuba, India, Mexico, USA
Fruit Research Station	Sangareddy, Andhra Pradesh, India	267	
BALITAN	Malang, Indonesia	239	Cultivars from India, Java, Pakistan, Philippines, USA
Subtropical Horticulture Research Station	Miami, Florida	200	Cultvated *M. indica* only
Horticultural Research Institute	Saharanpur, Uttar Pradesh, India	179	
MARDI	Kuala Lumpur, Malaysia	175	Duplicated at MARDI, Kluang; approx. half of accessions from Malaysia
Tropical Research and Education Center, Univ. Florida	Homestead, Florida	163	
Central Mango Research Station	Rehmankhera, Uttar Pradesh, India	139	

TABLE 3.2—*cont.*

Institution	Location	No. of accessions	Remarks
Regional Fruit Research Station	Anantharajupet, Andhra Pradesh, India	137	
Horticultural Research Center	Patharchatta, Uttar Pradesh, India	136	
U-Thong Field Crop Experiment Station	Subhanbari, Thailand	128	
College of Agriculture	Navsari, Gujarat, India	121	
Kasetsart Univ.	Thailand	120	
Fruit Research Station	Basti, Uttar Pradesh, India	119	
Regional Fruit Research Station	Vengurla, Maharashtra, India	115	
UH	Poamoho Farm, Oahu, Hawaii	110	Mostly commercial cultivars
BARI	Kazla, Bangladesh	107	101 local cultivars
Agricultural Research Station	Sabour, Bihar, India	104	
CENIAP	Maracay, Venezuela	103	*M. indica* only
CNPMF	Cruz das Almas, Bahia, Brazil	100	Cultivars from Brazil, India, USA
Koronivia Research Station	Nausori, Fiji	100	Some local cultivars
Estação Agrícola Nacional Quinta do Marquês	Oeiras, Portugal	100	Accessions from Guinea Bissau

Sources: Gulick and Van Sloten, 1984; field visits to the Tropical Research and Education Center, University of Florida, Homestead, and to the Subtropical Horticulture Research Station, USDA, Miami, Florida.

Note: See Appendix 2 for spelled-out names of institutions.

Curators of mango germplasm collections and breeders have been hampered by the large, and often confusing, array of cultivar names. To help sort out the plethora of varietal names, the Division of Fruits and Horticultural Technology of the Indian Agricultural Research Institute recently published an international checklist of mango cultivars (Pandey, 1986). The IBPGR also issued descriptors to assist germplasm curators sort out their collections, such as in the Seychelles, Philippines, and Thailand (IBPGR, 1989).

Field gene banks are labor-intensive operations that tie up large portions of land at research stations. Competing demands for land to set up test plots for other crops can jeopardize field gene banks of perennial crops, particularly if the germplasm collections are not linked to any ongoing research. Furthermore, field gene banks are attacked by diseases and pests, such as anthracnose, which has invaded the germplasm collection at the Subtropical Horticulture Research Station south of Miami, Florida. Mango seeds cannot be stored in seed gene banks because they are destroyed when dried and frozen. Research is under way to culture mango cells in vitro, and tissue-culture gene banks may ultimately reduce the cost of maintaining mango germplasm (Litz, 1984).

Wild relatives of mango are poorly represented in collections. Probably fewer than several dozen accessions of wild mangos are represented in all germplasm collections. While it would be helpful to include more wild material in field gene banks, such material is best conserved in situ.

THE CULTIVAR MOSAIC AND BREEDING EFFORTS

During mango's long association with humans, many selections of spontaneous seedlings have been made, which in the case of India, has led to the development of more than a thousand cultivars adapted to the diverse soils, climates, and cultures of the subcontinent (Mukherjee et al., 1968). Villagers at Sirsi in Karnataka along India's west coast cultivate at least seventeen varieties of mango (D. Flickinger, pers. comm.). Villages in many parts of Asia have a few prized mango trees that are known for their outstanding fruit.

Even in the New World, diverse forms of cultivated mango have arisen. In French Guiana, at least thirty varieties of mango were recognized in the late nineteenth century (Coudreau, 1883). Some sixteen mangos are recorded in Venezuela (Pittier, 1926). In the vicinity of Tenochtitlán alone in the Isthmus of Tehuantepec, Mexico, people appreciate nine varieties of mango (Coe and Diehl, 1980:89).

Local mango cultivars could prove useful for germplasm collections and future selections. In some areas, seedling mangos form extensive groves, while in other parts of the tropics and subtropics, grafted cultivars are planted in commercial orchards.

In addition to those found in India, a number of old and widely grown cultivars are found in Indonesia, Bangladesh, and Pakistan. In other countries, rural households usually rely on seedlings to propagate mango, and trees with insipid or very stringy flesh are often the result. Nevertheless, this "random pool" helps promote diversity in cultivated mango and serves as a basis for selecting superior forms. All the famous mango cultivars arose from spontaneous seedlings.

Widespread cultivars have been multiplied true to type by growing seedlings from maternal tissue of polyembryonic seeds, or if seeds are monoembryonic, by vegetative propagation. Various techniques for vegetative propagation of mango are used, including inarching and marcotting. With inarching, a scaffold is erected under the mango tree to be propagated, and seedlings in pots are placed next to shoots so that they can be wrapped together. Eventually they merge, and the shoot can then be cut away to be nurtured by the seedling rootstock. Inarching, with seedlings serving as rootstocks, has been employed for centuries; British botanists used the method to introduce selected mango forms to several colonies in the 1860s. This approach-grafting technique is labor-intensive, and other grafting and layering techniques have been developed, such as marcotting, in which twigs are wrapped with soil held in place by plastic so that roots form.

Scientific breeding of mango began in the West Indies in 1912 and in India in 1920 and is geared primarily to the export trade. A major concern for breeders is selecting mangos that bear fruit consistently. Mango trees are notorious for not producing fruit every year. Off-years can create cash flow problems for mango growers. When a bumper crop is produced, prices usually dip. Planting a mix of varieties with different fruit-producing cycles is one way to overcome this problem, but growers are interested in cultivars that fruit regularly. Some wild mangos (*M. indica* var. *mekongensis*) in Indochina apparently fruit twice a year and could be useful parents in crossing to improve yield consistency (Mukherjee, 1985). India has produced a promising dwarf cultivar, 'Amrapali', that also bears fruit regularly.

Apart from regular production, other criteria for mango breeders include outstanding taste, ability to travel and store well, uniform fruit shape, and compact tree shape to facilitate harvesting. Thick skins are normally required for export mangos. Desirable shape, skin color, and taste vary according to region, so selection and breeding goals often differ from one country to another. In Florida, for example, consumers prefer relatively large, round mangos with pink or purple blush. The elongated, yellow-skinned mango 'Manila' is popular in Veracruz, Mexico. The peach-colored 'Piña' mango is a favorite along the Pacific slope of Guatemala. The small, locally selected mangos of Ica, along Peru's desert coast, are highly prized in Lima. In tropical regions, people have learned to appreciate a variety of mango shapes, colors, and textures, whereas consumers in temperate countries are exposed to only a small sample of shapes and sizes.

Although mango breeding and commercial plantations are geared largely to the dessert fruit market, demand is growing for mango fruit drinks. Dessert cultivars do not necessarily yield the best mangos for juice production. The fruit juice industry is interested in cultivars that produce juice with superior flavor after processing. Local selections used for sucking, rather than eating, may prove useful for manufacturers of mango juice. The Punjab region of India and Pakistan is well known for its juicy mangos.

Mangos are attacked by many pathogens, particularly fungi, and resistance to them is a concern to breeders. The two most serious fungal diseases of mango are anthracnose (caused by *Colletotrichum gloeosporioides*) and powdery mildew (caused by *Oidium mangiferae*). Anthracnose triggers a range of symptoms, particularly in humid climates. One common symptom is disfiguring black splotches on the skin, thereby reducing the fruit's market value. Powdery mildew provokes the loss of flowers and immature fruit, thereby depressing yields, especially in India. After extensive screening, four mango cultivars in India have been found to resist powdery mildew (J. H. Gupta and Yadav, 1985). Bacterial canker is becoming a pressing disease problem, and a mealybug, *Rastrococcus invadens*, has recently invaded Africa, where it causes serious damage to mango and other tree crops. Sources of genetic resistance to these as well as numerous other diseases and pests are needed to reduce losses (D. K. Sharma, 1987).

Mango growers have historically relied on local selections or introduced cultivars to establish or rejuvenate orchards. Mango breeding has had limited success in developing cultivars that rival traditional selections in terms of yield or taste qualities. Breeding for disease resistance has begun in India, using cultivars as parents in controlled crosses.

Some poorly understood mango disease and pest problems need further elucidation before germplasm can be effectively screened for resistance genes. Flower malformation, for example, stems from several causes. A virus is suspected in some cases of deformed flowers, while in other instances a fungus (*Fusarium moniliforme*), transmitted by a mite, appears to be the culprit.

Wild relatives of mango have the same basic chromosome numbers as *M. indica*, but much more information is needed about their characteristics. Furthermore, crossing mango is not easy. For breeding purposes, mango pollination is carried out by hand, and thousands of meticulous pollinations must normally be performed to obtain just a few fruits (L. B. Singh, 1976). Also, progeny from crosses take many years to bear fruit, and some cultivars are self-incompatible. Some successful crosses have recently been achieved using house flies trapped in insectproof cages.

PROSPECTS AND CONSERVATION NEEDS

Some nontraditional uses for mango may further spur demand for good planting material. Mango kernel oil can be blended with cocoa butter, thereby reducing the cost of that ingredient in chocolate. Chocolate manufacturers in Switzerland, Italy, and Japan have expressed interest in exploring the potential of mango kernel oil (Battacharya, 1987). If the price is high enough to provide an incentive, juice companies could sell a by-product from their operations.

Demand for fresh mangos is increasing in Europe, North America, and the Gulf states of Southwest Asia. For example, imports of mango into the Netherlands tripled between 1975 and 1982 (Samson, 1986:220). Mexico is shipping ever-increasing quantities of mango to the United States; U.S. citizens now eat more mangos from Mexico than from Florida or Hawaii. In 1987, Mexico supplied two-thirds of the mangos eaten in the United States, up from one-third of the U.S. market in 1970 (R. J. Knight, 1988a; R. J. Knight et al., 1984). Several countries, such as Australia, Colombia, Côte d'Ivoire, Cuba, Fiji, Martinique (France), Mexico, Nicaragua, Peru, and Venezuela, are increasing mango plantings with a view to the export trade. Mango's genetic resources need to be better safeguarded and tapped to meet this burgeoning demand.

Considerable potential remains to exploit further the rich germplasm resources of mango and its near relatives both for the export trade and local markets. Progress in mango breeding hinges on upgrading existing gene banks and safeguarding gene pools of mango and its near relatives in the wild. Most of the existing mango germplasm collections are geared to the immediate interests of breeders rather than to conservation of the broad genetic diversity of mango.

Indigenous people have intimate knowledge about mango relatives and other forest resources, and such information could prove invaluable to germplasm collectors and mango breeders. Safeguarding wild populations of crops and their near relatives is only part of an overall conservation strategy. The rich cultural heritage of tribal peoples needs to be safeguarded as well. This does not mean trying to hermetically seal cultures from outside influences, only that native peoples should be allowed to develop at their own pace, while protecting their forest resources. All cultures change, but indigenous people can act as custodians of genetic resources in extractive reserves that remain under local control. A mix of in situ conservation strategies is thus called for: parks or nature preserves, extractive reserves, and agroforestry schemes.

In some cases, reserves will probably have to be proposed on the basis of sketchy field and herbarium work, given the constraints of time and resources. Too much delay in setting up parks could result in the loss of critical forest ecosystems. Reserves established too hastily may not contain much useful germplasm. The utility, and political safety, of reserves will be enhanced if they contain germplasm of crops or their near relatives and provide some environmental service, such as protecting water supplies for cities. The growing market for mangos in industrial nations could provide an incentive for governments in tropical Asia to set aside more reserves for wild mango and its close relatives.

CITRUS

The genus *Citrus*, comprising some dozens of species, has provided a cornucopia of fruits to freshen the palate, slake one's thirst, impart exquisite flavors to prepared foods, and treat ailments. Not surprisingly, then, citrus fruits and their myriad products are consumed in virtually every country and are thoroughly entwined in most cultures (Figures 3.8, 3.9).

The bright colors, tantalizing textures, enticing aromas, satisfying shapes, and appealing flavors of citrus fruits have inspired artists, have been incorporated into nursery rhymes ("Oranges and lemons said the bells of St. Clement's. . . ."), and have been absorbed into religious practices and beliefs. Citrus fruits are featured in many still life paintings, such as Cézanne's *The Kitchen Table*; Van Gogh's *Still Life with Coffeepot* (1888), *Still Life of Oranges and Lemons with Blue Gloves* (1889), and the endearing *Child with Orange* (1890); Matisse's *Harmony in Red* (1909); and Maurice de Vlaminck's *Still Life with Oranges* (1907).

Bengal quince (*Aegle marmelos*), a relative of the orange, is revered by Hindus. In Hindu culture it is sacrilege to uproot a Bengal quince, which is used to prepare sherbet and a preserve. Hindu mendicants sometimes adorn themselves with lemon garlands (P. Thomas, 1975). Citron is thought to be the "goodly tree" in the Old Testament, one of four used in the Feast of the Tabernacle.

Several species of *Citrus* are cultivated for fresh fruits and to make juice, marmalade, pies, or confections. Citrus fruits are rich in vitamin C, while the orange contains significant amounts of folic acid and vitamin B6. Some citrus oils, such as limonene, citrol, and bergamot, are used in the perfume industry and to flavor Earl Gray tea. In some countries, such as Brazil, orange peels are steeped to make a refreshing hot beverage. Citric acid is a useful byproduct of the lime juice industry.

Citrus fruits are major contributors to global commerce in tropical and subtropical fruits. The most important species include sweet orange (*C. sinensis*), sour orange (*C. aurantium*), lemon (*C. limon*), grapefruit (usually designated as *C. paradisi*, but possibly a hybrid between sweet orange and pummelo from Barbados around 1750), pummelo (*C. grandis*), lime (*C. aurantifolia*), and mandarin orange (*C. reticulata*). Marmalade is made from all cultivated species of citrus, but sour orange is grown almost exclusively for this purpose. A type of sour orange grown around Seville, Spain, is exported in large quantities to the United Kingdom. Lime is used extensively in India for making pickles.

FIGURE 3.8 Sweet oranges for sale in a market, Chichicastenango, Guatemala, March 1988.

Developing countries are gaining ground in the production of citrus for local consumption and for export. Citrus fruits are cultivated in over 125 countries, but the Mediterranean region, Central America, Florida, California, Brazil, and East Asia are the main producers. Citrus fruits are dooryard plantings or are cultivated in groves.

ORIGINS AND SPREAD

Little is known about the early domestication of citrus fruits. Most species were probably domesticated long ago in seasonally dry forests subject to monsoon rains. After a protracted period of simply gathering wild fruits, early citrus planters soon selected citrus forms with especially juicy pulp and appealing flavors. In areas of shifting cultivation, a farming method still predominant in many forested areas, introgression of genes from wild citrus to domesticated forms would have been common. When villagers abandoned an area, domesticated forms soon reverted to the wild. Some "wild" populations may have thus undergone selection in the past.

Citron (*C. medica*) reached the Mediterranean during the height of the

FIGURE 3.9 Sweet oranges on a market stall, Mangalore, India, December 1989.

Greek civilization from tropical or subtropical Asia (Figure 3.10). Citron could have been the *melea persika* of Theophrastus and the *persika mala* of Dioscorides. Arab traders brought lemons and sour oranges from India to the Arabian Peninsula in the ninth century. Lemons and oranges reached Sicily in A.D. 1002, and Spain and North Africa by the end of the twelfth century (Targioni-Tozzetti, 1855). Saint Domine is credited with planting an orange in the gardens of S. Sabina in Rome in 1200, and an orange grove still thrives on this Aventine Hill. Crusaders noted oranges, lemons, and citrons in Southwest Asia. Sweet oranges arrived in Europe in the fifteenth century.

From the Iberian Peninsula, oranges were taken to various parts of the New World. In 1493, Columbus took seeds of orange, lemon, and citron on his second voyage to the New World where they were planted on Hispaniola. Oranges were well established in Florida, California, and Louisiana by the mid-eighteenth century. In both the New and Old Worlds, new forms have arisen from seedlings or as sports from bud mutations. 'Wash-

1 cm

FIGURE 3.10 Citron (*Citrus med-ica*). (Illustration by R. Suralis Nur-biantoro, in S. Sastrapradja et al. 1980.)

ington' navel, one of the most important cultivars of sweet orange, arose in Bahia, Brazil.

Many early introductions of citrus to the New World and the Pacific soon became naturalized. A Catholic missionary marveled at the thick woods of orange trees in Brazil arising from discarded seed at the close of the sixteenth century (Crosby, 1986:151). Two and a half centuries later, Charles Darwin remarked on the density of spontaneous orange and peach trees on an island in the mouth of the Paraná River, derived from seeds washed downstream. Limes, sour oranges, and pummelos have become naturalized in Jamaica (Fawcett, 1920).

Sweet oranges are feral in the uplands of Tahiti, where they are relished by people and escaped pigs. In June and July, the Southern Hemisphere winter, Tahitians gather wild oranges in the mountains for sale along the coast. In Côte d'Ivoire, sour orange is reportedly spontaneous in the forest around villages, while in Guinea-Bissau, oranges have become naturalized around some Capuchin and Jesuit missions (Chevalier, 1912; Gomes e Sousa, 1930).

GENE POOLS

Citrus fruits belong to the family Rutaceae, which contains several hundred species from the tropics to temperate regions. This wide-ranging family is divided into five subfamilies, one of which, the Aurantioideae, includes genera that produce berrylike fruits with a pulp, such as *Citrus*. *Citrus* trees and shrubs occur naturally throughout Southeast Asia, the Indo-Malaysia region, and in southern China (Swingle and Reece, 1967).

The location of wild gene pools of most cultivated citrus species is unclear for two main reasons. First, the taxonomy is confused (H. C. Barrett and Rhodes, 1976), a problem with many domesticated perennial plants in the tropics. Between 16 and 159 species of *Citrus* are recognized, depending on the taxonomic system used (C. J. Hearn, pers. comm.). Second, fieldwork to locate truly wild representatives is difficult because they arose in primary forest, a habitat that has been drastically altered by human activities, particularly farming and logging (IBPGR, 1982). Wild limes have been reported in some forests remaining in northern India and the East Indies (Purseglove, 1974:498).

In 1981, a working group organized by the IBPGR identified a primary citrus gene pool consisting of lime, pummelo, lemon, citron, sweet orange, sour orange, and mandarin. A secondary *hystrix* gene pool was proposed containing C. *hystrix*, C. *ichangensis*, and C. *macroptera*, while a tertiary gene pool includes genera related to citrus, such as *Fortunella*, some of which can hybridize with *Citrus*.

BREEDING HISTORY AND CHALLENGES

All citrus fruits have been the focus of selection and breeding programs, particularly sweet orange, lemon, grapefruit, and mandarin orange. Some citrus fruits, such as lime, pummelo, citron, and liman-purat (C. *hystrix*), however, are little advanced from their wild progenitors.

Citrus breeding started in Florida in 1893, with California, China, the Soviet Union, Indonesia, the Philippines, and Zimbabwe following suit in the early part of the twentieth century. Japan, Italy, and Australia launched citrus breeding efforts after the World War II (J. W. Cameron and Soost, 1976).

Much of the research on citrus has concentrated on developing commercial material suitable for warm temperate areas, the subtropics, and to a lesser extent the dry tropics. Many forms adapted to the humid tropics are considered primitive and have been largely ignored by breeders. A rich array of traditional cultivars are tended in eastern Indo-Malaysia for lime, Indochina (sour orange), Malaysia and Thailand (pummelo), northern India to Indochina (mandarin orange), and Indochina and southern China (sweet orange), but they have been little exploited by breeders. Southern Thailand is particularly rich in pummelo types and is probably the center of origin for the species (Ueno and Akihama, 1988).

Many of the commercial cultivars arose from mutant sports or chance seedlings. But considerable scope exists for more controlled crossing to transfer desirable traits from one form to another. The important economic forms are mostly diploids and are interfertile. Even intra- and intergeneric hybrids are frequently fertile. These traits open the door to a multitude of possibilities for crop improvement.

Most citrus fruits are seedy, but fruits of some cultivars are seedless, or nearly so. Sexual seeds result in heterogeneous seedlings and cannot be used to perpetuate a cultivar. Pummelo (Figure 3.11) and citron usually reproduce sexually. Lemon, lime, oranges, mandarin orange, and grapefruit range from partly to fully asexually seeded.

Protoplast fusion promises to create an array of seedless citrus triploids for the fresh fruit trade (Grosser and Gmitter, 1990a, b, c). While the first hybrid triploid was obtained in the 1920s, only a few triploids have been produced because of poor endosperm development and embryo failure (Soost, 1987). Protoplast fusion is likely to increase the number of triploids, some of which may prove useful to growers.

FIGURE 3.11   Pummelo (*Citrus grandis*) is widely appreciated in tropical Asia. Northern Kerala, India, December 1989.

Rootstocks have also been the focus of breeding and selection. Traditionally, sweet oranges are budded on rootstocks of sour orange, sweet orange, rough lemon (*Citrus jambhiri*), *Poncirus trifoliata,* or crosses between citrus and *Poncirus.* In lowland areas of Thailand, *Feronia limonia*, a distant relative of citrus, is used as a rootstock because of its tolerance to poor drainage (Ueno and Akihama, 1988). Even when grafted on rootstocks, sexual or asexual seedlings of most citrus forms pass through a relatively long juvenile phase.

Characteristics high on the priority list for citrus breeders include fruit size, yield, flavor, time of ripening, juice quality, and seedlessness. Resistance to pests and diseases in both scions and rootstocks is also near the top of the agenda of citrus breeders. Cold tolerance remains a major objective of breeders working with scions and rootstocks for citrus plantings in temperate climates. In Florida, for example, 14 percent of the citrus acreage was destroyed by the Christmas 1983 freeze, and growers in the state were hit by two more severe cold snaps in January 1985 and December 1989 (Drummond, 1984a, b, 1985).

Most cultivars are clonally propagated by budwood to avoid the long juvenile phase of seedlings. Unfortunately, the movement of budwood increases the danger of spreading pests and latent diseases. To help address this problem, IBPGR and the FAO convened a meeting of specialists to develop a series of technical guidelines for the safe movement of citrus germplasm in November 1989. In vitro grafting has helped eliminate viruses and other pathogens from many citrus clones, and this technique is used in several countries, particularly Japan and Spain (S. R. Bhat, pers. comm.).

Citrus crops are attacked by numerous diseases, some highly localized, others virtually cosmopolitan. Some of the most serious diseases are caused by viruses or viruslike organisms. Tristeza ("sadness") is an aptly named disease that troubles citrus growers worldwide. Tristeza can trigger yellowing and wilting of leaves, twig dieback, and subsequent death of the tree. Fortunately, sweet orange, rough lemon, and some citrange (*C. sinensis* × *P. trifoliata*) and citrumelo (*C. paradisi* × *P. trifoliata*) rootstocks tolerate tristeza.

Another dieback attacks citrus in India, while other important diseases include leaf mottle yellows in the Philippines and citrus vascular degeneration in Indonesia. To reduce the chances of spreading viruses, viroids, and other bud-transmissable diseases, breeders normally distribute pathogen-tested budwood from nurseries to groves and apply insecticides to suppress vectors. Collections of virus- and viroid-free budwood are maintained in numerous countries, and most citrus-growing countries operate programs to register budwood.

An assortment of bacterial and fungal pathogens, nematodes, and insects attacks citrus. Control is mainly by sanitation, spraying, and pruning. Resistant or tolerant rootstocks are used against some major soil-borne problems, such as nematodes and phytophthora (R. Soost, pers. comm.). As citrus moves into southwestern Florida, growers are encountering tree-girdling termites and fungi that kill some young trees. The termites and fungi are more of a problem in the warmer, more humid area around Immokalee. Insulating wraps used to protect the trees against frost help the termites by providing them with a moist, dark environment to tunnel into the bark. Frost-tolerant citrus varieties would thus help overcome this insect problem. Genetic resistance would also reduce the cost of controlling other diseases and pests.

Several strains of citrus canker have been identified, thereby complicating the task of resistance breeding. Citrus canker appears occasionally in Florida, and millions of citrus trees and seedlings are torched in an effort to contain the disease (Plucknett and Smith, 1989). An apparently new strain of canker, dubbed the nursery strain, appeared on young trees of certain cultivars in several nurseries in central Florida during the mid-1980s.

Somatic hybridization is permitting breeders to overcome crossing barriers between citrus and its more distant relatives, thereby creating opportunities for developing hardier and more productive varieties. At the University of Florida's Citrus Research and Education Center, Lake Alfred, scientists are evaluating a number of interesting interspecific and intergeneric hybrids created by protoplast fusion (Grosser and Gmitter, 1990a, b, c). Developments in biotechnology thus underscore the need to conserve as wide a net in a crop's gene pool as possible.

GENETIC RESOURCE COLLECTIONS AND GAPS

Citrus cultivation is widespread, so many countries maintain orchard gene banks (Table 3.3). Many of the collections have not been fully described or evaluated. A list of descriptors has recently been published that can help sort out accessions in germplasm collections (IBPGR, 1988). It is difficult to gauge how comprehensive or unique collections may be without careful documentation. But excessive duplication may be burdening efforts to conserve citrus germplasm. Wild materials, as in the case of most ex situ collections, are poorly represented.

Much more collaboration is needed to avoid duplication. An IBPGR Working Group suggested in 1981 that regional collections might better serve the global citrus research and development community. Regional collections could be designated as follows: East Asia and South Asia for wild forms and old cultivars; Southeast Asia for wild forms and old cultivars; the Mediterranean for cultivars; Central Africa for wild forms. Many de-

TABLE 3.3

Citrus germplasm collections

Country	No. of institutes	Oranges, lemon, lime, mandarin	Citron, pummelo, grapefruit	Wild *Citrus*	Wild related species
Algeria	1	4	6	5	6
Argentina	5	2			
Australia	4	2	6	6	6
Bolivia	2	5	6	6	6
Brazil	5	3	5	4	5
China	2	4	6	6	6
Colombia	1	5			
Costa Rica	1	6			
Cuba	1	2			
Ecuador	1	5			
Fiji	1	6	6	6	6
France	1	1	4	1	4
Greece	1	5	6	6	6
India	2	5	6	5	6
Indonesia	3	4	5	4	6
Italy	1	4	6	6	
Jamaica	1	4	6		
Japan	5	3	5	3	5
Mexico	3	5	6		
Morocco	1	2	5	5	6
Nepal	1	6	6		
Nigeria	1	5	6	6	6
Philippines	1	6	5	6	
Seychelles	1	4	6	6	
South Africa	1	3	4	4	4
Spain	1	3	6	6	6
Thailand	2	4	4	4	6
Turkey	2	3	5	5	5
Former USSR	1	1			
USA	4	3	3	4	4

Source: IBPGR, 1982.

[1] 1 ≥ 750; 2 = 500–750; 3 = 150–499; 4 = 100–149; 5 = 50–99; 6 ≤ 50.

tails still need to be worked out, however, such as which countries could house effectively the regional collections. Such regional collections would not obviate the need to maintain other collections serving the immediate needs of national citrus research programs.

Citrus germplasm collections would benefit from the incorporation of new material from areas poorly represented in gene banks. Priority areas for collecting citrus germplasm include southern China for sweet orange, mandarin orange, and several near relatives, such as *C. honghoensis, C.*

hystrix, and *C. limonia* (Gmitter and Hu, 1990); Japan and China for *Poncirus* and natural hybrids with citrus; Southeast Asia for pummelo; Northeast India for wild and cultivated citrus; Japan, China, and the Mediterranean for old cultivars; and parts of Southeast Asia for wild relatives. Parts of east-central Africa may warrant collecting for *Citropsis* germplasm, particularly since deforestation is accelerating in that region with rapidly growing rural populations.

Future germplasm collecting will focus mainly on primitive forms and wild citrus progenitors. A number of wild and domesticated near relatives of citrus may be particularly useful to breeders in the near future and therefore warrant incorporation into more gene banks as well as preservation in the forest (Table 3.4). Wild relatives of citrus and some minor domesticated species contain genes for such valuable traits as resistance to diseases and tolerance to saline soils or drought (IBPGR, 1986; Samson, 1986:73; Swingle and Reece, 1967; Sykes, 1988). The remaining wild relatives are best conserved in their natural environments; only some citrus seeds can be dried and stored for long periods without adversely affecting their viability.

In association with national agricultural research systems and IBPGR, the Japanese have recently collected citrus germplasm in numerous coun-

TABLE 3.4

Some near relatives of *Citrus* with potentially useful genes

Scientific name	Common name	Distribution	Known traits
Clausena lansium	Wampee	S. China; widely grown in tropics	Cultivated
Eremocitrus glauca	Australian desert lime	Australia	Drought-tolerant
Feronia limonia	Wood apple	India; Sri Lanka; planted in tropical Asia	
Feroniella oblata		Thailand; Indochina	
F. lucida		C. Java	
Limnocitrus littoralis	Wild swamp orange	Bali; Java; Vietnam	Tolerates salinity
Microcitrus spp.		New Guinea; Australia	
Severina buxifolia	Chinese box orange		Tolerates salinity; immune to many diseases of citrus
Swinglea glutinosa	Tabog	Philippines	

tries of Southeast Asia and East Asia. Working with national counterparts, Japanese scientists collected 500 *Citrus* accessions, including cultivars and some related wild species, between 1983 and 1987 (Akihama et al., 1985). In 1984, the World Wildlife Fund (WWF) in Malaysia organized the collection of wild and cultivated citrus in areas of Sabah threatened by forest clearing for cacao and oil palm estates (D. T. Jones, 1985).

Our understanding of the distribution of the wild populations and near relatives of citrus is being furthered on several fronts. In 1986–87, David Jones of the University of Malaysia examined specimens in many herbaria and sought out living examples in the forests of Malaysia and Indonesia to better understand taxonomic nuances not readily discernible in pressed specimens and to verify distributions. The Botanic Gardens of the University of Malaysia, WWF-Malaysia, and the National Biological Institute of Indonesia have provided the institutional anchors for this ambitious survey of citrus gene pools in Southeast Asia.

Herbarium records have been employed by IBPGR to establish a database on localities, altitude, and vernacular names of seventeen aurantioid genera related to *Citrus*. Although many species are confined to tropical forests, several wild relatives of *Citrus* thrive in open, disturbed environments. Also, some species exhibit narrow ecological tolerances, while others appear less exacting in their growth requirements.

In 1981, Indian authorities established a gene sanctuary for citrus in the Garo Hills of Meghalaya, in northeastern India (B. Singh, 1981). This gene sanctuary is being established along the lines of a national park. The location for the citrus sanctuary followed examination of herbarium records and surveys of *Citrus* species throughout India. The 10,265-hectare Garo Hills sanctuary protects diverse populations of wild orange (*C. indica*), an endangered species. A buffer zone surrounds this gene sanctuary and provides local people with fuelwood and other resources, thereby minimizing human impacts on the core area, which contains the greatest concentrations of "ghost oranges," as the wild orange is known locally. Wild orange also grows in the forests of the Chase Hills, Nowgong (Assam), Rangpur and Dimapur in the Nuga Hills, and the Kaziranga forest reserve in Assam.

India deserves considerable credit for this far-sighted conservation effort, the first such reserve set up specifically to conserve genetic resources of a tropical perennial. More recently, Mexico has set up a reserve to safeguard a perennial maize, another bold move discussed in more detail in the section on avocado (see p. 129). India and Mexico are faced with rapidly growing human populations; indeed India is expected to surpass China as the world's most populous country early in the twenty-first century, so efforts to set aside wild areas are timely.

FUTURE NEEDS

Many citrus species that once were simply gathered in the wild have been domesticated in numerous areas for thousands of years. Now close to a dozen citrus and its near relatives are cultivated and have become integral parts of human cultures and economies in the tropics and subtropics. Breeders can draw on an expansive gene pool for improvement, both for local tastes and export markets. This rich gene pool has permitted breeders to increase the number of commercial cultivars available for growers in recent years. In many parts of the tropics, households in rural and urban areas continue to grow citrus from seeds, thereby maintaining variability. Genetic erosion is most serious with wild progenitors and near relatives, as forests come down. Also, old and primitive cultivars are being lost.

A broad range of strategies is called for to safeguard the genetic heritage of citrus, including field collections, seed gene banks for certain species, test-tube gene banks, and natural reserves. Some are in place, but much more needs to be done. More targeted collecting of citrus germplasm is needed, particularly in areas of origin, and more parks and reserves need to be set aside to safeguard wild germplasm of citrus as well as other crop gene pools.

PINEAPPLE

When Europeans stepped ashore in the Caribbean and the Americas, they were confronted by a strange new fruit: prickly on the outside like a pine cone, yet sweet and fleshy on the inside. Pineapple (*Ananas comosus*) does not taste or look like an apple, but Spanish, Portuguese, English, Dutch, and French explorers were intrigued by the juicy fruits and soon took them back home for subsequent introduction to their colonies and possessions in Africa, the Indian Ocean, Southeast Asia, and the Pacific.

Pineapple quickly spread from its native home in tropical America to other lands during the colonial period. It reached India in 1548, probably through the hands of Portuguese seamen at Goa (Chadha and Pareek, 1989). In 1558, pineapple was taken to the Philippines in Spanish galleons from Mexico, and to South Africa in 1660 (Stearn and Roach, 1989).

An enthusiastic welcome awaited pineapple in Europe and other lands where it was introduced. The prickly but tasty fruits were presented to the kings of France and England as delicious curiosities, and pineapple was soon incorporated in paintings and symbolism. In 1661, for example, 'Queen' pineapple from Barbados was presented to Charles II of England; this early fruiting clone is still cultivated in South Africa and Australia for the fresh fruit trade (Hooker, 1989:208). To prevent cold damage in Europe, growers cultivated pineapples in the new, heated glasshouses pioneered by the French and Dutch in the seventeenth century. These glasshouses, attached to dwellings of the well-to-do, were called orangeries because they were initially designed to keep citrus trees.

The edible portion of pineapple is a common mass of tightly packed individual fruits that form after the flowers have been fertilized. The aggregated "fruit" of pineapple is especially succulent, highly suited for eating fresh, for canning, and for preparing frothy and pulpy juices. Fruits are also candied and made into jam. Pineapple is used to flavor crushed ice sold by street vendors in many tropical countries and is sometimes used to make an alcoholic drink. Pineapple waste is fed to livestock and is used to manufacture a vinegar. In the past, long-leafed varieties have been cultivated to make silky cloth, particularly in the Orient, and to make cord for bowstrings (Kimber, 1988:93).

Although the nutritional value of pineapple varies somewhat among varieties, the fiber-rich fruits provide some carbohydrate and are a good source of vitamin C, carotene, and riboflavin. During the colonial period, the fruits were stowed on board ocean-going vessels because they helped prevent scurvy. Pineapple also contains traces of thiamin and nicotinamide.

In tropical countries, local people have generally selected sweet-tasting pineapples, whereas tarter forms are now preferred by commercial growers who cater to canneries. Pineapple is propagated vegetatively by several means, including planting suckers taken just below the soil surface, shoots from buds in the axils of leaves, slips from the fruit stalk, or *hapas* from the base of the fruit stalk. A common planting method is to lop off the crown of the fruit after harvesting and insert the leaf whorl into the ground. Slips can also be taken for planting. All these planting forms resist drying and can survive several weeks of storage or transportation.

Vegetative propagation allows interesting new forms to be easily "fixed" and multiplied. New variants from spontaneous mutations have been the basis of most pineapple improvement. Pineapple is a perennial herb that produces fruit in about 18 months after planting and can live for decades. In commercial plantations, the entire field is replanted after two or three ratoon crops.

ORIGIN AND SPREAD

Little is known about the domestication of pineapple (Pickersgill, 1976). Tupi-Guarani Indians in northern Paraguay have been credited with domesticating it (Collins, 1960), largely because southern Brazil and Paraguay contain the greatest diversity of the crop's near relatives. The great Brazilian botanist Adolpho Ducke speculated that *A. microstachya* in drought-plagued northeastern Brazil is the progenitor of pineapple (Ducke, 1946). Others have postulated the well-watered Guianas or the Amazon estuary as the home of pineapple.

It seems more likely that the terrestrial bromeliad was domesticated in the fringes of Amazonia, such as Mato Grosso or Goias, where the forest thins and becomes subject to pronounced dry periods. Pineapple is thus one crop domesticated away from the center of diversity for related wild species. Pineapple shows some tolerance to drought because of its reduced leaf area and specialized cells for water storage. The search for the ancestral home of pineapple is complicated by the fact that it is unknown in the wild. Reports of "wild" pineapple are probably documenting escapees.

Pineapple exhibits greatest variation in western Amazonia, particularly in Peru and Colombia. The fact that sun- loving pineapple is adapted to various soils, including infertile ones, suggests that pineapple's progenitor may have exploited disturbed sites. Its weedy nature would have brought it into close contact with humans, particularly farmers. Whatever the origin of pineapple, it appears to be an ancient cultigen.

Cultivated pineapple spread throughout the lowlands of central and northern South America, including the Atlantic Coast. Aboriginal groups, such as those belonging to the Arawak and Carib linguistic families, took the cultigen to the West Indies in precontact times; Columbus found the fruit growing on Guadeloupe in 1493 (Stearn and Roach, 1989). The antiquity of pineapple in the West Indies is suggested by the various uses and cultivars of pineapple on Martinique when Europeans arrived. The indigenous population on the island used one kind of pineapple for fresh fruit and to ferment, while another type, called 'La Pité' was grown for its leaf fibers to make bowstrings, among other products (Kimber 1988:93).

Provided that waterlogging is avoided, pineapple thrives in many soils and will fruit as long as temperatures do not fall below 20°C. Pineapple tolerates an unusually wide amplitude of rainfall (from 600 to 1500 millimeters per year), but yields are generally higher when the annual rainfall is more than 1,000 millimeters (Purseglove, 1975).

Although taken to other continents and warm archipelagos as early as the sixteenth century, pineapple lingered as a relatively minor fruit internationally until the twentieth century. Development of plantations for the fresh fruit market had to await the availability of refrigerated vessels for transport at the turn of the century. The Chinese started canneries in Singapore in 1886, and Hawaii followed suit in 1903. Canned pineapple is now produced in many countries and is available worldwide. Pineapple, probably introduced from Mexico, has been grown in Hawaii since 1813 (Crawford, 1937). It is now a major fruit crop, both for domestic markets in tropical countries and for export. Fresh fruit is the fastest growing segment of the pineapple export trade. Over sixty-five countries grow pineapple for domestic consumption and for export. Thailand, India, Indonesia, China, Brazil, Mexico, the Philippines, and the Hawaiian Islands are the principal pineapple producers.

WILD RELATIVES

Pineapple has seven near relatives in the genus *Ananas* and one more-distant cousin in the genus *Pseudananas*. All of pineapple's wild relatives are confined to lowland South America, where they grow in environments from moist forest to more open habitats. One related species, *A. bracteatus*, grows in wetter areas, often in partial shade; apparently cultivated at one time in Paraguay, the fibrous leaves of this species are used for making hammocks (Brücher, 1989:252). *A. erectifolius* is a denizen of hot, humid areas of northern Amazonia and the Orinoco; Waika and Makiritare Indians use this species for fiber, and some forms have spineless leaves. Although pineapple may not have come from forest proper, parts of its gene pool extends into the jungle.

Southern Brazil and nothern Paraguay contain a cluster of pineapple's relatives, including *A. ananassoides*, *A. fitzmuelleri*, *A. paraguayensis*, *P. sagenarius*, and *A. bracteatus*. *A. ananassoides* is drought-tolerant (Brücher, 1989:253). Pineapple's relatives are poorly known, and more species are likely be found with further field surveys.

SELECTION HISTORY AND BREEDING POTENTIAL

Pineapple is self-incompatible, whereas the wild relatives are partly self-sterile (Brewbaker and Gorrez, 1967). A pineapple clone produces fruits parthenocarpically, so they do not contain seeds. Thus, if an area is planted to one clone by vegetative propagation, crossing is not possible because of incompatibility. On the other hand, if diverse varieties are planted in the same field, cross-pollination can occur, and some fruits may contain seeds. Pineapple cultivars may be diploid, triploid, or tetraploid; the diploids and tetraploids can be used for crossing, but the triploids are sterile.

Indigenous peoples soon selected seedless forms, and presumably kept clones relatively distinct. In addition to seed-free fruit, other desirable properties selected for in early times included taste, juiciness, and color. Less prickly leaves for ease of handling are among the more recent selection criteria. Much of modern pineapple breeding has been geared to the canning industry, which has favored yellow-fleshed, cylindrical fruits.

Commercial pineapples largely trace their origins to materials sent to France from French Guiana in the early nineteenth century (Collins, 1960). These introductions were distributed throughout Europe, including England, which sent material to Florida, Hawaii, Jamaica, and Australia. W. H. Purvis obtained suckers of 'Smooth Cayenne' from the Royal Botanic Gardens, Kew, and planted them in Hawaii in 1885 (Crawford, 1937). Discovery of many smooth-leafed clones with different genes for spinelessness has opened up the possibility of developing superior pineapple varieties for a wide range of cultural environments (R. J. Knight, 1988b).

Pineapple breeding is still largely based on selection of mutant forms or progeny of crosses between cultivars. Pineapples are genetically heterozygous, and thus seedlings are variable and provide a rich array of materials to work with. Traditional screening and selection of seedlings is nevertheless slow; hybrid offspring take four years to fruit, and several generations of backcrossing to one of the original parents is usually required to amass the desirable qualities. Such procedures can take a quarter of a century before a suitable cultivar is produced.

Callus tissue culture would be a more efficient method of producing new cultivars, since heterozygosity can still be released by this method (Pannetier and Lenaud, 1976). By shortening the breeding cycle, scientists are more likely to produce tangible results in a shorter time, an important consideration for young scientists deciding on a career path to pursue. Also, commercial interests and farmers will be better served if new products and clones can be produced more quickly. By using callus tissue culture from slips of the syncarp and the crown, scientists can obtain somaclonal variants showing differences in spine and leaf color, waxiness, foliage density, leaf shape, and spininess (Wakasa, 1979). Tissue culture techniques can also be used for mass propagation of desirable selections (Mathews, 1979).

Hybridization has the advantage of widening the genetic base of pineapple production, which is excessively narrow on commercial plantations. Controlled crossing programs are particularly well developed in Hawaii, Puerto Rico, the Philippines, Malaysia, Australia, Brazil, and Guinea. In Hawaii, scientists are incorporating some of pineapple's near relatives in their breeding efforts. *A. ananassoides* has thus far contributed a distinctive flavor, tolerance to cool, moist conditions, and some disease resistance (IBPGR, 1986:29). India is considering using *A. ananassoides* and *A. erectifolius* for its pineapple improvement program (Chadha and Pareek, 1989).

In addition to desirable fruit characteristics, resistance to diseases is high on the list of priorities, particularly to Phytophthora and Ceratocystis fungal rots, fruitlet core rot caused by *Fusarium moniliforme*, wilt triggered by mealybugs (*Dysmicoccus brevipes*), and the yellow spot virus. At least eleven species of nematodes attack pineapple. Traditionally, crop rotation and chemical sprays have been used on commercial holdings to control pests and diseases of pineapple, but resistance breeding is needed to reduce costs and combat pests and diseases that cannot be controlled effectively with pesticides or agronomic practices. Resistances to mealybug wilt and *Fusarium* have been located by breeders in some cultivars and near relatives of pineapple (Samson, 1986:211).

The Genetic Resource Base

Breeders maintain pineapple germplasm in field gene banks, largely because they are interested in clones. The major collections of pineapple are held in Australia, Brazil, Côte d'Ivoire, Indonesia, Japan, Malaysia, Nigeria, and the United States. Smaller gene banks containing pineapple germplasm are tended in Cameroon, Colombia, Martinique, India, Mexico, Moçambique, the Philippines, Seychelles, South Africa, and Thailand. No collection contains more than 200 accessions, a limited number of samples considering the nutritional and commercial importance of the crop (IBPGR, 1986:30).

The National Clonal Germplasm Repository of the USDA Agricultural Research Service maintains one of the largest collections, with 164 accessions, at Hilo, Hawaii. This collection was started in 1988, mostly with materials previously held by the Pineapple Research Institute. The National Clonal Germplasm Repository contains several near relatives of pineapple, including *A. bracteatus*, *A. erectifolius*, *A. ananassoides*, and *Pseudananas*. Most of the accessions are distributed as tissue culture plantlets. Brazil, the ancestral home of pineapple, maintains only 109 accessions of *Ananas* at Cruz das Almas, Bahia (C. R. Clement et al., 1982).

Most ex situ collections are dominated by clonally propagated cultivars, and none of them contains sufficient material to represent the genetic diversity of pineapple or its near relatives. Given the importance of pineapple to many tropical regions, the precarious genetic base of breeding programs is worrisome. Germplasm collections must broaden their holdings to meet future challenges to the productivity of pineapple. Tissue culture techniques may eventually reduce the cost of maintaining pineapple germplasm and will facilitate the exchange of materials for gene banks and improvement programs.

With support from the IBPGR, efforts have begun to diversify germplasm collections of pineapple. A collecting mission has already scoured the Orinoco basin in search of novel material for gene banks. But all such efforts are hampered by our ignorance of the distribution and patterns of variation of primitive forms of pineapple and its near relatives. Much more fieldwork is needed, particularly to the remote areas of tropical South America, to collect material for study by taxonomists and breeders.

THE TASK AHEAD

Several steps can be taken to remedy the limited amount of genetic variability conserved in germplasm collections. First, it would be wise to collect and conserve as many primitive cultivars as possible. Some of the samples collected should be maintained in field gene banks for the immediate needs of breeders. More of the wild species need to be collected and their seeds stored. If more accessions are obtained, together with good data on their characteristics, the future of pineapple breeding will be on a firmer footing.

AVOCADO

Avocado (*Persea americana*) was domesticated several thousand years ago in Central America. Its natural range stretches from the frost-prone central highlands of Mexico to hot, humid jungles of northwest Colombia. Like the common bean (*Phaseolus vulgaris*), which also has an extensive natural distribution, avocado was probably domesticated in several parts of its ecologically diverse range.

The apparent early emergence of three distinct races of avocado—Mexican, Guatemalan, and West Indian (Figure 3.12)—supports the idea that avocado was brought into cultivation from genetically diverse and widely separated wild populations. The Mexican race (var. *drymifolia*) arose in highland Mexico and is characterized by small black or green fruits with thin skins (Figure 3.13). Mexican avocado matures relatively quickly, within six to eight months, and is the richest of all avocados with as much as 30 percent oil content. Mexican avocado can be grown at elevations up to 3,000 meters in the tropics.

The Guatemalan form (var. *guatemalensis*) also arose in the highlands of Central America, but farther south and generally at a lower elevation than the Mexican race. Guatemalan avocados have thick, woody skins and are relatively slow to mature, taking a year or longer before they are ready for harvest. The generally rotund avocados of Guatemala typically have an oil content of 8 to 15 percent. Most Guatemalan avocados are the size of a tennis ball, but some achieve the dimensions of cannon balls.

The larger West Indian form (var. *americana*) occurs in lowland forests of Central America and northwestern South America. L. O. Williams (1977) asserted that the West Indian avocado is derived from the Mexican form and originated in the lowlands of east-central Mexico, but a slender band of forest along the Pacific plain of Central America from Guatemala to Costa Rica is a more likely center of origin for the West Indian avocado (Storey et al., 1987). If the West Indian avocado had arisen on the east side

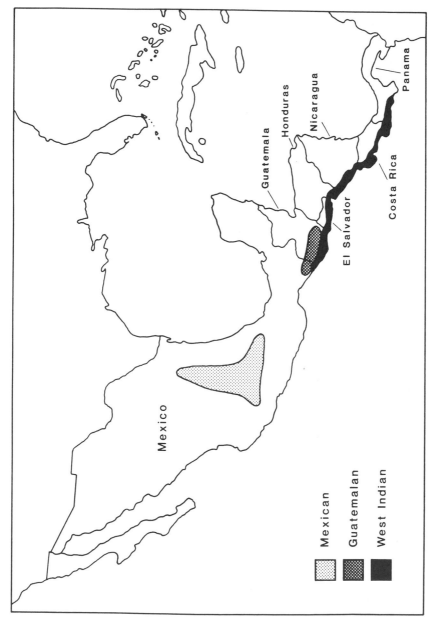

FIGURE 3.12 Possible centers of origin of the Mexican, Guatemalan, and West Indian races of avocado (*Persea americana*). (Adapted from Storey et al., 1987.)

113

FIGURE 3.13 Seedling Mexican avocados in the public market of Atlixco, Puebla, June 1988.

of the Central American highlands, it would surely have been taken to the West Indies before the colonial period. The leathery-skinned West Indian avocado matures in 6 to 9 months, weighs up to 1.5 kilograms, and has the lowest oil content (only 3 to 10 percent) of all avocados.

The creamy taste of nearly all avocados, coupled with the ability to fruit year round, made the trees attractive prospects for domestication. Avocado is much more than just a tasty snack food, however. Because of its high oil content, which accounts for its buttery smooth texture, avocado has a relatively high caloric value, a source of some concern for the diet-conscious, but a valuable asset in areas where food intake is marginal. The fruit is also rich in minerals and vitamins A, B, and E (Platt, 1962; Purseglove, 1974:196). In Central America, avocado is often eaten with maize tortillas, mashed to make guacamole, or cut up into soup. In Peru, *palta*, as avocado is called, is sliced in half and the cavity left after the seed is removed is filled with various diced vegetables or seafood. Peruvian *paltas* are large, and a meal in themselves (Figure 3.14). In Brazil, avocados are used mostly to make savory ice creams and milkshakes. In temperate countries, avocados are eaten as a first course at dinner or in salads. Avocado oil is blended into fine soaps and is used in the manufacture of cosmetics, dog food, and cooking oil.

By the time Spaniards arrived on the shores of the New World, avocado was grown from Mexico to Peru (Figure 3.15). West Indian avocados reached the Peruvian coast by at least 1800 B.C. (L. O. Williams, 1977) and appear to have been spreading across Amazonia when the Europeans arrived. In the sixteenth century, missionaries reported seeing and eating avocados in various parts of western Amazonia along the Magdalena, Upper Caquetá, Upper Napo, and along the Amazon upstream from its confluence with the Negro (Patiño, 1963a:220). Orellana's expedition down the Amazon in 1542 reported avocado (*aguacates*) in Paguana's tribal lands in the vicinity of the confluence of the Purus and the Amazon (Medina, 1988:203). Richard Spruce (Spruce, 1908b:376), a reliable botanist, mentioned the report of an Indian who found four jaguars (*Panthera onca*) feeding on wild alligator pear in the forest between the Uaupés and Japurá rivers in northwestern Amazonia in the latter part of the nineteenth century. If this report was accurate, the avocado tree would presumably have been a spontaneous seedling from discarded seed or a mature tree on the site of an abandoned village, suggesting that aborigines were cultivating avocado in that remote portion of the Amazon basin by at least the mid-nineteenth century.

The avocado probably reached Jamaica and Martinique in 1650 and 1694, respectively (Kimber, 1988:192), but despite avocado's exceptional nutritional qualities, its spread from the New World in historic times has been relatively slow. Avocado's relatively slow diffusion to the Old World may be attributed to the apparent lack of interest in disseminating the nutritious crop by botanic gardens. Although it was taken to southern Spain in 1601 (Purseglove, 1974:192), it evidently was not carried far by traders or missionaries.

Avocado emerged as a significant commercial plantation crop only in this century. Introductions to southern Florida in 1833 and southern California in 1871 did not lead immediately to widespread cultivation of the novel crop because the cultivars were unsuitable and people were generally not accustomed to the fruit. Germplasm collecting missions to Central America, experimental plantings, and advertising preceded large-scale commercial production. Only in the early 1900s was avocado planted extensively in southern California for market production.

The avocado planting boom in southern California began in 1911 with

FIGURE 3.14 West Indian (left) and 'Fuerte' (right) avocados on sale in a street market, Miraflores, Lima, July 1988.

the West India Gardens' release of 'Fuerte', a superb-tasting, cold-tolerant variety from Mexico. California soon became the leading commercial producer of avocados, a position only recently surpassed by Mexico. Avocado is planted on over 28,000 hectares in California, and the state usually produces over 150,000 metric tons of the delicate fruit a year (Bergh, 1976; O. Atkins, pers. comm.). A major expansion of the avocado area was made possible in the 1970s with the introduction of drip irrigation from Israel, which made the cultivation of hillsides feasible (Coffey, 1987). Avocado plantings in southern California are often on hillsides, where land is cheaper; slopes also help to avoid frost damage and improve drainage, an important consideration since avocado trees do not tolerate waterlogged soils. Recently, California has been sending avocados by air freight to Paris, Frankfurt, and Japan, where consumers are apparently willing to pay up to $10 for an avocado compared with 50 cents to $2 paid by aficionados of the savory fruit in the United States. Kenya and Israel also export avocados to Europe.

The drive to introduce avocado to American consumers was led by entrepreneurs in the Los Angeles area and illustrates the potential for developing a market for a new crop. Between 1890 and 1910, C. P. Taft, Joseph Sexton, and C. F. Wagner promoted the avocado in California and

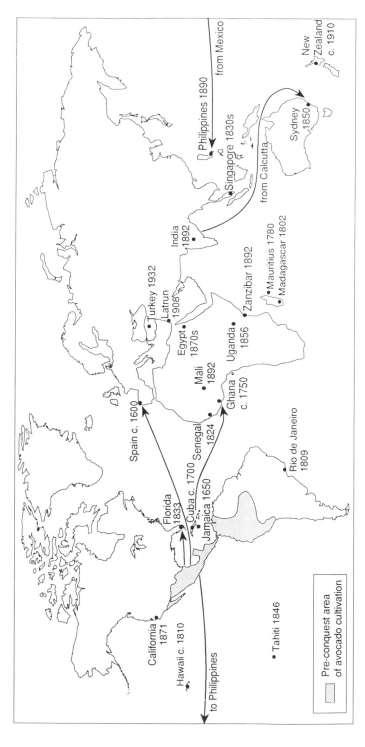

FIGURE 3.15 Spread of avocado from its area of origin in Central America, with dates of introduction, or first recorded presence outside of the American tropics. (Adapted from Bergh, 1975; Busson, 1965:507; Chapot, 1962; Coronel, 1983:22; Donadio, 1984; Friedman, 1936; Godfrey-Sam-Aggrey, 1969; G. A. Green, 1927; Milad, 1936; Morton, 1987:91; Neal, 1965; Pétard, 1986; Piang, 1936; W. Popenoe, 1934; Purseglove, 1974:193; Schroeder and Schroeder, 1982; Shachar, 1982; Wolfe, 1937.)

117

demonstrated its adaptability in coastal areas of the southern part of the state (Schroeder, 1967). Frederick O. Popenoe, owner of the West India Gardens, a private nursery in Altadena, California, which made way for housing in the 1920s, played a prominent role in popularizing the fruit among consumers and growers in the state. Popenoe dispatched professional plant explorers, such as Carl B. Schmidt and Roberto Johnson, to Central America in search of high-yielding and hardy avocados suitable for southern California (W. Popenoe, 1926; Schroeder, 1967).

The Office of Foreign Seed and Plant Introduction of the U.S. Department of Agriculture was also active in acquiring germplasm of avocado and other crops for testing in the United States. At the urging of the California Avocado Association, the Office of Foreign Seed and Plant Introduction, based in Washington, D.C., created a post for an avocado explorer, and Wilson Popenoe, one of Frederick Popenoe's three sons, was selected for the position. Wilson Popenoe scoured over 3,000 miles of highland Guatemala on horseback in search of promising avocados from September 1916 to December 1917, with a return visit in 1919 (W. Popenoe, 1918). Popenoe also collected avocado germplasm in highland Ecuador in 1921 (Figure 3.16). Popenoe cut budwood from promising avocado trees, wrapped them in moist sphagnum moss, and mailed them in tin tubes or wrapped in heavy, oiled paper, to Washington, D.C., where they were promptly grafted to seedlings in glasshouses on the mall. After the budwood had cleaved to the rootstock, it was sent to the Plant Introduction Garden at Chapman Field, Miami, to be grown outside on an old airstrip. Popenoe also mailed back thousands of miscellaneous avocado seeds to produce rootstocks.

Some of this painstakingly collected material has proved useful in the United States and elsewhere, particularly as parents in breeding (Bergh, 1975; N. J. H. Smith, 1986:30). For example, Popenoe selected 'Nabal', 'Benik', and 'Itzamna', all of which were once grown commercially in California, Hawaii, and Israel. Avocado germplasm collected by Popenoe in Guatemala has also furthered the avocado industry in Florida, the second leading producer of avocado in the United States, with 4,600 hectares under cultivation in 1988.

Avocado selections made in California and Florida have launched commercial avocado production in several countries. Chile's avocado export trade, geared principally to the United States, was made possible by the introduction of California cultivars beginning in 1930. Israel and South Africa have also benefited from the cooperation of avocado scientists in California and Florida. Israel's avocado industry was launched by 'Fuerte' and 'Dickinson', varieties introduced from California in 1924 (Morton, 1987:91). Of Israel's annual avocado production of 30,000 tons, 26,000 tons are sent to Europe; in December 1989, for example, 'Fuerte' avocados

FIGURE 3.16 Wilson Popenoe (right), a twenty-nine-year-old avocado explorer in Ecuador, 1921. (Courtesy of Hugh Popenoe.)

from Israel were selling for up to $2 each in London's gourmet stores. Florida selections have proved useful in several parts of the Caribbean and South America.

Mexico is now the leading producer of avocado, but thus far nearly all of the avocados are consumed locally. For over sixty years, quarantine regulations have prohibited the marketing in the United States of avocados from Mexico and other parts of Central America in an attempt to keep the avocado seed weevil (*Heilipus lauri*) out of California, Florida, and Hawaii (Hodgson, 1927). This measure has helped nurture the growth of the U.S. avocado industry.

Mexico is nevertheless vigorously expanding its production of grafted avocados, particularly 'Hass', with an eye to its growing urban market as well as Japanese and Western European consumers. Michoacan became Mexico's leading state in avocado production by 1970, largely because of extensive plantings of 'Hass' (Gallegos, 1983:20; Rodriguez, 1982:11). Farmers in Puebla and Guanajuato are also busy topworking *criollo* avocados to 'Hass', as well as planting freshly grafted 'Hass'. In June 1988, black 'Hass' and, to a lesser extent other improved selections such as green 'Fuerte' and 'Principe', were the only avocados encountered for sale in supermarkets in Mexico City, Puebla, Orizaba, and Celaya. 'Hass' is popular with supermarket produce buyers because this thick-skinned variety travels better than the delicate local forms. *Criollo* avocados are still found in public markets, but even there 'Hass' commands a higher price and sells briskly (Figure 3.17). This large-scale conversion to 'Hass' in Mexico, understandable from the viewpoint of urban consumers and farmers interested in boosting their incomes, is triggering serious genetic erosion among traditional avocados.

Brazil is also a major avocado grower, but production is based mostly on seedlings and is geared entirely to local consumption. São Paulo is the leading Brazilian state for avocados, where West Indian and Guatemalan crosses produce medium-sized fruits that truck well all over the country. The 2-million-hectare Cerrado region of central Brazil is thought to have untapped potential as a major avocado-growing area. To capture a greater share of the export market for avocados, Brazil and some other developing countries plan to plant greater areas to selections that produce relatively small avocados, weighing around 300 grams, with excellent taste and travel qualities.

Two principles emerge from the launching and spread of commercial avocado production worldwide. First, selections made in one climatic zone often do well in other areas with similar weather patterns. Thus avocado varieties developed in California generally thrive in the mediterranean climate of Israel and central Chile. Similarly, Florida's warm, moist summers match those of much of the Caribbean region and Brazil. The second prin-

FIGURE 3.17 'Hass' (left foreground) and 'Fuerte' (right) on sale in the public market of Atlixco, Puebla, Mexico, June 1988.

ciple is that progress with a crop, both scientific and commercial, is only possible with collaboration and the unimpeded exchange of germplasm.

The recent expansion of a market for avocados in Europe indicates some of the potential for future expansion of the crop. By 1987, the European Community (EC) was importing 100,000 metric tons of avocados a year. Between 1983 and 1988, avocado imports into the EC grew by 38 percent (*Spore*, March 1988, p. 1). Avocado recently supplanted pineapple as the number two exotic fruit in France, where banana still retains its position as the leading imported fruit. Eastern Europe and the former Soviet Union might also become significant markets for the crop. With the worldwide market for avocados expected to continue to grow, untapped potential exists for expanding avocado production in some areas while improving such attributes as yield, taste, and disease resistance. The ability to capitalize on the promising future rests to a large extent on the germplasm resources available to breeders.

WILD GENE POOLS

Wild populations of all three avocado races have been found in the forests of Central America and northwestern South America. Wild *P. ameri-*

cana is reported from widely scattered sites in Central America, but races are not always identified. Unfortunately, many reports are anecdotal with no voucher specimens, thereby complicating the task of establishing the validity and precise whereabouts of wild avocado populations.

The taxonomy of *Persea* is still in dispute, further complicating the wild gene pool picture. Gene flow is occurring between cultivated forms of avocado and at least one wild relative, further blurring the taxonomic lines. In some cases, "wild" avocados arise from seeds cast away by farmers or woodsmen after a meal. "Wild" avocados in groves, even in forest environments, may be escapees from abandoned villages or campsites (E. Anderson, 1950). Much of tropical America was densely settled in precontact times, and given the long history of avocado cultivation, it seems likely that some "wild avocados" could be feral. In general, characteristics of fruit of wild *P. americana* include small size and large seed in relation to flesh. From the viewpoint of breeders and gene bank curators, however, feral avocados are of interest because they are once again evolving under natural conditions.

Wild populations of *P. americana*, referred to as the basic primitive form (Bergh, 1976), have been found in lowland Honduras and along water courses on the slopes of Irazu Volcano, Costa Rica, at around 1,500 meters (W. Popenoe, 1927). Wild avocados have also been found in many forest types in other parts of Costa Rica. For example, wild *P. americana* have been noted on well-drained slopes at 400 meters in tropical wet forest of the Osa Peninsula; on a moderately drained site in premontane wet forest 30 kilometers southeast of San Isidro del General; at 830 meters on well-drained, acidic red latosols in Valle Escondido, 23 kilometers east of Turrialba; and in montane wet forest between 1,700 and 2,175 meters on the slopes of Volcán Barba, 7.5 kilometers north-northeast of Heredia (Holdridge et al., 1971:235, 360, 434, 484, 512). Wild *P. americana*, possibly vestiges from long-abandoned settlements, occur in the forest of Barro Colorado Island in Panama (Croat, 1978:414).

Primitive forms of the Guatemalan race of avocado occur, or at least occurred, at widely separated sites in Central America. In Guatemala, small and hard wild avocados, known as *aguacate de piedra* (rock avocado: C. A. Schroeder, pers. comm.), have been found in forest between Tecpan and Solola, as well as in the mountains of Zacapa, Chiquimula, Quetzaltenango, and Huehuetenango (Standley and Steyermark, 1949). In Guatemala, primitive forms of the Guatemalan race grow between 2,600 and 3,000 meters in fast-receding pine, oak, and cypress woodland that cloaks the rugged and cloud-capped Chichoy range (W. Popenoe, 1939). Wild avocados of the Guatemalan race have also been found near Tegucigalpa, Honduras.

Hard-shelled wild avocados, called *aguacate de mico* (monkey avocado) or *aguacamico* (in El Salvador), may belong to the Guatemalan race, and occur sporadically from southern Mexico to Turrialba Volcano in Costa Rica (Schieber and Zentmyer, 1981a). *Aguacate de mico* or *aguacamico* have been collected from such sites as Santa Maria de Ostuma in Nicaragua and the edge of El Boqueron's precipitous crater in central El Salvador. Wild *P. americana* with hard shells typical of the Guatemalan race, but with anise-scented leaves characteristic of the Mexican botanical variety, have been found in the mountains surrounding the Lancetilla Experiment Station in northern Honduras (Zentmyer and Schieber, 1982). Standley (1931) reported seeing these hard-shelled wild avocados in the forest around Lancetilla on well-drained sites between 260 and 600 meters; Wilson Popenoe (1927) considered this population of wild avocados similar to the one he located on Volcán Irazu in Costa Rica. Popenoe (1921) described the fruits of the wild avocado he found on Volcán Irazu as the size and shape of a small orange, with dark green hard shell, scanty flesh with a gritty texture, and a tinge of anise.

Ancestral forms of the Mexican race have been found in widely scattered locations of Mexico: near Volcán Orizaba, Veracruz, on the eastern slopes of the central highlands; in pine and oak forests of Nuevo Leon state in the northeastern part of the country; and in Michoacan state in the west (Storey et al., 1987). Herbarium records indicate the presence of apparently wild avocados in a broad range of forest environments in other Mexican states including Chiapas, Hidalgo, Jalisco, Mexico, Morelos, Oaxaca (where wild avocados are known as *aguacatillo*), and Tamaulipas (Figure 3.18).[1]

1. Habitat and location where herbarium specimens of apparently wild *Persea americana* have been collected in Mexico and deposited in the Herbario Nacional, Instituto de Biología, Universidad Nacional Autonómo de Mexico, Mexico City, are as follows: forest of *Carpinus, Liquidambar,* and *Turpinia,* Camino de Huauchinango a Xilocuautla, Puebla (Herbarium No. 13379); pine-oak woodland, El Alamo, Rio Balsas Valley, 6 kilometers east of Temascal, Charo Municipality, Michoacan (358365); 30 kilometers northeast of Teotitlan, road to Huautla de Jiminez, Oaxaca, mesophytic forest (445780); pine forest with tropical elements, about 1,500 meters, 21 kilometers south of Talpa, road to Cuesta, Allende Municipality, Jalisco (347378); mesophytic forest, 10 kilometers northeast of Nanchititla, between Los Bancos and Piedra Grande, Tejupilco Municipality, Mexico (295873); transitional between caducous woods and tropical forest, 1,420 meters, Ixtotento, 5 kilometers from Atzalan, road to Tlapacoyan, Veracruz (166743); temperate broad-leaf forest with conifers, 2,350 meters, Macuiltiangus, Oaxaca (291798); riparian vegetation at 1,600 meters, Almoloya de las Granadas, Tejupilco Municipality, Mexico (310138); oak forest at 1,200 meters, 20 kilometers northwest of Victoria, road to Jaumave, Cuidad Victoria Municipality, Tamaulipas (433215); mesophylic montane forest at 1,070 meters, 7 kilometers northwest of Tianguistengo, Tianguistengo Municipality, Hidalgo (375645); evergreen primary tropical forest, vicinity of Campto. Hnos. Cedillo (17°15′ N, 94°40′ W), Hidalgotitlan, Veracruz (207483); forest of juniper, pine, and oak at 2,000 meters, 19 kilometers north of Zimapan in direction of San Miguel mine, Zimapan Municipality, Hidalgo (366858); pine and oak forest at 1,950 meters, north of Cuernavaca, kilometer 64 of the Mexico-Cuernavaca freeway, Morelos (13574); slope with *Quercus, Dodonaea,* and *Calliandra* at 1,300 meters, Paraje de Mahbenchauk, Tenejapa Municipality, Chiapas (121413); slope with *Pinus, Quercus,* and *Liquidambar* at 2,800 meters, Paraje Shobleh, Tenejapa Municipality, Chiapas (104404).

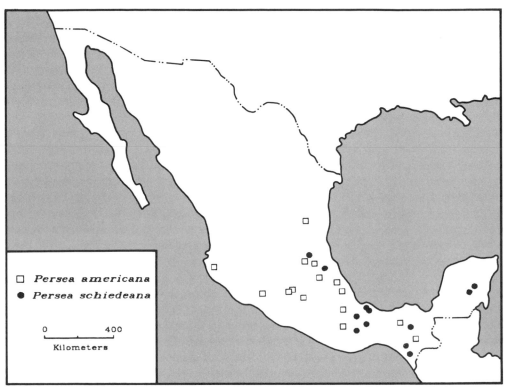

FIGURE 3.18 Occurrences of apparently wild avocado (*Persea americana*) and *chinene* (*P. schiedeana*) according to records in the U.S. Herbarium, Smithsonian Institution, and the National Herbarium, National Autonomous University of Mexico (see nn. 1 and 2).

Wild Mexican avocados have also been found in northern parts of Guatemala, where they are known as *matul-oj*. Wild Mexican avocados have been found in the high country of Guatemala including Cunen, in the remote northwest section of the country; in the Chuchumatanes range; at 1,600 meters near Malacatancito, Huehuetenango; near Itzapa and Parramos at 2,000 meters in the Chimaltenango Valley; and at 2,600 meters on the slopes of Agua and Acatenango valleys in the central highlands (Schieber and Zentmyer, 1974; Schieber et al., 1983). The ranges of wild Mexican and Guatemalan avocados thus overlap in places, and genes flow between these populations, with the pollen conveyed by insects.

A primitive form of the West Indian type of avocado was discovered in eastern Salvador in 1977 (Storey et al., 1987). Another possibly wild population of West Indian avocados is located between 300 and 1,300 meters on the slopes of Sierra Nevada de Santa Marta in northeastern Colombia (W. Popenoe, 1935). These wild avocados occur in groves of one hectare or more. Wild West Indian avocados also reportedly occur in the forests of northwestern Colombia.

With human populations growing at close to 3 percent a year in much of tropical America, the safety of many wild avocado populations is in doubt. The human population of Central America is likely to double within thirty years, and Central American forests are already being felled at the brisk rate of some 400,000 hectares a year (Tangley, 1987).

In the early part of this century, wild Mexican avocados were abundant on the lower slopes of Orizaba Volcano on the edge of the highlands east of Mexico City (W. Popenoe, 1920:14), but the present viability of this population, is not known. A few possibly wild avocados, with fruits the size of large olives, were noted in fields surrounding the village of La Perla at the foot of Orizaba Volcano in June 1988. The Lancetilla Valley, which at one time contained wild avocados of the Guatemalan race or an intermediate form with both Mexican and Guatemalan characteristics, has now been almost entirely cut over by settlers. An apparently insatiable demand for illicit drugs in North America and Europe has undoubtedly accelerated marijuana planting in the Sierra Nevada de Santa Marta, at the expense of food crops needed to feed Colombia's growing population and forests that contain wild avocados and other irreplaceable resources. Ecogeographic surveys of wild populations of avocados in that region would be difficult for security reasons.

NEAR RELATIVES

Much also remains to be learned about the status of avocado's near relatives (Appendix 3). Breeders are increasingly interested in some of the approximately eighty-five wild relatives of avocado as possible sources of disease and pest resistance, among other traits. Ecogeographic surveys of wild relatives of avocado are urgently needed in light of the rapid rates of forest clearing in Central America and many parts of South America. The IBPGR has ranked wild species of avocado a high priority for ecogeographic studies (IBPGR, 1985:14). The gene pools of several wild relatives of avocado have shrunk considerably in recent years, with many populations becoming isolated.

The majestic *P. schiedeana* occurs sporadically on wooded hillsides between 250 and 1,900 meters from southern and eastern Mexico to Panama (Appendix 3). In Mexico, *P. schiedeana* grows in warmer locations than wild avocados and appears to be concentrated in the eastern and southern part of the country (Figures 3.18, 3.19).[2] The tall, long-lived tree grows on a variety of soil types and drainage conditions, ranging from steep, porous slopes, to moist stream margins. The ability of *P. schiedeana* to withstand waterlogging has led to successful searches in germplasm of the species for resistance to root rot, one of the most serious avocado diseases.

The relatively large fruits of *P. schiedeana* are much appreciated by inhabitants of Central America, and the plethora of names given to the plant are one indication of its importance to local peoples.[3] In Veracruz, Mexico, and on some farms in Guatemala, *P. schiedeana* is deliberately spared when the forest is felled (Morton, 1987:102; Price, 1917). This large-fruited wild relative of avocado, reputed to have a distinctive coconut-like flavor (W. Popenoe, 1935, 1952), is also cultivated on a limited scale (Bergh and Ellstrand, 1987; Morton, 1987:102; W. Popenoe, 1935). This minor crop could become more important if selections were made for large, high-yielding and good-tasting fruits that travel well.

In Veracruz, Mexico, *P. schiedeana* is called *chinene*, and several old *chinene* trees were observed in backyards in the outskirts of Orizaba on the road to La Perla on 18 June 1988; at that time the trees, estimated to be at least fifty years old, had immature fruits. When *chinene* fruits mature in late June, July, and August, they reportedly command a higher price

---

2. Habitat and location where herbarium specimens of *P. schiedeana* have been collected in Mexico and deposited in the Herbario Nacional, Instituto de Biología, Universidad Nacional Autónomo de Mexico, Mexico City, as follows: primary caduceus forest at 1,200 meters, west side of Cerro Mastagaga (18°31' N, 95°9' W), approximately 13 kilometers northeast of San Andres Tuxtla, San Andres Tuxtla, Veracruz (232590); Cuauhtapanaloyan, Cuetzalan Municipality, Puebla (328396); secondary forest with *Trema micrantha*, Barrio San Pablo, Cuautlapan, Veracruz (30945); road to river 5 kilometers from Valle Nacional in direction of Arroyo Banco, Oaxaca (272750); rocky, open area with sandy soil and volcanic lava, Mata Larga, Veracruz (293068); urethral area, 13 kilometers south of Ucum., Quintana Roo (276751); calacareous slope with mesophylic secondary forest at 1,250 meters, 3 kilometers southeast of Ahuacatlan, Xilitla Municipality, San Luis Potosí (399244); tall evergreen tropical forest at 250 meters, lot 71, Cerro Lazaro Cardenas, Estación de Biología Tropical Los Tuxtlas (18°34' N, 95°4' W), Municipal San Andres Tuxtla, Veracruz (418020); tropical rain forest at 300 meters, 12 kilometers south of Palenque on road to Ocosingo, Palenque Municipality, Chiapas (397838, 397915); summit of Suspiro, Chiapas (202308). *P. schiedeana* specimens deposited with the U.S. Herbarium, Smithsonian Institution, were collected near Los Cartagos, Alajuela, Costa Rica, and in the Mexican states of Veracruz, Puebla, Oaxaca, Quintana Roo, San Luis Potosí, and Chiapas (vouchers 232590, 328396, 30944, 272750, 293068, 276751, 399244, 397838, 202308, 3018920, 2751902).

3. In Mexico, for example, the Nahuat speakers call *P. schiedeana pahua*, elsewhere in the country it is called *chinini*, or, in Veracruz, *chinene*. In San Luis Potosí, *P. schiedeana* is dubbed *aguacate de ardilla* (squirrel avocado), suggesting that squirrels may be dispersal agents for the species. In Guatemala, it is called *chucte* or *quillou* in Alta Verapáz, and *coyou* in Baja Verapáz, an indication of the rich cultural fabric overlaying much of Latin America (Schieber et al., 1984). In Belize, *P. schiedeana* is known as wild pear. Other local names for *P. schiedeana* include *coyó* and *yas* (Costa Rica), *chuti* (Honduras), and *aguacatón* (Panama).

FIGURE 3.19 Mexican states where *Persea schiedeana* has been found. (Data from herbarium records, UNAM, Mexico City.)

than *criollo* avocados in Orizaba. *Chinene* trees bear fruit every year in the Orizaba area and can be distinguished from Mexican forms of *P. americana* by their larger leaves, light-colored underside to younger leaves, and the lack of anise scent when the leaves are crushed. *Chinene* is propagated by seed, and the fruits are highly variable; some trees bear dark fruit when ripe, others remain green. In the vicinity of Orizaba, *P. schiedeana* serves as a shade tree in some small coffee groves.

The towering *P. schiedeana* is not always shown mercy by rural folk, however. At one homestead along a twisting gravel road leading from La Perla to Sometla on the flanks of Orizaba volcano in Veracruz, Mexico, a farmer cut down an ancient *chinene* in 1988 because it had grown so large that the long-leafed tree was shading out some of his other crops. In Guatemala, complete deforestation of Ipala Volcano, mainly for firewood, has eliminated a population of this species where it was once collected (Schieber and Zentmyer, 1981b). Another *P. schiedeana* population on Volcán Quetzaltepeque, Guatemala, is threatened by rampant deforestation. It is doubtful if any *P. schiedeana* survive in Lancetilla Valley, Honduras, where it once thrived on hillsides (Standley, 1930a). Populations of *P. schiedeana* in the Revantazón Valley of Costa Rica are now largely confined to steep ravines because of extensive clearing, principally for coffee and sugarcane estates (Schroeder, 1977). No more specimens of *P. schiedeana* are to be found in the vicinity of Yas, a town named after the species, located 12 kilometers east of Cartago. The area of primary forest in Costa Rica, the major habitat for *P. schiedeana* and its relatives, declined from 67 percent of the country's total area in 1940 to only 17 percent by 1983 (Sader and Joyce, 1988).

The picture also appears bleak for *P. nubigena*, another avocado relative with edible fruit. *Aguacate cimarrón*, as the small-fruited species is known in Mexico, has been extensively felled on the slopes of Orizaba volcano in order to procure its savory fruits (Schieber and Zentmyer, 1980). The threat is thus within as well as outside of the forest for *P. nubigena*.

In 1947, a precocious wild relative of avocado, *P. floccosa*, was discovered on the slopes of Orizaba volcano, in Veracruz, Mexico, by a germplasm collecting team organized by the California Avocado Society and the University of California (Schroeder, 1967). The status of this population, where wild populations of the Mexican race of avocado have also been found, is uncertain. On a visit to Orizaba volcano on 18 June 1988, approximately three-quarters of the volcano appeared to be in forest, but a more precise estimate was difficult because clouds capped the upper slopes. The lower slopes are being progressively cleared for crops such as maize and various flowers. Orizaba's rugged terrain may save some of the forest from the axe, but if the germplasm that remains is confined to rock outcrops or steep slopes, only a small portion of the gene pool remains; genotypes on better soils will be eliminated for agriculture or urban sprawl.

Fortunately, Central American governments are increasingly aware of the importance of preserving wild habitats, and this trend may help alleviate some of the pressure on the primary and secondary gene pools of avocado. Parks and reserves are being actively established in the region, particularly in Mexico, Costa Rica, and Guatemala. Mexico has set aside forty-four national parks, fourteen ecological reserves, and five biosphere reserves covering a variety of ecosystems (SEDUE, 1987). At least two of the parks are located in and around Orizaba volcano, which harbors wild avocado and as well as *P. floccosa*, *P. nubigena*, and *P. schiedeana*: the 19,750-hectare Parque Nacional Pico de Orizaba, and the 55,609-hectare Parque Nacional de Rio Blanco. The first park is confined to the upper reaches of Orizaba volcano and thus does not contain *P. schiedeana* or wild avocado. The second park surrounds the town of Orizaba, but its boundaries appear to have been violated by settlement and agricultural activities.

At least one of Guatemala's growing list of parks embraces wild avocado populations. A 675-hectare park near Purulha, Baja Verapáz, set up to preserve the endangered quetzal (*Pharomachrus mocinno*), the national bird of Guatemala, contains *P. donnell-smithii*, *P. steyermarkii*, and the rare *P. vesticula* (Schieber and Zentmyer, 1979). Several of avocado's near relatives, such as *P. donnell-smithii*, are important sources of food for the resplendent quetzal, a forest bird that does not tolerate captivity (Schieber et al., 1984; Skutch, 1983:146). The quetzal is a likely dispersal agent for wild avocado and its near relatives. Efforts to sustain the quetzal and to conserve wild populations of avocado's near relatives can be complementary goals. The interdependence of wildlife and flora underscores the need to focus on conserving ecosystems rather than individual species.

A desire to protect water supplies for Tegucigalpa, the largest city in Honduras, saved the unguarded La Tigra National Park, which was being invaded by loggers and farmers. The La Tigra forest contains populations of at least two wild relatives of avocado: *P. nubigena* and *P. vesticula* (Appendix 3). A workshop on forest conservation held in Tegucigalpa in late 1986, when the sprawling city was experiencing a severe water shortage, emphasized the importance of watershed management to supply water for urban dwellers and farmers (Tangley, 1987). Government officials took note, and now La Tigra National Park is effectively protected.

No reserve has yet been set aside specifically to safeguard *Persea* germplasm, but Mexico has provided an exemplary precedent by establishing a biosphere reserve to protect wild relatives of two other important crops, maize and beans. SEDUE (Secretaría de Desarrollo Urbano y Ecologia) has designated an in situ preserve for *Zea diploperennis*, a wild relative of maize with useful genes for disease resistance, in the Sierra de Manantlán in the State of Jalisco (Iltis, 1988; SEDUE, 1987:66). The 139,577-hectare Reserva de la Biosfera Sierra de Manantlán, managed in collaboration with scientists at the University of Guadalajara, also contains wild populations of the kidney bean (*Phaseolus vulgaris*: F. Cardenas, pers. comm.).

Although progress has been made in setting aside pristine areas in Central America, existing parks need to be better protected, and more reserves are clearly needed soon if appreciable portions of the wild gene pools of avocado and its near relatives are to be saved. More ecogeographic surveys of wild populations of avocado and other crops are urgently needed to provide concrete data for government officials as they grapple with hard decisions on where to establish parks and reserves.

TRADITIONAL AND MODERN VARIETIES

Traditional avocados are all seedlings, and since avocado is an outcrosser, the progeny are heterogeneous. Avocado seedlings reliably express only two characteristics from their parents: skin type of the fruit and hardiness (E. E. Knight, 1917). Thus, although only three races of avocado are recognized, each seedling is a distinct genotype. The variability of avocado progeny increases options for breeders but also complicates the task of selecting specific forms for gene banks.

Avocado trees require bees or other large flying insects for pollination and usually need to be fertilized by a neighboring tree. This dependence on other trees for pollination is due to distinct periods when the avocado flower is ready to be fertilized or is releasing pollen. In A-type avocado flowers, the stigma is receptive to pollen in the morning for a few hours, closes, and opens the next afternoon to shed pollen (Bergh, 1975). At that time, the stigma is no longer receptive. In B-type flowers, the stigma is receptive for only a few hours in the afternoon of the first opening, but not from the same flower since the stamens are not releasing pollen. Only in the morning of the second day are B-type flowers functionally male. To maximize seed set, then, breeders should interplant avocado trees of both flower types.

The pollination sequence of the two flower types functions well only when ambient temperatures are above 26°C; below that temperature, receptive times for pollen become irregular, thus large plantations of a single clone can bear heavy fruit crops. Fruit set is poor or nonexistent when temperatures dip below 15°C at flowering time. In traditional avocado growing areas, a single variety rarely dominates any large area so, neighboring trees, containing A and B flower types, normally fertilize backyard avocado trees.

Little information is available on the diversity and characteristics of traditional avocados in the tropics. In Guatemala, numerous morphological types are reported, including a sausage-shaped avocado and another as small as 2.5 centimeters in diameter in the markets of Quetzaltenango (Standley and Steyermark, 1949). Next to Guatemala and Mexico, Cuba has long enjoyed a reputation for superior avocados, many of which have been helpful to the Florida avocado industry (F. O. Popenoe, 1915). 'Catalina' is a pear-shaped and highly regarded avocado grafted in the region of Havana; its fame has prompted its inclusion in at least two germplasm collections (Appendix 4). The most common avocado in backyards of Pará state in Brazil is *abacate-de-pescoço* (the neck avocado: Cavalcante, 1976: 57), while in the neighboring state of Amazonas, similar, rather small West Indian seedlings dominate. In southwestern Brazil, early avocados are West Indian, while Guatemalan forms and hybrids bear later (Montenegro, 1960). In Trinidad and Tobago, six local avocados—'St. Ann's', 'St. Clair River', 'St. Joseph No. 1', 'St. Joseph No. 2', 'Lenegan', and 'Macqueripe No. 2'—were popular in the middle of this century (R. O. Williams and Williams, 1951:251). But judging from some of the cultivar names, they were selections made by research stations in the Caribbean. Information on their precise origins, their characteristics, whether they are cultivated today, and whether they are represented in gene banks, would be of interest to breeders.

Presumably, genetic diversity among cultivated avocados is more pronounced in Central and South America, but surveys of genetic variability in avocado should not overlook backyards and orchards in Africa and Asia. In Burundi, avocado trees of the West Indian form are occasionally used to fortify fences and stockades as well as for food; in the Kivu region of Zaire, West Indian avocados displayed a remarkable array of shapes and colors ranging from shiny green to red or black in Bukavu's roadside market in March 1989. The mostly West Indian avocados of Malaysia, for example, vary considerably in size, shape, and color (Lambourne, 1934). Avocado reached Mauritius in 1780 and Senegal in 1824 (Figure 3.15); at least two centuries have elapsed since the crop was first cultivated in the Old World tropics, ample time for new varieties to evolve in response to local ecological conditions and selection by people with varying taste and texture preferences.

The degree to which traditional avocados are being replaced by improved cultivars warrants investigation. Although commercial avocado production is concentrated in regions outside the crop's area of origin, selections emanating from private nurseries are infiltrating areas traditionally planted to locally produced seedlings. In Mexico, visits to avocado orchards in two distant areas in June 1988 revealed a trend to topworking with California or locally produced selections and to planting with grafted material. Small-scale nursery workers in Mexico are displaying a keen entrepreneurial spirit as they seize opportunities to meet a growing demand for high-yielding, grafted avocado material. A cottage industry supplying grafted 'Hass' avocados has emerged in the avocado growing parts of Mexico.

When Wilson Popenoe visited Atlixco, Puebla, in 1918, he saw no budded or grafted avocados (W. Popenoe, 1926). Seventy years later, three randomly selected avocado orchards in the vicinity of Atlixco were all planted to grafted material. 'Fuerte', was the most popular variety, followed by 'Hass', which has been planted in the broad, fertile valley of Atlixco since at least the early 1970s. 'Principe', a 'Fuerte'-like selection in the Colonia Ubera, a small nursery some 3 kilometers from Atlixco, was also evident in one orchard as well as in the public market of Atlixco. 'Fuerte', 'Hass', and 'Principe' accounted for about 90 percent of the avocados on sale in Atlixco's public market on 19 June 1988; small, black *criollo* avocados were relegated to a few sellers operating on the floor. The sellers handling *criollo* avocados were always different from those dealing with the thicker-skinned improved selections.

In the vicinity of Celaya, in the state of Guanajuato, the trend to topworking or planting with grafted avocado selections was also pronounced. While many groves in the 5- to 15-hectare range were still dominated by *criollo* avocados, others had a mix of Mexican seedlings, 'Hass', and 'Fuerte'. Owners of some groves planted to Mexican seedlings near the village of Comonfort expressed their intentions to switch to 'Hass' because of the greater earning potential of the California selection.

In Guatemala, one of the richest countries for avocado germplasm, genetic erosion of traditional avocados is also well under way. By 1981, close to a third of the avocado area was occupied by cloned material (Pérez and Salán, 1986:219). While seedling West Indian avocados are still common in backyards of Peru, commercial groves are increasingly turning to U.S. selections, particularly in the Ica area. 'Fuerte', a 1911 selection from Altadena, California, and the larger, rounder 'Nabal' (sometimes referred to as 'Naval'), collected by Wilson Popenoe near Antigua, Guatemala, in 1917, were the most commonly encountered varieties in supermarkets and public markets in Lima in July 1988. 'Fuerte' consistently commanded the highest price of any cultivar on sale in Lima, selling for around 50 cents a kilogram compared with half that amount for others.

It would be a tragedy if much of the traditional avocado material is lost before it can be surveyed, evaluated, and placed in field gene banks if warranted. Farmers should not be forced to plant older varieties when newer, and often higher-yielding selections become available. A strategy to conserve segments of the gene pool of traditional avocados could be accomplished by using select seedlings.

Native groups of Central and South America did an outstanding job of selecting avocados to suit different elevations and palates. Indeed, commercial avocado clones in developed countries are hardly any better than selections by indigenous people in tropical America (Bergh, 1976). Improved agronomic measures, such as irrigation and pest and disease control, largely account for superior yields on large-scale avocado plantations in temperate countries.

Spontaneous crosses have blurred racial distinctions in commercial avocados. Most avocado trees grown on modern plantations are a mix of Guatemala and Mexican ancestry or have Guatemalan and West Indian parents. For example, 'Fuerte', the single most successful commercial avocado variety, was the result of a chance cross between Guatemalan and Mexican avocado trees. Carl Schmidt took budwood from the 'Fuerte' tree growing in the backyard of a home in Atlixco, Mexico, in 1911 and sent it to the West India Gardens in Altadena, California (W. Popenoe, 1926). The date the original 'Fuerte' was planted is not known, but its owner, Alexander LeBlanc, a naturalized Mexican from France, noted that it first bore fruit in 1908 (Shamel, 1936). LeBlanc immediately recognized the unusually good taste of his 'Fuerte' fruit and kept all of it for his family. 'Fuerte', Spanish for strong, was thus named because the graft was so vigorous in the West India Gardens. It turned out that the variety was aptly named since it also withstands cold, and this sought-after trait undoubtedly came from its Mexican parent.

Given the global importance of 'Fuerte', Nigel J. H. Smith attempted to locate the house where Carl Schmidt collected budwood of the 'Fuerte' tree and to learn what has become of that famous original 'Fuerte'. Upon arriving in Atlixco at midday on Sunday, 19 June 1988, he proceeded by car to one of the two steep hills that overlook Atlixco. From that vantage point, it was easy to pick out the older part of the rapidly growing town, around the shaded main square. Working on the premise that older folk would be more likely to know of the whereabouts of LeBlanc's house, he approached a gray-haired barber at work by the main square. Although the barber had no idea of the location of LeBlanc's house and the 'Fuerte' avocado, his younger customer had driven by the house numerous times.[4]

LeBlanc's old home is several blocks from the main square. A ceramic plaque on the outside wall points out the historic importance of the home. But much of the house and yard are now in disrepair. LeBlanc died many years ago, and his children left for the opportunities of Mexico City. Squatters occupy most of the rooms; one occupant operates a homeopathic bone-healing service from a front room. Another room on the ground floor serves as a small library for schoolchildren at certain times of the year.

The original 'Fuerte' tree apparently died in the mid to late 1940s. A stump of the original 'Fuerte' tree was still standing in 1947 (C. A. Schroeder, pers. comm.). A copper plaque,[5] placed close to the tree by the California Avocado Society in 1938, is covered by weeds. The original 'Fuerte' tree lived for close to half a century, and its clones, still numbering in the hundreds of thousands if not millions, thrive in orchards around the world. It is particularly comforting to see 'Fuerte' flourishing near its birthplace, in the Valley of Atlixco (Figure 3.20).

Released in 1911, the superb-tasting 'Fuerte' accounted for more than three-quarters of the avocado area in California by the early 1940s (W. Popenoe, 1941). As late as the mid-1970s, 'Fuerte' still accounted for about half of California's productive avocado orchards, but by 1987, the

---

4. The current address of LeBlanc's residence in Atlixco, where the original 'Fuerte' avocado arose is 1102 Calle 3 Poniente. The former address of the house was No. 2, Calle Manuel Buen Rastro.

5. The plaque honoring the mother 'Fuerte' tree at 1102 Calle 3 Poniente, Atlixco, Mexico, reads in Spanish and English:

THIS TREE
HAS THROUGH ITS PROGENY PLAYED A MAJOR
ROLE IN THE DEVELOPMENT OF A NEW INDUSTRY
IN CALIFORNIA, U.S.A.
IN TESTIMONY OF OUR GRATITUDE AND
APPRECIATION THIS TABLET HAS BEEN PLACED
HERE BY THE
CALIFORNIA AVOCADO ASSOCIATION
1938

FIGURE 3.20 A grove of 'Fuerte' avocado with fruit in the outskirts of Atlixco, Puebla, Mexico, June 1988.

once dominant variety was responsible for less than a quarter of the state's avocado output (J. Shepherd, 1988). Other important avocado cultivars in California include 'Hass', 'Bacon', 'Zutano', and 'Reed'. 'Hass' was named after Rudolph Hass, a Wisconsin mail carrier who retired to California (R. Bergh, pers. comm.). Hass was given U.S. plant patent number 139 in 1935 for his variety that arose by accident when a graft failed and the rootstock grew out and produced abundant and good-quality fruit. The original 'Hass' seedling was still growing in a grove in La Habra Heights in 1988 (J. D. Sauer, pers. comm.). The dusky 'Hass' is mostly Guatemalan germplasm but also contains Mexican genes (Bergh, 1975).

'Hass' has replaced 'Fuerte' as the leading cultivar in California, where it accounted for three-quarters of the state's avocado production by the mid-1980s. 'Hass' dislodged 'Fuerte' as California's leading avocado cultivar because it is more adaptable, yields on average 50 percent more than 'Fuerte', and is a consistent bearer, unlike 'Fuerte', which tends to fruit erratically (Bergh, 1985).

The superior yield and fruiting characteristics of 'Hass' have not escaped the attention of avocado growers in other parts of the world. 'Hass' has emerged as the leading cultivar in the important Israeli industry (R. Bergh, pers. comm.). It is the principal variety in New Zealand and is second to

'Fuerte' in Chile (Morton, 1987:96). In Mexico, 'Hass' is catching on fast in various parts of the country, particularly in Michoacan, Guanajuato, and Puebla. Spurred by the high prices 'Hass' receives in markets ($1.50–$2.00/kg in June 1988), usually higher than any other avocado cultivar, Mexican farmers are increasingly topworking their seedling *criollo* avocados to 'Hass'. 'Hass' was the only avocado seen in supermarkets in Mexico City, Celaya, Orizaba, and Puebla in June 1988, and the dark-skinned cultivar is also common in public markets. This marked trend toward topworking and replanting with grafted 'Hass' is likely to continue to replace the genetic mosaic of traditional seedling avocados in Mexico. As in California, all commercial cultivars in Florida arose from private selections, particularly in Dade County (Appendix 4).

'Sharwil' (Redland Bay, Australia) is an avocado front-runner in Australia, while in Israel, 'Hass' and 'Ettinger' are the main avocado varieties. 'Ettinger', a 1965 Israeli selection, has a large genetic contribution of Mexican germplasm. Other Israeli selections include 'Ein-Vered', 'Entom', and 'Tova' (1971).

The diverse mosaic of traditional avocados and modern cultivars typical of avocado growing areas reveal four main patterns. First, diversity is much greater in traditional growing areas, which still rely on seed propagation. Second, varietal turnover is evident in commercial avocado areas, although it is not as rapid as that of cereal crops. 'Pollock', for example, is still grown in some areas more than eighty years after it was first grafted. 'Fuerte' is still pleasing customers worldwide after more than seventy years in production; for example, it was the most expensive cultivar (*palta punta*) on sale in markets in Lima, Peru, in July 1988. Third, market forces are shifting in response to consumer demands while the biophysical environment for avocado growing is also changing, thereby creating a constant demand for more-productive and hardier forms. Avocado breeders and private growers are thus interested in novel variation. Fourth, commercially important cultivars have arisen mostly in private orchards as a result of chance crosses, rather than as a result of breeders manipulating germplasm drawn from their collections. As breeding becomes more sophisticated, well-managed gene banks will clearly play an increasingly important role in breeding to boost and sustain yields worldwide.

GENE BANK COLLECTIONS

Avocado seeds are recalcitrant, as are the seeds of many tropical fruit trees, so germplasm cannot be stored as dried and frozen seed. Instead, avocado germplasm materials of potential interest to breeders are maintained as living plants. Avocado collections are scattered throughout the world in subtropical and tropical climates (Table 3.5).

TABLE 3.5

Germplasm collections of avocado (*Persea americana*) and its near relatives (*Persea* spp.)

Institution[1]	Location	No. of P. americana accessions	Near relatives Species	No. of accessions
FCAV	Jaboticabal, São Paulo, Brazil	343[2]	*indica*	1
CSIRO	Merbein, Victoria, Australia	294[2]	*indica* *schiedeana*	1 1
DICOF	Havana, Cuba	280[3]		
IPB	Los Baños, Philippines	246[2]		
Doi Muser Horticultural Research Station	Tak Province, Thailand	227[4]		
ARO	Bet-Dagan, Israel	211[3]	*Persea* spp.	7
USDA	Miami, Florida	204[4]		
UC	Riverside, California	200[3,4]	*Persea* spp.[5]	20
CIAB	Celaya, Guanajuato, Mexico	177[3]	*cinerascens* *indica* *schiedeana*	1 2 5
Plew Horticultural Research Station	Chantaburi, Thailand	136[2]		
IFAS	Homestead, Florida	116[4]		
USDA	Mayaguez, Puerto Rico	112[4]	*schiedeana*	1
Hope Gardens	Kingston, Jamaica	108[4]		
UH	Poamoho Farm, Oahu, Hawaii	107[4]		
INRA	San Nicolao, Corsica, France	75[4]		
CENIAP	Maracay, Venezuela	74[6]		
Grand'Anse Experiment Centre	Mahé Island, Seychelles	50[4]		
CEMSA	Santo Domingo, Villa Clara, Cuba	47[4]		
IRFA	Abidjan, Côte d'Ivoire	42[4]		

TABLE 3.5—*cont.*

Institution[1]	Location	No. of P. americana accessions	Near relatives Species	No. of accessions
IRFA	La Réunion, France	40[4]		
MIDINRA	Mazatepe, Masaya, Nicaragua	40[3]		
CPAC-EMBRAPA	Planaltina, Brazil	37[4]		
INIAP	Quito, Ecuador	36[4]		
EMPASC	Itajai, Santa Catarina, Brazil	32[4]		
ICA	Espinal, Tolima, Colombia	20[3,4]		

Sources: C. Campbell, pers. comm.; M. D. Coffey, pers. comm.; R. J. Knight, pers. comm.; Genú et al., 1987; Internal Management Memoranda IMM No. 6C1-6. 120-8, 4 June 1985, Institute of Food and Agricultural Sciences, University of Florida, Gainesville; IBPGR, 1986:50; Van Sloten and Gulick, 1984.

Note: Other small collections in Madagascar, Papua New Guinea, South Africa, Greece, Taiwan, Indonesia, Cyprus, and the Spanish Canaries are derived from those listed in the table.

[1]See Appendix 2 for spelled-out names of institutions.
[2]Spontaneous seedlings well represented.
[3]Local forms well represented.
[4]Largely commercial cultivars.
[5]*Persea borbonia, caerulea, cinerascens, donnell-smithii, indica, lingue, longipes, nubigena, schiedeana,* and *steyermarkii.*

National programs in developing countries maintain some of the largest avocado germplasm collections. Jaboticabal in São Paulo State, Brazil, has the biggest avocado gene bank, with over 340 accessions, but since these consist mostly of spontaneous seedlings it is difficult to estimate the range of diversity represented (Gulick and Van Sloten, 1984). Cuba has two avocado gene banks; one near Havana with 280 accessions that include commercial varieties and landraces from Central America and Cuba, and another at Villa Clara with 47 cultivars. Thailand has two sizable avocado gene bank collections indicating a growing interest in the fruit in Southeast Asia. The Doi Muser Horticultural Research Station in Tak Province, Thailand, has 204 cultivars from Hawaii and 23 varieties from Thailand, while the avocado germplasm collection at the Plew Horticultural Research Station in Chantaburi Province has 136 Thai varieties.

Several developed countries also hold sizable germplasm collections of avocado, a reflection of the importance of the crop both domestically and for export. The 294 avocado accessions maintained by the Commonwealth Scientific and Industrial Research Organization (CSIRO) at Merbein, Victoria, are mostly seedlings from Australia. The Israeli avocado germplasm collection at the ARO Volcani Center, Bet-Dagan, is more diverse, with 132 landraces and 47 rootstock cultivars from Latin America, 20 wild accessions from Mexico, and 12 commercial cultivars from Latin America.

The USDA Subtropical Horticulture Research Station in Miami, Florida, has been designated as the national depository for avocado germplasm in the United States. The Subtropical Horticulture Research Station maintains an orchard gene bank of some 204 avocado accessions and is also responsible for the 112 accessions, mostly duplicates, maintained on the grounds of the University of Puerto Rico at Mayaguez. The Tropical Research and Education Center of the University of Florida located near Homestead, Florida, has a collection of 116 avocado accessions, including many accessions not present in other gene banks (Appendix 4). Unfortunately, this University of Florida collection, which contains abandoned commercial varieties and material collected by Wilson Popenoe, has become infected with the highly damaging sun blotch viroid and is no longer available for germplasm exchange. Shallow soils on limestone bedrock have encouraged extensive interconnecting of roots near the surface in this germplasm collection, thereby making it difficult to isolate the infection. With competing demands for space on the research station, the future of this old and largely unique gene bank is uncertain. Tissue culture salvage of the collection would be difficult and costly.

Gene banks are safer when they are actively used by breeders and are cared for by full-time curators. The wild avocado collection maintained by the University of California at Riverside, and the gene bank of rootstock candidates, old commercial cultivars, and breeders' selections at the South Coast Field Station near Irvine, California, are secure because they are linked to two active breeding programs for cultivars and rootstocks, and scientists regularly oversee the safety and health of the germplasm collections. Avocado gene banks in Australia and Israel, and the small but well-tended avocado germplasm collection at the Cerrado research center (Centro de Pesquisa Agropecuária dos Cerrados, CPAC) near Brasília, also play an integral role in introduction and selection programs.

Many avocado germplasm collections, including the avocado gene bank at Homestead, Florida, are not tied to active breeding programs. Orchard gene banks are sometimes neglected because of intense pressure on limited resources such as staff, fertilizers, and irrigation water. The 177 avocado accessions held at the Centro de Investigaciones Agrícolas de El Bajío, near Celaya, Mexico, were neglected after the avocado breeder left for more training in early 1987. The orchard received no rain or irrigation water between October 1987 and mid-June 1988; consequently, several dozen *P. americana* accessions were near death in June 1988, and one of only two *P. indica* introductions had recently died. The summer drought that struck much of Mexico in 1988, combined with an accelerated drawdown of the aquifer for irrigation, caused one of the station's three wells to dry up, while the remaining two wells were hard pressed to service the station's other water needs. The two active wells at the El Bajío Station are already down to 200 meters, and pumping costs are rising. The continued drop in the water table in the Celaya area has led to shifting and structural damage to buildings in Celaya and jeopardizes the future of the avocado collection.

Commercial varieties and seedlings under observation dominate many avocado germplasm collections. It is a good idea to save old commercial varieties and some experimental lines. For example, the prostrate growth of an avocado maintained at the University of California's South Coast Field Station south of Los Angeles, would be ideal for patios and decks. Thus in addition to providing rich, lustrous foliage, this ornamental avocado would also provide generous and sizable fruit for urban folk.

Duplicates of some old commercial varieties, such as 'Booth 7' and 'Collinson' are fairly common in germplasm collections (Appendix 4). Some duplication is a good idea in order to avoid irreplaceable losses in the event that a germplasm collection is lost, to gain some idea how adaptable varieties are to different environments, and to bring avocado materials closer to breeders. Over 3,400 accessions of *P. americana* are held in collections throughout the world (Table 3.5), but few primitive or wild avocados are represented. No single collection represents more than a small part of the genetic diversity of avocado.

Coverage of traditional avocado varieties from Latin America and other tropical countries in gene banks is limited. Ideally, avocado germplasm collections should contain local avocados and wild material, basic resources for breeding programs. Sometimes wild specimens or traditional varieties are the only known sources for resistance to a particular disease or pest. The best collection of traditional Mexican avocados is maintained at the El Bajío Station near Celaya, Mexico, which contains 164 accessions from the states of Guanajuato, Nuevo Leon, Puebla, Veracruz, and Michoacan. Still, this 177-accession collection captures only a small sample of the genetic diversity of Mexican avocados. Less than 20 percent of the avocado accessions maintained by the USDA Subtropical Horticulture Research Station are accessions from seedling avocados in Latin America. Most of the traditional selections, such as 'Isham' and 'Egas', at the Subtropical Horticulture Research Station were collected by Wilson Popenoe in backyards of farmers and in coffee plantations in Guatemala between 1916 and 1917 and in Ecuador during 1927.

Some traditional avocados collected by Popenoe in Guatemala are apparently not in gene banks and are presumably lost. Among the cultivars collected by Popenoe that are missing from gene banks are 'Benik', collected at 1,700 meters near Antigua and once a commercial variety in Israel; 'Cabnal' (1,700 meters, Antigua); 'Cantel' (1,700 meters, Antigua); 'Coban' (1,400 meters, Coban); 'Kekchi' (Purulha, Baja Verapaz); 'Mayapan' (Purulha, Baja Verapaz); 'Panchoy' (1,600 meters, Antigua); 'Pankay', a hardy variety collected at 2,800 meters near Totonicapan; and 'Tumin' (1,600 meters, Antigua). The fates of other avocado introductions to the United States from Guatemala in the early part of this century, such as 'Chabil','Chisoy' (San Cristobal Verapaz),'Hunapuh', 'Kagua', 'Kanola', and 'Nimah' (Mazatenango) are also unknown (W. Popenoe, 1928).

It is unrealistic to "save" every interesting local genotype, plant introduction, or seedling spotted in a commercial nursery or backyards of traditional avocado growing areas. Millions of seedling avocados enrich the diets of people living in most parts of Mexico, and the task of deciding which trees are worth collecting for orchard gene banks is daunting. Over 700 avocado cultivars have been tried in the United States (Chandler, 1958:206), but less than a dozen are grown commercially today. Furthermore, only a small fraction of plant introductions make it in the ornamental trade or farming world anyway, and maintaining rejects could be seen as a wasteful use of scarce resources.

But today's rejects can be tomorrow's gold. It is particularly risky to discard plant germplasm when its source region is undergoing rapid ecological and cultural change. The early introductions of avocado from Central America were each unique. Given limited resources for germplasm collections, a difficult balance must be struck between holding on to the old and making room for the new.

Close to half a century ago, Wilson Popenoe (1939) emphasized the need to revisit periodically areas from which promising avocados had been collected in order to survey the new crop of seedlings in backyards, fields, and coffee plantations. Although taste is always subjective, several regions have earned reputations for particularly good-tasting avocados. In Mexico, Queretaro is famous for its *drymifolia* avocados, but urban expansion and topworking to 'Fuerte' and 'Hass' are shrinking the rich avocado gene pool in the area. By 1988, the booming industrial city of Queretaro had over one million inhabitants, and nearby Celaya had half a million city dwellers. Farther to the south at Atlixco, both Guatemalan, known locally as *pagua*, and Mexican avocados are planted side by side. The fertile plain of Atlixco is of particular interest to avocado breeders because of the spontaneous hybrids between Guatemalan and Mexican forms. Rapid changes are occurring in this rich avocado gene pool, however; some groves are being topworked to 'Hass', 'Fuerte', and 'Principe', while others have been cut down because of labor difficulties or to make room for the booming flower business, particularly roses. Atlixco's pleasant climate, clean air, and good road and rail connections to Puebla and Mexico City are attracting an increasing number of residents and businesses, resulting in urban sprawl into the surrounding countryside.

In Guatemala, Antigua and San Cristobal Verapáz are famous for their avocados. The Chinandega Valley between Corinto and Managua in Nicaragua reportedly has superior West Indian avocados. In South America, fine-tasting West Indian avocados are cultivated in the region of Santa Marta near Baranquilla on Colombia's Caribbean coast. On Colombia's Pacific coast, West Indian avocados are especially palatable around the port of Tumaco. The Chota Valley of Ecuador has unusually productive and savory Mexican avocados, as well as a smaller amount of West Indian forms (W. Popenoe, 1922). Remote valleys and plateaus in the Andes and other mountain chains in Latin America may well contain valuable germplasm of cultivated avocados. All avocado gene banks would greatly benefit from a fresh infusion of traditional material.

Australian collections of avocado germplasm are virtually devoid of traditional material and wild species. Australia is particularly keen on receiving more West Indian avocados for sources of salt tolerance and to expand production of the crop in the tropical north. Even commercial varieties are underrepresented in Australian avocado gene banks, which contain only two-thirds of the germplasm of improved types (Sedgley and Alexander, 1983).

Wild populations of *P. americana* and its near relatives are virtually absent in ex situ germplasm collections. Gene banks contain only thirty-nine accessions of wild *Persea* spp. (Table 3.5). The University of California at Riverside contains the best collection of near relatives of avocado, but only 20 accessions of some of the wild *Persea* species are maintained there. The hot, dry conditions of the east Los Angeles basin may not be conducive to growing all the close relatives of avocado. Other gene banks, in a variety of climatic and soil conditions, need to take on the responsibility for stocking accessions of wild species for evaluation and breeding.

The degree of evaluation at avocado gene banks varies widely. Fruiting characteristics of avocado accessions are relatively well studied, such as at the USDA Subtropical Horticulture Research Station in Miami and the El Bajío Station near Celaya, Mexico, but data on pest and disease resistance are usually lacking. One accession of *P. schiedeana* appeared to be immune to the leaf-damaging insect known as *agalla de hoja* (*Trioza anceps*) at the El Bajío station, while nearby *P. americana* accessions were heavily infested by the pest. This pest can attack Mexican avocados severely, raising numerous welts or bumps on leaves, thereby reducing yields (Torre, 1984:21). Not all *P. schiedeana* appear to resist *T. anceps*, however; several *chinene* were affected by the pest in the vicinity of Orizaba, Veracruz, in June 1988.

A large portion of avocado germplasm collections are composed of old commercial varieties whose characteristics are already well known, so evaluation efforts should focus on experimental lines and especially on primitive and wild material when the latter become more available. Germplasm should be evaluated both on research stations and in the field; growers are usually willing to make land available for such purposes.

Breeding Thrusts and Germplasm Resources

Major avocado selection and breeding programs operate in California, Florida, Hawaii, Israel, South Africa, Australia, and Chile. Mexico has recently greatly expanded its avocado research efforts. But apart from Mexico, little avocado breeding is under way in most developing countries, even in the native home of avocado in Central America. Although developing countries contain some of the larger avocado germplasm collections, those resources are underutilized.

All leading avocado cultivars arose as chance seedlings. Exact parents of modern or traditional varieties are rarely known. Hand-pollination has been largely abandoned because only a small fraction of the prolific flowers bear fruit. Labor costs and time constraints of scientists and technicians prohibit widespread employment of this breeding technique on avocados. More controlled crosses could introduce specific traits. Controlled crosses can be achieved by confining bees, which are effective pollinators, in screenhouses enclosing prospective parent trees.

Commercial avocado plantations are mostly based on grafting high-yielding material onto hardy rootstock, so avocado breeders have generally pursued a two-pronged selection strategy: superior fruiting characteristics, and rootstocks. The major breeding priorities for fruit include heavy yield, good shipping quality, sales appeal and consumer satisfaction, spreading habit to facilitate picking and to reduce wind damage, and disease resistance (Bergh, 1976; IBPGR, 1986:49). For rootstock, the priorities are root rot resistance, adaptability to saline and alkaline soils, and cold tolerance (Ben-Ya'acov, 1989; IBPGR, 1986:49). In California, drought tolerance has also emerged as an important attribute in screening avocado germplasm (B. O. Bergh, pers. comm.).

This parallel improvement strategy is likely to converge in the distant future as breeding programs attempt to use controlled pollination as a means of transferring desirable attributes to a single variety so that grafting scions onto rootstocks is unnecessary. Controlled crossing is more time-consuming but ultimately can be more rewarding. Assembling as many desirable qualities as possible in one cultivar reduces the costs of breeding and cultivating the crop.

The most important disease problem of avocado is root rot, caused by *Phytophthora cinnamomi*. This fungal pathogen was discovered on cinnamon trees in Sumatra in 1922 and has spread to over 70 countries (Zentmyer, 1985). Root rot infects over 1,000 plant species and was reported in

Puerto Rico and California in 1929 and 1942, respectively. By the late 1960s, root rot had severely damaged close to 2,000 hectares of avocado orchards in California (Kerr, 1970:198). By the late 1980s, root rot had penetrated most avocado groves in the state (R. Bergh, pers. comm.). In South Africa, root rot had seriously infected over 80 percent of avocado plantings by the late 1970s (Kotzé and Darvas, 1983; Wolstenhome, 1987). Root rot damage to avocado is increasing in Brazil (Donadio, 1987). The disease can wreak havoc in orchard gene banks; root rot destroyed an avocado germplasm collection maintained by the Escuela Agrícola Panamericana at Zamorano, Honduras.[6]

Use of chemicals to combat root rot in avocado plantations has only been partially successful and is costly (IBPGR, 1986:48). In commercial avocado orchards, waterlogging for as little as two or three days is often fatal to the trees. High levels of resistance to this widespread fungal pathogen are found in some near relatives of avocado, particularly *P. borbonia*. The pathogen has been found on roots of *P. schiedeana* in Alta Verapáz, Guatemala (Zentmyer, 1954). Some populations of *P. caerulea* are probably subject to heavy and prolonged exposure to the root rot fungus with no apparent symptoms; the species thrives in soil with a high water table close to San José's international airport in Costa Rica. Unfortunately, most wild *Persea* species are not graft-compatible with *P. americana* (Frolich et al., 1958; Schroeder, 1967).

One of avocado's closest relatives, *P. schiedeana*, is graft-compatible with *P. americana* and has thus been the main focus of resistance screening to root rot. In 1984, three selections of *P. schiedeana* (G755A, B, C) with moderate levels of resistance to root rot were released for use by the University of California at Riverside (Coffey, 1984; Coffey et al., 1988). These promising selections were obtained as seeds from a market in Cobán in central Guatemala in 1975. Isozyme analysis has shown that the G755 selection contains genes from *P. americana*, most likely from the Guatemalan race (Ellstrand et al., 1986).

One way to overcome the disease would be to transfer resistance genes from wild species to *P. americana*. All species of *Persea* are diploids, so no ploidy barriers complicate prospects for widecrossing within the genus. Avocado has been crossed with *P. floccosa*, but widecrossing attempts with other wild species have thus far proved unsuccessful (Bergh, 1976; Zentmyer and Schieber, 1987). New developments in biotechnology may eventually create fresh opportunities for widecrossing avocado (Bergh and Ellstrand, 1987).

---

6. Letter from Wilson Popenoe to Evelyn Smiley written in Antigua, Guatemala, on 5 May 1972 (courtesy of Hugh Popenoe).

Over two dozen other fungi attack avocado, causing varying amounts of damage (Weber, 1973:10). Fungal diseases for which genetic resistance would be useful include trunk canker (caused by *Phytophthora citricola*), Verticillium wilt (*Verticillium dahliae*), oak root rot (*Armillaria mellea*), Cercospora spot (*Cercospora purpurea*), anthracnose (*Colletotrichum gloeosporioides*), avocado scab (*Sphaceloma perseae*), powdery mildew (*Oidium* spp.), and algal spot (*Cephaleuros virescens*).

Another troublesome avocado disease for breeders and germplasm curators is sun blotch, which disfigures fruit and leaves and can stunt or kill avocado trees (Campbell, 1984; Zentmyer, 1984). This disease, once thought to be a result of severe sunburn (Coit, 1928), is now known to be caused by a viroid. The pathogen, transmitted by grafting, may have originated in California and probably spread to Israel on infected budwood (Schroeder, 1935). The disease appeared near Homestead, Florida, in the mid-1980s. By the late 1980s, sun blotch had spread to virtually all avocado growing regions, including Australia (M. Coffey, pers. comm.). No sources of resistance to the pathogen have thus far been found. Sun blotch–free, registered budwood is the only method currently used to control the disease.

A number of insects attack avocados, and not all of them can be enumerated here. Two general points need to be made from the breeding standpoint, however: insect pressure is expected to be particularly severe in avocado's native range, and new insect pests are constantly evolving or expanding their distributions. The greenhouse thrip (*Heliothrips haemorrhoidalis*), for example, started severely blemishing avocado fruits in southern California in the mid-1980s (Bekey, 1986). A biocontrol approach, if successful, would allow the continued cultivation of favored varieties. If no parasites or predators of the greenhouse thrip can be established to control the pest effectively, then resistance breeding would be especially desirable. Mexican avocados appear to be much less affected by the greenhouse thrip than other races, and sources of resistance may thus be found by tapping gene bank collections for pure Mexican varieties or even hybrid forms containing *drymifolia* germplasm. Resistance to the greenhouse thrip might also be found in Guatemalan germplasm; the pest apparently does not affect 'Reed', a Guatemalan avocado selected in Carlsbad, California, in 1960 (O. Atkins, pers. comm.). 'Reed' is a commercial cultivar in California and is maintained in at least two gene banks (Appendix 4).

As avocado breeding programs are implemented in tropical America, resistance to insect pests will be a major priority. One reason that traditional avocado varieties in Latin America are generally low-yielding is that many are severely damaged by insects (Campbell, 1984). Fewer insect pests of avocado are encountered outside of the crop's traditional growing

area. In the Cerrado region of Brazil, moth larvae (*Stenoma catenifer*) dis-
figure avocado fruit and provoke premature fruit drop. Larvae of beetles,
particularly those belonging to the genus *Heilipus*, feed on seeds and fruit
of avocado and tunnel into trunks of the trees.

Cold tolerance has always been a concern of avocado growers in temperate countries. The best sources of cold tolerance are Mexican avocados,
since they originated at higher altitudes in the tropics and their natural
range extends into the subtropics in northern Mexico. Mexican avocado
varieties typically tolerate temperatures down to −5°C for several hours
(Campbell, 1984). 'Knowles' and 'San Sebastian', Mexican varieties cultivated in California in the early part of this century, withstood temperatures as low as −6°C during the 1916–17 winter with only minor damage
to young leaves (Webber, 1918). 'Yama', another Mexican avocado, has
survived temperatures as low as −8°C without severe injury (Bergh,
1975).

Many of the avocado cultivars grown in California, Israel, South Africa,
and Chile for the export market contain *drymifolia* germplasm. Pure Mexican varieties are generally not suitable for North American and European
markets, but Mexican genes are responsible for the cold tolerance of several California varieties of mixed Guatemalan and Mexican germplasm,
such as 'Bacon', 'Zutano', and 'Fuerte', which can withstand temperatures
of −4.4°C, −3.3°C, and −2.7°C for four hours, respectively (Bergh,
1985).

Commercial avocado groves are generally irrigated, and the rising cost
and, in some cases, scarcity of water have some growers worried. Avocado
by nature is prone to moisture stress. During the 1960s, avocado growers
in southern California started switching from open furrow and overhead
spray irrigation to drip irrigation, microjets, and fan-jets to convey water
to their crop. Drip irrigation and microjet misters reduce evaporation
losses, but current commercial avocados still need a lot of water to remain
productive. Avocado orchards in California use between 1.5 and 3 acre-
feet of water per year (Kerr, 1970:341). Avocado growers in some parts of
the state were paying as much as $500 an acre-foot for irrigation water in
the late 1980s. Competing demands for water, particularly from industry
and growing urban areas, are sure to continue to drive up irrigation costs.
Arizona, for example, is expected to start receiving its full quota of water
from the Colorado River in the early 1990s. Much of the water for southern California comes from the Sierra Nevada Mountains and the Colorado
River, and further attempts to tap water supplies from northern California
are meeting with increased resistance. If southern California is successful
in obtaining more irrigation water, construction of more aqueducts will
boost costs. As oil prices have resumed their upward climb, pumping costs
have also soared. Even though the agricultural lobby still wields some

clout in California, irrigation costs are destined to rise. Farmers use 85 percent of California's water supply, and, as urban areas continue to expand, pressure on this resource will increase. The ability of avocado groves to thrive on less water would be a boon to growers in California, Israel, the Mediterranean region, and other parts of the world.

Drought tolerance is a relatively new concern among avocado breeders, so work has only just begun on screening germplasm that can still be productive with reduced irrigation. Some commercial varieties tolerate extreme heat and low humidity, but there is no evidence that they need less irrigation water. For example, 'Irving', a Mexican-Guatemalan cross, reputedly withstands the hot, dry summers of California, while in Morocco, 'Choquette' and 'Chavanier' withstand searing winds blowing off the Sahara (Bergh, 1975). 'Choquette', a Guatemalan and West Indian cross, originated from a seed planted in Miami in 1929 (Appendix 4). Wild species and most traditional avocados in the Third World thrive under variable rain-fed conditions and may prove to be useful sources of drought tolerance. Remarkably drought-tolerant wild avocados have been reported in Guatemala (R. Bergh, pers. comm.). Truly drought-tolerant varieties are unlikely to match the output of avocado orchards under optimal irrigation, but growers may have to live with lower yields if water costs rise dramatically.

Salt tolerance is another trait of increasing concern for avocado growers in some areas, such as Israel (Ben-Ya'acov, 1989). Unless irrigation is accompanied by good drainage and periodic flushing to remove salts that build up when water evaporates, the soil becomes progressively more saline and inhospitable to crops. As water costs rise, flushing becomes more expensive, so salt tolerance will become more and more desirable. The best sources of salt tolerance for avocado rootstock are West Indian varieties, but they do not tolerate cold. Cold-tolerant Mexican varieties, on the other hand, are especially sensitive to salts. A protracted breeding effort will be required to develop avocado rootstocks tolerant to salts under a variety of climatic conditions.

Another sought-after attribute of avocado rootstocks is dwarfing. Tall avocado trees are more prone to wind damage and are more costly to harvest and to treat with fungicides or pesticides. Florida is in the path of summer hurricanes, while in southern California powerful Santa Ana winds can dislodge appreciable quantities of avocados. Windbreaks can help alleviate this problem, but they take up space and can compete for water and soil nutrients. Near relatives and wild populations of *P. americana* are potentially good sources for dwarfing. Some genotypes of *P. schiedeana*, *P. floccosa*, and *P. nubigena* may also be useful sources of dwarfing for rootstocks (Bergh, 1975). Dwarfing rootstock is not a universally desirable trait for avocado growers. In some situations, tall avocado trees discourage stealing of fruit and provide generous shade.

It is time to take a closer look at wild populations of *P. americana* and its near relatives. The limited use of primitive and wild avocados in the past should not guide breeding policy in the future. Breeders have historically been reluctant to devote more time to wild species because they are difficult to cross with avocado and because little is known about their attributes. Avocado's near relatives represent a large reservoir of potentially useful genes. Evaluation of the attributes of wild *Persea* species, both in the wild and in field gene banks, would reveal possibilities for enriching the gene pool of the cultivated species.

If avocado plantations are to hold their own in Florida and southern California, for example, costs of production need to come down so that producers can make sufficient profits and withstand the pressure to sell out to land developers. Avocado groves in southern Florida are being bulldozed for subdivisions and ornamental nurseries; between 1984 and 1988, the avocado area in the state shrunk from 5,209 hectares to 4,600 hectares (R. J. Knight et al., 1984). Avocado growers in southern California remark that a climate propitious for avocado also attracts people and businesses; California's mild winters and long, warm summers are a magnet for retirees, suburbanites, and light industry. Avocado growers in the United States and elsewhere are keen for high-yielding varieties that produce reliably every year, resist diseases and pests, and require modest amounts of water, among other attributes. The constantly shifting demands of the avocado industry worldwide will have to be met, in large measure, by deploying genetic resources.

REALIZING THE POTENTIAL

Considering the importance of germplasm collections to the future of the avocado industry worldwide, more support for orchard gene banks is warranted. Not only do collections need to be broadened to incorporate more primitive and wild material, but many avocado gene banks would also benefit from fuller evaluation of their collections. In this manner, gene banks would be considered an even more valuable resource by breeders and thereby help ensure their survival.

The financial burden of maintaining avocado gene banks can be alleviated by the private sector. In the United States, growers' associations recognize their stake in the future of germplasm collections and thus contribute to their upkeep. The Florida Avocado Growers' Association donates funds for the avocado gene bank at the Tropical Research and Education Center of the University of Florida, near Homestead, Florida. The enduring California Avocado Society, established in 1915 as the California Avocado Association, annually provides $750 per acre for the avocado germplasm collection maintained by the University of California at its South

Coast Field Station near Irvine. In 1988, that expanding germplasm collection of 120 accessions occupied two acres.

The cost of maintaining orchard gene banks is small compared with the actual and potential benefit. The avocado germplasm collections maintained by the University of California at the South Coast Field Station and the Riverside campus probably cost under $20,000 to operate, including irrigation and staff time. This is a modest investment compared with the $200 million that the avocado industry generated in California in 1988.

Genetic erosion of wild populations of avocado and its near relatives in Central America and northwestern Colombia is the single most important threat to the genetic resources of the crop. More collecting of wild germplasm and setting aside reserves to safeguard avocado in undisturbed habitats are urgently needed. Mexico, Costa Rica, and Guatemala are leaders in setting up parks and reserves. Guatemala, for example, plans to set up over twenty parks in the near future. Botanical surveys in existing and planned parks are needed to ascertain whether gene pools of avocado and other crops are included within their perimeters.

The genetic resource picture for cultivated avocados is less precarious. Traditional avocado growing areas contain extremely heterozygous populations because of propagation by seed. A shift to a few high-yielding cultivars has occurred mainly in warm temperate regions, well away from the center of origin of the crop. Still, the avocado groves in traditional growing areas are becoming more homogeneous as they are topworked to superior varieties and grafted avocados become more available for fresh plantings.

Yield stability of commercial avocado growing areas in industrial nations could be threatened if pressures from wholesalers and retailers continue to focus on a single variety. Packers and supermarkets generally prefer reliable supplies of standardized fruits that travel well, have a consistent appearance, and a good shelf life. The danger of narrowing the genetic base of commercial avocado orchards is ever present because "obsolete" varieties can be readily topworked.

A single avocado variety is unlikely to produce optimal fruit year-round in a region, particularly if the region has a seasonal climate. Early and late-maturing varieties are desirable to prolong supplies. Also, by planting more than one variety, growers have a cushion in case there is overproduction of one particular variety and the price falls drastically. Also, varieties need to be developed to fit microclimates. Planting a highly desirable cultivar over vast areas could be a recipe for disaster, as has occurred with some other crops. In California, extensive topworking of old varieties such as 'Bacon' and 'Fuerte' in favor of 'Hass' could be short-sighted. In 1915, the nascent California avocado industry had eighty-six varieties planted in orchards (F. O. Popenoe, 1915); now only five cultivars account for nearly all of the avocado acreage in California, with 'Hass' alone accounting for more than three-quarters of the total. A shakeout of varieties is inevitable, but a handful of varieties dominating production of any crop can be risky, particularly with trees, where switching to backup varieties is more time-consuming and costly than it is for annual crops. For example, 'Hass', for all its virtues, is the least cold-tolerant of the commercial cultivars in California, and a hard freeze could exact a devastating toll; just such a freeze occurred at Christmas in 1990. That freeze destroyed approximately one-quarter of the avocado crop on the trees in California, and further economic losses were incurred because of poor fruit quality and shorter shelf life (B. Bergh, pers. comm.).

The argument that consumers prefer a standard fruit is true up to a point, but education and advertising can extoll the virtues of avocados with different shapes and skin colors at maturity. The most important factor is taste: good quality sells. Also, consumers are increasingly interested in culinary novelties and may well appreciate a range of avocados in the marketplace. Supermarkets could indicate the variety of avocado on display, as they generally do for apples and oranges, and help educate consumers about the merits of each variety. Certainly the job for breeders is made easier if they do not have to engineer all desirable traits, such as cold tolerance and disease resistance, into a single, standardized fruit type. The fact that the highly successful 'Hass' is dark at maturity, instead of green as is the once-dominant 'Fuerte', indicates that outstanding taste can overcome resistance to differences in fruit appearance.

Quarantine regulations are an obstacle to the exchange of avocado germplasm in some cases. Importation of avocado germplasm to Australia has been impeded by the problem of reliable indexing for sun blotch. A recently developed DNA probe for the viroid, however, promises to improve prospects for a fresh infusion of germplasm for avocado breeding programs in Australia and elsewhere (Sedgley and Alexander, 1983). Avocado germplasm for Australia still needs to be observed for a year for symptoms of black streak virus. The University of California, Riverside, is required by the USDA to quarantine avocado grafts for two years in glasshouses; avocado seeds are fumigated and released promptly. Most of the major diseases of avocado have accompanied the crop as it diffused from its origins. Sun blotch, though, warrants some concern, and more quarantine facilities, preferably away from avocado growing areas, would help reduce the chances of spreading diseases into new areas or introducing different pathotypes of already established diseases.

Given the pressing need to learn more about the genetic resources of avocado and the urgency in safeguarding wild germplasm, a flexible network approach is a viable way to further avocado research. A global avocado network linking existing avocado researchers could be established with modest funding from external donors to support workshops, publications, international nurseries, ecogeographic surveys of germplasm, and socioeconomic studies of avocado cultivation and consumption. International cooperation in avocado research is already under way. One example of such cooperation is the development and widespread adoption of the dot-blot test for detecting the sun blotch viroid (Wolstenholme, 1987). Such efforts need to be reinforced and expanded into other areas of avocado research.

Several tropical fruits are considered minor in terms of world trade but are nevertheless widely grown in the tropics. Guava, papaya, sapodilla, and passionfruit, for example, do not compare with oranges or mango as major trade items, but they enrich the diet of hundreds of millions of people in warmer climates. Guava, papaya, sapodilla, and passionfruit are indigenous to the American tropics and have become widespread within only the last few hundred years. European adventurers, traders, and missionaries have taken the fruits, used for fresh fruit, juice, and preserves, to Africa, Asia, and the Pacific. Guava is genetically diverse and now is being commercialized in many areas, while sapodilla is still little different from its wild progenitor.

GUAVA

Guava fruits, reaching well beyond the size of tennis balls in some cases, are relished in many tropical countries. Guava (*Psidium guajava*) can be a shrub or reach the dimensions of a 20-meter tree. Traditionally, guava is planted in dooryard gardens where the multibranched tree provides diffuse shade, thereby allowing other plants, such as medicinals, to thrive beneath. Known as *guayaba* in Spanish-speaking countries and *goiaba* in Brazil, guava fruits are eaten fresh and made into drinks, ice cream, and preserves (Figure 3.21). Brazilians are passionately fond of a dessert called Romeo and Juliette, which consists of a wedge of firm guava paste (*goiabada*) topped with a slice of white cheese.

Guava fruits have a distinctive, savory-fresh aroma that is thermostable, and thus survives processing. Guava is higher in vitamin C than citrus and it contains appreciable amounts of vitamin A as well. Guava fruits are also a good source of pectin, an enzyme used in making jam.

The precise range of wild guava is still not known, but it probably

FIGURE 3.21 Guava products sold in Latin America: guava paste (*goiabada*) from Brazil on the left; guava jam from Mexico on the right.

stretches through lowland parts of Central America and tropical South America. Given the long association between people and guava, it is probably impossible to establish with any certainty the precise former range of guava.

As were avocado and some other tropical perennial crops, guava was probably taken into cultivation, or at least semidomesticated, in several parts of its natural range. Both a camp follower and a spontaneous invader of open environments created by human activities, guava was a common and tempting candidate for domestication. Even before open field agriculture, Paleolithic hunters and gatherers in the New World probably favored wild guava when they burned thickets and grassland at the edge of forest in order to flush game. Isolated guava in the forest would provide seed sources for birds to disperse the sun-loving species into ash-rich, open areas. Guava may have been domesticated in Peru several thousand years ago. Guava remains have been found in Peruvian archaeological sites with beans, chili pepper, cotton, and squashes, all early cultigens (Hawkes, 1983). Abundant guava seeds have been found at El Paraíso, a preceramic site along the Peruvian coast, that was occupied between 1800 and 1500 B.C. (Quilter et al., 1991). It is not clear whether guava was cultivated at El Paraíso or whether fruits and wood were gathered from wild or managed stands. Guava wood was used extensively in construction and for fuel at the arid site. In Mexico, guava remains date from approximately 200 B.C. (Bray, 1976).

Portuguese traders took guava from Brazil to Goa in India, while the

Spanish took guava from Acapulco, Mexico, to the Philippines. From these beachheads in Southeast Asia, guava was picked up by other traders and carried farther afield. Henry Bricknell of the London Missionary Society brought guava to Tahiti from Brazil in 1815 (Hermann et al., 1989:20). Guava was reportedly introduced to Florida in 1847 and was common in the warmer parts of the state by 1886 (Morton, 1987:356).

Guava has spread widely in warm tropical areas with moderate to heavy rainfall because it thrives on a variety of soils, propagates easily, and bears fruit relatively quickly. The fruits contain numerous, ball-bearing-sized seeds that can produce a mature plant within four years. When guava is propagated vegetatively, it can produce even sooner.

Like many tropical fruit trees, guava has several uses. Folk cures employing guava were incorporated early into the Dutch colonial pharmacopoeia. In its Latin American homeland, guava leaves are chewed to treat mouth sores, and the bark is boiled to produce a remedy for dysentery. In Mexico, guava leaves are reputed to help relieve itching (Standley, 1924:1036). In Central Africa and Southeast Asia, tannin-rich guava leaves and bark are used for various medicinal purposes. In Zaire, people suffering from diarrhea sometimes resort to a decoction of guava leaves and bark to wash out their intestines (Bokdam and Droogers, 1975). In West Africa, mashed guava roots are mixed with water, and the concoction is drunk to alleviate dysentery (F. R. Irvine, 1930:355). Guava also makes excellent charcoal.

International markets are growing for guava juice and, to a lesser extent, preserves. North Americans and Europeans are eager for new taste sensations, and many are willing to try "exotic" fruit juices as an alternative to the morning glass of orange juice. A guava-passionfruit juice blend is sold in many supermarkets in the United States. In Hawaii and Florida, frozen guava puree for juice is also readily available. Commercial guava production is currently limited to only a handful of countries, such as Brazil, India, Pakistan, Mexico, Guyana, and the United States (Florida and Hawaii), but the planted area is expected to increase in the near future. Internal markets are also growing for processed guava products, such as juice and jelly, particularly in Latin America, where more than half the people now live in towns and cities.

GENE POOLS AND NEAR RELATIVES

Guava's gene pool is vast because of outbreeding (up to 40 percent), the fact that most farmers plant seed rather than clone the tree, and the existence of numerous spontaneous populations. Guava thrives in second growth, pasture, and other open habitats. It is so well adapted to disturbed habitats that it has become a weed in many areas, such as in Cameroon (J. D. Sauer, 1988:66), Fiji (Vitousek, 1988), Florida (Wolfe, 1937), Ghana

(Irvine, 1961:102), Guatemala (Brigham, 1887:368), Jamaica (Burke and Girvan, 1954:581), Martinique (Kervégant, 1937; Kimber, 1988:303), Mexico (Standley, 1924:1036), Nicaragua (Belt, 1888:203), Panama (Croat, 1978:661; Standley, 1927:26), St. Croix (Britton, 1918), St. Thomas (Britton, 1918), Samoa (Parham, 1972:46), Sri Lanka (Macmillan, 1935:264), Tahiti (Pétard, 1986:244), and Tonga (Yuncker, 1959).

Spontaneous guava can become a nuisance in pastures and can invade farmers' fields (Hawaii: Crawford, 1937; Tahiti: Pétard, 1986:35). In some cases, spontaneous guava forms extensive, almost pure stands, for example on gravel bars of the Terraba River in Costa Rica (P. H. Allen, 1956:306) and behind some beaches along Ghana's coast (F. R. Irvine, 1961:102). Guava is a ruderal plant in the Serra dos Carajás, Pará, Brazil, and near Cali, Colombia (N. J. H. Smith, field notes).

In warmer parts of Mexico, guava is spread readily by domestic animals, and spontaneous guava trees form dense thickets, called *guayabales* (Standley, 1924:1036). In Costa Rica, horses eagerly consume fruits that have fallen from spontaneous guava trees in pastures, and many of the small, hard seeds pass through the gut unscathed and start more trees (Skutch, 1980:166). Several monkey species are probably also involved in spreading guava seeds; along the Transamazon Highway, a pet capuchin monkey (*Cebus apella*) would head straight for a backyard guava tree and devour fruit when allowed out by its owners (N. J. H. Smith, field notes, 1974). In the eighteenth century, an artist accompanying the Brazilian naturalist Alexandre Rodrigues Ferreira depicted three species of monkey either holding or eating guava fruits (Figures 3.22, 3.23).

Several fruit-eating birds also feed on guava and disperse the seeds in rich droppings. In the coastal range of Venezuela, flocks of blood-eared parakeets (*Pyrrhura hoematotis*) feed avidly on guava.[7] Colorful and common blue-gray tanagers (*Thraupis episcopus*) eat guava fruits in backyards of Manaus, Brazil, and probably elsewhere in Amazonia.[8] Parrots in Amazonia were depicted feeding on guava fruits in the eighteenth century (Figure 3.24).

Wherever guava is taken, some local birds soon develop a taste for its fruit. Even on Bora Bora, a small, reef-ringed extinct volcano in the Society Island chain with very few land birds, stunted guava are common on rocky basalt ridges (Figure 3.25). Spontaneous clumps of guava probably outnumber "pristine" populations dispersed in shrinking forests.

7. Flocks of fifteen to twenty blood-eared parakeets, known locally as *perico cola roja*, were observed feeding on guava on the grounds of the Biological Station, Rancho Grande National Park, near Maracay, Venezuela, on 20 September 1971 (N. J. H. Smith, field notes).

8. Blue-gray tanagers were seen feeding on guava in Manaus, Amazonas, Brazil, in July 1976. On 8 July 1976, two adults were observed feeding guava to two fledglings; the parents subsequently flew away, and the young birds continued to feed by themselves (N. J. H. Smith, field notes).

FIGURE 3.22 White-lipped tamarin (*Saguinus labiatus*) with guava fruit along the Upper Amazon. (From Alexandre Rodrigues Ferreira, *Viagem filosófica pelas Capitanias do Grão Pará, Rio Negro, Mato Grosso e Cuiabá*, vol. 2, plate 134, Conselho Federal de Cultura, Rio de Janeiro, 1971.)

FIGURE 3.23 Saki (*Pithecia monachus*) with guava fruit in Amazonia. (From Alexandre Rodrigues Ferreira, *Viagem filosófica pelas Capitanias do Grão Pará, Rio Negro, Mato Grosso e Cuiabá*, vol. 2, plate 121, Conselho Federal de Cultura, Rio de Janeiro, 1971.)

FIGURE 3.24 Sun parakeet (*Aratinga solstitialis*) feeding on guava in Amazonia. (From Alexandre Rodrigues Ferreira, *Viagem filosófica pelas Capitanias do Grão Pará, Rio Negro, Mato Grosso e Cuiabá*, vol. 2, plate 100, Conselho Federal de Cultura, Rio de Janeiro, 1971.)

Several of guava's approximately seventy near relatives provide sufficient fruit to attract people as well as dispersal agents. In Brazil, rural folk gather the fruits of *P. incanescens*, while in Central America and Cuba, the fruits of *P. sartorianum* are appreciated for their spicy, subacid flavor.

Close relatives of guava have been domesticated in a few areas because of their unusual fruits. Fragrant strawberry guava (*P. cattleianum*) was first cultivated in coastal Brazil, while the Costa Rican guava (*P. friedrichsthalianum*) was domesticated in the highlands of Costa Rica. The Costa Rican guava, known locally as *cas*, is a common dooryard tree in San José, where the acidic fruits are used to make a popular drink (Landrum, 1991). Other relatives of guava that are occasionally cultivated include *P. guineensis* in tropical America and the West Indies; small-fruited guisaro (*P. molle*) in Central America; mountain guava (*P. montanum*) in the West Indies; pichiché (*P. sartorianum*) in Mexico; *P. multiflorum* and *P. sylvestre* in Brazil (McVaugh, 1956); and ararai (*P. acutangulum*) among the Siona and Secoya of the Ecuadorian Amazon (Vickers and Plowman, 1984).

FIGURE 3.25 Spontaneous guava (*Psidium guajava*) with fruit on a basaltic ridge behind Pointe Raititi, Bora Bora, Society Islands, French Polynesia, August 1989.

Guava's near relatives may be useful for products other than fruit. Extracts from the leaves of *P. acutangulum*, for example, contain a potent fungicide and also deter tobacco budworms (*Heliothis virescens*: Miles et al., 1990). This guava relative, which is native to western Amazonia, also has at least one compound in its leaves that inhibits *Xanthomonas campestris*, a bacterial pathogen that causes black rot on cabbage (Weber, 1973:78).

The status of wild populations of guava's near relatives is poorly known. At the end of the last century, mountain guava (*P. montanum*) was not common in Jamaica, and by now the forest tree is probably extremely rare (Fawcett, 1891). Widespread deforestation in the West Indies has probably greatly reduced other populations of mountain guava, which can reach 30 meters in height. Mountain guava produces an exceptionally hard wood with a striking grain pattern suitable for gunstocks and other specialty uses, so logging pressure has also undoubtedly taken its toll.

BREEDING EFFORTS AND GERMPLASM
COLLECTIONS

Several of guava's near relatives have properties that would lend themselves to wider cultivation or possible widecrossing work with guava to transfer desirable traits. Strawberry guava, known as *araçá da praia* in Brazil, can be grown in open areas or dense shade, breeds true from seed, and tolerates frosts to $-4°C$ (Samson, 1986:274). In India, crossing work has begun on *P. molle*, which is closely related to guava (IIHR, 1980). Research is under way to develop suitable rootstocks for guava cultivars. One objective is to select compatible rootstocks to induce dwarfing so that harvesting is easier. A wild relative of guava, *P. chinensis*, shows promise as a dwarfing rootstock.

Major selection aims for guava cultivars are high and reliable yields, bright pink or red-fleshed fruits with superior taste, reduced seed content, low wastage ratio, and firm fruits that withstand transportation. Mexican scientists have successfully developed guavas with especially high levels of vitamin C (Lakshminarayana and Moreno-Rivera, 1979).

Little breeding for resistance to diseases and pests has been undertaken. Most fungi and insects can be controlled by using disinfected nursery soil, destroying diseased plants, and applying chemicals. Some fifteen fungi attack guava, but only Fusarium wilt is a serious disease on acid soils in India. Fruit rot (anthracnose) attacks guava, but a number of local forms apparently resist the disease (J. P. Singh and Sharma 1981). In Florida, gill fungus (*Clitocybe tabescens*) kills guava trees in some areas (Ruehle, 1959). Canker and dieback affect guava branches, and fruit flies, scale insects, and mealybugs can also cause damage. Wild relatives of guava are known to resist some diseases, such as Fusarium wilt, as well as nematodes, but they have not yet been used in breeding efforts.

At present, only a handful of guava cultivars are propagated clonally. Cultivars are propagated vegetatively using a range of techniques, such as grafting, cuttings, or the promotion of suckers by pruning the roots. Micropropagation using tissue culture techniques could greatly increase the production of guava clones, although some snags still need to be ironed out (Amin and Jaiswal, 1987). Difficulties encountered with rapid multiplication of guava using in vitro techniques include seasonal effects on the establishment of tissue cultures and occasional genetic aberrations.

The wide interest in guava improvement is reflected in the numerous germplasm collections. Guava germplasm is maintained in orchard gene banks, and collections are generally geared to the immediate needs of breeders, rather than to representation of the broad spectrum of guava's gene pool. Typically, fewer than fifty promising local selections are held in guava gene banks. Brazil, for example, has collections of between five and twenty-five guava trees at five locations. The United States maintains only three guava germplasm collections, none of which exceeds forty-one accessions. The guava germplasm collection held at the Tropical Research and Education Center, University of Florida, near Homestead contains several near relatives of guava including *P. aromaticum* (1 accession), *P. cattleianum* (12), *P. coriaceum* (2), *P. cujavillus* (3), *P. cuneifolium* (1), *P. friedrichsthalianum* (3), *P. guineense* (3), and *P. polycarpon*. Cuba has a sizable guava gene bank, and a germplasm collection of over 100 trees is held in India. Guyana's guava improvement program is not backed by any germplasm collection.

Guava seeds can be dried and stored at low temperatures, but breeders generally do not use this cost-effective way of safeguarding germplasm because they want living genotypes for their experiments. Seed gene banks could be used to store large quantities of basic germplasm, including guava's wild relatives.

Seed gene banks would be only part of an overall germplasm conservation strategy for guava. Reserves are still needed for guava and its near relatives. Because of guava's weedy nature, multiple-use reserves that would include some clearing activities would be appropriate.

At present, the guava gene pool is not threatened to any great extent. While some forest populations may have disappeared, plenty of spontaneous guava occurs throughout the tropics. Also, some farmers and urban dwellers still plant guava seedlings, rather than use cuttings. Nevertheless, as breeding programs proceed, some germplasm will inevitably be lost as superior clones become more available. Vegetatively propagated cultivars are already displacing local diversity of guava in parts of Southeast Asia (Sastrapradja, 1975). Breeding programs are well under way in Brazil, Colombia (since 1961), the United States (Puerto Rico, Hawaii, and Florida), Mexico, Côte d'Ivoire, Nigeria, and India.

To assist breeders and germplasm curators in their work, scientists need to do much more surveying of the natural range of guava and its near relatives to provide a basis for systematic sampling. An interesting problem will be trying to separate truly wild guava from spontaneous populations in old second growth. It probably does not matter much, since guava was exploiting light gaps long before people penetrated the New World tens of thousands of years ago.

The density of wild guavas on a swampy plain near Aguasco, Nicaragua, was noted by a naturalist in the last century (Belt, 1888:203). Central America was probably more deforested than it is now when the Spaniards touched the shores of the New World in the late fifteenth century (C. O. Sauer, 1966). As human population levels fluctuated, cycles of forest retreat and expansion engulfed many crop populations and spontaneous seedlings, some of which survived as the forest reclaimed clearings.

Guava, avocado, cacao, and papaya are examples of crops whose once-domesticated populations have reverted back to a "natural" existence in forest environments in tropical America.

Given the bright prospects for expanding guava production in many parts of the tropics and subtropics, some form of coarse grid sampling across diverse ecosystems and subsequent seed storage is warranted. Sampling should be well planned to avoid too much onerous growing-out when seeds in storage begin to lose their viability. Guava is not likely to be high on the priority list of genetic conservation programs in most developing countries, so cost-effective methods of preserving germplasm are essential.

Prospects

Guava thrives in the tropics where rain falls during at least four months and temperatures remain above 16°C. Because of Guava's flexible ecological requirements it can be cultivated commercially as far as 27° N in Florida. Guava has the potential to be planted even more widely, and if genetic resources are tapped effectively, guava groves could produce higher yields in marginal environments. A well-tended guava orchard produces fruit within four to six years and can remain productive for half a century. Refinements in food-processing technologies, particularly at the local level, are likely to spur demand for guava.

PAPAYA

Papaya (*Carica papaya*) is a short-lived tropical plant that produces thin-skinned, easily digested fruits all year round. Also known as pawpaw or the melon tree, the ripe yellow- to orange-fleshed fruits are relished as a breakfast or dessert fruit and are a good source of vitamin C. Red-fleshed papayas are also rich in vitamin A, which is often deficient in the diets of people living in the tropics (Ochse et al., 1961:594). The soft fruits are also canned, crystallized, or blended to make nutritious juices or milkshakes. In some countries, immature fruits are stewed in cane juice to make a sauce or are boiled as a vegetable. In Java, a sweetmeat is prepared from the numerous flowers (Purseglove, 1974:45).

A meat tenderizer, papain, is extracted from papaya, particularly from unripe fruits. Papain is a proteolytic enzyme obtained from the dried latex of young papaya trees and is a significant export item for Tanzania,

Uganda, Zaire, and Sri Lanka (Becker, 1958). Latex is normally harvested by gashing immature fruits, but recent extraction methods use papaya leaves, thereby leaving unblemished fruits for marketing.

Papain is used to treat slipped disks and can prevent the need for painful and expensive back surgery. It is also employed to tan leather, prevent shrinkage of wool, and keep beer clear during the brewing process (Cobley, 1956:285). Papain's digestive properties are well known to the indigenous peoples of the American tropics.

Several folk remedies are prepared from papaya, both for humans and livestock. Papaya has been incorporated into the pharmacopoeia of rural folk in tropical America as well as introduced regions, such as Africa, Southeast Asia, and the Pacific. The seeds are eaten as a vermifuge and are apparently effective against intestinal helminths except hookworm. In Southeast Asia, papaya is used to stimulate the production of mother's milk, and the milky sap is applied to the body to help remove thorns. The fresh, tablemat-sized leaves of papaya and damaged or spoiled fruit with seeds are fed to pigs along the Transamazon Highway to fatten and tenderize them and to lower their worm loads. Indians tenderize tough meat by wrapping it in papaya leaves and boiling it. The leaves contain a glycoside, carposide, and the alkaloid carpaine. In Tahiti, a decoction is made from the male flowers to treat bronchitis (Pétard, 1986:239). The Ashanti in Ghana use papaya to treat gonorrhea and stomachache, and the powdered roots are eaten to relieve headaches (F. R. Irvine, 1961:87).

Papaya is a perennial herb that because of its height, which can reach 10 meters, looks akin to a soft-wooded tree (Figure 3.26). Plants can be either male, female, or hermaphrodite; the latter usually produce the most desirable fruit shapes. Pollination is probably by wind and small flies. Only when papayas begin to flower is the cultivator able to determine the sex of the plants and remove some of the male plants.

To avoid unwanted males, growers have selected a number of hermaphrodite and dioecious cultivars. Half of the seedlings of hermaphrodite cultivars are female, and the remainder are hermaphrodite if a hermaphrodite is used to cross with a female. If the hermaphrodite is self-pollinated or crossed to another hermaphrodite, it will give a 2:1 ratio of hermaphrodite to female plants. On commercial plantations in Hawaii, female trees are culled because the market prefers pyriform fruits produced by hermaphrodite plants rather than the female's rounded fruits. Most of the males of dioecious varieties are cut down.

Papaya has traditionally been propagated by seeds from selected fruits. Commercial cultivars, such as 'Sunrise', 'Kapoho', 'Waimanalo', and 'Sunset', are highly inbred hermaphrodite lines, and seeds collected from selfing of hermaphrodite plants breed true if flowers are protected from cross-pollination with other papayas.

FIGURE 3.26 Papaya (*Carica papaya*) interplanted with pineapple, Luzon, Philippines, 1986.

ORIGIN AND SPREAD

Of the twenty-one species of *Carica*, at least six have been domesticated. Tropical *C. papaya* is the most widely grown, but mountain papaya (*C. pubescens*), native to the high Andes near the tree limit, is cultivated throughout the Andes (NRC, 1989:254). Known locally as *chamburo*, mountain papaya produces small, golden fruits that are cooked before they are eaten (Goodspeed, n.d.:139). Mountain papaya is not widely cultivated outside its indigenous range, but it is grown in the Nilgiri Hills of South India (Ram et al., 1985). Babaco (*C. pentagona*) is grown in the Andean valleys of Ecuador for its fruits, which, like those of the mountain papaya, are cooked (Morton, 1987:346). Babaco may be a natural hybrid between *C. stipulata*, cultivated in southern Ecuador, and *C. pubescens* (Litz, 1986). Babaco is now cultivated in New Zealand, Australia, California, Colombia, and the Mediterranean. Other species cultivated on a minor scale include *C. monoica* and *C. goudotiana*. In Central America, the hard trunk of *C. dolichaula* is cut into sections in which grain is stored (Record and Hess, 1986:118).

Like the origin of many crops from lowland tropical forests, that of tropical papaya is obscure. Many useful plants were not depicted in ancient iconographies, and for many no archaeological record exists. The writings of early European explorers, conquerors, and missionaries can nevertheless be used to piece together the distribution, and in some cases, areas of high diversity of crop plants. Such early written accounts offer evidence that papaya was widely distributed in tropical America by 1500.

The fast-growing species was probably originally native to an area extending along the eastern flanks of the Andes from Ecuador to Venezuela (Brücher, 1989:224; Prance, 1984), although Central America is also proposed as the ancestral home for papaya (Purseglove, 1974:45; Storey et al., 1987). Over a century ago, a Brazilian and an English naturalist noted:

> Where Papayaceae most abound is on the wooded slopes of the Andes, both on the eastern and western sides, up to 8,000 feet elevation; and it is there that travellers and sedentary botanists may confidently expect to find not only materials for the more perfect elucidation of the species already partially known, but also many new species, which doubtless still remain hidden in the savage recesses of the oriental Andes. (Correa de Mello and Spruce, 1869)

Papaya probably had a hybrid origin involving several *Carica* species, such as *cauliflora, goudotiana, microcarpa, monoica,* and *pubescens* (Badillo, 1971).

By the time European colonists arrived, papaya had spread across Latin America, and many different forms, based on the shape and flavor of fruits, were recognized. In the vicinity of Orizaba, Mexico, papayas reach the proportions of a watermelon (Figure 3.27). Large papayas generally do not travel far because they are especially susceptible to damage. Although some regions may be known for certain papaya types, considerable local variation is often found because backyard papayas are all seedlings that do not usually breed true. Papayas sold in the public market of Itabuna, Bahia, Brazil, display a range of shapes from globular to cylindrical (Figure 3.28).

Papaya has also diversified on new continents and islands. In the Society Islands, a thick, cucumber-shaped variety with deep-orange skin, known locally as 'Carot', sells for almost $2 each (Figure 3.29). Tongans distinguish three main varieties of papaya, one with intriguing pink flesh (Yuncker, 1959).

The Spaniards coined the name *papaya* from the Carib *ababay*. The Spanish took papaya to the Philippines in the mid-sixteenth century and from there the quickly appreciated crop spread westward to South Asia and East Africa. Papaya reached Zanzibar by at least the eighteenth century and was reported in Uganda in 1874 (Purseglove, 1974:46). An ex-

FIGURE 3.27 Papaya (*Carica papaya*) fruit in the outskirts of Orizaba, Veracruz, Mexico, May 1988.

FIGURE 3.28 Papaya (*Carica papaya*) of assorted sizes and shapes on sale in Itabuna, Bahia, Brazil, January 1990.

FIGURE 3.29 'Carot' variety of papaya (*Carica papaya*), Bora Bora, Society Islands, French Polynesia, August 1989.

ample of papaya's esteem in introduced areas is the 80-franc stamp of Rwanda, which depicts a papaya tree, fruit, and female flower.

Papaya's popularity is also attested to by the hundreds of vernacular names ascribed to the succulent fruit; papaya is known by over a hundred names in Indonesia alone (Ochse and van den Brink, 1931). In the New World, papaya is also known by various names, including *melon zapote* in Mexico, *fruta bomba* in Cuba, *lechosa* (milky) in Puerto Rico, and *mamão* (big breast) in Brazil.

WILD POPULATIONS AND NEAR RELATIVES

Like guava (*Psidium guajava*), papaya is a "weed" by nature, but, unlike guava, it does not thrive for long in advanced second growth. The original populations of papaya no doubt exploited light gaps in the forest and along the margins of rivers and lakes. Papaya is a common volunteer plant, sprouting readily from seeds dropped by birds and discarded in kitchen wastes. Since domestication, papaya has readily become feral on sites opened by human activities, such as along roads and in abandoned fields. The gene pool of wild papaya has thus considerably expanded under human agency.

Papaya is spontaneous in many areas of the humid tropics and subtropics, particularly on more fertile soils. Papaya has become naturalized in the richer soils of Amazonia, such as the alfisols derived from weathered basalt near Altamira, Pará. Papaya also occurs as an escapee on rich volcanic soils of Colombia, St. Croix, St. Thomas, Tahiti, and Tonga (Britton, 1918; Britton and Wilson, 1924:605; Pérez-Arbeláez, 1956:259; Pétard, 1986:237; Yuncker, 1959:193). Limestone and sandy soils of Guam, Samoa, and the Florida Keys also support spontaneous papaya (Fosberg, 1960; Mayo, 1938:17; Parham, 1972:22). Other countries where papaya occurs as a welcome weed include the Golfo Dulce area of Costa Rica, Ghana, the Guianas, India, and Panama (P. H. Allen, 1956:154; F. R. Irvine, 1961:87; Standley, 1927:25; Van Roosmalen, 1985:70; Watt, 1908: 269).

Birds are the principal dispersal agents for spontaneous papaya. Papaya fruits contain hundreds of slick, pellet-sized seeds that are easily swallowed. Palm tanagers (*Thraupis palmarum*) and blue-gray tanagers (*T. episcopus*) were seen feeding on papaya in agrovila Nova Fronteira at kilometer 80 of the Altamira-Itaituba stretch of the Transamazon Highway in September 1973. These attractive garden and second-growth birds ingest seeds of papaya and other plants and disperse them. Also, papaya seeds may stick temporarily to their bills and feathers.

Although only six species of *Carica* have been domesticated, many of the wild species contain desirable characteristics that could be useful in breeding. Papaya has already been hybridized experimentally with five wild species of *Carica*. Scientists at the University of Hawaii, who developed the popular 'Sunrise Solo' variety of papaya, are examining the attributes of wild *Carica* species, particularly for disease resistance. Research is also being conducted using molecular analysis to understand more about the relationships between species of *Carica* and their variations. Fruits are gathered from at least seventeen wild species of *Carica* (Table 3.6).

BREEDING EFFORTS AND CHALLENGES

Papaya breeding started relatively early in this century, first in Hawaii in 1911 and then in South Africa in 1931. India started work on papaya crossing and selection in the 1950s; Colombia initiated a papaya breeding program at Palmira in 1963, and Trinidad and Tobago followed suit in 1965 (Morton, 1987:338). Côte d'Ivoire started papaya breeding in the 1970s. More recently, Mexico, Brazil, Taiwan, and the Philippines have

TABLE 3.6

Some wild or little-cultivated species of *Carica* whose fruits are harvested

Carica species	Local name(s)	Location
candicans[1]	*Mito*	Peru
cauliflora[4]	*Papaita*	Central America, Venezuela
cestriflora[1]	*Papaya de tierra fria*	Colombia
chilensis	*Palo gordo*	Chile
chysopetala[2]	*Chamburo, higacho*	Ecuador
chrysophylla[1]	*Chihualcan, higacho*	S.E. Colombia, Ecuador
digitata	*Mamão*	Brazil
erythrocarpa[3]	Not known	Brazil
frutifragrans	*Chamburo*	Colombia, Ecuador
goudotiana	*Tapaculo, papayuela*	Colombia
monoica[5]	*Col de monte, peladera, peladua, papaya de selva*	Andes
microcarpa[7]	*Tapaculco, lechosa de monte*	Venezuela, Colombia, Ecuador
peltata	*Papaya de mica*	Tropical America
pentagona[1]	*Babaco*	Ecuador
pubescens[3,6]	*Chamburo, papaya de olor*	Colombia, Ecuador
quercifolia[3]	*Figuera del monte, mamãozinho*	Tropical S. America
stipulata	*Sigalón, paronchi*	Ecuador

Sources: Badillo, 1971; Litz, 1986; Martin et al., 1987
[1]Eaten mainly in preserves.
[2]Fruits cooked.
[3]Eaten mainly in preserves or candied.
[4]Used as a vermifuge in Venezuela (Pittier, 1926:330).
[5]Cultivated in Ecuador and Peru.
[6]Cultivated in Peru and in northern Chile.
[7]Cultivated in Darien, Panama.

invested resources in attempting to raise the yield, quality, and disease resistance of papaya.

Research on papaya has been geared to increasing productivity of orchards in the hands of market-oriented farmers. Considerable progress has been made, but much basic work remains to be done. Genetic factors responsible for sex, earliness, and height at first fruiting have been elucidated, but the genetics of fruit color and the inheritance of fruit shape, size, and flavor are imperfectly understood. The latter traits are under multiple gene control, thus complicating breeding efforts.

Breeders have focused on developing improved types that can be propagated by seed. Some research has been carried out on grafting papaya on various rootstocks (Theakston, 1976). Also, the South African 'Honey Dew' cultivar is propagated by rooted cuttings.

'Solo', one of the most successful commercial varieties, was introduced to Hawaii in 1911 and selected there in 1919 (Morton, 1987:337). 'Solo' led to a series of superior cultivars, including 'Sunrise Solo' (Figure 3.30).

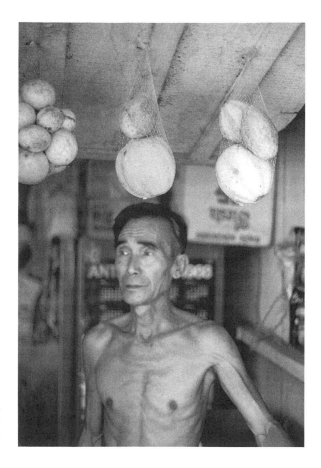

FIGURE 3.30   Japanese-Brazilian vendor with 'Sunrise Solo' papaya (*Carica papaya*), Rio Preto da Eva, Amazonas, Brazil, August 1988.

'Sunrise Solo' is an important cash crop in several parts of Brazil, such as in Bahia, near Manaus, and in the Bragantina zone east of Belém.

Pyriform 'Solo' fruit and its derivations are ideal for a single serving at breakfast or for dessert. Furthermore, their compact shape renders them less prone to damage during transportation and handling than larger varieties. Growers in Hawaii plant two or three seedlings in each hole to ensure a complete stand of hermaphrodite plants. Pear-shaped 'Sunrise Solo' is sweet, with firm, red-orange flesh. Although 'Sunrise Solo' thrives in hot, humid climates, the hermaphrodite flowers become male when plants are stressed.

Breeders tailor their objectives according to the intended market. In Hawaii, small fruits weighing less than 400 grams are exported, whereas the local market prefers fruits in the 500–800-gram range. Even larger papaya fruits, attaining up to 3 kilograms, are esteemed in some tropical countries. In some papaya improvement programs, enhanced papain production is

the major goal, while in others, higher-quality fruit is the main aim (Giacometti, 1987).

Papaya is attacked by several viruses and viruslike diseases, and commercial cultivars are generally highly susceptible to the pathogens. Papaya
mosaic and ringspot viruses are widespread and highly damaging. Viruslike diseases include bunchy top and decline. Ringspot, for example,
has limited commercial papaya production in the Caribbean, tropical
America, and Florida. Control measures include spraying insecticides to
reduce aphids that transmit the viruses and weeding out alternative hosts
of the pathogenic viruses. Sometimes, though, commercial papaya production comes to a halt because of viral diseases, as it did in Cuba (Roig and
Mesa, 1962). In a large country, such as Brazil, growers shift commercial
papaya production from one location to another in an attempt to stay one
step ahead of viral diseases (P. Alvim, pers. comm.).

To combat ringspot, the Tropical Research and Education Center of the
University of Florida developed 'Cariflora' in 1986. Derived largely from
dioecious lines obtained from a commercial grove, 'Cariflora' can produce
yields of 35 metric tons per hectare per year and is recommended for Florida and lowland areas of the Caribbean (Conover et al., 1986). India also
intends to plant the spherical-fruited 'Cariflora' (Chadha and Pareek,
1989). However, 'Cariflora' does not hold up well if papaya mosaic virus
or apical necrosis virus is present or if ringspot virus pressure is severe
(Campbell, 1986).

Monogenic resistance to papaya ringspot virus has been located in *C.
pubescens, C. stipulata,* and *C. cauliflora,* but none of these wild species
can be crossed with papaya by conventional breeding methods (Litz,
1986). In vitro techniques are now being used to surmount barriers to
widecrossing (Tsay and Su, 1985). Cell suspension cultures can readily
produce somatic embryos in papaya and related species (Litz and Conover,
1979; Vega and Kitto, 1988).

Other techniques, such as embryo rescue, inducing embryogenesis in immature zygotes, and microprojectile bombardment could be useful in incorporating exotic germplasm into papaya cultivars (Fitch and Manshardt,
1990; Manshardt and Wenslaff, 1989). John Sanford at the New York
State Experiment Station, Geneva, has recently genetically transformed papaya callus (R. Litz, pers. comm.), and DNA has successfully been introduced to papaya tissue cultures on high-velocity tungsten projectiles (Fitch
et al., 1990). Although the genes thus introduced where chimeric and only
for experimental purposes, the ultimate goal of this genetic engineering
work is to develop virus-resistant papayas.

Papaya can be multiplied rapidly in vitro (Litz and Conover, 1982; Litz
et al., 1985). Scientists at the Indian Agricultural Research Institute in
New Delhi are exploring the potential for rapid micropropagation of pa-

paya (Clemente, 1987). While progress has been made with micropropagation of papaya, some difficulties remain with establishing plants in the field after they come out of the laboratory. Various hardening techniques are being employed on papaya tissue cultures to boost their survival rate in the field.

Although papaya's close relatives undoubtedly contain sources of disease resistance, it will be easier in the short term to tap the heterogeneous but widely scattered gene pool of papaya in backyards and among spontaneous populations. Still, wild material may eventually prove to be the only source of polygenic resistance for some viruses and fungi.

Over fifty fungi attack papaya. Major fungal diseases include root rot, caused by *Phytophthora palmivora* in Hawaii and *P. parastica* in Brazil; root and trunk rot due to fungal infections by species of *Pythium*; and anthracnose, caused by *Colletotrichum gloeosporioides*, which spoils ripe and stored fruits by causing unsightly spots. Anthracnose can be reduced by dousing the fruit in hot water after it is harvested. Fungal diseases are normally controlled through rigorous orchard hygiene, chemical spraying, and crop rotation.

Several near relatives of papaya, *C. monoica*, *C. goudotiana*, and *C. cauliflora*, have been found to resist foliar diseases caused by fungi (*Acrosporium caricae*, *Asperisporium caricae*, and *Cercospora caricae*). In addition, *C. quercifolia* resists *Leveillula taurica* and *Ascochyta caricae*, both pathogenic fungi (Ullasa et al., 1983). Genetic resistance to fungi and other diseases would reduce operational costs and enable growers to keep orchards in production longer. The economic life of a commercial papaya plantation is only two to three years, compared with five or six years for isolated, backyard papayas.

Genetic resistance to insects and mites would also be desirable, especially since some insects are involved in disease transmission. Insecticides are currently the only control measures used to reduce pest damage in papaya orchards. Nematodes are checked by treating the soil with chemicals and by crop rotation.

Short stature, without losing productivity, would also be a desirable trait. Papayas in commercial groves are generally replaced after two to four years when they become too tall for easy harvesting. Shorter, high-yielding cultivars that can produce for two or three more years would reduce operational costs. Pest and disease resistance will become even more critical if papaya plants are to be kept in commercial production for longer periods.

The gap between potential papaya yield and actual productivity of commercial groves is vast. Papaya could annually produce as much as 100 metric tons per hectare, but plantations in the vicinity of Hilo, Hawaii, average only 38 metric tons per hectare per year (Morton, 1987:342; Samson, 1986:267). While it seems unlikely that yields of 100 metric tons per hectare per year will ever be achieved under farmers' conditions, scope exists for further raising the yield of papaya.

Germplasm Collections

Papaya germplasm can be conserved as seed as well as maintained as plants in field gene banks. Papaya breeders have assembled relatively small collections of *Carica* germplasm to further their work. Several collections have over 100 accessions, such as in India, Nigeria, Hawaii, and the Philippines, while Colombia and Mexico maintain collections with over 50 accessions. Smaller collections are held in Costa Rica, Cuba, Ecuador, Malaysia, Peru, and the Seychelles. All papaya collections have been built up in a haphazard manner, without any well-defined conservation strategy. Many collections have relied on plant introductions, rather than a systematic attempt to garner a wide range of local papayas. A recently published descriptor list for papaya will help curators characterize their germplasm accessions (IBPGR, 1987).

Existing papaya collections contain a reasonable amount of variation, but few of the wild *Carica* species have been collected or maintained in gene banks. Also, only a few samples of each wild *Carica* species are typically held in germplasm collections. Gene banks thus contain only a small fraction of the gene pool of the highly variable wild species.

To help ensure the future of papaya production worldwide, at least some collections should be built up from a coarse grid sampling of the papaya gene pool for ecogeographic variants. Such sampling should concentrate on Latin America, particularly in papaya's area of origin. Spontaneous populations in Africa, Asia, and the Pacific also warrant some collecting for unusual gene combinations. In addition, more wild species need to be collected and their natural history studied.

Papaya germplasm can be stored as seed, but regenerating the seed is expensive. Papaya seeds may not remain viable for decades under frozen conditions as do wheat and rice. When papaya accessions are regenerated, great care is needed to prevent pollen contamination from other samples, since the species is an outcrosser. Hand-pollination to maintain the genetic integrity of each accession drives up the cost of germplasm storage of papaya.

For the most part, no urgent "rescue" missions are needed to safeguard disappearing papaya germplasm in the wild. Collecting missions can be planned without the immediate threat of losing vast sections of the gene pool. Genetic erosion of papaya is not particularly serious, even in areas where commercial production is well under way. Papaya orchards still account for only a small proportion of papaya production worldwide. Backyard papaya trees vastly outnumber commercial varieties growing in orchards. Eventually, increased commercialization of papaya in such countries as the Philippines, Indonesia, India, Nigeria, and Côte d'Ivoire will mean the loss of locally adapted forms.

Genetic erosion of papaya's near relatives is also not worrisome at present because they exploit disturbed sites. As in the case of guava and other sun-loving species, however, not all human-induced changes in the landscape will favor fast-growing, successional species. Reservoirs and herbicide-treated pastures, for example, are not conducive to spontaneous papaya or its near relatives.

FUTURE PROSPECTS

Papaya is grown on a large scale in over thirty countries, and production levels have been steadily increasing since the early 1970s. Brazil and Mexico are major producers of papaya in the New World, while India, Indonesia, and Zaire are significant producers in the Old World. Much of the production is for local consumption, but demand for certain cultivars that travel well is growing in temperate markets. In London, for example, 'Sunrise Solo' papayas from Brazil were selling for $3.20 each at Harrods and $4 each at Fortnum and Masons in December 1989. In Florida, 'Sunrise Solo' flown from Hawaii sold for $1.19 each in November 1989.

Whereas market prospects are bright for papaya fruits, biotechnology may undercut the papaya plantations geared for papain production. Pharmaceutical companies are keen to clone the gene for papain in laboratories. If this feat is accomplished, most papain production would shift from developing countries to industrial countries. Such a move would be detrimental to the economies of several Third World countries, such as Zaire and Tanzania.

Nevertheless, papaya is a versatile plant, well- suited as a cash crop and for the nutritional needs of the rural and urban poor. Papaya thrives in small urban lots as well as in village settings. Like banana and guava, papaya provides abundant and relatively cheap fruit. Some of the commercial cultivars might do well in dooryard gardens, but traditional seedling papayas will probably continue to dominate production for local consumption. Papaya tolerates light shade and thus is well suited for some agroforestry systems. Demand is likely to grow for a range of more disease-resistant and highly productive cultivars suitable for commercial orchards and backyards.

SAPODILLA

Like papaya, guava, and pineapple, sapodilla (*Manilkara achras*) is a Neotropical fruit that has become widely planted in the warmer regions of the world. Sapodilla is a medium-sized evergreen tree native to the tropical forests of Central America, including Mexico and northern South America. More fruits are produced in areas where sapodilla has been introduced, such as the Philippines and India (Figure 3.31), than in its native range. Malaysia grows even more sapodilla than Venezuela, where the rich-tasting fruit is native and highly regarded. Other significant sapodilla-producing countries include Brazil, Cuba, Jamaica, Surinam, Netherlands Antilles, and Puerto Rico in the New World, and Indonesia, Cambodia, Sri Lanka, and Thailand in the Old World. The Spanish were instrumental in introducing sapodilla to Southeast Asia; after Spanish ships took sapodilla and silver from Mexico to Manila, the fruit spread to Indochina and other parts of Southeast Asia.

Technically, sapodilla fruits are berries. The tennis-ball-sized fruits are covered with rough, brown skin and contain from six to twelve seeds. The soft flesh somewhat resembles pear, but is creamier. Sapodilla is especially succulent because small, juice-filled veins thread throughout the pulp. Sapodilla is exceptionally sweet and a dessert fruit *par excellence*.

In addition to producing a fine fruit, sapodilla yields a latex (chicle), suitable for making chewing gum. Gum is obtained by boiling the milky latex that flows freely when the bark is gashed. The Aztecs enjoyed this masticatory, and sapodilla trees have been tapped for the export trade, although demand for chicle has slackened with the advent of synthetic gums (Dupaigne, 1979). Sapodilla's durable wood was employed in the construction of Mayan temples and is still used in buildings (Purseglove, 1974).

ORIGINS AND EARLY SPREAD

Sapodilla is thought to have first been cultivated in Mexico from the Isthmus of Tehuantepec south to the border with Guatemala, but the tree occurs wild as far south as Venezuela and was probably brought into cultivation various times in different parts of its extensive range. As in the case of avocado and banana, for example, highly useful species with wide distributions are likely to have caught the eye of early farmers and thereby presented themselves for domestication.

Efforts to understand the origins of sapodilla and other fruit trees of Central America are stymied by the paucity of archaeological data. Much of our current understanding of plant domestication in the region is based on archaeological finds in dry or semiarid zones. Sapodilla has been recorded along with maize, peppers, squashes, and beans in the archaeological strata of Tehuacán in Mexico, which span a period from 7000 to 1800 B.C. (MacNeish, 1967). Whether the sapodilla seeds found at the Tehuacán site are from wild or cultivated trees cannot be deduced.

FIGURE 3.31 Sapodilla (*Manilkara achras*) fruits for sale in a street market in New Delhi, India, December 1989.

Lowland Central America, particularly in Mexico, Guatemala, Belize, and Honduras, has witnessed a succession of impressive civilizations ranging from Olmecs (1500–100 B.C.) to the Classic Maya, whose empire collapsed around A.D. 1000. With the waxing and waning of major civilizations in this region, some plants would likely have been brought into cultivation, and then some of them, such as sapodilla and avocado, would have reverted to the wild as the areas around ceremonial cities became temporarily depopulated. It is a safe assumption that virtually all of the forests of Central America have been cleared at some point in the past. The current forest cover of the region represents a protracted recovery from precontact times when vast areas were cleared (D'Arcy, 1977).

Some present-day cultivars of perennial tree crops have thus been "brought back" from the wild, in some cases perhaps several times. Not all varieties of tropical fruit trees are thus necessarily linear descendants extending back to first domesticates.

Known as *chico sapote* in Guatemala, the cultivated fruit tree was taken to the West Indies in pre- Columbian times. Sapodilla probably penetrated the Caribbean from northern South America, where wild populations of the fruit occur and where it may also have been domesticated. Sapodilla was most likely taken by one or more migrating waves of people, such as the Saladoid culture that spread from the Orinoco through the Guianas and into the lesser Antilles between 900 B.C. and A.D. 600.

NEAR RELATIVES

Sapodilla has close to seventy near relatives, a large extended gene pool. None of the other species of *Manilkara* is cultivated, but several of them provide useful products. Three wild relatives are closely related: *M. sieben* from Trinidad, *M. spectabilis* (Costa Rica), and *M. bidentata* (South America).The latter species is a source of balata, a nonelastic rubber that has been used in the past for boot soles and machine belts; the fruits are also eaten, and the tree is a source of chicle. *M. sieben* was once a source of balata, as was *M. duriensis*, another Panamanian species. A number of species yield hard, heavy wood, such as *M. spectabilis*, which has been used for railroad cross-ties in Costa Rica. Several of sapodilla's near relatives produce fruits that are gathered in the wild, such as *M. balata* in Brazil and the Guianas, *M. bidentata*, *M. coriacea*, and *M. huberi* in Brazil, and *M. sieben* in Trinidad (Martin et al., 1987).

Synthetic rubber and natural rubber from *Hevea brasiliensis* have reduced the importance of sapodilla's near relatives as sources of balata, but at least some species of *Manilkara* could ultimately prove useful for upgrading sapodilla. Near relatives of sapodilla likely contain genes for resistance to diseases and pests that might prove valuable as sapodilla is grown in larger groves.

Breeding Efforts

Sapodilla improvement is at a very early stage and has thus far relied on selection of seedlings for vegetative propagation. Criteria for selection include precocity (some sapodillas fruit three years after planting), high and consistent yields, and superior fruit quality. Scientists based at the University of Florida's agricultural research station at Homestead and the USDA's plant introduction facility at Chapman Field, Miami, have screened and selected sapodillas for South Florida growers over the last fifty years (Campbell and Malo, 1973).

Related wild species have not been needed for breeding thus far, although some of sapodilla's near relatives have been tested as potential rootstocks. One of the more promising species that could be suitable as a rootstock for sapodilla is *M. kanki* from Southeast Asia, Indochina, and northern Australia. More distant relatives in the Sapotaceae, particularly *Madhuca latifolia* from northern India and several species of *Sideroxylon*, might also serve as hardy material upon which to graft superior forms of sapodilla.

Local populations of sapodilla are highly variable, a property that helps the species to adapt to a wide range of soil conditions. Sapodilla is still cultivated by planting seed in many parts of tropical America, so no large-scale erosion of the domesticated gene pool is under way at present. As more popular selections are cloned, though, sapodilla plantings are likely to become more uniform, as is the case in much of tropical Asia.

Wild sapodilla should be targeted for collection in view of the massive spasm of deforestation under way in Central America. After the indigenous population crashed in the sixteenth and seventeenth century following the rapid spread of introduced diseases, the forest retook a relatively open landscape and "swallowed up" numerous orchards and backyard trees. Some of these reverted populations of sapodilla need to be collected before they fall victim to the axe or chainsaw. It will be difficult to differentiate populations descended from these ancient orchards from trees that trace their origin to truly wild stock. Gene flow is likely to have occurred between old cultivated and wild forms. From the genetic resource viewpoint, however, such fine distinctions are probably not particularly important.

Conservation Needs and Prospects

Areas where sapodilla still occurs in the wild need to be identified to help justify the establishment of parks and reserves. Villagers could help in ecogeographic surveys by identifying where chicle is still gathered in the forest for local use. Although wild populations have not yet been used in breeding, they might prove valuable for rootstocks or for sources of resistance to diseases. At least a sample of the variability of wild populations of sapodilla are thus needed in gene bank collections. Whenever sapodilla is grown on a large scale, the incidence of diseases and pests increases. In India, for example, leafspot disease caused by *Phaeophleospora indica* can cause serious loss of fruit (IBPGR, 1986:127).

Seeds of sapodilla are recalcitrant, so they cannot be stored in seed gene banks. Field gene banks have neither sufficient space nor labor to grow the progeny of many diverse populations collected from different regions and environments. Criteria for incorporation of material into gene banks will thus have to be highly selective; a major goal should be to obtain samples from a range of ecosystems. Several countries maintain modest germplasm collections of sapodilla; the largest collections are in Costa Rica with 118 samples and the Philippines, where 77 accessions are kept. None of these collections is adequate for conservation purposes.

As tissue culture research advances and cryopreservation techniques are perfected, the size of field gene banks is likely to shrink. Curators and national agricultural research programs need to consider such changes on technological frontiers when devising plans for collecting and strategies for conserving crop germplasm. With careful planning, tissue culture germplasm collections and field gene banks should prove a valuable complement to in situ conservation. It will ultimately be cheaper to safeguard sapodilla germplasm in forests, rather than try to place even a small portion of the wild variation of the species in test tubes or in extensive field gene banks, which few institutions can afford to maintain. If sapodilla breeding intensifies in the future, such wild populations in forests will be invaluable sources of germplasm for improvement programs and gene banks.

Farmers are interested in new selections of sapodilla, so vegetative propagation is expected to become more common at the expense of the more traditional method of sowing seeds in backyards. Cloning is particularly common where sapodilla has been introduced, such as India, where inarching and marcottage are practiced (Samson, 1986:303).

PASSIONFRUIT

Various species of *Passiflora*, climbing vines native to the American tropics, have been domesticated to eat as fresh fruit or to make refreshing juices. Yellow passionfruit (*P. edulis* var. *flavicarpa*) is the most widely planted species in the warm, humid tropics. This perennial climber probably arose as a mutant of the purple form (var. *edulis*) somewhere in a vast area from Brazil to northern Argentina. The yellow gelatinous pulp of yellow passionfruit is mixed with water and sugar to make drinks, sherbet, ice cream, jams, jellies, and salad dressing.[9]

Passionfruit juice is popular in many tropical countries and has been available in many supermarkets in North America for a few decades. Apart from its distinctive flavor, which is gaining converts in industrial countries, passionfruit juice contains some protein, fat, and carbohydrates, and vitamins A and C. Growing consumer preference for natural fruit drinks, rather than carbonated colas, has spurred interest in developing more productive varieties of passionfruit.

Durable passionfruit has the advantage of traveling well and can be stored for up to a month, although the thin-skinned fruits become increasingly wrinkled after harvest (Figure 3.32). The attractive purple form (var. *edulis*) is grown in cooler parts of the tropics, such as in highland Kenya and Australia, but is lower yielding than the yellow form. Purple passionfruit has an appealing, lustrous skin, and along with kiwi and grapes, provides a cool color for fruit salads. Purple passionfruit's intriguing appearance has gained it a place in restaurants in such cosmopolitan cities as London and Amsterdam. In December 1989, purple passionfruit from Kenya was selling in London markets at the equivalent of 50 cents apiece.

The yellow passionfruit grows on a variety of soils, provided drainage is adequate. The billiard-ball-sized fruits contain dark seeds that can produce mature plants within a year. Passionfruit vines are trained along fences or on trellises (Figure 3.33) and can yield good crops for up to a decade. In commercial plantings, the vines are usually replaced after about five or six years. Passionfruit is a versatile component of some agroforestry systems because it provides shade for tree seedlings and, in turn, tolerates partial shade.

CULTIVATED AND WILD RELATIVES

Of the approximately 400 species of *Passiflora*, some 40 produce sizable fruit, 24 are harvested, and 11 are cultivated. Of the 11 cultivated species, 5 attain regional significance (NRC, 1989:287). Lowland giant granadilla (*P. quadrangularis*) has fruit as long as 30 centimeters and has become naturalized in several regions after introduction, such as in Southeast Asia and Australia. Giant granadilla produces edible tubers and grows wild in Nicaragua (Belt, 1888:69). Orange to ochre-colored granadilla (*P. ligularis*) sells briskly in markets in the highlands of Central America and in the Andes, where it is eaten fresh (Figures 3.34, 3.35). Granadilla is also a

---

9. A full-page British Airways advertisement in a recent edition of the *Economist* features a first-class menu with passionfruit and almond dressing as an option for one of the salads. Swissair serves passionfruit sherbet between courses in first class on some of its flights.

FIGURE 3.32 Yellow passionfruit (*Passiflora edulis* var. *flavicarpa*), known locally as *maracujá*, on left, and granadilla (*P. ligularis*), on right, for sale in Lima, Peru, July 1988.

cash crop in highland Kenya, and the colorful fruits adorn fruit bowls in Nairobi hotels. Deep-orange granadilla juice resembles yellow mombim (*Spondias mombim*) and is relished in Kenya, among other countries. In Peru, *P. ligularis* is fed to infants to aid their digestion. Water lemon (*P. laurifolia*) is cultivated in the lowlands of northern South America and is found wild or spontaneous after cultivation on Ilha de Maracá, Roraima, in northern Amazonia (W. Milliken, pers. comm.). The banana passionfruit, or *curuba* (*P. mollissima*), is native to the Andes, where it was domesticated in preconquest times. The sweet calabash, or *chulupa* (*P. maliformis*), is found in parts of the Caribbean and northern South America and is cultivated in upland Jamaica, Brazil, and Ecuador (Morton, 1987:334). Sweet calabash is probably wild in the Andes.

With further selection and breeding, all of the eleven cultivated species of *Passiflora* could be more widely grown. Banana passionfruit and the little-known curubejo (*P. popenovii*) of the upper Amazon in Ecuador, are particularly promising candidates for breaking into new markets (NRC, 1989:290).

Wild species of *Passiflora* have been little studied. The large number of

FIGURE 3.33 Farmer with yellow passionfruit (*Passiflora edulis* var. *flavicarpa*) trained on a trellis on Careiro Island, near Manaus, Amazonas, Brazil, in 1972.

passionfruit's near relatives grow in many habitats and therefore contain an assortment of adaptations that could prove useful for breeding. On Guam and the Marianas Islands, for example, *P. foetida* must be drought-resistant because it thrives on extremely thin soils resting on porous limestone on Guam and on shell ridges and sand banks in the Guianas (Fosberg, 1960; Van Roosmalen, 1985:356). Native to tropical America, *P. foetida* appears to grow in a range of environments, such as on mounds and along the brackish coast of Marajó Island at the mouth of the Amazon (Huber, 1898). The wide-ranging species has been naturalized on Java for close to a century, since it was introduced to the Bogor Botanic Garden (then Buitenzorg) and subsequently dispersed by birds (Backer, 1910). The species is a common weed in Hawaiian sugarcane fields.

Several wild passionfruit species are gathered in the wild and might prove suitable for domestication or crossing with the cultivated forms. Near Manaus in the Brazilian Amazon, rural folk gather *maracujá suspiro* (*P. nitida*) from second growth in abandoned fields and along the margins of rivers and roads during the rainy season. The highly prized fruit of *P.*

FIGURE 3.34 Granadilla (*Passiflora ligularis*) for sale in the public market, Chichicastenango, Guatemala, March 1988.

FIGURE 3.35 Granadilla (*Passiflora ligularis*) for sale in a public market, Lima, Peru, July 1988.

nitida is the same size as yellow passionfruit but has an orange-yellow skin. *Maracujá suspiro* was on sale in Manacapuru, Amazonas, in February 1978. The Chácobo of the Bolivian Amazon collect wild *P. coccinea* (Boom, 1989). In Ecuador, *P. caudata* is eaten by locals (Gentry, 1986).

Passionfruit can be propagated by two to three internode cuttings, but it is usually grown from seeds, so considerable variation is maintained in farmers' fields. Passionfruit has also escaped from cultivation and occurs as a spontaneous plant in some parts of the humid tropics, such as in the Blue Mountains of Jamaica (Fawcett, 1891; Fawcett and Rendle, 1926). In Hawaii, escaped passionfruit are common climbers along roadsides and waste places. Such "weedy" populations of passionfruit might also contain useful genes.

THE ROAD TO IMPROVEMENT

The gene pool of *Passiflora* might better be envisaged as a vast ocean of genetic resources that has hardly been dipped into for valuable traits. Few of the eleven cultivated species have received much attention from the handful of breeders working with passionfruit, and the distribution, natural history, and characteristics of the wild species are still poorly understood.

No true commercial cultivars of the yellow passionfruit have been developed, although certain "strains" are recognized. Breeders are hampered by the self-sterility of many plants and by low pollination rates. In Florida, hollow logs are recommended for passionfruit groves to encourage carpenter bees, which serve as pollinators. Hand-pollination is usually needed to achieve high yields in the subtropics.

In the Neotropics, hummingbirds as well as large bees pollinate passionfruit. Passionfruit flowers are perfectly designed for pollination by hummingbirds since the latters' foreheads brush the protruding anthers while they drink nectar. Most species of *Passiflora*, whether in open habitats or in the deep shade of forest, probably coevolved with the many species of hummingbirds in tropical America. On Barro Colorado Island, for example, a near relative of passionfruit, *P. vitifolia*, is pollinated by hummingbirds (Foster, 1985). In the Old World, pollination is presumably mostly by large bees and possibly sunbirds, the ecological equivalents of hummingbirds.

In the subtropics, labor costs can be an impediment to raising passionfruit yields. Hand-pollination can pay for high-value crops, such as vanilla, but is marginally profitable for passionfruit in many cases. Improved fertilization rate is thus a major goal for breeders. Passionfruit productivity is currently within the 5–20 metric tons per hectare per year range, or about 1 kilogram per plant. If more germplasm was evaluated, breeders could eventually raise this yield ceiling.

In addition to striving to reach the yield potential of passionfruit, breeders are concerned with several serious diseases and pests. Sources of resistance to several fungal diseases, such as brown spot (caused by *Alternaria passiflorae*), Fusarium wilt and crown rot (*Phytophthora* sp.), are needed to boost yields and reduce operating costs. Water lemon (*P. laurifolia*) has shown promise as a disease-resistant rootstock, but genetically resistant varieties of passionfruit also need to be bred.

Apart from larvae of heliconid butterflies, herbivorous insects hardly bother passionfruit vines, probably because they contain defensive compounds, such cyanogenic glycosides and cyanohydrins. Ants may also play a role in defending *Passiflora* vines. An unidentified *Passiflora* with red flowers in a forest reserve (Base 4) maintained by Eletronorte along the right margin of the Tucurui Reservoir was fiercely defended by ants. Ants are attracted to passionfruit by nectaries on sepals and on young leaves (Belt, 1888:224).

The yellow form of passionfruit is more disease-resistant than the purple form, and the two color types have been crossed in an attempt to transfer desirable traits. Widecrossing has been achieved with at least thirteen species of *Passiflora*, and more species could be tapped in this manner, particularly for disease resistance (IBPGR, 1986).

PASSIONFRUIT'S BLOSSOMING FUTURE

Passionfruit is catching on as a popular drink in both industrial and developing countries. In Puerto Rico, for example, passionfruit cultivation was virtually nonexistent in 1976, but production has now reached at least 3,000 metric tons. Puerto Rico's passionfruit harvests are converted into numerous locally made drinks (NRC, 1989:287). Commercial passionfruit plantings are well established in Brazil, South Asia, South Africa, New Zealand, Australia, Western Samoa, and in Hawaii. New Zealand, Kenya, and Sri Lanka plan to markedly increase passionfruit plantings for the export trade (Chadha, 1989). In established passionfruit growing areas and regions slated for expanded production, farmers are keen to acquire higher-yielding, yet hardy, passionfruit selections.

Germplasm collections of passionfruit are few and narrow in scope. The largest germplasm collection contains only about 100 accessions, and only a handful of institutions maintain collections of more than a few dozen samples. Simple freezers are a cost-effective way to safeguard seeds of primitive cultivars and wild forms of *Passiflora*. Larger germplasm collections would be a prerequisite for more intensive breeding efforts.

Fortunately, genetic erosion of cultivated and most wild species appears to be minimal. Wild and domesticated species occur in a range of habitats from second growth to deep forest. Many species of *Passiflora* are favored by forest disturbance since they exploit light gaps. Nevertheless, more information is needed on the precise ecological niches of the wild species; some of them may not tolerate significant forest disturbance, or pollinators may be adversely affected by clearing. In spite of intensive collecting efforts, *P. sumatrana* has not been found in Sumatra since the beginning of this century, possibly because of habitat disturbance (De Wilde, 1989). A combination of pristine preserves, extractive reserves, and agroforestry systems would help conserve the wild and cultivated gene pool of one of the premier tropical fruits for making juice. Brazil has over 200 hundred species of *Passiflora*, while the Guianas contain 78 species in the genus; both regions are thus a high priority for in situ conservation of *Passiflora* germplasm (Feuillet, 1989; Oliveira et al., 1988).

CHAPTER 4

Regional Fruits

Whereas the temperate zone has produced few fruit crops of any significance, tropical forests have produced over a dozen major fruit crops that enter international trade and many more of regional or local importance. Tropical Asia and the Neotropics are the richest regions for domesticated tropical fruit trees (Arora, 1985). Surprisingly, Africa and Madagascar contain only a limited diversity of tropical fruit trees. Africa has less tropical forest than Asia or Central and South America and may have been more radically altered by human activities in the past.

Despite concentration in certain centers, an often sizable group of indigenous fruit tree species flourishes with great variation in each tropical region. Only a small fraction of the tropical fruits have gained international prominence, and many more could enter regional and possibly international commerce if their properties were better known and more selections were made. Many of the minor tropical fruits have underexploited economic potential (Martin, 1984). Mangosteen (*Garcinia mangostana*) is one of the world's best-tasting fruits, but its cultivation has been mostly confined to its native region, Southeast Asia. Japanese-Brazilian farmers near Belém in the Brazilian Amazon are starting orchards of the apple-sized fruits, and canned mangosteen is now being exported to the United States from Thailand.

Over 100 species of tropical fruits are cultivated locally in Southeast Asia alone, and at least 100 additional species of fruits are gathered from the diverse forests in the region. A similar situation prevails in tropical America. Of the many tropical fruits with potential for wider cultivation, only a handful of the minor ones are sampled here. We focus on some of the fruits that are now gaining converts in new areas, are on the verge of conquering new markets, or provide multiple uses, such as food, fodder, fuelwood, and medicine. Our coverage of minor tropical fruits includes durian, rambutan, several species of *Annona*, African plum, and Indian jujube. We concentrate on current plant breeding and germplasm conservation efforts.

Although the minor or regional fruits discussed here may not enter world commerce, they are nevertheless important locally. Many of the regional or minor fruits are a valuable source of vitamins and oil. To what degree any of them could become more significant on a global scale is difficult to predict. In the last several hundred years, various minor fruits have been introduced to new areas but have not taken off. Sometimes novel fruits are rejected by local people because of taste or other attributes.

More research is needed by social scientists on criteria used by people in accepting or rejecting new crops. A growing literature addresses why farmers adopt or eschew agricultural technologies, such as machinery or high-yielding varieties, but few scientists have focused on factors that affect the decision to take on an entirely new crop. If more information were available on the needs, aspirations, and limitations of farmers, as well as market conditions, it might be possible to predict which of the hundreds of minor fruits are worth extra research effort.

DURIAN

Durian (*Durio zibethinus*) is widely cultivated in tropical Asia, especially in Thailand, Malaysia, and Indonesia. In dense, lowland forests, the bombacaceous giant can soar to 40 meters, but in orchards and backyards durian normally reaches only 10 to 15 meters. Durian is adapted to the humid tropics with an annual rainfall of at least 2,000 millimeters. The conical trees produce round or oval fruits with prominent spines. The fruits vary considerably in size, but usually weigh 1 to 8 kilograms and can often be seen stacked for sale in street markets like prickly cannonballs (Figure 4.1). Each fruit contains from five to over twenty date-sized seeds, which are sometimes roasted and eaten (Fairchild, 1939:74). The seeds are surrounded by a fleshy aril that has the consistency of cream cheese and is highly regarded in Southeast Asia and in parts of India and Sri Lanka.

Durian's rich pulp has a strong aroma, sometimes regarded as objectionable, particularly by outsiders. Once people try the fruit, however, they are usually won over by its distinctive flavor and creamy texture. The pale green pulp is eaten raw, cooked as a vegetable, or frozen or dried for later use. Indonesians ferment the pulp for a side dish (*tempoya*) or mix the fleshy arils with rice and sugar to produce *lempong* (Ochse et al., 1961:573). The seeds can be boiled or roasted and are used in confections.

FIGURE 4.1 Durian (*Durio zibethinus*) fruits for sale in a street market, Tak Fa, Thailand, 25 June 1986.

Premier fruits are sweet and have a deep cream color. Durian pulp is a good source of carbohydrates and also contains significant amounts of protein and vitamins B and C.

Durian was probably domesticated on Borneo and taken early to the Malayan Peninsula and Thailand. Durian does not appear to have reached New Guinea, but has long been cultivated in parts of Burma. Although the seeds have short duration, Malays nevertheless took them along on trips and helped spread cultivated forms of durian (Burkill, 1935). In ancient Burma, kings had runners bring durian fruits from afar if none were available locally. Durian is cultivated sporadically along the warm, humid coasts of India and on Sri Lanka.

Natural History and Near Relatives

Of durian's twenty-seven near relatives, nineteen are found on Borneo and fourteen are endemic to the island. Eleven species of *Durio* are found on the Malaysian peninsula, where five species are endemic (Kostermans, 1958). Sumatra has seven species of *Durio*, but none of them is unique to the lenticular island.

All species of *Durio* produce fruit, some at the bottom of the trunk, and at least six of durian's near relatives are harvested in the wild. Fruits of the following durian relatives are relished and gathered in the wild: lai (*D. kutejensis*) in Borneo; kerantongan (*D. oxleyanus*) in Malaysia; lahong (*D. dulcis*) in Borneo; tabelak (*D. graveolens*) in Malaysia; and durian manjit (*D. grandiflorus*) in Borneo (M. Jacobs, 1988:142; Soegeng-Reksodihardjo, 1962).

Several of durian's near relatives have become established around settlements, either because they are deliberately planted or because of discarded seeds that germinate. The superior fruit quality of *D. kutejensis* has made it worthwhile for people to domesticate the species in northern and eastern Borneo (Soegeng-Reksodihardjo, 1962). Kerantongan (*D. oxleyanus*) is cultivated sparingly on Borneo. Both species are thus indicators of human occupation. In addition to providing fruit, many of durian's close kin are logged to make furniture and other wood products.

The strong odor of the fruits of durian and its near relatives is designed to attract animal dispersal agents. In this regard, *Durio* is similar to frutão (*Pouteria pariry*) in the forests of the Brazilian Amazon. In the rainy season, frutão drops its soccer-ball-sized fruits, which invariably split open and release a strong smell that permeates the forest understory for at least 50 meters from the base of the tree. Hunters often stop by to snack on the generous yellow pulp surrounding frutão seeds, much as hunters and gatherers have done with durian's near relatives on Borneo, Malaysia, and Sumatra for hundreds of thousands of years (Lee, 1980). On Borneo the

Kantu' know that a long period of drought is about to be broken when the durian drops its fruit (Dove, 1985:145). By falling in the wet season, seeds have a better chance of germinating.

A number of animals are involved in dispersing seeds of *Durio*. An English name for *D. zibethinus* is civet-cat tree (Ochse et al., 1961:569). Nineteenth-century naturalists noted that orangutans feed on durian fruits (Beccari, 1986:201; Wallace, 1986:66, 71). The pervading smell of durian also attracts elephants, tigers, deer, rhinoceros, and monkeys (Hawkins, 1986: 107). Several of these animals may ingest *Durio* seeds while feeding on the arils, thereby dispersing them.

At least one bat species (*Eonycteris spelaea*) pollinates durian and its near relatives (Corner, 1964:206; Whitmore, 1990:70). The various species of *Durio* flower only once or twice a year, so other food resources are necessary at other times of the year to maintain pollinators for durian and its near relatives. Extensive forest reserves are needed to secure the animal communities responsible for fertilizing durian and for pollinating and dispersing durian's near relatives.

Some of durian's near relatives are threatened by the increased pace of logging in Southeast Asia, particularly in Borneo and Malaysia where so many endemic *Durio* species are found (Choke, 1973). More and better enforced reserves are needed to protect the wider gene pool of durian.

Cultivated durian is not known in truly wild form and may be a natural hybrid that was subsequently selected for cultivation. Some hybrids between species of *Durio* produce fertile offspring, and Borneo is the most likely origin of durian because so many *Durio* species grow there (Hambali et al., 1989; B. Seibert, 1989). Many spontaneous stands of durian arise from seeds of cultivated forms (Corner, 1951). Feral *D. zibethinus* has been recorded in the forests of Sulawesi, Moluccas, and Sumatra. Durian has presumably been cultivated a long time since it is widespread, highly variable, and escapees have become established at widely scattered locations. On Borneo, durian patches survive in the forest long after settlements have been abandoned (Beccari, 1986:166).

CONSERVATION AND DEPLOYMENT OF GENETIC RESOURCES

Throughout its extensive range, local inhabitants have selected superior forms of durian. The richest array of durian forms are found in areas where the fruit is most popular, such as Thailand, Malaysia, Indonesia, southern Philippines, southern Cambodia, and Vietnam. 'Chanee' and 'Kanyao' are two famous selections from Thailand. Genetic erosion of local seedlings, triggered by the dissemination of some clonally propagated cultivars, is particularly advanced in Thailand, Malaysia, Indonesia, and Vietnam (S. Sastrapradja, 1975).

A variety of propagation techniques are used. Durian is often propagated by sowing seeds, which must be done within two weeks, otherwise the seeds lose their viability. In commercial orchards, selections are propagated by grafting onto seedlings of the cultivated species or occasionally onto those of *D. malaccensis*, a wild relative. Under orchard conditions, durian fruits seven to eight years after planting the seeds and can produce for over sixty years (T. B. Wilson, 1954). Grafted durian fruits within four to five years, another reason why market-oriented farmers often prefer to graft superior selections. In southern India, some farmers use *Cullenia excelsa* as a rootstock for durian (Tidbury, 1976).

The Dutch published a list of durian cultivars in 1886 and again in 1931, but the cultivars listed do not breed true to type (Ochse and van den Brink, 1931). In the late 1970s, the International Board for Plant Genetic Resources (IBPGR) assisted national agricultural systems in Thailand, Indonesia, and Malaysia in their efforts to systematically collect, characterize, and conserve local selections. Field gene banks were established in those countries to maintain durian accessions. Thailand has the most comprehensive field gene bank for durian because it includes some near relatives. The Philippines also maintains a collection of 200 durian accessions. These gene banks are an important asset for future breeding efforts.

Disease problems, such as root rot (caused by *Phytophthora palmivora*), have historically been tackled by cultural practices rather than by deployment of resistant varieties. Until recently, little systematic breeding has been tried with durian. Some colonial agricultural services and national agricultural services have maintained limited selection programs for the fruit since the 1920s, but only within the last few years have scientists attempted crossing durian with any of its near relatives.

Indonesian and Japanese scientists have begun widecrossing durian with some of its near relatives to improve fruit quality and performance (Hambali et al., 1989). Fertile hybrids have been achieved between *D. zibethinus* and *D. graveolens*, and these are under observation. Eventually, superior varieties of durian may be bred that resist a broad range of diseases and pests and are adapted to different cultural and ecological conditions. Perhaps durian varieties with a less pungent odor may be developed for non-Asian markets. A cultivated near relative of durian, lai (*D. kutejensis*), is aromatic, but the smell is not offensive (Soegeng-Reksohihardjo, 1962).

The private sector is likely to play an increasingly important role in multiplying highly desirable cultivars of tropical fruits and trees. Nurseries in Jakarta and Bogor, for example, are already distributing at least one durian hybrid that apparently arose spontaneously (Hambali et al., 1989). Commercial nurseries in Hawaii, Malaysia, and Singapore are able to air freight clones of tropical fruit trees to customers in many tropical and subtropical countries.

With growing Asian communities in North America and elsewhere, the export prospects for durian are bright. Whole frozen durian fruits and frozen pulp are flown to Chicago and San Francisco to cater to the Asian community in those cosmopolitan cities. If the fruits are quick-frozen, durian can be imported into the United States without quarantine barriers. It may be feasible to establish plantations of durian closer to North America, such as in the Caribbean and Central America. Durian fruits will probably be confined to ethnic markets for the foreseeable future. It would be difficult to conceive of odorous durian fruits gracing the shelves of mainstream supermarkets in Canada, Europe, and the United States.

The domestic market for durian in tropical Asia is booming. In Thailand, durian is probably the most prized tropical fruit, and it also commands a high price on Java. As cities grow in Southeast Asia, durian is likely to become an ever more attractive cash crop. India is also keen to explore the possibilities of growing durian on a larger scale (Chadha and Pareek, 1989). Considerable scope exists for providing a range of flavors and textures to suit local preferences. If the forests in which they grow survive, wild relatives of durian, with their often fine-tasting fruits, may one day make a much larger contribution to the diets of city as well as rural folk in Southeast Asia.

RAMBUTAN

Rambutan (*Nephelium lappaceum*), a relative of the lychee (*Litchi chinensis*), is one of the most popular fruits of Southeast Asia. The evergreen tree reaches 25 meters in height and produces distinctive ovoid fruits covered with soft, hairlike spines. *Rambut* in Malay means "hair," hence the name of the fruit. A thin, leathery red or yellow skin covers the subacid or sweet flesh. The translucent pulp is normally eaten by breaking the skin at one end and squeezing until the seed and attached flesh are ejected. The firm yet jelly-like pulp is an excellent source of vitamin C and carbohydrates and also contains some vitamin B.

In addition to the appealing fruit, rambutan provides other useful products. The handsome red wood is hard and suitable for many types of construction. The supple fruit walls are high in tannin and are used medicinally. Leaves and roots of rambutan are also employed in various folk remedies. A red dye is extracted from the fruit and leaves of rambutan to color batik cloth.

Trees fruit about five years after sowing seed. Progeny from seed are often highly variable, so farmers and villagers have propagated some superior forms vegetatively since at least the 1930s (Lambourne, 1937). Bud grafts, inarching, and marcotting are used to clone outstanding rambutans. Rambutan is grown in mixed stands in villages and farms as well as on small-scale plantations for home consumption and sale in cities (Figure 4.2).

In some areas, such as Malaysia and Indonesia, rambutan produces two crops annually, whereas in the Philippines it usually fruits only once a year (Lambourne, 1937). Weather conditions a few weeks before flowering apparently influence fruiting behavior, but other factors, such as agronomy and soil nutrients also play a role in how often rambutan produces a crop (Whitehead, 1959).

ORIGIN AND SPREAD

Rambutan is native to Malaysia and is widely cultivated in the lowland, humid tropics of Southeast Asia, provided there is no pronounced dry season. Arab, Indian, and Malay traders carried rambutan far from its native range, and it has been cultivated since early times from Zanzibar to the Philippines. Throughout its extensive range, locals have favored rambutans with sweet taste and a partially free stone because they are easier to eat. Selection for these qualities has been difficult to maintain because some trees produce only male flowers, and much chance crossing takes place

FIGURE 4.2 Rambutan (*Nephelium lappaceum*) on its way to market in Banjarmasin, Kalimantan, Indonesia, January 1985.

(Lim, 1989). In spite of this difficulty, people in some areas, such as Java, have managed to develop distinct rambutan "races."

NEAR RELATIVES

Borneo is the center of diversity for *Nephelium*. The genus comprises twenty-two species, sixteen of which are found on the equatorial island (B. Seibert, 1989). Borneo contains eight endemic species of *Nephelium* and is therefore a high-priority area for conserving the larger gene pools of rambutan and durian.

Several of rambutan's near relatives are collected in the wild and are cultivated, albeit on a more limited scale. Pulasan (*N. mutabile*) is wild in the Philippines and is cultivated for its especially sweet fruit in west Malaysia and Indonesia. Unfortunately, pulasan fruits usually have large stones, but some seedless forms have been selected by farmers and villagers. Another relative of rambutan, *N. eriopetalium*, is occasionally cultivated in Indonesia and its generally sour fruits are collected sporadically in Malaysia. Rural people in parts of India to Thailand and southern China cultivate *N. langana* for its acidic, lychee-like fruits. Several other species of *Nephelium* are used for wood or medicinal purposes.

VARIABILITY AND BREEDING CHALLENGES

Rambutans vary greatly because of extensive cross-pollination. Nevertheless, variation is not continuous, and some local forms or varieties can be recognized. Clones have been selected intermittently for over fifty years in Malaysia, Indonesia, and the Philippines. High yield, exceptional fruit quality, especially for canning in recent years, and leaf characteristics have been high priorities for breeders. High yielders include 'Zamora' in the Philippines and 'Kepala besar' in Malaysia.

On the whole, rambutan is remarkably free from major diseases. Several fungi attack it, causing sporadic damage. For example, powdery mildew (*Oidium nephelii*) sometimes provokes premature fruit drop in Java. In Malaysia, a canker fungus (*Dolabra nephaliae*) can affect vegetative parts of the tree. No breeding for resistance to fungal diseases has been attempted because some local varieties are already resistant and produce high-quality fruit.

The main pests of rambutan are fruit-eating bats and certain caterpillars. Nocturnal vigilance is necessary to ward off the bats, whereas severe outbreaks of caterpillars are controlled by chemical sprays. Breeding for resistance against caterpillars is feasible, but little if any screening has been attempted to find resistant rambutans.

Much more research is needed to select high-yielding and disease-resistant clones and suitable rootstock. Although local forms have been named and listed from time to time, the literature suggests that there are probably no more than fifty favored selections. A better list of descriptors would aid future efforts to characterize rambutans.

Rambutan can be budded relatively easily, thereby facilitating the propagation of desirable selections. Tissue culture techniques could also help in the mass production of good-quality selections for planting in backyards and commercial groves. Some trees are bisexual, and cloning allows growers to avoid "useless" male rambutan trees.

In the near term, a more systematic search for desirable rambutans in backyards and orchards will pay more dividends than a protracted breeding effort. Diversity of local cultivars is particularly pronounced in Thailand, Malaysia, and Indonesia. Unfortunately, genetic erosion of this rich gene pool is accelerating as more people turn to superior clones to increase production and fruit quality (S. Sastrapradja, 1975). The dilemma of cloning superior forms and the loss of traditional varieties are common to several tropical fruit trees, such as avocado and mango.

More germplasm collecting is needed in view of the rapid loss of local forms in some areas and the limited scope of ex situ gene banks. Rambutan seeds cannot be dried and frozen for storage without destroying them, so germplasm must be maintained in situ or in orchard gene banks. The major rambutan germplasm collections are held at Bogor and Lembang in Indonesia; Kuala Lumpur, Malaysia; Los Baños, Philippines; and at Chantaburi and Bangkok, Thailand. Over 700 accessions are held in these ex situ field collections, and probably over 60 percent of the samples are unique.

PROSPECTS

Rambutan, already an important regional fruit in Southeast Asia, is conquering new markets and could be used more extensively for industrial uses. The seeds of rambutan are fatty, consisting mainly of oleic acid. The extracted fat sets at normal ambient temperatures to a hard white substance with many potential uses. The Indian agricultural research system is interested in exploring the potential for increased rambutan production (Chadha and Pareek, 1989). Rambutan commands a high price in up-scale fruit and vegetable markets in London; at Harrods, rambutan from Madagascar was selling for $14/kg in December 1989. Rambutan may also become increasingly common in some North American stores.

Although not well known outside Southeast Asia, rambutan has been introduced to such countries as Australia, Brazil, Cameroon, China, Honduras (1927),[1] Liberia, Mexico, Panama, Seychelles, Trinidad (1937: R. O. Williams and Williams, 1951:202), and the United States (Florida, Hawaii, Puerto Rico). The potential of rambutan is thus widely recognized, even if its cultivation is still essentially limited to Southeast Asia. No ecological barriers prevent rambutan from becoming established in hot, humid areas of tropical Africa and America, so rambutan may one day become a pan-tropical fruit.

ANNONACEOUS FRUITS

The large tropical and subtropical family Annonaceae has provided us with some of the world's most exquisite, but often little known fruits. Close to a dozen species in three genera (*Annona*, *Rollinia*, and *Cananga*) have been domesticated in mountains and lowland forests of tropical America, the Caribbean, and Malaysia. At least nine species of *Annona* were domesticated in the New World long before the arrival of the Europeans (Appendix 1). The original distribution of many of these species is not known precisely, in part because people have been responsible for spontaneous populations of the fruit trees around villages and temporary camps.

Some fruits of wild or spontaneous *Annona* are collected in tropical forests, such as ilama (*A. diversifolia*) and soncoya (*A. purpurea*) in Central America, *A. scleroderma* in parts of Mexico and Guatemala, and pond apple (*A. glabra*) and mountain soursop (*A. montana*) in Venezuela and the Caribbean. The hard, black seeds of *Annona* contain an oil that can be used in paints and insecticides (Samson, 1986:275).

Cherimoya (*A. cherimola*) is grown in highlands and is prized for its smooth, custardlike pulp. Sweetsop (*A. squamosa*), soursop (*A. muricata*), bullock's heart (*A. reticulata*), and ilama (*A. diversifolia*) arose in the tropical lowlands of Latin America.

Sweetsop is widely cultivated in tropical Brazil, particularly in Bahia, where it is known as *pinha* and is typically planted as a dooryard tree. Sweetsop is eaten fresh and is a reasonably good source of vitamin C (Martin et al., 1987:18). Although sweetsop is native to tropical America, India is the largest producer of the fruit (Samson, 1986:276). In India, desirable sweetsops are propagated by inarching. Young budwood can easily be grafted, and rooted cuttings are another way to multiply superior selections.

Aptly named, bullock's heart is cultivated sporadically in various parts

---

1. Wilson Popenoe introduced rambutan to the Lancetilla Experiment Station, Honduras, in 1927 (letter to Evelyn Smiley, Miami, Florida, from Wilson Popenoe written at Antigua, Guatemala, 5 January 1974, shortly before his death; courtesy of Hugh Popenoe.

of tropical America, but nowhere achieves the importance of soursop or sweetsop. Introduced to a few locations outside the New World tropics, such as the Society Islands (Figure 4.3), bullock's heart is strictly a backyard tree planted largely for domestic consumption.

Mountain soursop (*A. montana*) is indigenous to the West Indies and northern South America. Atemoya, a manmade cross between soursop and cherimoya, is grown commercially in various tropical and subtropical regions, such as South and Central America, the Philippines, Israel, New Zealand, southern Florida, and South Africa.

Biribá (*Rollinia deliciosa*) is grown sporadically in Amazonia (Figure 4.4). Apparently wild populations of biribá around Ariquemes, Rondônia, Brazil (Cavalcante, 1988:58), are threatened by accelerating deforestation by ranchers and farmers. Ylang-ylang (*Cananga odorata*) occurs in the forests of Malaysia, and is cultivated in Réunion, the Philippines, and Java (Purseglove, 1974:626). Unlike the other domesticated species in the family Annonaceae, ylang-ylang is cultivated for its essential oils, obtained from the flowers.

FIGURE 4.3 Bullock's heart (*Annona reticulata*) in a backyard in Bora Bora, Society Islands, where it is known as *coeur de boeuf*. August 1989.

FIGURE 4.4 Biribá (*Rollinia deliciosa*) from the Camatiã tribe, near São Paulo de Olivença, Amazonas, Brazil.

Although all the cultivated annonaceous trees have some potential for future development, we focus on two species with a particularly promising future: cherimoya and soursop. Cherimoya does well in parts of the highland tropics and subtropics, while soursop is well adapted to the humid, lowland tropics. If their genetic resources were better understood, these species could help boost the income of farmers in many parts of the tropics and subtropics.

CHERIMOYA

Cherimoya is one of the most prized fruits in the highlands of tropical America and is a good source of thiamine, riboflavin, and niacin (NRC, 1989:229). The fruits are heart-shaped, conical, or oval and weigh up to 2 kilograms (W. Popenoe, 1920:163). The green fruits are covered with a thin skin, either smooth or covered with round lumps. The flesh is generally white, creamy, and slightly acid; a much-prized variety with red pulp is reported from highland Guatemala (Brigham, 1887:368). Peruvians, Guatemalans, and Mexicans esteem cherimoya as a premier dessert fruit, and it accordingly fetches a high price in rural and urban markets. Cultivated throughout the cool, dry valleys of the Andes and highland Central America between 1,200 and 2,000 meters, grapefruit-sized cherimoya is also grown to a limited extent in Australia, Sri Lanka, the Canary Islands, Madeira, and southern Spain.

Controversy has surrounded the origins of cherimoya. De Candolle (1902) asserted that the fruit originated in the Andes of Peru and Ecuador. Early in this century, Gabriel Alcocer of the Mexican Natural History Museum pointed out that Cobo found cherimoya in Guatemala City in 1629 and sent seeds to friends in Peru, where it was supposedly unknown. Wilson Popenoe (1913) speculated on a Central American origin for cherimoya and suggested that the species became naturalized in the Andes. However, Popenoe (1920:164) later modified his position in the light of archaeological evidence from Peru in the form of terra cotta vases apparently modeled after cherimoya fruits. Cherimoya was most likely domesticated in highland Ecuador or Peru, and then gradually disseminated in precontact times as far north as Mexico and south to Chile. Apparently wild cherimoyas can be found in the Loja area of southern Ecuador (M. Holle, pers. comm.).

Although cherimoya trees do not tolerate excessive heat or cold, they are relatively hardy and produce well in cool, relatively dry areas of the highland tropics. Cherimoya usually fruits after about four years, and some selections produce up to 300 fruits per tree. Cherimoya is pollinated by insects, and because annonas are obligate outcrossers, considerable variation is often found among cherimoya planted from seed.

To overcome wide disparities in fruit shape and productivity, some growers have turned to grafting superior selections. This practice is particularly common where cherimoya has been recently introduced. In parts of the Andes, some large-fruited selections are grafted, thereby helping growers increase their income. Unfortunately, though, this move has extinguished a few local forms, especially in Ecuador and Peru (M. Holle, pers. comm.). 'Cumbe' is a favored cherimoya cultivar in Peru. Cherimoya is grafted onto a variety of rootstocks, including cherimoya itself or near relatives such as *A. glabra*, *A. reticulata*, *A. squamosa*, or *A. montana*. For additional drought resistance, cherimoya is grafted on *Rollinia emerginata*, or in Africa an indigenous species, *A. senegalensis*, can be used for the same purpose. Enough cultivars have been released to warrant more careful identification of clones. Recent work has demonstrated the usefulness of isozyme markers in separating cherimoya genotypes (Ellstrand and Lee, 1987).

Soursop

Soursop (*A. muricata*) is one of the better known annonaceous fruits since it is widely cultivated in the lowland tropics of Central and South America. In Spanish-speaking countries, *guanábana* is used to make fruit drinks, sherbets, and ice cream. Known as *graviola* in Brazil, soursop fruits can reach the size of a pineapple (Figure 4.5), and are relished for their distinctive, sweet-sour flavor. Soursop fruits contain appreciable amounts of vitamins B1, B2, and C (Samson, 1986:275).

Of all the cultivated annonaceous fruits, soursop has perhaps the greatest potential for further development in the lowland, humid tropics. But more research is needed to develop highly productive varieties that also resist diseases and pests. In areas served by all-weather roads and electricity, small to large-scale plantations can be established to supply juice and ice cream factories. In general, markets are not oversupplied with soursop; demand usually exceeds supply. Export markets could be developed for soursop juice and flavoring for tropical ice creams and fruit juices, although the English name for the fruit might deter some sweet tooths in

FIGURE 4.5 Soursop (*Annona muricata*) in a field gene bank maintained by the Agricultural Research Center for the Cerrado (CPAC, Centro de Pesquisa Agropecuária dos Cerrados), part of Brazil's agricultural research system (EMBRAPA, Empresa Brasileira de Pesquisa Agropecuária) near Brasília. August 1988.

North America and Europe. A name change and aggressive advertising in up-scale markets could create new opportunities for soursop growers.

PESTS AND DISEASES

The most serious insect pests of *Annona* are eurytomid wasps, whose larvae bore into the fruits, and caterpillars belonging to the families Stenomidae, Sphingidae, Geometridae, and Noctuidae. Mealybugs (*Pseudococcus* spp.) attack sweetsop, especially in India, and soursop is infested by a moth (*Cerconota* sp.) and a wasp (*Bephrata* sp.) in the Caribbean (Peña et al., 1990; Samson, 1986:277). Soursop is also attacked by a variety of insect pests. Commercial growers place blue, translucent plastic bags around soursop fruits to reduce pest damage, but such practices are labor-intensive and drive up costs. Anthracnose, the only major disease of cherimoya, is controlled by chemical sprays.

If *Annona* fruits are to be promoted in tropical and subtropical lands, breeders will need a much more secure resource base to develop hardy cultivars for future challenges. Any outbreak of a new disease, or a more virulent form of an existing pathogen, could wreak havoc for growers and greatly diminish genetic variation. At present, biocontrol research and cultural practices are the only efforts under way to combat disease and pest pressures (Peña et al., 1990).

Conservation and better evaluation of genetic resources will help smaller production areas for cherimoya and other annonaceous fruits, such as sub-Saharan Africa, the Cape Verde Islands, and India. Little breeding work is being carried out to develop hardier, more productive varieties of annonaceous fruits. The Zill nursery at Boynton Beach, Florida, began an *Annona* improvement program in 1984, by interbreeding varieties and selections (Mahdeem, 1990). Eventually, superior material will be propagated vegetatively for growers in southern Florida. Similar efforts are needed in tropical countries.

CONSERVATION AND COLLECTION
IMPERATIVES

The gene pools of even the most important annonaceous fruits have not been systematically surveyed or collected. The sixty-year-old Zill nursery has dispatched a plant breeder to Belize, Guatemala, and Yucatán in search of promising *Annona* fruits cultivated in backyards, but scientific surveys are needed to better understand variation in domesticated and wild forms. Germplasm of cherimoya and its relatives can be stored as seed, thereby reducing maintenance costs for ex situ collections.

In spite of the relative ease in storing *Annona* germplasm in seed gene banks, few institutions have established ex situ collections. Moreover, the few ex situ germplasm collections of cherimoya contain only a small portion of the extant variation. A few working collections of cherimoya are maintained in Australia and the United States. Ecuador has a field gene bank with twelve cherimoya accessions, while forty and fifty accessions are held in Costa Rica and Peru, respectively. Brazil, the United States and Mexico maintain only fourteen selections of soursop, while Costa Rica holds forty accessions in a field gene bank. Only three significant collections of sweetsop have been assembled in the Philippines, Seychelles, and the United States. All germplasm collections of annonaceous fruits are fragmentary, and strategies need to be drawn up to guide collecting missions and subsequent efforts to store germplasm over the long term.

Ethnobotanists would be especially helpful in surveys of genetic variability of the cultivated species of *Annona*. Unusual forms are sometimes maintained by indigenous people, selections that might prove suitable for wider cultivation. Local people can also identify other uses for cherimoya and its cultivated relatives, such as medicinal qualities that could prove helpful additions to the pharmacopoeia of distant lands.

The annonaceous fruit species are currently the focus of research for biomedical applications. This family is especially rich in alkaloids, some of which may help arrest the growth of certain cancers and promote healing of damaged livers (Kowalska and Puett, 1990). Potentially useful compounds for medicine have been found in both domesticated and wild species of the Annonaceae, further underscoring the need to conserve natural areas.

AFRICAN PLUM

The long and intimate association between people and tree species selected for cultivation or semidomestication in tropical Asia and America also prevails in the hot, humid parts of Africa. The tropical forests, savannas, and highlands of the Great Lakes Region account for most of the crops domesticated in Africa (Harlan et al., 1976). Many trees in Africa have been protected or cared for in the wild and have become major cultivated plants, such as the oil palm. Other economic trees remained essentially wild or were planted only on a limited scale.

Some of these minor perennial crops, such as African plum (*Dacryodes edulis*), African breadfruit (*Treculia africana*), and the shea butter tree (*Butyrospermum paradoxum*), are nevertheless important locally. Many of the minor perennial crops were brought into cultivation in the transition zone between dense forest and broad, open savannas.

African plum is planted in southern Nigeria, Cameroon, and the Congo for its nutritious fruit, which has a high oil content. The round or conical fruits are dark blue when ripe and also contain a range of amino acids (Busson, 1965). The fruits, which measure 4–12 centimeters by 3–6 centimeters, are boiled in saltwater, fried, or roasted over charcoal. African plum was probably more widespread in the past, ranging from the Gulf of Guinea to Uganda and central Angola (Aubreville, 1962). African plum has a wider ecological tolerance than oil palm and coconut, and is well suited for plantations as well as backyards.

African plum could be cultivated more widely again, since it adapts well to differences in day length, temperature, rainfall, soils, and altitude. Oil sales from African plum could provide a valuable source of revenue in low-income, rural areas of the continent and possibly other tropical regions (Ngatchou and Kengu, 1989). Fortunately, the Institute of Agronomic Research of Yaoundé, Cameroon, is exploring the untapped potential of the little-known African plum. But more such efforts are needed to boost African plum from relative obscurity to an important source of revenue and food.

African plum could produce 7–8 tons of oil per hectare, compared with some 3 metric tons per hectare for oil palm (Giacomo, 1982). Multipurpose trees are especially important in areas where land may be scarce and farmers do not have a lot of space to plant a mix of trees for food, income, construction materials, and firewood. African plum provides oil for cash income, fruit for the table, and a perennial pharmacy for a variety of ailments ranging from ear infections to fevers and oral problems.

In forest, African plum can reach over 40 meters high, but under cultivation it rarely grows taller than 10 meters. The tree is pollinated by bees, and the flowering time and duration depend on latitude and genotype. Some African plums are early, while others flower late and may produce blossoms continuously for several months. Trees are male, female, or hermaphrodite. Males may produce a limited number of female flowers, and thus some fruit.

The Gene Pool for Future Development

The genus contains eleven species, mostly found in West or Central Africa. Gabon is the center of diversity for *Dacryodes*. Other relatives of African plum are found in Indo-Malaya and South America. Several near relatives of African plum produce edible fruit that are gathered in the wild (Aubreville, 1962; Lam, 1932).

African plum has been divided into two botanical varieties, one with large fruit, the other with smaller ones. But African plums exhibit great diversity since they are raised from seed, and boundaries of the proposed botanical varieties can be blurred. No artificial selection has been attempted with African plum, thereby contributing to a continuum of fruit sizes rather than clear-cut botanical races.

Selection is nevertheless possible, and although previous attempts at vegetative propagation have failed, superior genotypes probably could still be cloned. If growers can be assured of high yield, they might be encouraged to establish small orchards or even large plantations.

No germplasm collections have been initiated for African plum, nor have any strategies been drawn up for gene banks or in situ reserves for the species. Seeds of African plum are recalcitrant, so ex situ collections will entail orchard gene banks. One problem facing germplasm collectors is that African plum shows a very plastic response to the environment. Trees collected from diverse geographical areas and habitats would have to be later screened under controlled conditions to determine more accurately patterns of genetic variation.

INDIAN JUJUBE

Indian jujube (*Zizyphus mauritiana*) is a fruit tree of the home garden, particularly in India, where the fruit is highly prized. Also called Malay jujube or Chinese date, the medium-sized tree has drooping branches that bear fruits like small, glossy crabapples or large dates. Indians know the tree as *ber*, and its fruits are consumed locally and taken to markets in towns and cities. An affordable fruit, Indian jujube is widely appreciated among the Indian population.

Indian jujube was domesticated in India, where it still grows wild in seasonally dry, subtropical and tropical forests of northern and central India, Sri Lanka, Burma, and as far east as Java (Watt, 1908:1143). The thorny tree was taken early to China, where diverse cultivated forms are found. Indian jujube is hardy, able to survive cold blasts of air descending from the Himalayas in winter, as well as extreme heat and dryness before the onset of the summer monsoon.

The highly variable fruits may be subacid or sweet, with a juicy or mealy texture. When ripe, the fruits generally turn yellow, then red or reddish brown. At maturity, many of the fruits are crisp and juicy. Fruits are normally left to overripen so that they begin to shrivel and the pulp softens; at this stage they emit a strong odor and are relished. The pulp is a good source of carbohydrates and vitamins A and C. Jujube butter is prepared by squeezing cooked fruits through a sieve. Jujube jelly is made from unripe fruits. Dried fruits are savored as a delicacy, incorporated in bread, and made into paste. China exports dried jujube to several countries, including India, which is currently unable to satisfy its own demand for the product. The kernels of Indian jujube are also eaten and are high in protein.

In addition to food, Indian jujube is used for a wide variety of purposes. In folk medicine, the fruits are believed to purify the blood and to aid digestion. Powdered roots and leaf poultices of Indian jujube combat a number of ailments in South and Southeast Asia. The tree is a valuable source of fuelwood and fodder for livestock. Branches of Indian jujube make effective fences, and the wood is hard enough for agricultural implements (Parker, 1986:19). The bark is used for tanning (Watt, 1908:1143). The leaves are also employed to rear lac insects, which produce a resin used in the manufacture of shellac and varnishes, and are fed to silkworms (Cowen, 1984:89).

The importance of jujube in the diet and culture of Indians is exemplified by the degree to which the life-supporting tree has inserted itself into the religious life of Hindus. Known as *badari* in Sanskrit, a grove of Indian jujube trees in the Himalayan foothills was chosen for the hermitage of two saints, Nara and Narayana, the latter an incarnation of Vishnu (Gandhi and Singh, 1989:63). This site, Bandrinath, is a sacred pilgrimage center. Indian jujube appears in several Hindu epic stories, such as *Ramayana*, in which the tree fails to secure the dress of Sita and after confessing to Rama, droops in shame. A Punjabi legend tells of a leper who dipped his feet in a pool while clutching the branch of an Indian jujube tree and was cured.

WILD POPULATIONS AND NEAR RELATIVES

Indian jujube occurs wild on the semiarid plains of India. Fruits of wild Indian jujube are used mainly for sherbet because they tend to be less sweet than cultivated forms (Parker, 1986:19). Although wild Indian jujube is confined to dry tropical forest, often in scrubby and heavily grazed environments, many near relatives grow in moist tropical forests, such as *Z. vulgaris* and *Z. rugosa* in southern Indian.

At least 100 species of *Zizyphus* are found in the tropics and subtropics, but only a few are used by people. Christian tradition looks upon *Z. spina-christi*, indigenous to the Near East, as the tree used to fashion the crown of thorns worn by Christ shortly before his crucifixion (Zohary, 1982). Talmudic literature mentions the species as a source of edible fruits. Chinese literature of the fourth century B.C. records areas where the Chinese jujube, *Z. jujuba*, can be grown (Needham, 1986). Chinese jujube has been traded by Arabs for centuries and is used to prepare cough lozenges and syrups.

GENETIC RESOURCES AND POTENTIAL

Increased attention to Indian jujube is warranted because of its food and medicinal value (Campbell, 1984). More resources for research on Indian jujube would be a wise investment because it provides sustenance, medicine, fuelwood, and fodder to many densely settled rural areas with rapidly growing populations in South Asia and parts of Southeast Asia. A multipurpose tree, Indian jujube is well-suited to agroforestry by smallholders in the subtropics and seasonally dry tropics.

Breeders have plenty of material to work with in future efforts to improve fruit quality and yield, among other characteristics. Great diversity is to be found in both domesticated and wild Indian jujube (Watt, 1908: 1144). The extensive distribution of Indian jujube in the wild and under domestication increases the chances that locally adapted forms have arisen that contain different potentially useful genes. Indian jujube grows in a wide range of environments and is likely to have evolved defenses against a host of pests and pathogens, including different races of pests and pathotypes.

Several hundred varieties have probably arisen under cultivation, some of which are perpetuated by bud grafting. Local selections by farmers would probably be fertile ground to launch any improvement efforts.

A limited amount of Indian jujube germplasm is maintained in India, but nowhere is the species given high priority for research. A first step would be a survey of local cultivars, including a detailed description of their attributes and uses. Samples of all cultivars would be taken during this first step, but not all of them would necessarily be conserved. Survey results should help to target specific cultivars for conservation. If samples were collected during the initial survey, a second, follow-up mission would not be necessary, and costs could be reduced.

Several complementary conservation efforts should be pursued. More field gene banks need to be established, with some duplication in tissue culture collections. Field surveys in existing parks and reserves are needed to determine if wild forms are found within their boundaries. If wild populations are not safeguarded, then extractive reserves should be established that allow local people to harvest the trees and other products without destroying the resource. Until more is known about the near relatives of Indian jujube, they are unlikely to be of much use to improvement efforts.

CHAPTER 5

Rubber, Oils, and Resins

Several perennial trees of moist tropical forests are so important for latex, oil, or resins that they have been domesticated and widely planted. Here we highlight the significance of rubber, oil palm, and resins and balsams from tropical trees. Industrial applications for such crops not only provide cash income for growers but also generate jobs for urban dwellers in Third World cities. The widespread use of perennial crops for industrial products in both developing countries and industrial nations means that all citizens have a stake in the survival of tropical forests.

The rubber tree is native to Amazonia and has become a major plantation crop in Southeast Asia. Citizens from virtually every country have directly or indirectly benefited from natural rubber. Natural rubber is used in such products as aircraft tires, radial tires for cars, and latex gloves. Rubber plantations in tropical Asia, the major source of natural rubber, are genetically uniform and thus are vulnerable to catastrophic failure from introduced diseases. Many potential sources of resistance to diseases, among other desirable traits, are found in the wild gene pool of rubber in the Amazon basin, which is shrinking as the forest cover dwindles. Extractive and Indian reserves, among other conservation strategies, hold some hope for preserving at least some portions of rubber's genetic variation.

African oil palm has long provided a valuable source of food for peoples of West and Central Africa. In this century, oil palm has emerged as a major underpinning to the agricultural economies of Malaysia and Indonesia. Several Latin American countries, particularly Brazil and Colombia, envisage oil palm as an increasingly important crop in the humid lowlands. Not only is palm oil an important food in some regions, but oil extracted from the fruits and kernels of oil palm is a good substitute for diesel oil, and the palm may thus become a significant alternative fuel in some countries.

For oil palm to be successful in the American tropics, however, more disease-resistant varieties are needed, particularly against spear rot. Fortunately, a near relative native to the American tropics promises to help catapult oil palm into a major crop in Latin America because it resists spear rot and its short stature can facilitate harvesting of the fruit bunches.

Whereas oil palm and rubber are relatively well known and are major export commodities, a large assortment of gums and resins that also find industrial applications are collected in the tropics. Technically, resins are mostly mixtures of either terpenoids or phenolic compounds, whereas true gums are polysaccharides produced in response to bacterial infection (Langenheim, 1990). In practice, the terms *resin* and *gum* are used interchangeably. Although the economic importance of such products pales in comparison with that of natural rubber or oil palm, gums and resins nevertheless deserve closer attention because they can satisfy the needs of the growing industrial base in developing countries as well as generate foreign exchange.

A vast array of gums and pectins are obtained from wild and domesticated plants, many of them perennial trees in the tropics. The fact that most trees with copious resin occur in the tropics and subtropics suggests that the sticky compound plays a defensive role against pests and some diseases (Langenheim, 1990). Gums and pectin act as cementing material between plant cells and, in some species, ooze from wounds. Gums and pectins are soluble in water and are mainly polymerized carbohydrates that form colloidal gels, suitable for use in confectionery, food, beer, and ice cream. Such compounds are also employed in the textile and paint industries and to emulsify lotions, creams, and medicines.

Some gums have been exploited from prehistoric times. Gum arabic, derived from wild species of *Acacia* in the parched landscapes of North Africa, has been an article of commerce for thousands of years. In Australia, wattle gum is obtained from another species of *Acacia*. Gum tragacanth is produced by herbaceous and shrubby species of *Astragalus* in Southwest Asia and has long been used for medicinal purposes. Whereas the above gums come from arid areas, many other gums are produced by perennial plants of the tropical forests.

Gums and pectins are common, but supplies and demand on the world market fluctuate sharply. Alternative sources are readily tapped to fill the demand, thereby discouraging major investments to promote international commerce in gums and pectins from the tropics. Vegetable wastes from various food-processing industries provide similar substances. Thus gums and pectins from wild and domesticated tropical plants will probably remain mostly a local resource. No international bodies have formed to promote tropical gums and pectins, unlike the International Rubber Research and Development Board.

In contrast to gums and pectins, resins are not soluble in water and must be dissolved in various solvents before they can be used industrially. The international market for resins is stronger than for gums and pectins. Resin is produced in special ducts by some plants in response to stress, such as attack by predators or bruising.

Resins are used to produce varnishes and as sizes for the plastics and rubber industries, as well as for medicines and certain soaps. Pines exude an oleoresin that, after steam distillation, produces turpentine and rosin. In the past, turpentine was used mostly as a solvent, but more recently it has been employed mainly as a source of terpenoids for the chemical industry. Rosin is used directly or, after chemical modification, in adhesives, size for paper, rubber products, and various coatings.

Pine oleoresin is complex and can be processed in a number of ways. Many pine resin products are used in medicines. For example, pine resins are used in liniments. Destructive distillation of pine oleoresin produces tar or pitch. Tar is an ingredient in some antidandruff shampoos.

Papermakers consume most of the rosin as size to prevent ink from smearing during printing. Rosin is also used as a drying agent in paints and in adhesives and floor coverings. Resin oil, obtained by dry distillation of rosin, is employed for carbon black in lithography as well as in inks and axle greases. Medical disorders treated with resins include skin problems and ringworm.

Turpentine, obtained from pines, was once used extensively as a solvent for paint, but synthetic solvents have now largely replaced the natural compound. Nevertheless, turpentine is still used for special paints, varnishes, and polishes. Oil of turpentine is employed to purge intestinal worms, as a diuretic, and to treat flatulence and cystitis. Turpentine is applied externally to alleviate rheumatism and arthritis, and the fumes are inhaled to treat bronchitis.

Turpentine is a valuable source of refined chemicals that are used in perfumes, in household disinfectant and cleaning liquids, and in food processing. The refined chemicals are still used in the production of adhesives and sizes for paper and textiles. Turpentine can also be chemically manipulated to render such scents as citrus, cinnamon, menthol, and licorice.

Although most of the world's resin traded on international markets comes from pines in temperate countries, many tropical and subtropical trees produce copious amounts of resin, some of which have been domesticated. More than two-thirds of the resin-producing plants occur in the tropics (Langenheim, 1973). Numerous species in the Burseraceae, Anacardiaceae, Guttiferae, Styracaceae, and Euphorbiaceae synthesize resins, but members of the Leguminosae and Dipterocarpaceae are particularly noted for their copious resin. We profile balsam (*Myroxylon balsamum*) from tropical America and some tropical pines.

As in the case of gums, resins have been employed by ancient civilizations and cultures for incense, medicine, and embalming. In Egypt in the age of the pharaohs, for example, mummies were protected from moisture by coating their cases with resins. Myrrh, mentioned in the Bible, is a complex resin from species of *Commiphora* in eastern Africa. Benzoin is an oily resin from *Styrax* species of Southeast Asia, and various lacquers are prepared from resins obtained from trees in Southeast and East Asia. Copal and amber resins, both fossilized, are used for varnishes, particularly for furniture, and for jewelry. Copal and amber are obtained from a wide variety of trees in the Congo basin, Tanzania, tropical America, Southeast Asia, and Europe. Resins, then, are still widely employed for a variety of purposes in many countries.

A number of resins contain essential oils and are regarded as oleoresins. Some oleoresins are aromatic, owing to the presence of benzoin or cinnamic acid. The aroma-producing oleoresins are referred to as balsams. Oleoresins are insoluble in water, but can be mixed with oils and spirits. Turpentine, extracted as crude turpentine or pitch from a range of temperate and tropical pines, is one of the more commonly used oleoresins.

Oleoresin extracts from pines, still called "naval stores" because of their early use for caulking boats, have undergone a series of processing refinements. At first, resin was simply tapped from pine trees and used directly. Later, pine resin was distilled in the United States by heating chips splintered from the stumps of felled trees. In the 1960s, naval stores were also obtained as a byproduct from the pulp and paper industry.

Tapped resin is still important in many countries. The amount of tapped resin compared with that produced from pulp processing depends on the availability of labor. In China, for example, tappers still account for most of the resin production. Europe, the former USSR, the United States, China, India, and Mexico are the major producers of natural rosin. Honduras and Turkey also export modest quantities of resin. Scandinavia and the United States are major paper manufacturers and consequently hold a dominant position in resin production.

At the moment, the developed world dominates the production of oleoresins and essential oils (ITC, 1986). As costs of extraction and processing increase in industrial nations, Third World countries could seize the opportunity to enhance export earnings by better developing indigenous sources of resins and balsams. By tapping their own resources, some developing countries would be able to cut their imports of resins and balsams (C. L. Green, 1989).

RUBBER

Latex tapped from the Amazonian forest tree *Hevea brasiliensis* produces the best-quality rubber, whether from other latex-bearing trees or synthetic products (Figure 5.1). Because of its superior latex, *H. brasiliensis* now accounts for virtually all of the world production of natural rubber. Although rubber emerged as a major commodity on world markets in the early part of this century, it was never a major resource for Amazonian tribes (Galvão, 1979). In the eighteenth and nineteenth centuries, the Omagua, a once powerful tribe inhabiting the Upper Amazon, manufactured a variety of rubber products for domestic use and trade, including waterbottles, balls, boots, and elastic bands (Hemming, 1987:271; Métraux, 1963a). Several tribes in northwestern Amazonia had rubberball games (Steward, 1963). Such uses attracted the attention of early European visitors to Amazonia (Imle, 1978). The British took rubber back and found that it could erase pencil marks, hence the name "rubber."

Yet the earliest and principal use of rubber trees in Amazonia has always been for food, particularly in the northwest part of the basin (Schultes, 1956). Seeds of rubber and some of its close relatives can be safely eaten after prolonged soaking or boiling to remove cyanic poisons (Schultes, 1977a). Some of the denser "wild" stands of rubber in Amazonia may be due to artificial enrichment by indigenous peoples to increase food supplies (Dean, 1987:45). The mottled seeds of *H. brasiliensis* are still used for fish bait by rural folk along the Amazon (N. J. H. Smith, 1981a:63).

Charles Goodyear's discovery of the vulcanization process in 1839, which hardens rubber, thereby making it more durable and tolerant of heat and cold, set the stage for large-scale exploitation of the resource. Two Englishmen, Hancock and Broding, also made the same discovery in 1842 (Santa-Anna Nery, 1885:194). The bicycle craze of the 1890s spurred demand for rubber, and the mass-production of automobiles in the early twentieth century created an even larger market for the product (Weinstein, 1983:8).

Until 1910, most of the world's rubber supply came from trees tapped in the forests of the Amazon basin. In addition to several species of *Hevea*, latex for making rubber was obtained from *Castilla* in Amazonia. Early "competitors" of *Hevea* rubber, such as manicoba rubber (*Manihot glaziovii*), a relative of cassava, Bombay rubber (*Ureola elastica*), and Indian rubber (*Ficus elastica*) were planted to a limited degree in several tropical countries. But the superior qualities of rubber from *H. brasiliensis* soon undercut other sources of wild and planted latex.

By 1920, most rubber was coming from plantations of *H. brasiliensis* in Southeast Asia, particularly on the Malayan Peninsula. This radical trans-

FIGURE 5.1 *Seringueiro* in the Brazilian Amazon tapping a rubber tree (*Hevea brasiliensis*). (Illustration by Percy Lau, in *Tipos e aspectos do Brasil*, Instituto Brasileiro de Geografia e Estatística, Rio de Janeiro, 1975, p. 55.)

formation of the rubber industry was made possible by introducing rubber seedlings to Southeast Asia, where abundant and inexpensive labor was readily available to tend plantations, and by the development of a better knife and improved cutting methods, which did less damage to rubber trees. Henry Ridley, one of the early directors of the Singapore Botanic Gardens, designed and tested the new tapping knife and urged growers to plant rubber on a large scale. Rubber is cultivated on 1.9 million hectares in Southeast Asia, and 500,000 smallholders account for nearly two-thirds of the planted area (Zakri et al., 1987). In Thailand, virtually all rubber production is in the hands of smallholders (P. W. Allen and Jones, 1988).

Southeast Asia currently provides over 95 percent of the world's natural rubber. Rubber generates over $3 billion in export earnings and is particularly important to the economies of Malaysia, Indonesia, Thailand, and Sri Lanka. Indonesia will rival Malaysia as the leading producer by the year 2000, and China has recently expanded production in the southern part of the country and now ranks fourth in world production of natural rubber (P. W. Allen and Jones, 1988). Brazil, the native home of rubber, imports the product, but hopes to become self-sufficient and ultimately an exporter of rubber. Major rubber plantations are being established in Mato Grosso do Sul, and if hardy and high-yielding varieties are developed for the Amazon region, Brazil, Peru, Colombia, Venezuela, and Bolivia could become significant exporters of the commodity.

Germplasm Base for Plantations

The germplasm base of the extensive rubber plantations of Southeast Asia is extremely narrow. The hundreds of millions of rubber trees planted there are genetically very homogeneous. This worrisome genetic "simplicity" can be traced to limited introductions of seedling stock and stringent quarantine regulations that have historically halted germplasm exchange efforts.

Henry Wickham's now legendary efforts to secure seeds of the rubber tree in Amazonia in 1876 were by no means the first attempt to obtain seeds or seedlings of *H. brasiliensis* for planting abroad. Several consignments of rubber seeds from Amazonia in the preceding decades either failed to germinate or the seedlings perished. Rubber seeds were sent to India from Brazil in 1873, but the resulting plants failed to survive in Sikkim's climate. A second shipment of seeds in 1875 were no longer viable when they reached India.

In 1876, Henry Wickham and his assistant, Robert Cross, changed the course of rubber's history when they collected 70,000 rubber seeds from the Amazon (Figure 5.2) and legally exported them to the Royal Botanic Gardens, Kew, near London. Only 2,397 seeds of that consignment germinated at Kew. In the same year, Kew sent 1,919 of those seedlings to Sri Lanka (then Ceylon), where they were planted at Heneratogoda, a satellite garden of Peradeniya, part of a network of botanic gardens established by the British during the colonial era. In 1877, Kew sent twenty-two rubber seedlings to the Singapore Botanic Gardens, and these formed most of the stock for the rubber plantations of Southeast Asia (N. J. H. Smith, 1986: 22). Peradeniya also sent some rubber seeds, derived from the same Wickham and Cross consignment, to its sister botanic garden in Singapore (Schultes, 1984b).

Fewer than two dozen seedlings and a few additional seeds is a shaky foundation for a global industry. The full genetic potential of a species is unlikely to be contained in such a limited founding stock. But the story gets worse. The 70,000 seeds collected by Henry Wickham and Robert Cross in 1876 probably came from only twenty-six rubber trees (Schultes,

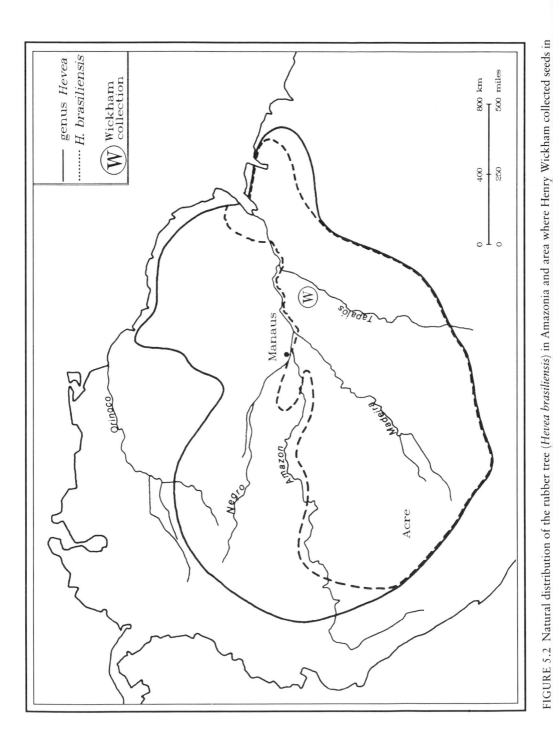

FIGURE 5.2 Natural distribution of the rubber tree (*Hevea brasiliensis*) in Amazonia and area where Henry Wickham collected seeds in

1984b). Even though rubber is an outcrosser, being fertilized by pollen carried by insects such as midges and thrips, slightly over two dozen female parents represents only a miniscule sample of rubber's gene pool.

Seed stock for the world's rubber industry thus passed through a tight bottleneck that surely excluded many valuable genes. Furthermore, Wickham and Cross confined their collecting to a very limited area of the natural range of rubber near Boim on the lower Tapajós river. This restricted sampling of rubber trees was due to the fact that a steamship, a rare sight along the Amazon in those days, was waiting to return to England, and rubber seeds had to be assembled quickly for shipment since they deteriorate rapidly. Attempts to send rubber seed to Kew prior to 1876 had failed because the seed embryos died waiting for passage or during the several-week trip by sailboat to England. Unfortunately, the area where Wickham and Cross collected seeds in 1876 is one of the poorest ecotypes for rubber.

Once rubber plantations became established in Southeast Asia, Brazil prohibited the export of rubber germplasm in an unsuccessful attempt to regain its former preeminence as a supplier of natural rubber. Nevertheless, some small rubber seed shipments arrived in Southeast Asia, but they were poor yielders and thus had little overall impact on the productivity of plantations (Schultes, 1977b; H. Tan, 1987). Introductions of rubber seed to Java from Brazil in 1896 did not enrich the germplasm base of the region's plantations since the resulting trees turned out to be poor specimens, and progeny were not distributed (Purseglove, 1974:150).

Major Breeding Thrusts and Challenges

Rubber breeding, as opposed to simple selection of chance crosses, began in 1919 when a Dutch scientist, Maas, undertook controlled pollination experiments in Indonesia (S'Jacob, 1931). In Malaysia, L. E. Morris began a program to breed *H. brasiliensis* by artificial pollination in 1928 (Gilbert et al., 1973). Efforts to boost yield and other qualities of rubber by controlled crosses and simple selections are now under way in close to a dozen countries. In 1980, eleven national research institutes were working on rubber in addition to six rubber research stations operated by private plantation interests (World Bank, 1982:22). The Rubber Research Institute of Malaysia (RRIM) accounts for three-quarters of global research expenditures on natural rubber. Sri Lanka is also investing considerable resources in rubber breeding with some promising results (R. E. Schultes, pers. comm.). Brazil devotes a sizable effort to rubber research, particularly at the Agroforestry Research Center (CPAA, Centro de Pesquisa Agroflorestal da Amazônia) near Manaus, but South American leaf blight (SALB), caused by the fungus *Microcyclus ulei*, continues to prevent the establishment of commercially viable rubber plantations in the Amazon.

Increased yields and disease resistance are the two primary goals of rubber breeding programs. But other high priorities for breeders of the crop include different composition of the latex, adaptability to various soils, ability to produce high yields even in areas with suboptimal rainfall, shorter trees to reduce storm damage, and softer bark (R. E. Schultes, pers. comm.). The desire for wind-fast rubber trees has arisen because sturdiness is negatively correlated with yield; selection for highly productive trees has increased vulnerability to high winds, particularly evident during typhoons (Wycherley, 1969, 1976:80). Drought tolerance is another sought-after trait by rubber breeders in China, India, and Vietnam (IBPGR, 1984:4).

SALB is the single most important disease of rubber. Native to the Amazon basin, this fungal pathogen does not destroy wild rubber populations because the trees are dispersed in the forest, and many escape infection. Furthermore, varying degrees of resistance to the disease undoubtedly exist in wild rubber populations. When rubber trees are crowded together, as in plantations, the virulent pathogen triggers massive and repeated defoliation, eventually killing the trees.

SALB has effectively prevented the establishment of profitable rubber plantations in the New World (Chee and Wastie, 1980; Holliday, 1970:4). Rubber tapped from wild trees in Amazonia accounted for 94 percent of world rubber production in 1908, but by 1928 the region accounted for less than 3 percent of global production of natural rubber (Galey, 1979). The Ford Motor Company attempted to establish rubber plantations at Fordlandia and Belterra along the lower Tapajós in the Brazilian Amazon in 1928, but to no avail. Successive SALB outbreaks largely doomed Ford's extensive rubber plantings, which included over 4 million rubber trees (Cruls, 1939). In an effort to save his Amazonian enterprise, Ford's scientists and technicians attempted three-component budding, in which crowns with some resistance to *M. ulei* were grafted onto high-yielding trunks, which in turn were budded on different rootstock; costs soared, and production was not competitive with output from tropical Asia (Dean, 1987:102). Although the old Ford rubber plantations along the Tapajós are still tapped (Figure 5.3), latex production is minimal.

Aerial spraying with fungicides can control the disease (Alvim, 1977), but the relatively high cost of such measures has discouraged extensive planting of rubber in Amazonia. Also, the pathogen would likely evolve resistant strains to fungicides, thereby necessitating periodic changes in chemical formulations and further raising costs.

In an attempt to flee the ravages of SALB, Brazilian planters have shifted efforts to plant rubber to southern and southeastern Brazil. While the open savanna landscapes of central Brazil checked the spread of the pathogen for a long time, SALB was recorded in São Paulo State in 1960, and it has also turned up in Southeastern Brazil (Dean, 1987:126).

FIGURE 5.3 Rubber tappers returning from work from the Belterra plantation, Pará, Brazil, 1970.

Genetic resistance to the pathogen needs to be bred into high-yielding rubber varieties. If SALB were to gain a foothold in Southeast Asia, the results would be devastating to the economies of Malaysia, Indonesia, Thailand, and Sri Lanka. At least two decades normally elapse between the time promising material is spotted in early trials until it is ready for release as clones to rubber growers; efforts thus need to be stepped up to incorporate sources of resistance to the pathogen in Southeast Asia before the disease arrives.

Single-gene resistance to *M. ulei* may not hold up long, particularly since several strains of the pathogen occur. Resistance genes for microbial pathogens in plants are often race-specific (Bennetzen et al., 1988). Multiple-gene resistance to SALB, as well as other fungi and bacteria, is thus called for. Further research is needed to locate disease-resistance genes in the germplasm of rubber and its near relatives, and to understand how those genes are expressed.

SALB is only one of dozens of fungal and bacterial diseases that attack rubber (Weber, 1973:465). For example, powdery mildew (*Oidium hevea*), is a serious problem in some rubber groves in Sri Lanka, northern Vietnam, and Malaysia (Wycherley, 1969). Black stripe, caused by *Phytophthora palmivora*, damages rubber plantations in India and parts of Central America (Purseglove, 1974:165). Genetic resistance to these and other fungal diseases is eagerly sought by rubber breeders.

Cold tolerance is emerging as a major priority with rubber breeders in China, Vietnam, and Brazil. Southern China and northern Vietnam, areas targeted for increased rubber plantings, are prone to cold snaps (P. W. Allen and Jones, 1988). The heavy investment in rubber plantations in Mato Grosso do Sul in Brazil could be threatened by periodic cold fronts that penetrate the southwestern portion of Amazonia during the Southern Hemisphere winter. Temperatures in the Brazilian territory of Acre can drop to the $5°-7°C$ range at $9°$ S (A. T. Guerra, 1955:53; Schmidt, 1942). Temperatures in Mato Grosso do Sul, some $6°$ farther south would likely approach freezing during such a *friagem*, as cold fronts are known in Brazil. During a cold front in July 1975, for example, Cuiabá at $15°$ S recorded a low of $3°C$, and light frost was recorded on some plants (Willis, 1976). Cuiabá is only 150 meters above mean sea level. As many as five cold air masses have pushed into warm, humid Amazonia in a year (I. A. Guerra, 1957). Cold fronts in Amazonia are channeled in the western portion of the basin by the Brazilian shield and can be felt as far north as the vicinity of Manaus, just three degrees south of the equator, where temperatures as low as $14.3°C$ have been recorded (Brinkmann et al., 1971; Ribeiro, 1976).

Rubber does not grow naturally in frost-prone areas, so extending plantings into areas subject to periodic cooling is risky. Acre, the Beni, and Madre de Dios in southwestern Amazonia are likely to prove the best areas to search for cold-tolerant rubber.

The productivity of rubber plantations increased spectacularly between 1910 and the 1960s, in large part because of bud-grafting of superior clones. Rubber yields on rubber plantations in Southeast Asia increased nearly fivefold by the 1940s (R. J. Seibert, 1948). In the 1970s, some Asian rubber plantations were achieving yields approaching 1,000 kilograms a hectare, a remarkable seventeenfold increase from 1910 yield levels. The increase was achieved by applying ethanol to tapping panels (Schultes, 1977a). The improved productivity of rubber plantations in Malaysia is reflected in the country's increased exports of natural rubber, which rose from 931,611 tons in 1965 to 1.6 million metric tons by 1983 (Jain, 1988:10). Few crops have experienced such rapid yield increases due to breeding and improved agronomy and harvesting techniques in such a short time. To maintain such momentum, growers need fresh germplasm to boost and stabilize rubber yields.

GENETIC RESOURCES

Rubber is a highly variable species in the wild, a major reason why the taxonomy of *Hevea* has been confused for so long. Disputes have arisen as to whether *H. brasiliensis* has crossed frequently with its wild relatives. Some claim that spontaneous hybridization is relatively common in the wild, whereas Schultes (1987a) suggested that such events are infrequent. Rubber has reportedly crossed with seringueira barriguda (*H. spruceana*) along the Amazon and Rio Negro (Baldwin, 1947a; R. J. Seibert, 1948).

Although the impressive variability of wild rubber has stirred disagreements among taxonomists and geneticists, such variation is a boon to breeders. The best rubber trees in terms of quality and yield are found in Acre, the Madre de Dios in Peru, and the Beni in Bolivia. 'Acre fina' consistently fetches a higher price than the superior rubber produced on plantations of Southeast Asia. Acre rubber is coveted for its outstanding strength and elasticity, qualities needed for such applications as surgeons' gloves (Schultes, 1970). Some rubber trees in the vicinity of Leticia in the Colombian Amazon exhibit an unusually high degree of resistance to SALB (Schultes, 1987a). Other important qualities, such as drought tolerance and insect resistance, may be found in rubber trees in other parts of the species' extensive range (Figure 5.2).

Rubber has nine near relatives (Table 5.1), all of which can interbreed (Schultes, 1956; Wycherley, 1976:78). Several wild relatives of rubber have already been used in breeding programs, while others have been entirely ignored. Only the tip of a genetic pyramid has been utilized thus far.

Breeders have been wary of using near relatives of rubber in breeding programs because of yield and latex quality considerations. Latex quality

TABLE 5.1
Distribution and habitat of rubber's near relatives

Hevea species	Distribution	Habitat
benthamiana[1]	Mainly N of Amazon, Upper Orinoco, Lower Madeira	Flooded forest
camargoana	Marajó Island	Grassland
camporum	Amazonas, Pará, Brazil	Grassland on sandy soils
guianensis[1]	Amazonia, Guianas, Upper Orinoco	Forest on well-drained soils
microphylla	Middle and upper Rio Negro	Flooded forest
nitida	Upper Rio Negro	Scrub on sandy soils
pauciflora	Amazonia, Guianas	Forest on well-drained soils
rigidifolia	Upper Rio Negro	Scrub on sandy soils
spruceana	Middle and lower Amazon	Floodplain forest

Sources: Aublet, 1775:873; Baldwin, 1947a, b; Ducke, 1939; Schultes, 1945, 1970, 1977b, 1990; Teixeira and Secco, 1989; Wisniewski and Melo, 1986; Wycherley, 1976:78.
[1]Supplies a useable latex.

is judged according to the elasticity and durability of the resulting rubber. Latex of most of rubber's wild relatives has now been examined and tested in laboratories. As expected, *H. brasiliensis* consistently produces the best rubber. Rubber's wild relatives, in descending order of latex quality, are as follows: *H. benthamiana, guianensis, pauciflora, rigidifolia, nitida, spruceana*, and *microphylla* (Wisniewski and Melo, 1986). Both *H. benthamiana* and *H. guianensis* produce commercially acceptable latex.

Little is known about the full range of genetic resources of rubber's wild relatives, but they harbor resistance genes to several diseases. *H. benthamiana*, for example, resists SALB, and its latex is of sufficient quality to warrant tapping, such as along the Lower Rio Negro (Wisniewski and Melo, 1986). Attempts to transfer resistance from *H. benthamiana* to *H. brasiliensis* have met with limited success because although the resulting hybrids resist the disease, they are modest yielders. Also, the stability of many supposedly resistant clones in Brazil is doubtful, probably because at least four races of SALB attack rubber trees (Chee and Wastie, 1980; Dean, 1987:148). Nevertheless, rubber clone IAN 6158, developed near Belém with *H. benthamiana* germplasm in the 1950s, has exhibited strong resistance to SALB. A clone of IAN 6158 planted in 1978 at CPAA near Manaus was still vigorous a decade later when surrounded by other rubber genotypes that were clearly suffering from *M. ulei* attack. A more protracted breeding effort is required in order to break linkages with undesirable genes. Biotechnology developments, particularly emerging recombinant DNA techniques, may facilitate this process.

Another wild relative of rubber, seringa da serra (*H. pauciflora*), has exhibited strong resistance to SALB. A clone of *H. pauciflora*, PA 31, has been successfully crown-budded on rubber, and this combination shows some promise in producing high yields in areas where SALB prevails (Hashim, 1983). Hybrids between *H. pauciflora* and *H. brasiliensis* resist SALB, but they produce poor latex. However, spontaneous introgression of *H. pauciflora* genes into rubber populations has apparently occurred in the Rio Negro watershed, and that broad river system may harbor wild rubber trees with varying resistance to SALB (R. J. Seibert, 1948).

Wild relatives of rubber contain genes for resistance to a host of other diseases and insect pests. For example, *H. rigidifolia* resists Phytophthora leaf fall and dieback caused by *Botryodiplodia theobromae*, two serious fungal diseases (Schultes, 1977b). In addition to resistance to SALB, *H. pauciflora* is immune to, or tolerates, several diseases. When target leaf spot was first seen attacking rubber in Peru in 1942, breeders were unable to find sources of resistance to the fungal pathogen, *Thanatephorus cucumeris*, in germplasm of *H. brasiliensis*. Fortunately, both *H. rigidifolia* and *H. benthamiana* show some immunity to the disease, which attacks numerous crops. A *brasiliensis* × *benthamiana* hybrid has been used as a top-bud where target leaf spot is prevalent, such as near Manaus, where it is known as *mancha areolada*.

Other useful traits undoubtedly abound in close relatives of rubber, such as adaptability to a wide range of soil conditions (Table 5.1). The development of high-yielding rubber varieties that thrive on diverse soil types is a high priority among rubber breeders. *H. nitida* shows promise for tolerance to sandy soils as well as high winds (Schultes, 1988:258). A subspecies of this rubber relative, var. *toxicodendroides*, thrives on the thin, coarse soils atop sandstone hills in the Colombian Amazon. The short stature of this subspecies, reaching only about 3 meters, makes it an ideal candidate for crossing with *H. brasiliensis* to improve resistance to high winds.

GENE BANKS

Rubber seeds are recalcitrant and cannot be dried and frozen for safeguarding in a seed gene bank. Rubber germplasm must thus be conserved as trees in field gene banks and in wild habitats. Only a handful of countries maintain sizable gene banks for rubber, with Brazil and Malaysia containing the largest collections. CPAA near Manaus in the Brazilian Amazon has the most comprehensive collection of *H. brasiliensis* from many parts of the 5-million-square-kilometer basin. The rubber gene bank at CPAA started in 1982 and by 1988 had 1,300 genotypes, 60 percent of which are improved clonal materials. CPAA also maintains fifty samples of *H. pauciflora*, two *H. guianensis*, several *H. benthamiana*, and one each of *H. nitida*, *H. camporum*, and *H. camargoana*. Brazil generally exchanges rubber germplasm on a bilateral basis, with the understanding that the recipient country must check with EMBRAPA before it passes any materials to other countries.

A major problem with the rubber collection at CPAA is that it is severely attacked by SALB as well as other diseases and insect pests. CPAA is an excellent place for screening germplasm for resistance to *M. ulei*, but is a difficult location for maintaining a wide assortment of rubber germplasm. Many introductions succumb, including some genotypes of rubber's wild relatives that are brought into a "hot bed" of pathogens and pests. Brazil's national agricultural research program (EMBRAPA, Empresa Brasileira de Pesquisa Agropecuária), the parent body for CPAA, is seeking another site for the rubber germplasm collection, possibly in the drier cerrado region of central Brazil.

Malaysia also holds a sizable collection of rubber germplasm (Table 5.2), but since it has relied on introductions, is less comprehensive than the collection in Brazil. Furthermore, the Malaysian collection is geared more to short-term breeding objectives than to encompass rubber's broad ge-

TABLE 5.2

Accessions to the rubber germplasm collection held by the Rubber Research Institute of Malaysia (RRIM)

Year	Material	Origin	Source
1877	22 seeds	Pará	Wickham
1951–52	1,000 seeds[1]	Acre, Rondônia	Bilateral exchange
1951–52	100 clones	Various	Ford collection
1966	100 clones[2]	Various	Schultes collection
1980	100 seeds[2]	?	?
1981	35,000 seeds	Acre, Mato Grosso, Rondônia	EMBRAPA/IRRDB expedition

Source: IBPGR, 1984.
[1]Includes 20 seedlings of *H. spruceana* and 10 seedlings of *H. pauciflora.*
[2]From the following *Hevea* species: *bethamiana, guianensis, nitida, pauciflora,* and *rigidifolia.*

netic diversity. Most of RRIM's germplasm is composed of seedlings of material gathered during a joint EMBRAPA/International Rubber Research and Development Board (IRRDB) expedition in the southern part of the Brazilian Amazon in 1981.

IRRDB, based in Brickenbury, United Kingdom, helps distribute rubber germplasm to its member countries, which include Brazil, Cameroon, China, India, Indonesia, Côte d'Ivoire, Malaysia, Mexico, Sri Lanka, Thailand, and Vietnam. Nonparticipants in IRRDB are not eligible to receive germplasm collected on IRRDB-financed missions, but member countries account for 95 percent of the world's rubber production.

Although valuable, other rubber germplasm collections contain a much narrower range of germplasm. A collection at Turrialba, Costa Rica, has some of the better-yielding, partially resistant selections from the former Ford plantation along the Tapajós River. Turrialba also has several hundred rubber seedlings from the prime Madre de Dios area (IBPGR, 1984:4). Guatemala also has several hundred trees from the Madre de Dios in Peru on the former Firestone plantation at Clavellinas. The Summit Gardens in Panama have a mixture of clones resistant to SALB and some high-yielding material from Asia. This old rubber breeding garden could harbor a wealth of seedlings resistant to SALB with high yield potential.

During World War II, the USDA dispatched several botanists to Amazonia in search of promising material to establish plantations in the New World. Between 1943 and 1947, Russel J. Seibert obtained more than 300 selections from elite rubber trees in the Madre de Dios. Richard Evans Schultes scoured large areas of the Colombian and Brazilian Amazon and made more than 200 selections of superior *H. brasiliensis* for breeders. Most of these selections are lost or neglected (Schultes, 1977b). After

World War II, trade in natural rubber resumed from Asia and interest in furthering plantations in the American tropics cooled.

A major problem with rubber gene banks is that they are relatively costly to operate. A field gene bank for 5,000 rubber clones would occupy approximately 40 hectares, a sizable portion of land for most research stations. The farther a gene bank is located from urban areas, the more difficult it will be to secure qualified personnel to work with the collection. Land is certainly cheaper away from towns and cities, but if a field gene bank is isolated from places where scientists and technicians work, it will not be as well maintained or properly used.

Rubber gene banks need to be monitored constantly for diseases and pests. At CPAA near Manaus, Brazil, for example, the trunks of young rubber clones are painted with a bactericide to improve survival rates. IRRDB plans to establish and evaluate two rubber breeding collections, one in Malaysia and the other in Côte d'Ivoire, at a cost of $200,000 per year (IBPGR, 1984:9).

One way to reduce the cost of ex situ conservation of rubber germplasm would be to employ tissue culture techniques. That way, backup materials could be maintained in small test tubes in a laboratory instead of on extensive plots of land. Unfortunately, tissue culture techniques for rubber are a long way from being perfected. IRRDB and the rubber research institute (IRCA, Institut de Recherches sur le Caoutchouc) at Montpellier, France, are investigating ways to store rubber germplasm in vitro.

IRCA's Paris office and CPAA near Manaus are exploring the potential for identifying genetic markers to determine genetic variability in species of *Hevea*. Such techniques can cut down on redundancy within collections and assist efforts to broaden variation within them.

Quarantine operations no longer impede significantly the exchange of rubber germplasm. For Brazilian materials destined for Asia, *Hevea* seeds or budwood are first inspected and treated at the point of departure. Further inspections and treatments are performed at the Tun Abdul Razak Laboratory on IRRDB's grounds in the United Kingdom, and upon arrival in Malaysia and Côte d'Ivoire. Intermediate quarantine facilities for rubber are provided by IRCA's tropical fruit research station on Guadeloupe; budwood taken from material quarantined there is kept under glasshouse observation in Malaysia for six months (IBPGR, 1984:10). The USDA Subtropical Horticulture Research Station in Miami formerly provided a rubber quarantine service and might be able to do so again if the need arose.

In consultation with rubber experts worldwide, the International Board for Plant Genetic Resources (IBPGR) has drawn up a list of priorities for collecting *Hevea* germplasm (Table 5.3). Wild relatives of rubber are poorly represented in germplasm collections, so all species of *Hevea* are

TABLE 5.3

Priorities for collecting *Hevea* germplasm

Hevea species	Priority area(s)	Reason
	Top Priority	
brasiliensis	(a) Purus and Madeira river basins	Settlement and development
	(b) N. Bolivia and Madre de Dios	Superior germplasm
benthamiana	Central Amazon around Manaus	Material from this area poorly represented in gene banks; resistance to South American leaf blight (SALB)
guianensis	Peru, Bolivia, S. Amazonas, Guianas	Potential of material from these areas unknown
pauciflora	West of Manaus	Potential unknown
camporum	Amazonas, Pará, Brazil	Potential unknown
microphylla	Brazilian-Colombian border	Potential unknown
	Lower Priority	
nitida		Minimal threat from human settlement or development
rigidifolia		Minimal threat
spruceana		Minimal threat

Source: IBPGR, 1984:8.

targeted for further collecting. How much germplasm can be collected over the next few years will depend on the availability of funds and on policies regarding germplasm conservation. As of 1988, for example, Brazil has halted its own rubber germplasm collecting program pending resolution of where the gene bank will be ultimately located. Germplasm of plants or animals can be collected in Brazil only with permission from the national genetic resources program (CENARGEN, Centro Nacional de Recursos Genéticos e Biotecnologia), usually on a partnership basis. Until the rubber gene bank at Manaus is relocated, further rubber germplasm collecting trips in Brazil are unlikely. It is hoped that this matter will be resolved soon because some populations of *H. brasiliensis* and its near relatives are threatened by settlement and development projects.

IBPGR has designated rubber as a crop deserving highest priority for future ecogeographic surveys (IBPGR, 1985:14). The Rio Negro watershed in Amazonia, comprising over 2 million square kilometers, is a high priority for collecting since it contains most of rubber's near relatives as well as a high degree of variability within each species (Schultes, 1945).

IN SITU CONSERVATION

Efforts to conserve rubber and its near relatives in their natural forest habitats are urgently needed (IBPGR, 1985:20). The Rio Negro watershed is still relatively sparsely settled, so populations of rubber and its near relatives in that portion of the Amazon basin are not in immediate danger of extinction. The Negro river system has not been densely settled because of the infertile sandy soils that prevail in the region and the low productivity of fisheries, partly a result of the extremely acid conditions of the Negro and its dark tributaries. Furthermore, a number of parks and nature preserves have been set up in the Rio Negro area. These include the 2.2-million-hectare Pico da Neblina park straddling the Brazil-Venezuela border, which was established in 1979; the Jau national park inaugurated in 1980 with 2.27 million hectares; and the Maracá-Roraima Ecological Station, which encompasses 920 square kilometers (S. W. Barrett, 1980; IBDF, 1982:15). Botanical surveys of these national parks have not been made, but a limited amount of botanical work is under way in the Maracá-Roraima Ecological Station.

The sizable national parks in the Rio Negro system are a valuable addition to Brazil's growing list of parks. They should be well protected to safeguard plant and animal resources, because forest clearing in the Negro watershed is accelerating. The Manaus-Caracaraí-Caracas road has been open for nearly a decade and is attracting settlement in the eastern part of the Negro watershed (Figure 5.4). Also, work on the Northern Perimeter Road (Perimetral Norte), initiated in 1974 to open up border areas of the Brazilian north, has started again, and this 5,000-kilometer highway is expected to entice settlers. Mineral prospecting may uncover valuable deposits of gold or other precious metals in the Negro watershed, thereby acting as a magnet for development.

Acre, a Brazilian state in southwestern Amazonia, is home to the highest concentration of, and reputedly the best-yielding, wild rubber trees. A sizable proportion of the rubber collected in the Amazon basin has always come from this corner of South America's equatorial rain forest. In the early 1950s, Acre accounted for a third of rubber produced in the Brazilian Amazon, even though the state accounts for less than 5 percent of Brazil's northern region. In 1979 and 1985, Acre accounted for 47 and 45 percent, respectively, of rubber tapped in the Brazilian Amazon (Castro Soares, 1956:97; IBGE, 1981:217, 1986:284). Acre, then, should be a high priority for collecting rubber germplasm with high yield potential and would be an important area to establish in situ reserves for *H. brasiliensis*.

Acre contains some of the most productive rubber trees but it is also a pioneer area undergoing rapid development. In the mid-1980s, a road was

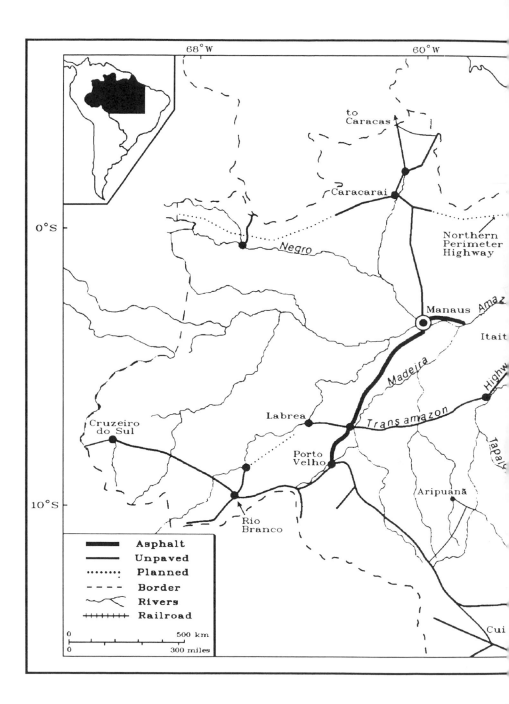

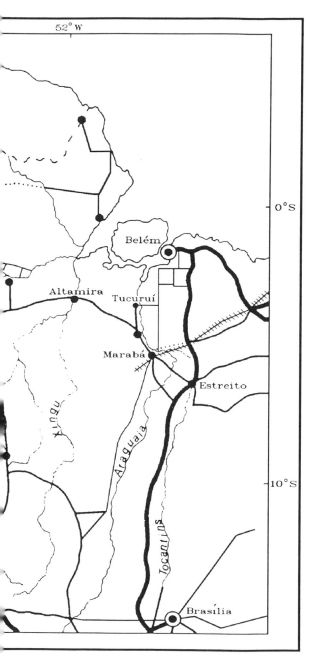

FIGURE 5.4 Highway system in
the Brazilian Amazon.

227

opened linking Porto Velho in Rondônia with Rio Branco, the capital of Acre. This road is being extended to Cruzeiro do Sul. The Porto Velho–Rio Branco road is being asphalted, a move that is accelerating settlement and agroindustrial development. Cattle ranching, and to a lesser extent sugarcane planting, are the two main agricultural enterprises being promoted. Both entail destruction of the forest and the loss of wild gene pools of rubber and other perennial crops.

Attempts are being made to conserve parts of the forest as "extractive reserves"; such conservation could help safeguard some wild rubber germplasm if they are viable. (The role of extractive reserves in conserving plant genetic resources is discussed more fully in Chapter 11, p. 438) The debate over wise use of pioneer territory in Acre involves a resource of concern to rubber producers and consumers throughout the world. Several sizable extractive reserves have recently been established in Acre, but their success ultimately depends on whether the people who live in them can obtain a substantial portion of their income from forest products. If they cannot, the forest may have to be cut down.

Rondônia, the second largest producer of rubber in Brazil, is also experiencing rapid settlement. Rondônia is the focus of the Polonoroeste development scheme, which is attracting large numbers of settlers from central and southern Brazil. Deforestation rates in Rondônia are among the highest in Amazonia, and genetic resources of a number of crops are undoubtedly being lost. Some of the best rubber groves are making way for cattle ranches because their land titles are sometimes more regular and therefore easier to purchase (Dean, 1987:149). Of more than 14,000 seedlings established in Malaysia from three states in the Brazilian Amazon, those from Rondônia showed the greatest vigor and plant height (Mohd, 1984). The wave of migrants to Rondônia increased from about 65,000 people a year during the 1980–83 period to some 165,000 settlers annually between 1984 and 1986 (Mahar, 1989). Fortunately, two large areas have been set aside for conservation in Rondônia: the 764,801-hectare Pacaas Novos National Park, and the 286,150-hectare Jaru Biological Reserve, both set up in 1979.

Two other parks or reserves that have been set up in Amazonia probably contain some wild rubber. The 1-million-hectare Amazon national park on the left bank of the Tapajós above Itaituba, created in 1974, is sure to contain some *H. brasiliensis*, although gold mining has now surpassed latex gathering in this part of the Tapajós watershed. The 385,000-hectare Trombetas biological reserve, set aside in 1979, along a northern tributary of the Amazon is likely to contain several species of *Hevea*. Even if these parks and reserves are adequately protected, however, only a small part of rubber's genetic variability would likely be represented. Several species of *Hevea*, such as *H. camporum* and *H. camargoana*, are probably not present in any park or reserve.

Realizing the Potential

All commodities suffer periodic swings in prices in response to shifts in supply and demand, but the price prospects for rubber through 1995 are favorable (World Bank, 1983:56). In the late 1980s, for example, Indonesia reaped the benefits of natural rubber's competitive edge in export markets (*Economist*, 13 August 1988, p. 60). Beyond 1995, prospects for natural rubber remain bright because increased petroleum prices will likely boost natural rubber prices as well (Grilli et al., 1980:6). The fortunes of natural rubber are tied to a large extent to the price of petroleum, the raw material for natural rubber's major competitor—synthetic rubber. As petroleum prices climb in the future, so will the incentive to plant rubber.

Some three-quarters of world rubber production is destined for tires (Schultes, 1977a), so technological changes in the tire industry will have a considerable bearing on the fate of commercial rubber plantings. When supplies of natural rubber from Southeast Asia were disrupted during World War II, industrial nations developed synthetic rubber and some foresaw the demise of rubber as a viable commercial crop. In the late 1950s, synthetic rubber overtook natural rubber consumption and by the mid-1980s, synthetic rubber accounted for close to two-thirds of rubber use worldwide (Grilli et al., 1980:20).

Still, no artificial products can match the extraordinary resiliency and durability of natural rubber, and demand for the product continues to grow. Radial tires, once the domain of racing and sport cars, are now virtually standard equipment on family sedans in Western Europe and North America (S. C. Tan, 1984:4). Radial tires were first marketed in Europe in 1948. Some twenty years lapsed before they made significant inroads in the U.S. tire market (Brice and Jones, 1988). The superior performance of radial tires is largely due to their much greater content of natural rubber than conventional cross-ply tires. Natural rubber now constitutes at least 40 percent of the rubber in radial tires for automobiles (Gentry, 1986). Although the world output of automobiles is not expected to grow much over the next decade, the proportion of vehicles equipped with radial tires is expected to grow substantially. Trucks also use tires with a considerable proportion of natural rubber to withstand better the stress of heavy loads and high, sustained speeds. Aircraft tires have always contained a large proportion of natural rubber. Most airlines anticipate an increase in the size of their fleets over the next decade in response to a marked increase in air traffic.

The global concern about AIDS has resulted in an explosive demand for latex gloves. More hospital personnel now use latex gloves to reduce the chances of acquiring the usually fatal disease. Latex gloves have now appeared in other quarters to help protect employees. Dentists and dental technicians, for example, routinely wear latex gloves while working with patients. And maids in many hotels and motels in North America don latex gloves when servicing bathrooms. Most latex gloves used in North America at least are imported from Malaysia.

With the recent development of epoxidized natural rubber (ENR) by scientists in the United Kingdom and Malaysia, the prospects for natural rubber are expected to be even more buoyant. In the ENR process, rubber is treated with peracids, a class of chemicals that can be made cheaply in bulk quantities, so that it has improved capacity to absorb shocks (*Economist*, 15 November 1986, p. 107). Engine mountings and vehicle suspensions are two potential uses for ENR. This form of modified rubber adheres well to plastics, thereby providing a flexible composite for conveyor belts and shoes. ENR is essentially impermeable to gases and liquids, making it particularly suitable for vapor and oil seals. ENR, when combined with some synthetic rubber, produces tires superior to those made from either natural or synthetic rubber.

A growing market for high-quality tires and the great potential for ENR products are generating optimism about the economic future of rubber. Natural rubber is finding increased uses for vehicle suspension and chassis systems, bridges, and earthquake-resistant buildings (Brice and Jones, 1988). A large array of other products could be developed from rubber, such as hormones and other chemicals. Given sufficient research and marketing studies, rubber trees could become veritable factories for a range of products.

The earliest use of rubber, for food, also deserves a second look. Rubber plantations could be set up to provide food or livestock feed in addition to other products. Although rubber seeds are toxic, they can be rendered safe to eat by simple household procedures still used in parts of Amazonia today (L. C. Wheeler, 1977). If a market cannot be developed for rubber seed as human food, livestock may benefit from being fed seed cake. After oil is expressed from rubber seed, the cake contains about 30 percent protein (Purseglove, 1974:154). In Sri Lanka alone, some 7,000 metric tons of rubber seed cake are discarded annually after the oil is extracted (L. C. Wheeler, 1977).

Another traditional use for rubber, as fish bait, provides further opportunities for devising rubber plantations with multiple uses. Some highly esteemed fish in Amazonia, such as tambaqui (*Colossoma macropomum*), feed avidly on seeds of *H. brasiliensis* and *H. spruceana* (Goulding, 1980:79; N. J. H. Smith 1981:63). Aquaculture could be integrated with rubber plantations, thereby further brightening prospects of the crop. Rubber produces seeds seasonally, so plantations could provide supplemental food, rather than a mainstay, for aquaculture schemes.

Growth in demand for rubber is coming not only from industrial countries, but from developing countries as well. Between 1961 and 1980, the overall share of world consumption of rubber doubled in Third World countries (World Bank, 1983:54). The ultimate success or failure of attempts to meet growing demand for natural rubber will have a major impact on the lives of millions of rural families in the Third World and on the economies of several developing countries (World Bank, 1982:11).

How much producers will be able to capitalize on the expected growth in demand for natural rubber as petroleum supplies dwindle depends to a large degree on planting rates. In the late 1980s, rubber planting worldwide lagged behind replacement needs by roughly one half. Rubber planting rates need to be stepped up by at least 100,000 hectares a year just to maintain current levels of production (World Bank, 1982:35). The useful life of a rubber tree on a plantation is about thirty years, so a little over 3 percent of existing plantings need to be renovated every year in order to maintain the same planted area. Only Thailand is expanding its rubber area. India, Brazil, the Philippines, and several West African countries hope to expand their rubber plantings. Nigeria, Cambodia, and Vietnam need to rehabilitate old plantations that have suffered neglect over the last few decades. Clearly, demand will remain strong for high-yielding, disease-resistant rubber germplasm—a major challenge for rubber breeding programs worldwide. To help ensure a reliable supply of latex and other products of rubber, producers and the rubber industry share a responsibility in providing some of the resources needed for gathering, conserving, and evaluating the genetic resources of this versatile crop.

OIL PALM

Oil palm (*Elaeis guineensis*), native to the tropical forests of West and Central Africa, is also widely grown on plantations and on smallholdings in Southeast Asia and in the humid lowlands of Latin America. Known best as a large-scale plantation crop, oil palm is a successful cash crop for farmers with as little as 4 hectares in Malaysia (World Bank, 1988). Oil palm is the most productive crop for vegetable oil and is also a good source of vitamin A, the highest source of that vitamin among vegetable oils (Cobley, 1956:127; Rajanaidu, 1990). In Bahia, Brazil, orange-colored palm oil is an ingredient in many regional dishes, and, considering the dearth of vegetables, the savory oil makes a valuable contribution to the nutritional health of the region's growing population.

Two distinct oils are derived from *E. guineensis*. Palm oil is extracted from the fleshy mesocarp that covers a nutlike endocarp. Kernel oil is extracted from the nut after the hard casing has been broken open. Hundreds or even thousands of the fruits are attached to broomlike bunches (spadix). Hooked knives attached to long handles are used to cut down the bunches from trees (Figure 5.5). Fruits are loosened from the bundles by steaming and turning them in a drum. The softened fruits are then pressed

FIGURE 5.5 Harvested oil palm fruits at the DENPASA plantation east of Belém, Pará, Brazil. August 1988.

to extract the oil. In Togo, the pressed fruits are dried and fashioned into cakes for cooking fuel. In some small-scale oil palm factories in Pará, Brazil, the fruit stalks and endocarp shells are fed into boilers to generate steam for loosening the fruits.

In Malaysia, the potash-rich residue from such boilers is routinely recycled back onto plantations to help enrich the soils. In other areas, such as Brazil, labor and transportation costs discourage such nutrient recycling.

Palm oil is used in sauces and baked goods, as shortening, to fry food, and to make soaps, candles, and margarine. Palm oil is also an ingredient in ice cream, coffee whitener, confections, polishes, and cosmetics, and is employed in the tin-plate industry. Palm kernel oil is used to make glycerine soaps and is a common substitute for cocoa butter in the manufacture of chocolate (Svarstad, 1987).

In West Africa, an alcoholic drink is prepared by fermenting sap tapped from the flower stalks of oil palm, similar to the tradition of making toddy wine from *Borassus flabellifer* along the Malabar coast of India and in Sri Lanka. Wine from oil palm is a locally lucrative business and a significant source of vitamin B (Hartley, 1988:8). Some people cut down oil palms to tap the sap and visit the dead trunks later to garner mushrooms (F. R. Irvine, 1930:173). Oil palm provides heart-of-palm, and the generous fronds are used for thatch.

In 1964, palm and soybean oil accounted for approximately equal shares of the world trade in vegetable oils and fats, but by 1986, palm oil shipments worldwide were three and a half times greater than soybean oil on a tonnage basis (Hatje, 1989). Palm oil production more than doubled in the 1970s, and by the mid-1980s, palm oil accounted for nearly half the global trade in vegetable oils (Gascon et al., 1989).

The market for oil palm in industrial countries has declined somewhat in recent years because of health concerns. The soybean lobby in the United States has been at the forefront of the campaign to discourage consumption of palm oil. Palm oil contains highly concentrated fats, and such fats have been found to raise blood cholesterol levels (Vles and Gottenbos, 1989). Palm oil does not contains as many saturated fats as kernel oil, and neither contains as much as coconut (*Cocos nucifera*) oil (Purseglove, 1975:480). Some cookie and breakfast cereal manufacturers now label their products with phrases such as "no tropical oils."[1] Polyunsaturated vegetable oils, such as from maize, soybean, and safflower are thus replacing palm oil and coconut oil in some markets. Vegetable oils based on crops grown in temperate countries are sometimes touted as "cholesterol-

1. See, for example, "Cookies, the Heart Can Love: Bowing to Consumers, Foodmakers are Abandoning Tropical Oils," *Time*, 23 January 1989, p. 71; "Tropical-oil War Still Heating up," *Gainesville Sun*, Gainesville, Florida, 27 January 1989, p. 9C.

free," even though all vegetable oils, including palm and coconut oil, are free of cholesterol. Less-saturated palm oil could be a high priority in breeding and selection programs in the near future. American oil palm (*E. oleifera*), a near relative of oil palm, has higher levels of unsaturated fatty acid than its African cousin (Ooi et al., 1981; Rajanaidu, 1983).

For markets in the Third World, however, saturated fats are not such a concern because calories are frequently deficient in the diet and there are often shortages of vegetable oils. Furthermore, the main producers of unsaturated vegetable oils are industrial countries, particularly the United States and Western Europe, and developing countries need to develop their own sources of vegetable oil whenever feasible.

Dramatic shifts in oil palm production have occurred within the last half century. The relative importance of West Africa as an exporter of palm oil has declined, but Malaysia has greatly expanded production and has surpassed West and Central Africa. In 1900, for example, Nigeria produced virtually all palm oil traded on the world market, and the commodity accounted for 88 percent of its export earnings (Harrison Church, 1960:106).

Malaysia began exporting palm oil in 1926 and now dominates global exports. In 1940, Malaysia had a little over 30,000 hectares in oil palm; by the mid-1970s oil palm plantations covered 400,000 hectares, and by 1980 one million hectares were in oil palm (Hartley, 1988:28; Purseglove, 1975:508). Oil palm production in Malaysia has grown from 34,544 metric tons in 1965 to 3 million tons by 1983 and 6 million tons by 1990 (Jain, 1988:10; R. Rajanaidu, pers. comm.). Malaysia now accounts for two-thirds of the palm oil traded on the world market (Rajanaidu, 1990).

Some of the dramatic expansion of oil palm plantings in Malaysia has come at the expense of tropical forest. In the Jengka triangle, for example, 40,000 hectares of forest were cleared to plant 26,000 hectares of oil palm and 13,800 hectares of rubber (World Bank, 1988). On the other hand, oil palm is a good crop for rehabilitating degraded areas, as on Sumatra, where oil palm has been successfully established on abandoned farmlands taken over by rank *Imperata cylindrica* grass.

In addition to the main producing areas of Malaysia and West Africa, oil palm plantations have also been established on a commercial scale in Angola, Brazil (Bahia, Pará), Cameroon, China, Colombia, Costa Rica (near Golfito, Osa Peninsula), Ecuador, Guatemala, Honduras, Indonesia, southern Mexico, Panama, and Thailand. Palm oil exports generate close to $3 billion annually for developing countries (Rajanaidu, 1990). Several countries plan to expand oil palm production, such as Brazil and Papua New Guinea. The governments of France and the United Kingdom are assisting Papua New Guinea's efforts to plant more oil palm. Yet, the European Community (EC) levies a 12 percent tariff on Malaysian palm oil; the income from this tax helps subsidize rape seed producers in the EC.

Origin and Spread

Oil palm is native to forested portions of West and Central Africa, particularly along rivers. Six thousand-year-old archaeological remains in the savanna region of West Africa indicate that oil palm and tropical forest once extended farther north than at present. Oil palm may have been domesticated first in Chad (Portères, 1962), but incipient domestication probably occurred over a much broader area. Village sites in humid regions of West and Central Africa have undoubtedly been enriched with oil palms for millennia, so its natural distribution has been blurred by human agency (Ascenso, 1966).

Long before the arrival of Europeans, oil palm was taken to the Congo basin and East Africa. People took oil palm to Sudan at least 5,000 years ago (Clark, 1976). Tribal migrations or intergroup exchanges spread oil palm across Africa, as inferred from distinct patterns of variation and close correlation between them and ethnic groups (Meunier, 1969). Oil palms are now characteristic of the Ruzizi plain in the hot, down-faulted valley bottom north of Lake Tanganyika. Africans probably took oil palm to Madagascar in the tenth century (Figure 5.6). Oil palm is also cultivated sporadically along the southern coast of Kenya and on Pemba and Zanzibar, where it is thought to have been established in the tenth century (Hartley, 1988:6; Wiedner, 1962:101), although it was probably introduced there earlier.

Slave traders took oil palm for food to the New World, but until recently it has been cultivated to any extent only in Bahia, Brazil. There the propitious climate and large community of Afro-Brazilians, particularly of Yoruba extraction, have favored the trees. It is possible that oil palm and banana reached the New World before Europeans, but both crops would have had to be brought by people. Palm fruits do not float, and banana is propagated vegetatively. No evidence has yet emerged to suggest that such long-distance travel occurred in pre-Columbian times.

The Dutch introduced oil palm to Southeast Asia. In 1848, the Bogor (then Buitenzorg) Botanic Garden received four oil palm seedlings, two from Amsterdam Botanic Garden and two from Réunion (then Bourbon). The Singapore Botanic Gardens obtained oil palm seeds from Java around 1870 and also helped diffuse oil palm throughout the Malayan Peninsula and into Sumatra.

The Genetic Foundation

The initial introduction of oil palm to Bogor laid the foundation for the oil palm industry in Southeast Asia. Progeny from the Bogor introduction were planted initially as ornamentals on tobacco estates around Deli and Medan on Sumatra between 1880 and 1900. As the market for oil palm

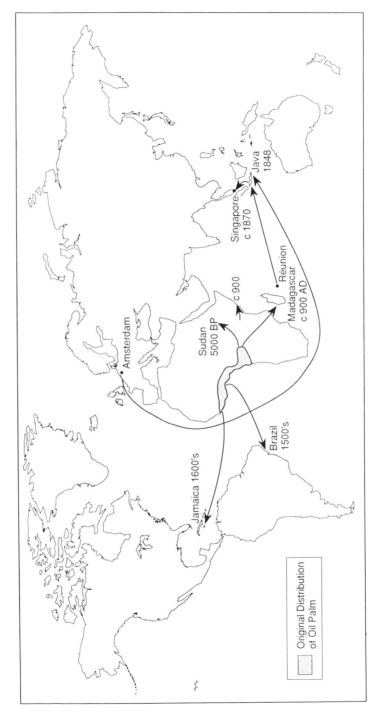

FIGURE 5.6 Origin and spread of African oil palm (*Elaeis guineensis*). (Adapted from Busson, 1965:507; Hartley, 1988:6; Purseglove, 1975:482; Wiedner, 1962:101.)

expanded in Europe, the exotic palm was cultivated on a plantation scale starting in 1911. The first oil palm plantations in Malaysia were established in 1917.

The four palms introduced to Bogor produced thick-shelled fruit of the *dura* type. In *dura* types, pulp typically constitutes 35 to 50 percent of the fruit. Fortunately, the original introductions were of superior genetic material, for they ultimately led to highly productive cultivars, including the 'Deli' types from Sumatra (Hardon, 1976). Although highly productive, the genetic base of the oil palm industry was exceptionally narrow (Arasu, 1985). All four oil palms introduced to Bogor are thought to have come from one tree (Hartley, 1988:20).

The genetic base to improve oil palm production in Central and West Africa was also narrow. Only ten open-pollinated fruit bunches of small-kerneled *tenera* types were used to launch oil palm breeding at Yangambi in Zaire (Hardon et al., 1987). One of these bunches was taken from a high-yielding palm from Eala, called *Django* (the best), and progeny from this bunch contributed over 70 percent of selections up to 1934. In Nigeria, most of the breeding material was obtained from four *tenera* palms and several *dura* types. Oil palm breeders in Côte d'Ivoire employed material from just thirty-eight palms from Benin and four in-country selections.

Gains in oil yield have been impressive. Oil yield jumped approximately 23 percent during the first selection cycle from plantings in the latter half of the nineteenth century. For much of the twentieth century, yield increases have averaged between 10 and 15 percent per eight- to ten-year generation (Hardon et al., 1987). More recently, little additive variation for yield improvement has been noted in current breeding populations, thus underscoring the need to broaden the crop's genetic base.

Limited breeding collections proved to be an adequate springboard for oil palm improvement in Africa and Asia, but long-term improvement requires fresh infusions of germplasm. Overall, the genetic base of oil palm plantations is still relatively narrow, considering the commercial and dietary importance of the crop (Rajanaidu, 1987). Future breakthroughs in yield and in disease resistance are likely to hinge on a sizable contribution of genes from wild or semidomesticated populations and American oil palm.

WILD GENE POOLS AND NEAR RELATIVES

Oil palm occurs in disturbed forest and along rivers and streams, both in its native range in West Africa and in some introduced areas. Oil palm is a successionary species, and its gene pool has thus expanded as farmers clear land and create more open habitats for oil palm to germinate (Figure 5.7).

FIGURE 5.7 Oil palm (*Elaeis guineensis*) in old successionary forest near Ibadan, Nigeria, June 1983.

The tall palm is usually spared when swidden fields are cleared and is often a conspicuous tree in second growth (Buchanan and Pugh, 1955:103; Busson, 1965:509). Spontaneous populations of oil palm thrive along the coast of Bahia, Brazil, and on Jamaica (Grisebach, 1864:522). In Brazil, black vultures (*Coragyps atratus*) feed avidly on oil palm and may be involved in their dispersal. Majestic oil palms do not survive in mature forest and are thus another tropical perennial favored by slash-and-burn farming (Hardon, 1974; Harrison Church, 1960:105; Hartley, 1988:3; Purseglove, 1975:485).

African oil palm has only one close relative, American oil palm (*E. oleifera*), which generally grows in damp sites from southern Honduras to the Amazon basin (Croat, 1978:173; León, 1987:58). Known as *caiaué* in Brazil, *coquito* or *palmiche* in Costa Rica and Nicaragua, *corozo* or *noli* in Colombia, and *obe* in Surinam, *E. oleifera* favors the margins of rivers and streams that are periodically flooded (Figure 5.8). In Amazonia, American oil palm is apparently confined to the Brazilian states of Amazonas and Roraima, where it occurs sporadically along some silt-laden rivers. In Costa Rica, the palm forms dense groves in freshwater swamps inland from sand dunes along the low-lying Caribbean coast.

FIGURE 5.8 American oil palm (*Elaeis oleifera*) on the floodplain of the Amazon River. Ilha do Risco, 15 kilometers downstream from Itacoatiara, Amazonas, Brazil, March 1977.

In the Magdalena Valley, Colombia, fruits from American oil palm are fed to pigs, and oil is extracted from the kernels. Some farmers in the broad and deep Andean valley have selected progeny from superior specimens for planting (Hartley, 1988:85). Oil is extracted from American oil palms in the Perlas Islands in Panama Bay (P. H. Allen, 1956:184). On Careiro Island near Manaus, Brazil, American oil palm fruits are occasionally fed to pigs and chickens.

American oil palm also serves a number of medicinal uses. On Careiro Island near Manaus, for example, old-timers assert that the oil discourages insect bites when smeared on the body (E. Barcelos, pers. comm.). Some Indian groups also anoint themselves with oil from American oil palm to ward off insects (Plotkin and Balick, 1984). Folk curers (*curanderos*) in Colombia employ the oil to treat stomach inflammation, and the oil is also used in Brazil and Colombia as a hair tonic. In Amazonas State, oil from *E. oleifera* is sometimes swallowed to alleviate whooping cough (Van den Berg and Silva, 1986). For the most part, though, American oil palm appears to be little used by rural peoples in Amazonia today or in the last century (Spruce, 1908a:479).

Although American oil palm is only sporadically harvested or planted at present, indigenous people may have altered its distribution. Some tribes may have found uses for the fruit and could have planted some individuals. America oil palm was probably planted at various locations in the past for its medicinal properties. Along the Urubu, Madeira, Negro, and Amazon rivers in Brazil, for example, *caiaué* is strongly associated with anthropogenic black earth, sites formerly occupied by Indian villages (Barcelos, 1986). In Amazonia, American oil palm is almost invariably found associated with deposits of broken pottery left by indigenous groups, and the palm is considered an indicator species for archaeological sites (Balée, 1988; De Blank, 1952). American oil palm may have been introduced to Amazonia long ago by indigenous peoples. In Amazonia, American oil palm is distinguished by its large fruit (Hartley, 1988:87; Rajanaidu, 1983), another indication that the thick-stemmed palm may have been introduced by people. Also, *E. oleifera* populations along the Amazon and Madeira rivers show little interpopulation variation (Ghesquière et al., 1987).

American oil palm may have been taken long ago from its ancestral home along the Pacific side of Central America to the Atlantic coast, where larger-fruited forms are found. From the Atlantic side of Central America, the palm could have been introduced to northwestern Colombia, which also has large-fruited forms (Rajanaidu, 1983).

American oil palm may have penetrated Amazonia along the Rio Negro or via headwaters in the Guianas (Figure 5.9). American oil palm appears to have been introduced to Surinam, where it is locally abundant on white

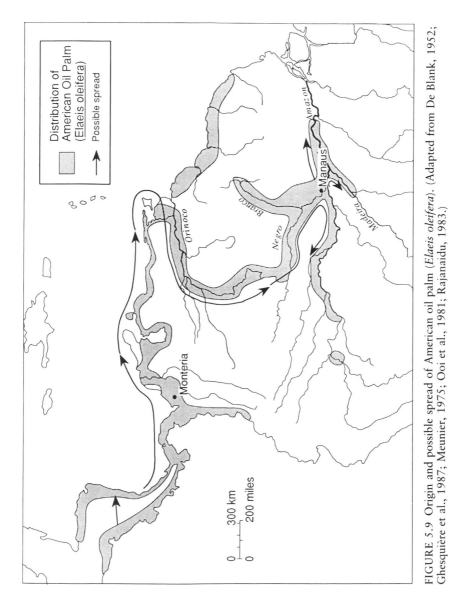

FIGURE 5.9 Origin and possible spread of American oil palm (*Elaeis oleifera*). (Adapted from De Blank, 1952; Ghesquière et al., 1987; Meunier, 1975; Ooi et al., 1981; Rajanaidu, 1983.)

sand savannas near Zanderij and Stoelmanseiland in the Marowijne watershed (Boer, 1965). Alternatively, seeds of the palm could have been carried across the Andes, since a population of American oil palm has recently been located near Iquitos in the Peruvian Amazon (E. Barcelos, pers. comm).

American oil palm was probably cultivated and harvested to a much larger degree in the past. Many of the peoples who once cared for the plants have long since gone. With the introduction of African oil palm to the New World and the arrival of coconut on the Pacific side of the American isthmus around the time of contact, other, more productive vegetable oils have become available. Although American oil palm is no longer planted and tended as in the past, its importance for the improvement of oil palm production has increased dramatically as a source of resistance to diseases.

Little systematic collecting and genetic evaluation of American oil palm have been attempted. American oil palm grows on a variety of soils from anthrosols to sandy soils and ill-drained swamps. Some natural or spontaneous groves are disappearing as settlers prepare land for agriculture and cattle raising. In Costa Rica, for example, cattle ranchers consider the palm a nuisance (Hartley, 1988:87). American oil palm generally prefers periodically flooded sites, which may provide some respite from development pressures, although in the Crete and Monteria area of northwestern Colombia, drainage of swamps for cattle raising and agriculture has destroyed extensive groves (Rajanaidu, 1983). Populations of American oil palm in better-drained areas are more vulnerable to expanding agricultural frontiers.

THE DRIVE TO COLLECT GERMPLASM

African oil palm has been relatively well surveyed in the field, but even for this important crop, only a broad-brush picture of its genetic variation has emerged. Prior to the 1970s, germplasm collecting missions in Africa focused on superior fruit forms and potential oil yield. The first systematic prospecting to gauge oil palm variability was undertaken in 1974 by the Palm Oil Research Institute of Malaysia (PORIM) and the Nigerian Institute for Oil Palm Research (NIFOR). The purpose of this mission and subsequent expeditions, supported in part by the International Board for Plant Genetic Resources (IBPGR), was to collect random samples of oil palm groves. The Institut National pour l'Étude Agronomique du Congo Belge (INEAC) in Zaire and the Institut de Recherches pour les Huiles et Oléagineux (IRHO) in Côte d'Ivoire also mounted expeditions to collect germplasm of oil palm in the 1970s. IBPGR also supported a cooperative effort between PORIM and agricultural research institutions in Costa

Rica, Honduras, Nicaragua, Peru, Colombia, Surinam, and Panama to sample populations of American oil palm.

The impetus for collecting came from two sources. First, the planned clearing of large tracts of land for development jeopardized stands of oil palm. Palm groves with possibly unique or rare genes of potential value have already been cleared in many parts of Africa (Arasu and Rajanaidu, 1975). Second, the practice of thinning palm groves to intercrop with food plants removes some towering, older palms that could be useful "mother trees" in future breeding efforts.

In 1984, IBPGR convened an Oil Palm Working Group to assess progress in germplasm collecting and to provide guidance for future collecting efforts. Angola, Zaire, the rift valley around Lake Tanganyika, parts of Cameroon and Côte d'Ivoire, seven other locations in Africa as well as Bahia, Brazil, were all accorded high priority for germplasm collecting. In 1984, IBPGR, PORIM, and a private British-Dutch firm, Unilever, collected around Lake Tanganyika and on Madagascar. All together, some 50,000 seeds, representing a wealth of genetic material for conservation, evaluation, and breeding, were gathered in the field between the mid-1970s and 1990.

Further collecting efforts are planned in some areas. The African Oil Palm Development Association (AFODA) plans to collect oil palm germplasm in West Africa to assist breeders in its member countries of Benin, Cameroon, Côte d'Ivoire, Ghana, Liberia, and Nigeria. Although such a collection mission would duplicate to some extent previous gene hunting expeditions mounted by PORIM in Malaysia, AFODA may well obtain some unique materials. Also, some redundancy in gene banks is useful for easier access for breeders and for safety. AFODA recognizes the need to institute exchange programs with existing oil palm gene banks to expand the variation in all germplasm collections.

CONSERVATION EFFORTS

The Malaysian government has agreed to safeguard much of the material collected in the 1970s and 1980s in a field gene bank. This "universal" gene bank, which will eventually cover several hundred hectares, is open to all oil palm breeders, and materials will be available free of charge.

Working collections of oil palm germplasm are also maintained by various public and private companies in other countries. At an experiment station on the Urubu River near Manaus, Brazil, for example, 474 accessions of oil palm are maintained by CPAA (Centro de Pesquisa Agroflorestal da Amazônia), formerly known as CNPSD (Centro Nacional de Pesquisas de Seringueira e Dendê). This 49-hectare field gene bank contains 246 accessions from spontaneous oil palm in Bahia, 178 accessions of

American oil palm from Amazonas State, and 50 accessions of *E. guineensis* from Africa and Southeast Asia.

Oil palm seeds cannot be preserved by conventional drying and freezing methods, but embryo culture and freezing pollen are likely to reduce the costs of maintaining ex situ germplasm collections in the future. Oil palm embryos can be extracted from seed and, once the proper techniques are perfected, stored for long periods under cryopreservation (Grout et al., 1983). Pollen preservation provides a backup method for conserving part of the variation, but when preserved pollen is used to fertilize mother plants, the progeny will not be the same as the parents. Further, germplasm curators would have to be careful to maintain maximum variability as germplasm holdings are regenerated.

Field gene banks and depositories for embryos and pollen serve as genetic reservoirs rather than conservation collections. Although ex situ collections can capture only a small slice of oil palm's extensive gene pool, they satisfy many of the foreseen needs of breeders if they contain sufficient variation. To conserve the broad spectrum of variation in oil palm will require in situ conservation.

Information is needed about variation under natural conditions to conserve "wild" populations in situ. In general, oil palm shows great intrapopulation variation but less variability between sites or areas. Malaysian scientists have verified this greater variation between individuals than between populations in well-designed field trials of collected material (Rajanaidu et al., 1982).

This finding does not necessarily mean that a sizable portion of forest environments in one area will embrace most of the genetic variability of oil palm. Different populations of oil palm in its vast range must be maintained in order to secure the gene pool. As long as slash-and-burn farming and gathering of products from forest and second-growth communities predominate as the main means of securing food, oil palm will be safe. If on the other hand, more intensive land use patterns emerge to feed Africa's rapidly growing population, many populations of oil palm could be threatened.

In situ conservation is needed to complement field gene banks for oil palm and many other tropical perennial crops. Three main steps are required to provide a sound basis for in situ conservation efforts. First, a survey should be conducted and patterns of variation assessed throughout oil palm's range. Second, suitable areas for reserves need to be demarcated, taking into account the need to conserve gene pools of other species. Finally, and most important, a management strategy for the reserves must include provision for disturbance so that oil palm can regenerate. In the Congo basin, for example, local people recognize the need to periodically thin forests that contain oil palm so that the palms can survive (Briey, 1920).

Protected areas that permit some human activities, such as national forests for timber extraction or extractive reserves, would be ideal for conserving oil palm germplasm. Fieldwork is urgently needed to provide economic and cultural information for devising and operating oil palm reserves. The newly created International Program for Tropical Tree Crops of the International Fund for Agricultural Research (IFAR) is currently stressing this need.

BREEDING CHALLENGES AND GENETIC RESOURCES

Given oil palm's genetic heterogeneity, it is not surprising that limited populations were adequate for decades of improvement work. Because cultivated oil palm is generally little removed from its wild state, mass selection has paid enormous dividends. Owing to the great variation in yields, harvest index, and size and thickness of fruit, dramatic advances were possible by crossing different materials. For several decades, such progress was not hindered by major diseases.

Oil palm breeding is relatively recent, and much room exists for improving planting stock (Zeven, 1972). Improvement work began in Indonesia and Malaysia in the early 1900s, followed by the efforts of Belgians in the Congo basin in the 1920s. In 1920, scientists began evaluating oil palm from various parts of the Congo basin at Boyeka, a satellite station of the Eala Botanic Gardens (Anonymous, 1924). Research focused on developing high-yielding palms to replace the tall, low-yielding wild palms growing in the region. In 1939, five INEAC scientists working at the Yangambi station in northern Zaire unraveled enough of the genetics of oil palm to produce superior hybrid varieties after only six years' work (Eicher, 1989:8).

INEAC's path-breaking research spilled over to West Africa, where scientists initiated oil palm breeding in Côte d'Ivoire and Nigeria in the 1950s. With assistance from Portugal's Overseas Mission for Agronomic Studies (MEAU, Missão de Estudos Agronómicos do Ultramar), Guinea-Bissau launched an oil palm breeding program in 1960 (Ascenso, 1966). Efforts to improve oil palm in the New World are more recent, in spite of the fact that the palm has been there for at least two centuries.

A major contribution to oil palm enhancement was the understanding of the three botanical fruit types: *dura, tenera,* and *pisifera.* The three types differ in the thickness of the shell and pulp (mesocarp), and therefore in their value for palm oil production. *Tenera* is preferred to *dura,* the original type introduced to Southeast Asia, because the oil-producing part of the fruit makes up 60 to 90 percent of the weight. Less than 60 percent of the *dura* fruit is capable of producing oil.

The basis for the difference in oil-producing capacity of the fruit was discovered in the early 1940s. A single gene determines the thickness of the pulp. This gene exists as two alleles, Sh + and Sh −; oil palms homozygous for Sh + have *dura* fruits, while those homozygous for Sh − bear *pisifera* fruits (Beirnaert and Vanderweyen, 1941; Tudge, 1988:115). *Pisiferas* have no shell but are generally sterile. Heterozygous *teneras* are preferred by planters because they are fertile and produce fleshy fruits.

After this discovery, breeding shifted from the selection of individuals and families of individuals to selection of *dura* and *pisifera* lines that could be crossed to give high-yielding families of *tenera*. Fortunately, the breeding programs in Malaysia, Indonesia, and Côte d'Ivoire were able to provide the different parents: *Deli dura* from Malaysia and Indonesia, and *tenera* and *pisifera* from Côte d'Ivoire and the Congo.

The genetics of oil palm have been studied for over thirty years, but research has been particularly intensive in the last few years. Sustained basic research has helped breeding efforts greatly. Most economically important traits are under polygenic control, rather than the influence of single major genes. Oil palms thus express continuous variation, which is modified by the environment (Hardon et al., 1985).

Oil palm yields vary considerably according to disease pressure, agronomic management, and germplasm deployed. Proper crop management is well known to improve yield; in Nigeria, weeding around oil palms increases the fruit harvest (O. T. Faulkner and Mackie, 1933:97). But genetic potential can be further tapped to improve yields in different soils, climates, and pest environments. One lesson learned from breeding for high yield is that tolerance for competition is an important consideration when identifying highly productive genotypes for breeding (Hardon, 1974). Individual palms with high yields may not fare so well under the more crowded conditions of commercial plantations.

Yield potential of oil palm is great, and further breeding will surely push up yields. In West Africa, annual yields of fruit bunches on plantations are in the 7–15 metric tons per hectare range, whereas in Malaysia and Sumatra, where oil palm production is backed by intensive research, yields are typically twice as high. Some Nigerian palms have exceptional yield potential; they could produce as much as 12 tons of oil per hectare per year, more than double the current average oil yield of 5 tons (Rajanaidu, 1990).

Another reason to step up breeding efforts is that the genetic basis of most of the commercial oil palm plantations remains narrow (Hartley, 1988:250). Increased efforts are especially needed to develop more disease- and pest-resistant material. As the area of oil palm in large plantations has increased in the last few decades, particularly in parts of Latin America and Malaysia, diseases and pests have become more prevalent. New diseases may appear or minor pathogens suddenly may become more serious as more uniform plantings of oil palm spread. Diseases of particular concern include Fusarium wilt (caused by *Fusarium oxysporum*) and bacterial bud rot in Zaire; dry basal rot (caused by *Ceratocystis paradoxa*) in Nigeria; Ganoderma trunk rot and crown disease in Asia; and spear rot in the American tropics (Hartley, 1988:580). The Commonwealth Agricultural Bureaus International (CABI) is supporting research to differentiate pathogenic forms of Fusarium wilt, a helpful step in attempting to locate sources of resistance to the disease.

Spear rot has emerged as a major threat to oil palm production in Latin America within the last decade. Also known as sudden wither because of the rapid mortality the disease provokes, spear rot lingered for over a decade as a minor disease, but became more noticeable by the mid to late 1980s. Known as *amarelecimento fatal* in Brazil, the etiological agent of the disease is unknown, but it could be a viroid or mycoplasma-like organism, similar to the pathogen that causes lethal yellowing of coconut.

Spear rot appears to be transmitted by insects. If that proves the case, insecticidal spraying would not necessarily offer a solution because pesticides would destroy pollinators, mainly beetles (*Elaeidobius kamerunicus* and *E. subvittatus*, Curculionidae; *Mystrops costaricensis*, Nitidulidae) and thrips (*Thrips hawaiiensis*). In the Peruvian Amazon, Endrin has been sprayed on the soil around oil palms in an effort to control spear rot (Watson, 1985:49). A Brazilian company, Empresa Amazonense de Dendê (EMADE), manages some 2,000 hectares of oil palm near Tefé, but spear rot is jeopardizing plans to increase production at that site.

In 1985, the disease struck the largest oil palm plantation in the Brazilian Amazon operated by Dendê do Para, S.A. (DENPASA) in the Bragantina zone east of Belém. By 1990, the disease had penetrated 400 hectares of DENPASA's plantation. The only recourse at present is to remove oil palms at the first sign of crown dieback. More than 20,000 hectares of oil palm have been planted in the Huallaga Valley in the Peruvian Amazon since 1967 (Pulgar, 1987:134), but they are also likely to succumb to spear rot. Spear rot could hinder the Peruvian government's efforts to find viable alternatives to coca in the Huallaga Valley.

American oil palm may hold the key to combating spear rot, as well as several other diseases. American oil palm has been used in crossing work with its African cousin to lower the stature of oil palm. Fortunately, *E. oleifera* resists spear rot, so interspecific hybrids will likely become more common on plantations. In Colombia, oil palm plantings devastated by spear rot have been replanted with hybrid *E. guineensis* × *E. oleifera* materials, and plantations are once again productive since they are not affected by the quick-acting disease (Hartley, 1988:34).

Widecrossing work using the two species has been under way for over twenty years (Hardon, 1969). Current emphasis is on improving the yield of interspecific hybrids. Malaysian, French, and Brazilian scientists are evaluating *E. guineensis* × *E. oleifera* hybrids for promising material.

Other oil palm diseases that could be tackled with genetic solutions include species of *Ganoderma* that cause basal stem rot in Indonesia and Malaysia and seedling blight triggered by *Curvularia eragrostidis* in Malaysia. In Côte d'Ivoire, vascular wilt disease (caused by *Fusarium oxysporum*) and oil palm leafminer (*Coelaenomenodera elaeidis*) are notorious problems on some oil palm plantations (Hartley, 1988:31). American oil palm has exhibited tolerance to *Ganoderma*, and seedlings of the palm resist Fusarium wilt (Rajanaidu, 1983)

Nurseries are particularly susceptible to a range of diseases. Among the more serious diseases of young oil palms are blast (caused by *Pythium splendens* and *Rhizoctonia lamellifera*); freckle (*Cercospora elaeidis*); and anthracnose (*Botryodiplodia palmarum*, *Melanconium elaeidis*, and *Glomerella cingulata*). Genetic resistance is thus needed against a wide range of diseases and pests.

Numerous other diseases afflict oil palm. Breeding for disease resistance is a complicated task, because a disease currently confined to one region could become pantropical. In spite of quarantine precautions, pathogens and insect pests are sometimes transported across oceans or other ecological barriers. All oil palm growing countries therefore have a stake in enhancing the germplasm base of breeding programs.

PORIM in Malaysia and NIFOR in Nigeria are keenly aware of the need to enrich the germplasm resources of their oil palm breeding programs. Both institutions have taken major steps to garner a wider array of the gene pool and to screen and use novel germplasm. Wisely, PORIM and NIFOR recognized the need to collect wild material, rather than just rely on selections in fields and on research stations.

With the development of tissue culture propagation of oil palm, the need to conserve and evaluate genetic resources becomes even more urgent. Unilever in the United Kingdom began oil palm tissue work at Colworth House, Bedford, in 1968 (Tudge, 1988:115). Unilever focuses on propagation of root material, while other institutions, such as IRHO in France, prefer to work on leaf tissue for micropropagation. By the late 1980s, laboratories in the United Kingdom and France were shipping superior clones in tissue culture form to Brazil and other developing countries for field trials. DENPASA in Pará, Brazil, for example, is testing oil palms developed in vitro in the United Kingdom.

While clonal material promised to improve yields by up to 30 percent (L. H. Jones, 1987), tissue cultures do not always regenerate true to type. The need to go through a callus phase has resulted in an unpredictable percentage of abnormally fruiting palms (J. Hardon, pers. comm.). Until this somaclonal variation is better understood and can be controlled, direct planting of tissue cultured clones for commercial production will be attempted only on a limited scale. Also, planting large blocks to a single clone could increase disease and pest problems. As a safety precaution, Unilever is recommending that at least ten clones be planted in an area (L. H. Jones, 1987).

To become self-sufficient in planting material of high-yield varieties, several developing countries, such as India and Brazil, are investigating the potential of in-house micropropagation of oil palm selections. In India, scientists at the Central Plantation Crops Research Institute (CPCRI) at Kasaragod, Kerala, are producing oil palm plantlets from somatic embryo lines and are studying them to see if any mutations have occurred as a result of in vitro procedures (Figure 5.10). Research on tissue culture of oil palm is also under way at the Bhabha Atomic Research Center in Bombay and at PORIM in Malaysia (ICAR, 1988).

FIGURE 5.10 Oil palm (*Elaeis guineensis*) raised from tissue culture at the Central Plantation Crops Research Institute (CPCRI), Kasaragod, Kerala, India. December 1989.

PROSPECTS

The long-term prospects for oil palm are bright. Several Latin American nations, such as Brazil, Ecuador, Venezuela, Colombia, Costa Rica, Honduras, and Panama, plan to increase oil palm production to meet growing domestic demand for vegetable oils as well as for export (Alvim, 1989; Bromley, 1981; EMBRAPA, 1986). In Asia, India is keen to expand oil palm plantings, particularly along the Malabar Coast and on the Andaman and Nicobar Islands to meet a national shortage of vegetable oil that has prompted substantial imports (ICAR, 1988). Although oil palm is usually planted in monocultural stands, it is intercropped, and such practices are likely to be more common in the future as agroforestry systems assume more important roles in sustainable development. Near Santa Isabel in the Bragantina Zone of the Brazilian Amazon, for example, several small-scale oil palm growers have interplanted peach palm (*Bactris gasipaes*) for fruit production.

Oil palm is particularly suited to the more fragile soils in the humid tropics, which are unsuited to continuous cultivation of annual crops. Oil palm plantations are often carpeted with a thick leguminous cover, which helps check soil erosion. The substantial and relatively permanent biomass of oil palm plantations means that evapotranspiration rates probably approach those of the original forest. Oil palm plantations thus are unlikely to contribute to disruption of rainfall patterns or to global warming.

An added factor in oil palm's favor is that it can serve as a substitute for diesel oil, a valuable property that has been recognized for over half a century (Dalziel, 1948:505). In 1912, Rudolf Diesel suggested that vegetable oils would one day rival petroleum as power sources for engines (Quick, 1989). Palm oil can be used to fuel diesel engines after it has been heated for several hours between 30° and 40°C in the presence of methanol or ethanol (Alvim and Alvim, 1980). The alcohol can be partially recovered afterward.

While there is insufficient space to plant oil palm to supply the fuel needs of all trucks, locomotives, and ships currently using diesel oil, oil palm could be used to reduce oil imports for many developing countries in the humid tropics. Furthermore, oil palm is a renewable energy resource with annual yields up to 10 to 60 barrels of oil per hectare (Duke, 1986). Brazilians envisage palm oil as a substitute for diesel fuel in the agricultural sector; it could be used to power tractors, irrigation pumps, processing machinery, and vehicles to transport harvests (Alvim, 1989; Alvim and Alvim, 1980; *Folhin* [Folheto Informativo, EMBRAPA, Brasília], 1 February 1991, 4[71]).

Vegetable oils are often in critically short supply in developing countries, to the detriment of local diets and foreign exchange reserves. The shortage of vegetable oils in the Third World has precipitated two recent international actions: interest in promoting rapeseed oil and coconut production. Rapeseed is a temperate zone crop and thus imposes an import burden for developing countries. Coconut is ubiquitous in the lowland and coastal tropics, but coconut oil is higher in saturated fats than oil palm. Other crops for vegetable oils that can be grown in the tropics and subtropics include soybean, groundnut, maize, and cotton seed. These alternatives require either prime agricultural land and thus compete directly with food production or irrigation and the heavy use of pesticides to obtain good yields.

Oil palm has been cultivated and tended on a traditional scale for thousands of years, and only in this century has the crop emerged as a major export commodity. Some feel that dozens of other, typically more localized crops, could be exploited in the same way. While some scope may exist for a few new vegetable oil crops, they will likely have to have special properties to conquer a place on supermarket shelves or in industry. The cost of developing new or minor crops must be balanced against the potential for further improvement of existing crops. Market forces will often be the determining factor, both at the local and global scale. Oil palm has emerged as a major vegetable oil crop for good reasons, and most of the agronomy and breeding methods have been well researched. Furthermore, new uses will undoubtedly be found for palm oil, such as a fermentation substrate for the production of proteins and antibiotics (Gascon et al., 1989).

Oil palm thus remains one of the best options for improving vegetable oil production in the tropics. Oil palm is an intricate part of the cultural fabric of hundreds of millions of people living in tropical Africa, Asia, and America. In spite of concerns about "tropical oils" in some temperate countries, demand for palm oil and kernel oil is likely to remain strong.

BALSAMS

The legume family, well known for its protein-rich pulses, also encompasses many wild and semidomesticated plants used for medicine, fruits, ornamental purposes, fodder, and for balsam. Balsam is obtained from *Myroxylon balsamum*, which is native to the tropical forests of the New World. Balsam occurs in a wide range of forest life zones, from subtropical dry forest to rainforest (Duke, 1981). Gray-barked balsam trees reach as high as 35 meters and bear single-seeded winged fruits that are dispersed by wind. The genus occurs naturally from southern Mexico to tropical South America (Dillon, 1980).

Balsam is used for flavoring, as a fragrance, and in medicine. The hard, strong wood of Peru and Tolu balsams is similar to red mahogany in color. In addition to its rich color, balsam wood has a pleasant, long-lasting scent and a fine grain that polishes well and is thus ideal for cabinet work. Seeds of balsam are used to flavor *aguardiente*, a potent, clear rum commonly consumed throughout Latin America.

Two chemically and physically distinct balsams are tapped from *M. balsamum* according to various pharmacopoeias. Balsam of Peru, also called black balsam or Indian balsam, is obtained from *M. balsamum* var. *pereirae*. Balsam of Tolu, also known as quinoquino, is collected from another subspecies, *M. balsamum* var. *balsamus*.

Peru balsam contains between 50 and 64 percent of cinnamene, a volatile oil. Cinnamene consists of benzoic and cinnamic acid esters and traces of other constituents, such as styrene, vanillin, and coumarin. Tolu balsam contains from 75 to 80 percent resinous materials, mainly toluresinotannol, cinnamic and benzoic esters; 7 to 8 percent volatile oil (primarily benzyl benzoate); 12 to 15 percent free cinnamic acid; 2 to 8 percent free benzoic acid; and vanillin.

Tolu balsam is native to Argentina, Paraguay, Brazil, Bolivia, Peru, Colombia, Venezuela, and Panama (Dillon, 1980; Duke, 1981). The tree is particularly abundant along the Magdalena and Cauca valleys in Colombia and in the northwestern part of the mountainous country. Tolu balsam has been cultivated successfully on Sumatra and in Sri Lanka, India, as well as some of the West Indian islands.

Peru balsam is so named because it has historically been shipped from Peru's main port of Callao, even though it is native to Central America (Record and Hess, 1986:298). Peru balsam is indigenous to the Pacific coast of Central America, especially in El Salvador, Nicaragua, Guatemala, with outlying populations in Belize. The balsam coast extends along the Pacific seaboard of El Salvador from the port of Acajutla to the Bay of Jiquilisco (Hanbury, 1863). Peru balsam has been introduced to southern Florida, Sri Lanka, India, and West-Central Africa, where it is cultivated on a limited scale (Duke, 1981). The Singapore botanic garden served as a springboard for introducing balsam to tropical Asia. Peru balsam apparently tolerates alkaline and nutrient-poor soils.

Although one polymorphic species of balsam is currently recognized, at least six species of *Myroxylon* have surfaced in the literature at various times. Taxonomic confusion is always an impediment to genetic conservation work. Early literature in the colonial period of Latin America refers to four balsams: *Hoitziloxitl, Huaconex, Maripenda*, and a balsam "from the province of Tolú" in northwestern Colombia. On the basis of seed coat morphology, taxonomists reduced the number of valid *Myroxylon* species to *M. peruiferum*, from Colombia to Bolivia and Brazil, and *M. balsamum*, from Colombia to Venezuela and Mexico. At one time *Myroxylon* was in the genus *Myrospermum*, but, since its separation, *Myroxylon* is now generally considered a monotypic genus.

HISTORICAL USES

The earliest mention of Peru balsam is in a work of Nicolas Monardes published in 1565. Monardes referred to *balsamo* and noted that it was produced by a tree in New Spain that the Indians called *xilo*. Monardes described two methods for extracting balsam, one by cutting the bark and the other by boiling the branches in water. At that time, balsam was in demand for its valuable medicinal properties.

The Incas used balsam to stop bleeding and to promote healing and cultivated the tree in their gardens (Mansfeld, 1959). After conquest, balsam was exported to Europe for use in perfumes and medicine. Popes Pius IV and Pius V issued bulls in 1562 and 1571, respectively, authorizing Catholic clergy to use Peru balsam for anointing oils; copies of these papal edicts on Holy Chrism are preserved in Guatemala and in the Vatican. The bulls emphasized that destruction of balsam trees was sacrilegious.

Hoitziloxitl balsam was so named by the Mexicans for its abundant resin; it was cultivated in the Huaxtepec gardens, established by Montezuma I south of Mexico City in 1466. In Central America, balsam was also tapped to provide tribute to indigenous leaders (Hanbury, 1863).

CURRENT EXPLOITATION AND MARKETS

Balsam can be obtained at any time during the year, although the dry season is the preferred collecting period. In El Salvador, the harvest begins in December and can continue until May. Balsam is obtained from the wild and to a lesser extent from planted trees (Record and Hess, 1986:298). In Mexico, balsam is cultivated in Chiapas, Guerrero, Oaxaca, Veracruz, and in the northeastern part of the country (Linares and Bye, 1987). Once established, balsam trees tolerate heavy tapping and can live for a century.

Methods of extracting balsam differ according to variety. Tolu balsam is tapped in a manner similar to rubber, by carving a groove in the bark and collecting the exudate in a cup hung below the incision. Balsam of Peru is obtained by beating four patches on the side of the tree with wooden mallets to loosen the outer bark. In this manner, a "window" (*ventana*) is opened to the inner bark. Enough connecting bark is left around the panels to allow the tree to survive (Hale, 1911).

After about a week, balsam flows from the bruised panels and is soaked up with cotton or linen cloths, which are periodically replaced as they become saturated. After some eight to ten days, the flow abates, but can be stimulated again by warming the tree trunk with torches. Each year, new windows are opened on the tree trunk, usually about a meter above the previous square or oblong scar. Balsam collectors erect platforms as they work their way up to the top of the tree.

The balsam-soaked cloths are first boiled then placed in a small, netted press and squeezed. The expressed balsam settles to the bottom of the container in which the cloths were first boiled. The water is then poured off, together with any impurities, leaving the balsam. This method is known as the *panal* (cloth) or *trapp* (rag) process (Duke, 1981).

Another extraction method used on Peru balsam trees is the *cascara* (bark) process, which renders a lower-grade product (Duke, 1981). Balsam is obtained from bark removed during the cloth process or from bark removed specifically to obtain balsam. In the latter case, collectors build a fire at the base of the tree and let it burn for about ten minutes so that it scorches, but does not kill, the tree.

The technique of burning under tropical trees to obtain a harvestable product is not confined to balsam. As we discuss in Chapter 9, Brazil nut gatherers often set fire to the base of Brazil nut (*Bertholletia excelsa*) trees at the end of the dry season to clear space for collecting the nuts when they begin to fall within a few months and to increase the yield. Fire may promote flowering of Brazil nut trees as well as the production of resin, as evidenced by resin rings seen in some felled Brazil nut trees.

After some eight days, the heat-treated balsam bark softens, and collectors cut away segments of the bark with machetes. The bark segments are then crushed or cut up and boiled to extract the resin. The balsam is then poured into calabash gourds or tin cans for transportation to dealers. Before export, the balsam may be further purified to increase its value.

In its crude form, viscous Peru balsam is dark brown, but appears transparent in thin layers. Peru balsam emits a pleasant odor, reminiscent of cinnamon when fresh but more like vanilla when aged. Balsam of Tolu is a translucent yellow or reddish brown that darkens and hardens with age (Leung, 1980). Up-market perfumes are scented with Tolu balsam because its delicate, hyacinth-like scent blends well with floral and oriental compounds (Duke, 1981). Tolu balsam's fragrance also has a hint of vanilla. Balsam is soluble in alcohol, acetone, benzene, and solvents containing chlorinated hydrocarbons.

Balsam is used for a wide variety of medicinal purposes. Peru balsam has mild antiseptic properties and is employed to help heal skin lesions. In addition to its antibiotic action, Tolu balsam is an expectorant. In the tropics, people use balsam to treat such conditions as wounds, indolent ulcers, scabies, ringworm, diaper rash, hemorrhoids, anal pruritus, bedsores, tuberculosis, and invertigo (Levingstone and Zamora, 1983; Linares and Bye, 1987). Tolu balsam is used in cough remedies and as an ingredient in compound benzoin tincture. In folk medicine, both balsams are reportedly employed to treat cancer. In Mexico, the scented stems of Peru balsam are used to alleviate asthma and rheumatism. Balsam is also used in dentistry, particularly for dental impressions. However, some people are allergic to balsam.

In the cosmetics and food industries, balsam is used as a fragrance in feminine hygiene sprays and as a fixative or fragrance in soaps, detergents, creams, lotions, and perfumes. Balsam is a minor flavoring in some beverages, ice cream, candy, baked goods, gelatins, chewing gum, and puddings.

THE GENETIC RESOURCE BASE
FOR IMPROVEMENT

Apart from a few introductions to the Old World, Peru balsam is little cultivated, mainly as a shade tree and in backyards. Tolu balsam is grown as an ornamental and is cultivated in parts of Argentina and Brazil for its fine wood. Most balsam thus comes from wild trees. Some of these "wild" trees may have been tended, and in some cases planted, in the forest to increase the density of stands, particularly along the balsam coast.

Peru balsam tolerates a broad range of environmental conditions and is thus widely adapted. Little breeding or selection is needed to obtain material suitable for planting in many parts of the tropics. Also, the abundant, wind-dispersed seeds of Peru balsam allow it to establish spontaneous populations quickly, as near Kandy in Sri Lanka.

Little is known about patterns of genetic variation in the two botanical varieties of balsam. Higher-quality resins may be found in certain trees, which could prove useful under plantation conditions and in small-scale agroforestry systems. Desirable selections can be propagated by cuttings, thus "fixing" the qualities for all progeny.

No germplasm collections have been assembled for balsam. Balsam seeds are probably orthodox, so they could presumably be stored in conventional seed gene banks. Nevertheless, the seeds are relatively large, and their size could pose some problems with drying and freezing them for germplasm storage. The seeds of Peru balsam are 15 to 18 millimeters long, while those of Tolu balsam are slightly smaller. Experiments would quickly determine whether desiccation and storage under reduced temperatures destroy their viability.

A thorough inventory of all in situ reserves in tropical America would indicate to what extent balsam populations are at least nominally protected. As will be discussed in a later chapter on conservation strategies, floristic inventories in the tropics are still in their infancy. Information about what plants and animals are in existing reserves and parks would help guide efforts to establish complementary reserves and gene banks.

The global nature of such an undertaking would take large-scale coordination and cooperation among countries, as well as substantial funding from industrial countries.

THE RATIONALE FOR ACTION

Trade in Peru and Tolu balsams is still vigorous, and artificial or synthetic compounds have failed to dislodge the complex compounds from a wide variety of uses in cosmetics, medicine, dentistry, and food preparation. Given the widespread use of balsam both within tropical America and in importing countries, a concerted effort to assess the genetic resources of balsam is justified. Balsam is already an integral part of some agroforestry systems, both at the village level and as a shade tree on coffee farms. If highly productive and precocious material is selected, plantations of balsam in tropical America could also be economically and ecologically viable. As more tropical countries acquire domestic industrial capacity, the internal demand for balsam is likely to grow.

As with many tropical perennial crops, time is of the essence. The forests of Central America, where Peru balsam and many other valuable crops are found, are rapidly disappearing. Populations of Peru balsam are undoubtedly being lost before they are even surveyed for their genetic properties.

TROPICAL PINES

Although pines are often associated with colder climates, some of the 70 to 100 species in the genus *Pinus* inhabit mountainous terrain and coastal plains in the tropics. Pines are cultivated on a vast scale in temperate and tropical countries, mainly for their timber or pulpwood. Of the thirty or so widely planted tree species in forestry, over twenty are coniferous, and most are pines. Pines also produce resin, which has long been used for a variety of industrial purposes. Both wild trees and those cultivated in plantations are tapped for resin.

The primary centers of diversity for *Pinus* are in North Asia and North America, but as the genus spread south, new species arose in warmer climates. In the Americas, the natural distribution of pines reaches as far south as Nicaragua, where *Pinus caribaea* reaches 12° N near Bluefields (Parsons, 1955). In Southeast Asia, one species has crossed the equator.

The sizable trade in resinous products from pines has traditionally been dominated by temperate countries, particularly in North America and northern Europe. Production is now shifting to pine-growing areas with lower labor costs, such as southern Europe, particularly Portugal, and to China. Production of pine resin could expand much farther in tropical and subtropical countries to meet the growing demand for resin worldwide.

Forestry and Plantations

Tropical pines are ideal candidates for afforestation and reforestation schemes because they provide a range of products useful for developing countries. One advantage of working with pines is that the scientific knowledge base is generally greater than for other forestry species suitable for tropical areas.

Pioneer work in the United States and Europe has helped catapult pines into the forefront of advanced breeding and selection work. Fortunately, many of the technologies and research methodologies related to temperate pines are readily transferable to tropical species. The main goals for most pine improvement programs have been improved growth rate to shorten the cutting cycles, straighter and more finely branched trees, better wood quality, disease and pest resistance, and tailoring genotypes for particular climatic and edaphic conditions (FAO, 1970; R. Faulkner, 1976).

Interest is growing in the sustainable use of natural and seminatural forests for a range of products. Although some confusion lingers about what is meant by sustainability in agricultural research and development, it is not a buzzword that will die down and be replaced with another fad. However vague the concept, sustainability is here to stay as a guiding principle for development. But sustainable use of tropical forests and crops derived from such environments will need to be underpinned by more research so that management strategies can be devised that allow regeneration of valuable trees and fuller exploitation of domesticated species.

The International Tropical Timber Organization (ITTO), headquartered in Japan, has recently issued guidelines for sustainable management of tropical forests (ITTO, 1990). If member countries adopt and enforce the guidelines, more enlightened management strategies will help ensure the survival of wild gene pools of timber and resin trees.

Two tropical pines are ideally suited to management of natural stands as well as plantation systems. Caribbean pine (*P. caribaea*), native to Central America and parts of the Caribbean, and Tenasseim pine (*P. merkusii*), indigenous from Burma to Indochina, have received a great deal of attention by foresters for timber production. Both species produce good-quality resins, an attribute thus far little exploited. Conservation of the genetic resources of these warm-climate pines will be essential if they are to realize their full potential.

Caribbean Pine

One of the richest centers of diversity for pines is in Mexico south to Guatemala, Honduras, and El Salvador. The most widely planted species from this center of diversity is Caribbean pine. Caribbean pine thrives in a variety of environments ranging from stream banks to upland marshes and well-drained sites. The quick-growing pine is typically found in pine-hard-wood formations, including marginal areas, up to 300 meters, but in some cases grows on hillsides up to 1,000 meters. Fire appears to be a factor in the maintenance of many Caribbean pine stands (Johannessen, 1963:23; Parsons, 1955).

Caribbean pine has long been exploited for naval stores. Pine pitch was exported from Realejo, a colonial port in Nicaragua, to Peru in the last half of the sixteenth century and early seventeenth century (Radell and Parsons, 1971). English traders obtained pitch from Caribbean pine and other species for the Royal Navy in the early colonial period.

In Honduras, resin from the savanna-inhabiting pine is still used locally as a waterproofing agent (Johannessen, 1963:90). Caribbean pine is tapped by removing bark from part of the trunk and gouging out a hole at the bottom to collect the resin. Country folk burn splinters of the pitchy wood as a substitute for candles.

Caribbean pine is widespread in the Bahamas, Caicos, western Cuba, Honduras, Guatemala, and Nicaragua. Not surprisingly, botanical vari-eties have arisen in response to different soils and climates. In the lime-stone and coral Bahamas, a distinct form is recognized (var. *bahamensis*), while var. *hondurensis* in Honduras is adapted to savannah. Slash pine (*P. elliotii* var. *densa*), native to Florida, is closely related to Caribbean pine, and both are widely planted for pulpwood. Another closely related species is *P. occidentalis* in Haiti and the Dominican Republic.

In this century, deforestation in Central America has accelerated alarmingly, thereby endangering wild gene pools of Caribbean pine and other tropical pines. According to a 1980 survey sponsored jointly by the Food and Agriculture Organization and the United Nations Development Program, the estimated annual loss of forest (not counting logged areas) in countries containing populations of Caribbean pine was as follows: Gua-temala (2%), Honduras (2.3%), and Nicaragua (7.7%). In no part of its range can Caribbean pine be considered safe.

The need for reforestation in Central America and parts of the Carib-bean is widespread. In some parts of the region, deforestation is approach-ing a critical stage (P. J. Wood et al., 1982). Watersheds need better pro-tection to safeguard germplasm, provide reliable water supplies, and to supply forest products, such as fuelwood. Degraded forests in Guatemala and Cuba, which contain Caribbean pine among other valuable species, need rehabilitation and better management. Caribbean pine is particularly well suited for plantations and as a purveyor of multiple products in a managed forest setting. Multilateral aid organizations increasingly realize the important role that tropical forests, agroforestry, and plantations play in providing a broad range of products (Burley, 1986).

The yields of existing timber plantations need to be raised and sustained, which can be accomplished in part by using improved germplasm. Caribbean pine yields on the Jari plantation in the Brazilian Amazon have been rising steadily, in part because of the deployment of improved genotypes. As of December 1989, Jari had 42,830 hectares in *P. caribaea* var. *hondurensis*, planted mostly on sandier soils of the 1.6-million-hectare property (Figure 5.11). The best sources of seed have come from wild populations of Caribbean pine in the vicinity of Guanaja, Honduras; Alamincamba, Nicaragua; and Mount Pine Ridge, Belize (M. Albuquerque, pers. comm.).

GERMPLASM EVALUATION AND CONSERVATION

A network approach has helped generate results from the high priority accorded to tropical American pines by a working group of the International Union of Forest Research Organizations (IUFRO). The Oxford Forestry Institute (OFI) in the United Kingdom has been working closely with national forestry programs in Central America to further research on the genetic potential of Caribbean pine as well as *P. oocarpa*, *P. patula*, *P. tecunumanii*, *P. pseudostrobus* (known as piño triste in Guatemala), and *P. maximinoi* for almost thirty years (Barnes, 1988). In addition, OFI works on *P. kesiya* from Southeast Asia. The institute has organized teams to explore natural populations of these pines, undertake taxonomic studies, assess natural variation, collect seeds, and bring them back to Oxford for storage and widespread testing. In addition, a Central America and Mexico Coniferous Resources Cooperative Program (CAMCRORE) has been operational since 1980 with assistance from the United States.

By employing a germplasm screening network of close to a thousand trials in sixty to seventy countries, scientists are able to identify promising material suitable for various countries and environmental conditions. Trials have been assessed by researchers at OFI, and certain entries have been recommended and breeding strategies devised for the varying conditions of countries participating in the international pine nurseries. Many of the most successful provenances have been resampled on an individual tree basis in the wild for further breeding work. Scions of the best entries in trials are propagated and bulked in greenhouses in Britain for distribution to regional centers throughout the tropical world. In this manner, developing countries can draw superior materials to enrich their own breeding programs.

The degree of international research collaboration on tropical pines is much greater than for most other tropical perennial crops. OFI has played a key role in this pantropical effort and thus provides a role model for other organizations interested in furthering work on genetic resources of cultivated woody plants of the tropics. A major factor in the collaborative

FIGURE 5.11 Caribbean pine (*Pinus caribaea* var. *hondurensis*) on third cutting cycle at Jari, Pará, Brazil, January 1990. The three-year-old plantation has been treated with herbicide.

venture coordinated by OFI is the close attention paid to seed collection, herbarium studies, reproductive biology, wood and chemical analyses, and the maintenance of databases.

Concern for the plight of genetic resources of Central American pines spurred the creation of a regional collaboration between national governments and industry to promote germplasm conservation and evaluation. CAMCRORE has fostered the collection and storage of conifer germplasm in national gene banks established by the eight countries belonging to the cooperative (Zobel et al., 1987:323). Such activities have paved the way for fuller exploitation of species such as Caribbean pine.

The quality of turpentine and resin from Caribbean pine is exceptionally high, exceeding that from *P. palustris* in the south and southeastern United States. The latter species has dominated turpentine and resin production since the pioneer days of the United States. In Brazil, *P. caribaea bahamensis* produces more resins than the *hondurensis* variety of Caribbean pine as well as slash pine or *P. oocarpa* (Zobel et al., 1987:379).

Although great potential exists for developing oleoresin production from Caribbean pine and other pines in tropical America, progress will hinge in part on how the debt crisis is managed and on governmental economic policies. The FAO is helping Central American governments to formulate forestry sector plans, but a substantial investment is needed to rationalize further exploitation and conservation of Caribbean pine and other tropical pines. To cover the costs of institution building, development of forestry resources, implementation of sound management plans, and technical assistance, Central American countries will need annual investments on the order of $5.6–$7 billion until the turn of the century (FAO, 1988).

TENASSEIM PINE

Tenasseim pine (*P. merkusii*) occurs in Burma east to Indochina, and Sumatra, but is not found naturally in Malaysia (Krüssmann, 1985:222). Outlying populations of Tenasseim pine are found in the Philippines. Tenasseim pine is the most tropical of all the pines and thrives between 50 and 2,000 meters on a variety of soils (Stein, 1978). The Dutch pioneered plantations of Tenasseim pine during Indonesia's colonial period.

Although Tenasseim pine is not as variable as Caribbean pine, a distinct form, *tonkinensis*, is recognized in Vietnam (Mirov, 1967:553). Also, the oleoresin of Philippine populations of Tenasseim pine differs since concentrations of 3-carene are lower there than in other parts of the pine's range. Distinct differences are noticed between Tenasseim pines from mainland Asia and the islands of Southeast Asia. Presumably, founder populations on islands of Southeast Asia followed different evolutionary trajectories since they were isolated from gene pools on the mainland.

Reaching the Horizon

Tenasseim and Caribbean pines are ideally suited for industrial use because they produce high-quality oleoresin in addition to timber. Resin can be tapped before the trees are felled for pulp. Or the resin can be extracted in the sulfate pulping process (Zobel et al., 1987:380). Furthermore, Caribbean pine, *P. patula*, and *P. oocarpa* adapt well to many tropical areas outside their native ranges. Several tropical pines could spearhead industrial expansion in developing countries by providing naval stores. All developing economies require a range of processed products from lumber and the pulp and paper industries.

Caribbean pine especially is already a well-established plantation species, so agronomic practices are reasonably well honed. Furthermore, much ground has already been covered on understanding variation in several tropical pines and evaluating their genetic resources (FAO, 1974 to date; Keiding and Kemp, 1977; Kemp, 1975). At least fifteen unrelated, diverse parental genotypes may suffice to maintain the vigor of a newly launched plantation, but over the long term 300 to 400 breeding parents are advisable to reduce vulnerability to environmental threats (Zobel et al., 1987:218). In practice, few if any breeding programs with forestry species achieve this ideal. But the wider the gap between the ideal and current operations, the greater the danger that productivity will decline.

Other less commonly planted tropical pines, such as *P. patula* and *P. tecunumanii* from Central America and the Caribbean, could eventually move to the forefront of planted pines in the tropics (Zobel et al., 1987:50). In the Great Lakes Region of Africa, for example, *P. patula* is an important source of fuelwood at higher locations and is used extensively for erosion control.

Tropical farmers and foresters need to forge symbiotic relationships in order to develop appropriate technologies. By better understanding the historical importance of tree species in the cultures they are dealing with, professional foresters and development planners can design and implement more effective forestry programs. Farmers and urban folk are eager for new trees or varieties that can supply them with food or cash income. With sensitivity to the needs and constraints of local people and their environment, foresters can play an increasingly important role in ecodevelopment (Golley, 1983).

The industrial importance of trees from tropical forests clearly demonstrates that the economic value of perennial crops in the humid tropics reaches far beyond the primary products, such as fruits or nuts. The extraction and refinement of secondary products have provided industry with a wide variety of chemicals. Consumers benefit from such compounds when they purchase radial tires for their cars, board a plane, varnish a boat or deck, wash themselves with soap, and clean their homes with disinfectants.

Historically, most plantations and industry in developing countries have been geared to the production and export of primary products. In the future, more processing will be done in the Third World, thereby generating more jobs for the growing ranks of people entering their working years.

CHAPTER 6

Daily Bread

Although the basic staples for humankind are normally considered to be the major cereal or tuber crops, several cultivated perennials supply a substantial amount of carbohydrates for people in certain tropical regions. Bananas and plantains, breadfruit, peach palm, and sago palm are "pillars of life" for large numbers of rural and urban folk in the humid tropics. All of these crops have been cultivated for millennia and could make an even more important contribution to rural and urban economies in developing countries with further support for breeding backed by genetic resource conservation.

Wild populations of sago palm in Southeast Asia and peach palm in western Amazonia are threatened by deforestation. Wild progenitors and relatives of banana and breadfruit are also endangered by land use changes. For all such crops, in situ reserves are needed, as well as some scientifically planned backup conservation in gene banks.

Some perennial staples in the tropics are both important sources of foreign exchange as well as backyard crops for domestic consumption and for local markets. A large array of banana and plantain varieties satisfy varying culinary tastes in tropical areas. Commercial plantations are based on a few dessert varieties for the export market, and are thus highly vulnerable to disease and pest attacks. In spite of the mosaic of traditional cultivars grown by small-scale farmers, diseases make serious inroads from time to time on banana and plantain production on small farms. Black Sigatoka, for example, threatens bananas in the Great Lakes Region of Africa, an area where millions of people depend on these fruits and on vitamin B–rich banana beer for sustenance.

Breadfruit is an especially important food item in Polynesia, where dozens of varieties have been selected with varying fruiting times. Besides featuring as the main ingredient in many repasts, various parts of the breadfruit tree are also employed to treat numerous ailments in Polynesia. Breadfruit has been taken to other tropical regions, such as Africa, the Caribbean, and Latin America. Breadfruit could achieve greater importance if more suitable, higher-yielding varieties were developed. Breadfruit is particularly well suited for backyards in rural and peri-urban areas, since it provides fodder for livestock, carbohydrate-rich food for the family, and generous shade for the yard.

In the New World, peach palm was domesticated long ago for its oily, farinaceous fruits. Peach palm fruits are boiled before they are eaten and are a good source of vitamin A. Recently, some peach palm plantations have been established in parts of Central America to supply heart-of-palm for the export trade. Scope exists for improving selections for traditional food production as well for the luxury export trade in heart-of-palm.

Sago palm products also enter world commerce, although on a much smaller scale than dessert bananas. Starch extracted from the stems of sago palm is an important food item in various parts of Southeast Asia and Polynesia. Sago palm thrives in swampy locations, and is thus an important crop for boosting food production in some marginal areas of the Asian tropics.

Our sample of bananas, plantains, peach palm, and sago palm by no means exhausts the range of perennial crops in the tropics that provide significant sources of carbohydrates. In parts of Ethiopia, for example, ensete (*Ensete ventricosum*) has long served as a basic staple (Brandt, 1984; Hemsley, 1868). The banana-like plant does not produce an edible fruit, but the pseudostem and roots are eaten. Ensete roots are cooked over hot stones in Kaffa Province and in other areas, such as the highlands bordering the Rift lakes. The pseudostems are buried for a year or more. After they have fermented, the stems are dug up and made into porridge or bread. Ensete is grown in backyards of Addis Ababa and is sold in the city's main market.

BANANAS AND PLANTAINS

Bananas are considered a dessert fruit in industrial nations, but in many parts of the humid tropics bananas and plantains are a major source of calories. In equatorial Africa, for example, bananas and plantains are a significant staple (Samson, 1986:142). Bananas tend to be sweet when ripe and are usually eaten fresh, whereas plantains are generally larger, have a higher starch content, and are usually boiled or fried (Figure 6.1). This distinction does not always hold, however, since some plantains ripen into a sweet-tasting fruit, while a few banana varieties, the so-called cooking

FIGURE 6.1 Plantain 'Beyaco' in a market, Lima, Peru, July 1988.

bananas, which include some red-skinned forms, are eaten fresh or boiled. Bananas and plantains are also made into chips, flour, and puree. In parts of Southeast Asia, the terminal male flower is eaten as a delicacy.

In addition to providing significant amounts of carbohydrate, bananas and plantains are also rich in potassium, a nutrient important for the proper functioning of the heart. Bananas are a good source of vitamins C and B6, whereas plantains contain high levels of vitamin A, a vitamin often deficient in the diets of people in the tropics. In Central Africa, bananas are used extensively to make home-brewed beer, which contains vitamin B.

Bananas and plantains provide numerous additional benefits. Banana leaves are used for thatch and to wrap food. Cattle are fed the trunks, or more technically, pseudostems, in some countries, such as Rwanda. Bananas and plantains are also important in some religious ceremonies. Among the Waika of the Upper Rio Negro in Amazonia, for example, the ashes of deceased tribal members are mixed with plantain soup and eaten by relatives (Mee, 1988:132). In India, green plantains have long been eaten as a folk remedy for stomach ulcers. Recently, scientists at the University of Aston in Birmingham, United Kingdom, have identified the active ingredient in plantain that stimulates cell growth in the stomach lining, which may help heal lesions there. This discovery underscores the importance of ethnobotany for identifying new medicines and other valuable traits.

Bananas and plantains are thoroughly entwined in cultures throughout the tropics. In India, arches are fashioned with young banana trees and mango leaves to welcome guests, particularly for religious festivals. The Garo tribe of northeastern India offer pseudostems of plantain with the first harvest of rice or millet to their gods (B. Singh, 1981:139). Another indication of the importance of bananas in India is their frequent appearance in Hindu art (Isacco et al., 1982). Banana trees appear in *Krishna Played upon His Flute* (painted book cover, Bundi, ca. 1750); *A Messenger Leads Radha to Krishna* (illustration to the Gita Govinda, Bundi, eighteenth century); and *The Meeting of the Eyes* (Devgarh, ca. 1780). In the *Mahabharata*, a Vedic epic story written around 1000 B.C., a banana garden on the banks of the Kuberapashkarni is the home of the monkey god, Hanuman. Some Buddhist sculptures depict banana leaves (Gandhi and Singh, 1989:49). In New Guinea, banana leaves are used in rituals to entice rain (Powell, 1976).

Bananas are ideal for small- or large-scale agroforestry schemes. Indeed, over 90 percent of global banana production comes from smallholdings. The vegetatively propagated crop offers generous shade for the establishment of other crops, such as cacao. In the Altamira area of Brazil's Transamazon Highway, for example, bananas provide valuable income during the four to five years before the cacao seedlings bear fruit. Once cacao is old enough, the bananas are cut down for mulch. Bananas are often grown in clumps in small farm plots and in dooryard gardens.

When space is sufficient between banana plants, other crops can be grown alongside. In the Kivu province of Zaire, for example, banana shades coffee bushes on small farms. In the vicinity of Gisozi and Mwokora in highland Burundi, banana is frequently intercropped with taro, beans (*Phaseolus vulgaris*), peas, or cabbages. In Rwanda, banana is also intercropped with sweet potato. In the wetter parts of West Africa, plantains are intercropped with oil palm, African pear (*Dacaroides edulis*), African breadfruit (*Pterocarpus africana*), citrus, cassava, cocoyam, beans, and vegetables (G. F. Wilson, 1987).

Bananas are also well suited to large, uniform plantations. Extensive commercial plantings of banana are found in Central and South America, particularly in Ecuador, Colombia, Honduras, Guatemala, Costa Rica, and Panama. Bananas are an important export crop for several Caribbean islands, particularly Jamaica and the Windward Islands. In Africa, Cameroon and Côte d'Ivoire export bananas to Europe, while in Asia, the Philippines benefit from a sizable export trade in bananas to Japan, the Middle East, and Hong Kong. The Philippines, Dominica, Grenada, St. Lucia, St. Vincent, Martinique, and Guadeloupe derive close to half their export earnings from bananas (Chadha, 1989; *Economist*, 27 October 1990, p. 72). Israel exports bananas to Europe.

Four points are worth stressing with respect to the banana export trade and rural and urban poor in developing countries. First, many small farmers are involved in the export trade. Farmers close to good transportation and processing facilities may plant 10 to 30 hectares of the desired clones for the export trade. Second, 10 to 20 percent of bananas grown for export are rejected and are trucked to local markets. Third, the export trade provides substantial employment opportunities. Banana production and processing are labor-intensive, requiring at least one person for every 2 hectares of planting (Samson, 1986:146). Pests and diseases that attack commercial plantings also damage many traditional banana and plantain varieties, so resistance breeding can benefit smallholders as well as company and cooperative plantations.

The fortunes of banana production and trade have profound implications for the diets and livelihood of hundreds of millions of people. Dozens of developing countries rely heavily on banana exports to generate foreign exchange. Guatemala received $38.4 million from banana exports in 1987. The value of banana exports from Jamaica, St. Lucia, Dominica, Grenada, and St. Vincent increased from $39.1 million in 1977 to $194.6 million in 1986 (*Economist*, 6 August 1988, p. 10).

Demand for bananas on the world market is likely to increase, thereby creating more opportunities for employment and income for farmers in the Third World. Per capita consumption of bananas in the United States has jumped 43 percent since 1970, and demand for bananas in Western Europe is likely to soar as protective barriers are removed when the European Community is fully integrated in 1992.

The ability of banana producers to respond to market opportunities, both domestic and abroad, rests largely on the productivity of the varieties they deploy. A widespread disease or pest outbreak could seriously undermine banana production, driving up costs for consumers and depressing the economy of rural communities. In some cases, agronomic practices alone are insufficient to control diseases; genetic resistance is thus called for.

ORIGINS AND HISTORY

Bananas evolved in Southeast Asia, and most cultivated clones are triploid. Some bananas trace their origin exclusively to the diploid *Musa acuminata* (*AA*). Other cultivated bananas are derived from crosses between *M. acuminata* and *M. balbisiana*, with the latter species supplying the B genome (Simmonds, 1976). Five major groups of bananas are thus found: the *AA* and *AB* diploids and the triploids *AAA*, *AAB*, and *ABB*. A few natural tetraploid bananas (*AABB*, *AAAB*, *ABBB*, and *AAAA*) are grown in Southeast Asia, while breeders have also developed some *AAAA* bananas.

Edible *M. acuminata* evolved several thousand years ago. Parthenocarpy, the ability to set fruit without pollination, arose with *M. acuminata* somewhere in its broad range stretching from the Malaysian archipelago to New Guinea (Simmonds, 1976). Diploid seedless bananas (*AA*) were spotted by farmers, and suckers were eventually removed for planting. The seedless trait was relatively easy to preserve since the crop is propagated vegetatively.

N. G. Vakili (1967) argued that *M. balbisiana* also produced edible forms, but if so, the species does not appear to have been cultivated for its fruit. Prior to the evolution of parthenocarpy, wild bananas were probably collected for snacks, but their pellet-sized seeds and small fruits evidently did not warrant domestication efforts. Only when seedless fruits arose spontaneously did farmers take up banana planting. Parthenocarpy may have developed at several places in *M. acuminata*'s extensive range, and various cultural groups may have seized the opportunity to domesticate the sun-loving plant.

As domesticated forms of *M. acuminata* spread north into the range of *M. balbisiana*, more chances for spontaneous hybridization were created. The distributions of the two species overlap slightly along the interface between monsoonal and equatorial climates, but contact zones between *M. acuminata* and *M. balbisiana* increased dramatically as the former species penetrated drier areas as a result of human agency. Spontaneous crossing between the two species thus began, or at least accelerated, several thousand years ago. The resulting triploid hybrids were more vigorous and higher-yielding than domesticated *M. acuminata*, so the latter soon began to lose ground.

Another line of banana domestication has led to the Fe'i group. Fe'i bananas developed parthenocarpy and sterility independently of *M. acuminata*. Fe'i bananas are diploids, with a basic chromosome number of ten, unlike *M. acuminata* and *M. balbisiana*, which have a basic chromosome number of eleven. Another distinguishing characteristic of Fe'i bananas is that the fruit stalks generally protrude upward, rather than hanging down (Figure 6.2). Starchy Fe'i bananas are boiled or baked before they are eaten.

Thought to be derived from *M. maclayi* and possibly other related species, Fe'i bananas originated in New Guinea and spread to the Philippines and the Pacific as those islands were colonized in prehistoric times. Fe'i bananas may have been a basic staple in New Guinea 9,000 years ago (Golson, 1977). Although the Fe'i group is not as important as other bananas on a global scale, plump Fe'i bananas are a significant food in the Society Islands and the Marquesas. Fe'i bananas are also cultivated in Melanesia, the Cook Islands, and Hawaii.

Fe'i bananas are an essential component of feasts and other special occasions in the Society Islands, where they are considered a prestige food. In August 1989, Fe'i bananas fetched twice the price of bananas on Tahiti, Moorea, and Bora Bora. The rich orange colors of Fe'i bananas attracted the attention of Paul Gauguin, a French impressionist painter who visited Tahiti and the Marquesas in the late nineteenth century. Three of Gauguin's canvasses feature Fe'i bananas, suggesting their importance in the Society Islands a century ago: *Les Bananes* (1891); *Ia orana Maria* (1891); and *Passage de Tahiti* (1892).

During the historical periods of exploration, numerous voyagers wrote about bananas. The elongated fruits were variously called bananas, Adam's fig, or plantains. Portuguese, Spanish, French, Dutch, and British sailors and explorers reported seeing bananas in many parts of the Old World tropics (Figure 6.3).

In 1350, Marignolli described bananas in the Garden of Adam in Sri Lanka. Captain May recorded them in the Nicobar Islands in 1591, as did travelers to India around the same time. Between 1560 and 1593, explorers such as Ramusio and Sir Richard Hawkins wrote of bananas in many parts of Africa and the Americas. When bananas reached Africa is uncertain, but they are thought to have arrived via Madagascar. Indone-

FIGURE 6.2 Fe'i bananas at the head of a valley near Afareaitu, Moorea, Society Islands, September 1989.

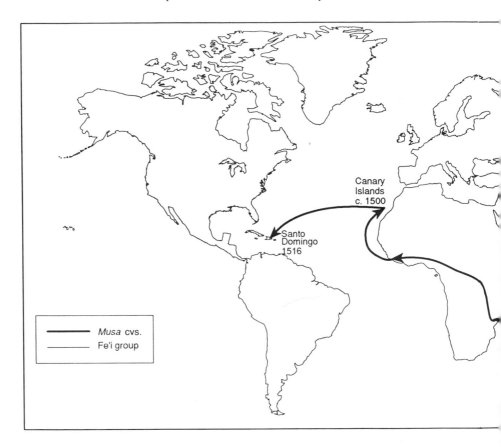

sians brought bananas to Madagascar as early as A.D. 300, and from there the crop was taken to East Africa (Davidson, 1969:23). Banana cultivation then spread up the chain of rift valley lakes and over the watershed divide into the Congo basin.

Reports of bananas in the New World during the early colonial period are controversial. The Musaceae family is native to the New World as well as Africa and Asia, but the genus *Musa* is not indigenous to the Neotropics. Nevertheless, banana-like *Heliconia* plants, better known in the ornamental trade as "lobster claws," thrive in moist, open sites in the tropical lowlands of South America. Reports of bananas in the New World may thus be cases of mistaken identity. Nevertheless, it is possible that West Africans took banana to the New World prior to the arrival of Europeans. This issue may be eventually resolved if *Musa* pollen is found in prehistoric archaeological sites in the New World, such as on Marajó Island in the mouth of the Amazon.

It appears more likely, however, that bananas were brought to the New

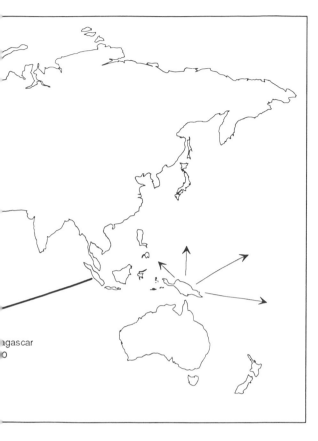

agascar
o

FIGURE 6.3 Spread of bananas (*Musa* cultivars) from their areas of origin. (Adapted from Davidson, 1969:23; Purseglove, 1975:349; Simonds, 1976.)

World by Europeans. Portuguese voyagers brought bananas from West Africa to the Canary Islands around 1500 (Figure 6.3). The subtropical Canary Islands served as a staging ground for exchanging several crops between the Old and New Worlds. Bananas were taken from the Canary Islands on Spanish galleons to Santo Domingo in 1516 (Kimber, 1988: 109). Once in the Caribbean, the relatively care-free crop soon spread to other islands. Also, a few further introductions were made by colonial powers directly to Central and South America.

TRADITIONAL CULTIVARS

Hundreds of banana and plantain varieties have been selected over the millennia by observant farmers. Banana has long been an important food, as evidenced by Buddhist sculptures that depict banana leaves and Buddhist scriptures that mention banana drinks (Gandhi and Singh, 1989:49); so ample time has passed for new varieties to have been selected. Most of the varieties were chosen from naturally occurring mutations. Farmers and villagers maintained interesting novel types that fit some local need or fancy. Estimates on the number of banana and plantain varieties range from 100 to 500 (Chandler, 1958:445; De Langhe, 1987; Samson, 1986:155; Simmonds, 1976). Given the long history of banana and plantain cultivation and its diffusion throughout the tropics and subtropics, the actual number of genetically distinct banana and plantain cultivars is probably close to 500.

The greatest diversity of traditional banana cultivars is found in tropical Asia, where the crop has the longest history. Some fifty banana varieties are grown in Tamil Nadu, India, while at least 25 banana cultivars are found in Sri Lanka (Chandler, 1958:445). Filipinos relish over seventy-five banana cultivars tended on thousands of widely scattered islands (Valmayor and Pascua, 1985).

An impressive diversity of bananas is also found in parts of Africa. In Rwanda, for example, eighty-five local banana cultivars were collected between 1984 and 1988 (Nyabyenda, 1988). At least thirty banana cultivars are recognized in Uganda (Johnston, 1902:97). More surveys in developing countries are likely to reveal a greater diversity of cultivars than was hitherto expected.

Some traditional varieties contain traits of potential interest to breeders; in Malaysia, one banana variety has a stalk over 2 meters long that contains an average of 2,000 fruits, more than five times the number of fruits typically found on export varieties.

A considerable array of new banana varieties has also arisen in Africa, Latin America, the Caribbean, and the Pacific. The mutation rate of bananas is apparently relatively slow, averaging one or two per million plants. African, American, Caribbean, and Polynesian farmers have done a remarkable job of recognizing and preserving mutations stemming from a limited number of introductions.

In the New World, the Atlantic seaboard of Nicaragua and Costa Rica has a particularly rich assemblage of banana and plantain cultivars. In the mid-1800s, a well-traveled naturalist remarked on the numerous banana varieties grown by Nicaraguans (Belt, 1888:67). The Bribri, Cabecar, and Chiripo groups of Costa Rica typically cultivate some eight varieties of banana and from four to thirteen varieties of plantain (Johannessen, 1970). Farmers in the vicinity of Tenochtitlán in the Isthmus of Tehuantepec, Mexico, cultivate at least eight varieties of banana (Coe and Diehl, 1980:89).

In the Brazilian Amazon, the Kayapó cultivate fourteen varieties of banana and plantain (Posey, 1988). In Colombia, twelve main banana cultivars are widely grown, and others are probably cultivated in the Amazon lowlands (Pérez-Arbeláez, 1956:530). On Tahiti, one farmer reported that his grandfather could name twenty-seven varieties of banana, four of which have apparently disappeared in the last twenty-five years.

Secondary diversity in Africa and Latin America suggests that when bananas were brought to new lands, people eagerly accepted the high-yielding crop, and it spread quickly. As bananas diffused across various soil and climatic zones, different cultures selected some of the new forms arising from spontaneous mutations.

Some traditional selections remain idiosyncratic and of localized importance. The Bribri of Costa Rica, for example, maintain two unique clones; 'Krum macra' has attractive brown fruits, while 'Dorocrum' has a striking black pseudostem (Johannessen, 1970). Other indigenous selections eventually gained wide acceptance. 'Sucrier' has penetrated deep into the Old and New World because many cultures appreciate its particularly sweet flesh. Known by various names (Table 6.1), this thin-skinned, fat banana measures no longer than an adult's finger. The diploid 'Sucrier' is less hardy and productive than triploid bananas, but has survived because of its exceptional flavor.

In Central Africa, several medium-sized banana varieties are cultivated exclusively to make beer. Banana varieties for beer production are called 'Mbidde' in Uganda and 'Makandili' in eastern Zaire. Mild banana beer, known locally as *pombe* in Burundi, Rwanda, the Kivu Province of Zaire, Uganda, and Western Kenya, is consumed by adults and children alike. *Pombe* is prepared by placing bananas in a pit for a week so that they ripen and begin fermenting. The soft bananas are then peeled and mashed in a wooden trough with circular pads of grass added to provide better consistency. Ground, roasted sorghum is added, as well as some banana beer, and the brew is left to ferment for a day or two. Banana beer spoils within a few days, so households typically borrow from each other if stocks run low. Banana beer is made only in Central Africa and probably arose from the habit of making beer from sorghum (*Sorghum bicolor*), an indigenous crop. Banana is such an important food crop in Rwanda that it is featured on the country's 20-franc coin.

'Red' bananas are pantropical and are eaten fresh or boiled. Restricted to backyards, 'Red' bananas are fat, medium-sized fruits. The triploid (*AAA*) bananas are called 'Banana roxo' in Brazil and 'Banana morado' in Costa Rica, Guatemala, and Panama, where they are marketed in towns and cities. Another distinguishing feature of these handsome bananas is that they resist leaf spot.

Plantains probably originated in southern India and are relatively high-yielding triploids (*AAB*) with good levels of resistance to Panama disease and Sigatoka. While plantains are most variable in their area of origin,

TABLE 6.1

Synonyms of the *AA* 'Sucrier' banana

Synonym/cultivar	Location
'Banana de oro'	Guatemala*
'Banana de ouro'	Brazil*
'Bocadillo'	Colombia
'Camela'	Kivu, Zaire*
'de Rosa'	West Indies
'Figue sucree'	West Indies
'Honey'	West Indies
'Klue kai'	Thailand
'Lady's Finger'	Hawaii
'Pisang mas'	Malaysia, Indonesia
'Sagale nget-pyaw'	Burma

Sources: Stover and Simmonds, 1987:113;
*N. J. H. Smith, field notes.

secondary areas of diversity are found in tropical America and in Africa, especially Zaire. At least fifty-six plantain varieties are known in West Africa and and fifty-one in Cameroon (Stover and Simmonds, 1987:124).

Plantains are eaten boiled, steamed, or fried and are a basic staple in many parts of Africa and Latin America, such as in Venezuela. In southern Venezuela, the Yanomamo rely on plantain as their main staple. In Brazil, 'Horn' plantains are grown in the Amazon and Northeast, where they are known as 'Banana comprida' (long banana) and 'Chifre de boi' (bull's horn). In Veracruz, Mexico, the same, or a very similar, variety is called 'Platano macho', and a smaller, 'French' type plantain is known as 'Platano Dominico'.

The greatest diversity of Fe'i bananas occurs on Tahiti, in the extreme eastern fringe of their range. Tahitians eat some thirteen varieties of Fe'i bananas, whereas on other islands in the Pacific and Southeast Asia, only a few cultivars are typically found (Table 6.2).

Many more cultivars of this ancient crop probably existed in the past, but the number of cultivars has declined in this century. Population crashes of some Pacific Islands as a result of aboriginal contact with Europeans and the spread of banana pests have all taken their toll on Fe'i bananas. British, French, and Spanish explorers landed approximately a dozen times on Tahiti in the latter part of the eighteenth century to barter, and in some case linger, with Tahitians (Moorehead, 1967). Such contacts unleashed a number of illnesses among the indigenous population.

Samuel Wallis arrived in Tahiti in 1767 and was reportedly ill much of the time (Ferdon, 1981). Louis Antoine de Bougainville, commander of two ships, dropped anchor in Taipahia Bay in 1768, followed by Captain

TABLE 6.2
Cultivars of the Fe'i bananas

Cultivar	Location
'A'ata'	Tahiti, Moluccas
'Afara'	Moluccas
'Aiuri'	Tahiti
'Arutu' ('A'aia')	Tahiti
'Autafun' ('Utinifun', 'Wagalovo')	New Ireland
'Borabora'[1]	Hawaii
'Chongk'	New Hebrides
'Daak'	New Caledonia
'Fe'i'	Tonga, Society Islands
'Ha'a'	Tahiti
'Ha'i'	Hawaii
'He'e'	Hawaii
'Huetu'	Marquesas
'Kokokongai'	New Ireland
'Mahani'	Tahiti
'Mo-pa'	New Guinea
'Ngalas'	New Ireland
'Oeoe'	Tahiti
'Oviri'*	Tahaa (Society Islands)
'Paru'	Tahiti
'Pisang tongkat langit'	E. Indonesia
'Polapola'	Hawaii
'Porek'	New Guinea
'Poti'a'	Tahiti
'Puputa'[2]	Samoa
'Purani'*	Bora Bora
'Ravoro'*	Bora Bora
'Rureva'	Tahiti
'Soa'a'	Samoa
'Soanga'	Tonga
'Soaqa'	Fiji
'Sula sula'	Samoa
'Tati'a'	Tahiti, Moluccas
'Toro aiai' ('Arapoi')	Tahiti
'Uatu'	Cook Islands
'Urtuk'	New Ireland
'U'ururu'	Tahiti

Sources: Stover and Simmonds, 1987:105, 109–111; *N. J. H. Smith, field notes, 1989.
   [1]Possibly two or three cultivars involved, two of which appear to be identical to 'Aiuri' and 'Afara'.
   [2]Resembles 'Poti'a' of Tahiti.

James Cook in 1769, and Don Domingo Boenechea in 1772. Several deck-hands on Boenechea's *Aguila*, which had sailed from Callao, reportedly brought influenza to Tahiti (Howarth, 1983). Cook returned to Tahiti in 1773–74, while in 1774, Boenechea brought back from Peru two Tahitians whom he had taken on his first trip to the Society Islands. Cook made a third trip to Tahiti in 1777, shortly before he died in a skirmish with Hawaiians, and William Bligh arrived in 1788. A few months after Bligh's ill-fated departure in 1789 with breadfruit seedlings, John Henry Cox visited Tahiti for almost a month. Captain Edward Edwards spent two weeks at Matavai Bay in 1791 in pursuit of mutineers on the *Bounty*. In the same year, George Vancouver stopped for a month at Tahiti with two ships, and Bligh returned in 1792 to collect more breadfruit seedlings. William Broughton stopped at Tahiti on his way to Hawaii and Japan in 1795, and Captain James Wilson dropped off missionaries in various parts of the Society Islands in 1797. Tahiti was clearly at a Pacific crossroads in the eighteenth century, with tragic consequences for the health and customs of the indigenous peoples.

Introduced diseases triggered a rapid decline of the human population on Tahiti from an estimated 140,000 at the time of contact with Europeans to under 5,000 within three generations (I. Cameron, 1987:36). Population thinning also occurred in some other parts of the Society Islands, such as Moorea, where forest-covered temple sites, known locally as *maraes*, attest to formerly dense farming communities in the interior valleys and along the coasts.

Unique varieties of Fe'i bananas and other crops may have perished along with Polynesians even before the eighteenth century as a result of introduced diseases such as tuberculosis, smallpox, syphilis, and the common cold. The Spanish, in particular, were probing the south Pacific in the sixteenth century. Ferdinand Magellan was the first European to cross the Pacific in 1519–20. Alvaro de Mendaña left Callao, Peru, for two trips across the Pacific in 1567 and 1595. Pedro Fernandes de Quiros also operated out of Callao in search of spices and other riches in the wide Pacific in the late sixteenth and early seventeenth centuries. Although none of these adventurers reached Tahiti, Quiros stopped at the Marquesas in 1585 and probably infected other Polynesian peoples with introduced diseases, who may in turn have carried pathogens to other islands (Howarth, 1983:44).

Plant diseases are another reason why Fe'i bananas are less common than in the past. The cultivation of Fe'i bananas on Tahiti has declined in this century because of pests and diseases (Pétard, 1986:118). In 1989, Fe'i bananas in the Terre Amo (vallée Teehu) of southern Tahiti were observed suffering from severe attacks of the banana root borer (*Cosmopolites sordidus*). Farmers report that Fe'i are more susceptible to diseases and pests at lower elevations, which is one reason they tend to be grown in the upper reaches of valleys. But even in the highlands of Tahiti, Fe'i bananas are becoming increasingly shaded out by an aggressive weedy tree, *Miconia magnifica*. Miconia was introduced to Tahiti as an ornamental, where it is known locally as turtle shell after the tree's broad, carapace-shaped leaves. This fast-growing melastomaceous tree sets abundant wind-dispersed seed, and it has gained a foothold in many valleys on Tahiti, where it shades out the native vegetation. Miconia proliferated in 1983 after a powerful cyclone flattened parts of the highland forest on Tahiti. Clearly, Fe'i bananas warrant special efforts for collecting before more forms disappear.

MODERN CULTIVARS

In contrast to the plethora of traditional banana cultivars, only a handful dominate the banana export trade. Export varieties are relatively thick-skinned, are cut when still green, and after being dunked in a chemical bath to retard browning caused by fruit rots, are placed in plastic-lined cardboard boxes. After arriving at port in refrigerated cargo holds, the bananas are gassed with ethylene to promote ripening and reach market shelves within one to three weeks of picking. Agronomic, transportation, and marketing constraints greatly limit the number of banana varieties entering world trade.

The banana export trade, developed over the last hundred years, has relied exclusively on a few triploid *AAA* clones. Until the early 1960s, 'Gros Michel' (Figure 6.4) and its shorter mutant, 'Highgate', dominated the banana export industry. 'Gros Michel', known by various names in different regions (Table 6.3), originated in Southeast Asia and was introduced to Martinique in the early 1800s and to Jamaica around 1835. 'Highgate' arose in Jamaica, while 'Cocos', another 'Gros Michel' mutant, comes from Honduras.

'Gros Michel' is now grown in limited quantities, mainly in northern Ecuador, Mexico, and Rwanda, and then only for local use. This obsolete variety, known locally as 'Platano Tabasco' in Mexico, was encountered in the public market of Puebla in 1988 (Figure 6.5). 'Gros Michel' was served as a dessert banana in some restaurants in Rwanda in 1990. The demise of 'Gros Michel' in the 1960s was triggered by Panama disease, caused by at least four races of *Fusarium oxysporum cubense*. Fungicidal applications against this variable pathogen are ineffective. The answer, it seemed, lay in locating genetic resistance. Fortunately, the Cavendish group, many of which have been around for as long as 'Gros Michel', appeared to be the answer.

The Cavendish group, like 'Gros Michel', is triploid *AAA* and travels well. But the Cavendish group is generally higher-yielding and, most im-

FIGURE 6.4 'Gros Michel' bananas at a United Fruit Company plant, Tiquisate, in the Pacific lowlands of Guatemala, in 1965. Commercial banana production has since largely switched to the Caribbean side of Guatemala. (Courtesy of Nixon Smiley.)

portant, resists Panama disease. Cavendish bananas are a closely related group of cultivars all derived from a tall parent, 'Pisang masak hijau' from Southeast Asia. The four principal cultivars of the Cavendish group are 'Dwarf Cavendish', 'Giant Cavendish', 'Grand Nain', and 'Pisang masak hijau' ('Lacatan' in the American tropics).

The Cavendish group of bananas is difficult to track sometimes because of the proliferation of local names for the same clone. 'Dwarf Cavendish', for example, is known by at least forty-four names (Table 6.4). The same local name can refer to genetically different clones. In Hawaii, for example, 'Lady's Finger' is a synonym for the diploids 'Sucrier' and 'Ney poovan'.

'Dwarf Cavendish' is the most widespread of the Cavendish group. 'Dwarf Cavendish' not only withstands storm damage better than many of

TABLE 6.3

Synonyms of the *AAA* 'Gros Michel' banana

Synonym/cultivar	Location
'Anamalu'	Sri Lanka
'Au malie'	Samoa
'Avabakor'	Papua New Guinea
'Banano'	Colombia
'Bluefields'	Hawaii
'Disu'	Papua New Guinea
'Fi'a palagi'	Samoa
'Gros Michel'	West Indies
'Guaran'	Puerto Rico
'Guineo'	Colombia
'Guineo gigante'	Puerto Rico
'Habano'	Colombia
'Jainabalavu'	Fiji
'Klue hom tong'	Thailand
'Makanguia'	French West Indies
'Pisang ambon'	Malaysia, Indonesia
'Platano roatan'	Orizaba, Mexico*
'Raimbaud'	French West Indies
'Tabasco'	Puebla, Mexico*
'Thihmwe'	Burma

Sources: Stover and Simmonds, 1987:114; *N. J. H. Smith, field notes.

its peers, but it also tolerates cool conditions. 'Dwarf Cavendish' therefore penetrates the subtropics and higher elevations in the tropics.

The narrow genetic base of export bananas renders them particularly susceptible to catastrophic outbreaks of disease. Commercial banana plantations are also especially vulnerable to the rapid spread of diseases and pests because the plants are planted close together, usually between 1,600 and 2,000 per hectare, although densities reach as high as 4,400 per hectare in India (K. L. Chadha, pers. comm.). Agronomic practices only partially alleviate the many pests and diseases that attack large, genetically uniform commercial plantations. Much unrealized potential exists for tapping gene pools to combat threats to banana production worldwide. Indeed, diverse and carefully evaluated germplasm collections are likely to prove invaluable in future efforts to raise and sustain banana yields.

WILD GENE POOLS

The highly diverse family Musaceae is divided into five series (Table 6.5). Most cultivated bananas belong to Eumusa, the largest series. The

FIGURE 6.5 'Gros Michel', known locally as 'Platano Tabasco' on sale in Puebla, Mexico, May 1988.

Fe'i bananas belong to the Australimusa series, which also contains abaca (*M. textilis*), a relatively recent domesticate in the Philippines used to make sturdy Manila hemp. Ensete (*Ensete ventricosum*), cultivated in Ethiopia for its starchy pseudostem, also belongs to the Australimusa series.

The Eumusa series is of more immediate interest to banana breeders because it contains *M. acuminata* and *M. balbisiana*. Both species are common in forest openings, such as along water courses, clearings, and forest margins. Polymorphic *M. acuminata* exhibits greatest diversity at the eastern edge of its range (Simmonds, 1966). Five main subspecies of the highly variable *M. acuminata* are recognized—*banksii, burmannica, malaccensis, microcarpa,* and *simaea*—some of which overlap. Three outlying populations of *M. acuminata* are found on Pemba, off the east coast of Africa; Hawaii; and Samoa, where the *banksii* subspecies occurs. These outlying populations were probably established by humans since long-distance dispersal by birds, wind, or sea currents is unlikely (Simmonds, 1962).

TABLE 6.4

Synonyms of the *AAA* 'Dwarf Cavendish' banana

Synonym/cultivar	Location
'Ai keuk heung ngar tsiu'	Hong Kong
'Ana'	Brazil
'Ananica'	Brazil
'Banane gabou'	Seychelles
'Basrai'	India
'Bhusawal'	India
'Binkehel'	Sri Lanka
'Camayenne'	Guinea
'Camburi pigmeo'	Latin America
'Canary Banana'	Pantropical
'Caturra'	Brazil
'Cavendish'	Hawaii
'Chinese'	Hawaii
'Chuoi duu'	Indochina
'Dwarf Cavendish'	Australia
'Dwarf Chinese'	Pantropics
'Enano'	Latin America
'Fa'i palagi'	Samoa
'Giuba'	Somalia
'Governor'	West Indies
'Guineo enano'	West Indies
'Jahaji'	India
'Jainaleka'	Fiji
'Johnson'	Canary Islands
'Kabuli'	India
'Kaina vavina'	Papua New Guinea
'Kinguruwe'	Tanzania, Zanzibar
'Klue hom kom'	Thailand
'Malindi'	Tanzania, Zanzibar
'Maouz siny'	Egypt
'Mauritius'	India
'Moris'	India
'Moz hindi'	Egypt
'Nanukehel'	Sri Lanka
'Nyoro'	Kenya
'Paca vazhai'	India
'Pachawara'	India
'Pandi'	Sri Lanka
'Petite naine'	West Indies
'Pisang badak'	Indonesia
'Pisang serendah'	Malaysia
'Poot'	Philippines
'Tampohin'	Philippines
'Vamanakeli'	India
'Wet-ma-lut'	Burma

Sources: K. L. Chadha, pers. comm.; Stover and Simmonds, 1987:114.

TABLE 6.5

Taxonomic division of the family Musaceae, along with basic chromosome numbers, number of species in each series, and some of the species of *Musa* in each series

Series	Basic chromosome no.	No. of species	*Musa* species
Australimusa	10	5–7	*jackeyi* *maclayi* *textilis* *angustigemma* *peekelii*
Callimusa	10	6–10	*coccinea*
Eumusa	11	13–15	*acuminata* *balbisiana* *basjoo* *itinerans*
Ingentimusa	14	1	*ingens*
Rhodochlamys	11	5–7	*laterita*

Sources: Adapted from Simmonds, 1976; Stover and Simmonds, 1987:2.

Another isolated population of M. *acuminata* may exist on Sri Lanka, where it grows in certain foothills alongside M. *balbisiana* (Kajale, 1989). Seedlings of both species are common weeds after burns, thus slash-and-burn agriculture may have favored the survival of hybrid forms. By creating an extensive contact zone between weedy bananas and cultivated forms, introgression has enriched the cultivated gene pool of bananas on Sri Lanka and elsewhere in tropical Asia.

Second-growth and ruderal areas are important reservoirs of variability for the enrichment of some crops. Unlike many other perennial tropical crops, bananas may benefit from human activities, particularly deforestation, because they can create larger gene pools for wild relatives of bananas. As human population pressure builds, however, fewer patches of second growth are left in fallow; instead, they may be converted to permanent fields or used for other purposes. In such cases, it is difficult for wild populations of *Musa* to become established. Also, deforestation can force monkeys and domestic pigs to focus on wild bananas for food, as happened in Karnataka State along India's Malabar Coast, with devastating results (S. R. Bhat, pers. comm.).

The other parent of some bananas, M. *balbisiana*, is considered less variable, but it undoubtedly contains valuable traits. It is thought to have conferred drought tolerance to bananas, since it occurs in monsoon climates, which have a pronounced dry season. Presumably, even greater drought tolerance can be located in M. *balbisiana* populations at the drier, and more marginal, parts of its range.

Other species in the Eumusa series could also prove useful in breeding programs. For example, *M. itinerans* grows as scattered individuals, rather than in clumps. This isolated growth habit could be useful for certain agroforestry configurations that are not suited to the bunched growth of most bananas. Another species in the Rhodochlamys series, *M. laterita*, shares the nonclumping growth habit. Other species in the Rhodochlamys series are currently of ornamental interest only. At the moment, the Callimusa series is also mostly of interest to the nursery trade; the handsome *M. coccinea*, for example, bears brilliant scarlet bracts.

The Australimusa series, which includes the Fe'i bananas, are native to a relatively restricted area that includes the Moluccas, New Guinea, New Britain, New Ireland, and the Solomon Islands. The fact that Fe'i bananas are on the decline should not relegate them to oblivion from the viewpoint of breeders and geneticists. Fe'i bananas could well make a modest comeback in certain areas if some breeding and selection are undertaken. Other members of the Australimusa series also warrant attention. On New Ireland, for example, bananas of the wild *M. peekelii* have an intriguing orange skin and bright yellow flesh. The little-studied Australimusa series probably contains several as yet undescribed species in the Solomon Islands and the New Hebrides (Stover and Simmonds, 1987:103).

Wild relatives of banana, especially *M. acuminata* and *M. balbisiana*, would be best conserved by protecting forest environments in Indochina and Southeast Asia. Although human activities create more openings, not all such clearings are left to natural succession. It is therefore important to retain relatively extensive areas of forest with naturally occurring light gaps suitable for species of *Musa*.

More research is needed on the natural history of banana's wild relatives. For example, breeders would benefit from information on whether subspecies of *M. acuminata* are more than ecotypes and whether ecoclines exist (De Langhe, 1987). Other questions urgently needing research include the degree of heterozygosity in wild relatives and how much interpopulation variation is found within each wild *Musa* species.

GENE BANK COLLECTIONS

Edible bananas generally do not set seed, so field gene banks are used to assemble germplasm collections of cultivated forms. Tissue culture would greatly reduce the cost of assembling and maintaining germplasm collections of banana. Banana plants were first produced by in vitro shoot-tip culture in Taiwan in the early 1970s (Vuylsteke, 1989:2).

Experience with tissue cultures of banana in the Philippines in the 1980s indicated that not all banana cultivars responded easily to in vitro conditions. Another impediment to the conversion of banana germplasm collections to test tubes and flasks is the relatively high rate of somaclonal variation. Up to 5 percent of banana tissue cultures undergo somaclonal variation (Stover and Simmonds, 1987:188). Such variation can be a useful source of new variation for breeding, but it undermines the task of preserving distinct genotypes.

Several institutions are investigating in vitro storage of banana germplasm. The International Network for the Improvement of Banana and Plantain (INIBAP) headquartered in Montpellier, France, is supporting research on tissue culture of bananas at the Catholic University, Leuven. This Belgian university is a principal transfer center for banana germplasm moving between tropical countries. One objective of research on tissue culture of bananas at the Institute of Plant Breeding at Los Baños, Philippines, is to assess the feasibility of duplicating accessions of bananas held in a field gene bank. In New Delhi, the National Bureau of Plant Genetic Resources (NBPGR) has 100 accessions of bananas in tissue culture, and NBPGR plans to duplicate all nonseeded *Musa* accessions in vitro by 1994. CATIE (Centro Agronómico Tropical de Investigación y Enseñanza) near Turrialba, Costa Rica, maintains thirty-eight tissue culture accessions of *Musa* (Jarret et al., 1986). When kept at 18°C, banana tissue cultures can remain for eighteen months before a shoot-bud needs to be transferred to a fresh medium. Cryopreservation might eventually prove the most practical way for the long-term conservation of cultivated bananas since it will eliminate the need to reculture and should prevent somaclonal variation. More research is needed on cryopreservation of bananas and plantains because *Musa* cultures exposed to temperatures below $-20°C$ die with current techniques (Vuylsteke, 1989:31).

As research on various methodologies for conserving banana germplasm continues, the issue as to how many cultivars need to be held in germplasm collections remains. Experts attending meetings sponsored by the International Board for Plant Genetic Resources (IBPGR) in 1977 and 1982 concluded that no more than 250 cultivars are needed to represent the crop's diversity in Southeast Asia, and not more than about 100 are needed for Africa (J. T. Williams, 1987). Approximately 350 cultivars are considered sufficient for the Old World, but no estimates are available for the Neotropics.

To address the need for more comprehensive germplasm collections of bananas, IBPGR has designated two field gene banks with regional foci. The banana gene bank at Davao on Mindanao, Philippines, concentrates on Southeast Asian material. The Davao gene bank, maintained by the Philippine Bureau of Plant Industry, contains 281 accessions, mostly primitive cultivars. The regional gene bank at Ekona, Cameroon, contains 116 accessions, mostly banana cultivars from Africa. The Ekona collection has been fully evaluated for resistance to nematodes and a few insect pests and diseases. At both gene banks, banana cultivars are checked for redundancy and classified.

In addition to the two regional gene banks, thirty-six banana germplasm collections with more than thirty accessions are found in over two dozen developing countries (Table 6.6). Most of these collections are for the immediate use of breeders rather than for conserving the broad genetic diversity of *Musa*. Some of these collections, however, bring together a diverse range of banana germplasm.

Three nationally controlled germplasm holdings are of particular interest. The banana collection assembled by the United Fruit Company in Honduras is currently maintained by FHIA (Fundación Hondureña de Investigación Agrícola), a Honduran research foundation supported in part by a grant from the U.S. Agency for International Development. This is the largest banana collection, with over 500 accessions, and contains an impressive assortment of primitive material and wild species (Table 6.6). This diverse banana gene bank underpins the world's most advanced banana breeding program.

The Banana Company of Jamaica, a parastatal organization, safeguards a valuable banana germplasm collection of 356 accessions near Kingston. The Jamaican banana gene bank contains wild species, primitive cultivars, fifteen Cavendish mutants, and some breeders' lines. The Banana Company's collection has also been evaluated for resistance to Panama disease and yellow Sigatoka. In Papua New Guinea, the Department of Primary Industry maintains a banana collection at Laloki, near Port Moresby. The latter collection has been largely duplicated at the regional banana gene bank at Davao, Philippines.

Other collections concentrate on local cultivars. The Bogor Botanic Garden, for example, maintains a small collection of traditional banana varieties and some wild species at its 40-hectare satellite garden at Cibinong (N. J. H. Smith, 1986:38). Small banana collections are also found in Malaysia, Thailand (duplicated at Davao), India, Tanzania, Zaire, Kenya, Uganda, Madagascar, Nigeria, Côte d'Ivoire, and the Windward Islands. Several developed countries also maintain banana germplasm collections, such as Australia, France (in Martinique), and the United States (Hawaii, Puerto Rico). Together these minor collections contain a valuable array of different banana varieties.

Except as source materials for multiplication and distribution to farmers, banana germplasm collections have been little used. Most breeders are interested in working with export varieties, so local cultivars have been largely neglected. Better selections could be made from germplasm collections to suit the needs of rural poor in developing countries. Also, a wider array of banana varieties might appeal to consumers in North America and Western Europe, particularly if they can be quickly transported with

TABLE 6.6
Germplasm collections of *Musa*

Institution	Location	No. of accessions	Wild material	No. of accessions of wild species
FHIA	La Lima, Honduras	516	*acuminata* *balbisiana*	93 12
Banana Co. of Jamaica	Kingston, Jamaica	356	54 wild spp.	
LBN	Bogor, Indonesia	333	Wild spp. from Indonesia	
Kasetsart University	Thailand	323	*acuminata* *balbisiana* *ornata*	
Kawanda Research Station	Uganda	308		
INERA	Kisangani, Zaire	300		
BPI	Davao, Philippines	281	*acuminata* *balbisiana* *itinerans* *violascens*	10 1 1 1
DPI	Laloki, Papua New Guinea	234	*acuminata* *balbisiana* *maclayi* *lolodensis*	11 1 4 1
IIHR	Bangalore, India	200		
CNPMF	Cruz das Almas, Bahia, Brazil	181	Wild *acuminata*	8
IRFA	Martinique	140	Australimusa 3 spp. Callimusa Rhodochlamys 2 spp.	1
Kerala Agricultural University	India	133		
CEMSA	Santo Domingo, Cuba	129		
IRFA	Abidjan, Côte d'Ivoire	129	*acuminata*	5
MARDI	Malaysia	128		
IRA	Ekona, Cameroon	116	Australimusa Callimusa Rhodochlamys	2 BB 3 1 2
Tamil Nadu Agricultural University	India	112		
UH	Kapaa, Hawaii	105		

TABLE 6.6—*cont.*

Institution	Location	No. of accessions	Wild material	No. of accessions of wild species
Pacific Tropical Botanic Garden	Kauai, Hawaii	100		
TARI	Chiayi, Taiwan,	97		2
Banana Research Center	Hajipur, India	96		
IITA	Ibadan, Nigeria	93		
IAC	Campinas, São Paulo, Brazil	90		
NIHORT	Ibadan, Nigeria	71		
Lowlands Agri- cultural Experiment Station	Kainantu, Papua New Guinea			
Jember University	Indonesia	63		
ICA	Tibaitata, Colombia	62	Some wild species	
Agriculture Research Institute	Maputo, Mozambique	61		
Department of Agriculture	Thailand	60		
WINBAN	St. Lucia	57	11 wild relatives	
FONAIAP	Maracay, Venezuela	53		
Koronivia Research Station	Nausori, Fiji	52		
University of Puerto Rico	Mayaguez, Puerto Rico	52		
BALITAN	Malang, Indonesia	45		
Highlands Agri- cultural Experiment Station	Aiyura, Papua New Guinea	41		
CATIE	Turrialba, Costa Rica	40		
CENRADERU	Tananarive, Malagasy Republic	39		
TFRS	Alstonville, New South Wales, Australia	32	Wild *acuminata* *balbisiana* *ornata* *velutina*	1 1 1 1
INIAA	Tarapoto, Peru	32		

Sources: Gulick and Van Sloten, 1984; N. J. H. Smith, 1986:41.

little damage from subtropical and tropical areas. While the export of bananas from the Philippines to Japan, its major market, has recently declined, the proportion of local cultivars in shipments has increased dramatically. Japanese consumers have acquired a taste for two Filipino cultivars, 'Senorita' and 'Latundan' (Segura, 1986). Exports of 'Senorita' and 'Latundan' from the Philippines to Japan doubled between 1983 and 1984. A more diverse array of export varieties would help reduce pest problems for farmers and plantation owners. 'Latundan', for example, has an *AAB* genome, unlike the *AAA* triploids that dominate the export market.

More gene bank accessions of indigenous banana varieties are needed in view of the widespread genetic erosion of traditional cultivars in Southeast Asia (S. Sastrapradja, 1975). The loss of traditional cultivars closes future options for new marketing ventures or for breeding. The job of assembling collections of traditional banana and plantain varieties is complicated by insufficient information on the characteristics, distinctiveness, and distribution of traditional cultivars. Less than half of the world's banana cultivars have been adequately described or studied (Stover and Simmonds, 1987:95). Pockets of unusual diversity of bananas and plantains surely await discovery in the tropics. The large number of synonyms of traditional varieties needs more sorting out, particularly for plantains. The appearance of a banana cultivar may vary in different locations in response to changes in soil and moisture conditions.

Chemotaxonomy will help germplasm curators and breeders streamline their operations. Laboratory techniques, such as electrophoresis and restriction fragment length polymorphism (RFLP), would help establish a clearer picture of the genetic diversity of plantain and banana varieties. At Kasetsart University in Bangkok, scientists used chemotaxonomic techniques to uncover only forty-two distinct cultivars in their large banana collection (Valmayor and Pascua, 1985). More such facilities for chemotaxonomy are needed in the humid tropics.

Wild species of banana tend to be aggressive weeds when added to field gene banks. With rare exceptions, curators are thus reluctant to include more than a few wild species of *Musa* in their germplasm collections (Table 6.6). Seeds of wild species of *Musa* can be stored in conventional seed gene banks, thereby avoiding the weediness problem in field collections. Before much progress can be made in collecting wild banana seeds, however, extensive field surveys are needed to map variation. IBPGR has initiated a field survey of wild bananas on Papua New Guinea and surrounding islands, but vast areas remain to be investigated.

BREEDING PROGRAMS AND CHALLENGES

Banana breeding started in the early part of this century and has since had a roller coaster ride. The first banana breeding programs began in

1922 at the Imperial College of Tropical Agriculture in Trinidad and at the Banana Research Station in Jamaica. Banana breeding continued on Trinidad until 1960, when such efforts were transferred to Jamaica (Mendenzez and Shepherd, 1975; Simmonds, 1966; Stover and Simmonds, 1987:172). By 1980, Jamaica had phased out banana breeding because of financial constraints. Recent support from Canada's International Development Research Centre has revived banana breeding on Jamaica.

The United Fruit Company launched a banana breeding project at La Lima, Honduras, in 1959. Wisely, the company initiated the program by first sponsoring a germplasm collecting trip to centers of banana diversity in Southeast Asia and the Pacific. Some 850 banana accessions were obtained during the collecting mission, which lasted from 1959 to 1961. After private-sector support for banana breeding in Honduras ended in 1983, FAO, and more recently the U.S. Agency for International Development and Canada's International Development Research Centre, have stepped in to sustain breeding efforts.

A more recent breeding effort is under way at the International Institute of Tropical Agriculture (IITA) in Nigeria. At its Onne Research Station in eastern Nigeria, IITA scientists are screening banana germplasm for resistance to black Sigatoka, a serious disease of bananas and plantains in Africa and other tropical regions. Also, INIBAP is encouraging research on major diseases and on the collection and utilization of *Musa* germplasm.

The main approach to banana breeding has been to cross edible, disease-resistant *AA* diploids with disease-susceptible *AAA* triploids, primarily 'Highgate'. Such crosses produced an array of *AAAA* tetraploids. Tetraploid bananas can be high-yielding, but often suffer from a short shelf life. In the 1970s, primary tetraploids were selected and backcrossed to disease-resistant *AA* diploids to produce *AAA* triploids. These triploids generally have more-compact foliage, stronger petioles, higher yields, and are more resistant to disease. Furthermore, triploid cultivars developed for export can be marketed three weeks after cutting, whereas tetraploids generally last only two weeks after harvesting.

Wild species have been little used in banana breeding because of their inferior agronomic features. Still, wild relatives of bananas warrant more attention because of their resistance to a number of diseases, among other favorable characteristics. Accessions of *M. acuminata* subspecies *burmannica*, obtained during United Fruit–sponsored collecting missions in Southeast Asia have demonstrated resistance to races 1 and 2 of Panama disease, as well as black Sigatoka (De Langhe, 1987). Two other *M. acuminata* subspecies, *banksii* and *malaccensis*, have been used in crosses with 'Highgate' for their bunch characteristics and long fingers (Stover and Simmonds, 1987:176). The two subspecies have also shown resistance to race 1 of Panama disease.

Continued breeding efforts are needed because of the ceaseless onslaught of existing diseases and pests, as well as the periodic appearance of new races and pathogens. Today, the banana industry appears to be on the eve of a another major varietal turnover at least as dramatic as the demise of 'Gros Michel' as a commercial variety because of the devastating spread of Panama disease in the late 1950s and early 1960s.

Black Sigatoka, a devastating leaf disease caused by another fungus (*Mycosphaerella fijiensis*),[1] is the culprit this time. Yellow Sigatoka (*M. musicola*) has been a problem in some banana-producing areas since the turn of the century, but black Sigatoka is a much more serious pathogen. Black Sigatoka was first detected in the Sigatoka Valley of Fiji in 1963 and reached Zambia by 1973 and Gabon by 1979. The virulent new pathogen began damaging banana plantations in Honduras in 1972 and reached Costa Rica in 1980 (Hall, 1985:169). By 1981, black Sigatoka had spread south to Colombia and Venezuela (Samson, 1986:176; Stover and Simmonds, 1987:283).

'Gros Michel' and the Cavendish group of bananas are highly susceptible to black Sigatoka. Many traditional bananas and some 'Horn' *AAB* plantains are also vulnerable to infestations of black Sigatoka. For example, black Sigatoka is drastically reducing the number of banana trees around Kampala, Uganda, an area known for its diversity of cultivars; as might be expected, the price of bananas is climbing in Uganda, a worrisome trend considering the importance of bananas in the regional diet. The export industry as well as domestic consumption of bananas and plantain will continue to suffer if black Sigatoka is not checked. Chemical control for black Sigatoka is three to six times more expensive than for yellow Sigatoka. In Central America and Colombia, fungicides need to be applied up to forty-five times a year to control the disease (Stover and Simmonds, 1987:293). On commercial plantations in Central America, black Sigatoka drove up the cost of banana production by 25 percent in 1984. An estimated $60 million is spent annually attempting to control black Sigatoka in Central and South America (Brownlee, 1989).

To help combat the global threat of black Sigatoka, scientists are busy screening *Musa* germplasm for resistance to the disease. The Honduran program is using *M. acuminata burmannica* as a source of resistance to black Sigatoka, but this subspecies has inferior bunch characteristics. One progeny of a *burmannica* cross, SH2989, displays reasonable agronomic qualities and is being used in further breeding efforts (Stover and Simmonds, 1987:178). Some types of cultivated bananas also exhibit resistance to black Sigatoka. Many cooking bananas (*ABB*), particularly 'Bluggoe',

---

1. The cause of black Sigatoka is sometimes designated as a subspecies of *Mycosphaerella musicola*, and other authorities differentiate black leaf streak (*M. fijiensis*) from black Sigatoka (*M. fijiensis* var. *difformis*). Gowen (1988) suggested that *difformis* arose in Central America. Until the taxonomy and distribution of *Mycosphaerella* are better understood, the term black Sigatoka is used here to include black leaf streak and black Sigatoka.

'Pelipita', 'Cardaba', and 'Saba', resist the disease (Stover and Simmonds, 1987:183).

The task of finding acceptable replacements for the Cavendish group is complicated by the narrow genetic base of the parents used in current breeding programs. Banana breeders generally use 'Highgate', which became available in 1960, as the female parent in most of their crosses. Under plantation conditions, the triploid 'Highgate' does not normally set seed because it is male sterile. If pollen is available from diploid forms of *Musa*, 'Highgate' can be induced to set a limited amount of seed. The Cavendish group of bananas is seed sterile, and so they cannot be used as female parents. With the existing approach to breeding, genetic diversity can be introduced only from the male side.

Recombinant DNA techniques may eventually break this breeding barrier with bananas. As of 1989, protoplast fusion and subsequent regeneration of plants from cells had not been achieved with banana. In the short term, somaclonal variation and spontaneous or induced mutation are more likely to generate disease-resistant material. In Taiwan, for example, some banana plants that mutated during tissue culture show promising levels of resistance to race 4 of Panama disease (Stover and Simmonds, 1987:188). On Mindanao in the Philippines, the Del Monte Corporation is testing two somaclonal mutants for resistance to Panama disease under field conditions (Buddenhagen, 1986).

Although widespread preoccupation with black Sigatoka has tended to eclipse Panama disease, the latter is still a serious threat to bananas in many areas. Panama disease attacks 'Silk' bananas (*AAB*); in Brazil, the disease forces plantings of 'Banana maça' farther and farther from cities. The widespread race 1 of Panama disease is not a problem for the Cavendish group of bananas, but race 4 attacks that group. Race 4 is currently confined to Australia, the Philippines, Taiwan, the Canary Islands, and South Africa (Dale, 1988). The Cavendish bananas are thus highly vulnerable to two relatively new pathogens: race 4 of Panama disease and black Sigatoka.

Panama disease and black Sigatoka are the major diseases of banana, but numerous other diseases are of regional or local significance. Some diseases can be controlled by agronomic practices, while breeding ultimately offers the best hope for other afflictions of banana. Bacterial wilt, caused by *Pseudomonas solanacearum*, is a serious problem in the Philippines and parts of Latin America. The pathogen is spread by planting infected suckers, contaminated tools, and flower-visiting insects. No effective chemical control is available. Only strict sanitary precautions and lopping off the male inflorescences can prevent the disease from becoming established in a plantation. Resistance to bacterial wilt has been located in 'Pelipita', a triploid *ABB* cooking banana (Samson, 1986:177).

Several viral diseases attack bananas and plantains. Bunchy top virus, first noted in Fiji in 1890, devastated the Australian banana industry when it arrived there in 1913 (Thurston, 1973). The disease is endemic to Southeast Asia, Fiji, and Taiwan. Bunchy top virus has become established in Burundi, Rwanda, Zaire, Gabon, and Egypt and may seriously undercut food production in a continent already struggling to feed itself. Bunchy top has also been detected recently in Hawaii. The vector for bunchy top virus, an aphid (*Pentalonia nigronervosa*), is present in banana-producing areas throughout the tropics, and conditions are thus ripe for the disease to spread to Latin America and the Caribbean. The Cavendish group is highly susceptible to bunchy top virus (Stover and Simmonds, 1987:314).

Over 200 insect pests have been recorded on banana (Purseglove, 1975: 370). Banana borer (*Cosmopolites sordidus*) is one of the more damaging insects. This borer, native to Southeast Asia but now found throughout banana-producing areas in the tropics, tunnels into banana corms in large-scale plantations and backyard patches alike. Banana borer, first recorded as a pest in 1885, is partly responsible for the decline of some Fe'i bananas during the last fifty years (Ochse et al., 1961:390; Stover and Simmonds, 1987:108).

Varietal turnover to combat one disease can sometimes unleash another pest. The burrowing nematode *Radopholus similis*, for example, became a problem only after Cavendish bananas replaced 'Gros Michel' (Shepherd, 1968). Nematodes are controlled by chemical applications in Costa Rica and Panama; elsewhere the practice is too expensive (Gowen, 1988; R. Ploetz, pers. comm.). Breeders' work is thus never complete.

The harvest index could be improved for most cultivated forms of bananas and plantains. Only 10 to 14 percent of the 'Gros Michel' biomass is fruit, compared with about 33 percent for 'Grand Nain'. Some plantains achieve a harvest index as high as 57 percent (Stover and Simmonds, 1987:77). The harvest index of most bananas and plantains clones is generally below 45 percent, compared with 50 percent or more for improved varieties of such crops as sugarcane, citrus, and semidwarf wheats and rices. An average 10 percent increase in the edible proportion of banana plants worldwide would greatly increase food for local people and foreign exchange earnings from exports.

To combat black Sigatoka, Panama disease, and many other pests, as well as to improve the harvest index, breeders must broaden the genetic base of banana breeding. The breeding program in Honduras has made substantial progress in this regard. Scientists in Honduras are building on the traditional approach to banana breeding by including 'Manqueño', an *AAB* clone from Ecuador, as the female parent (Rowe, 1987). In Bahia, Brazil, the national agricultural research system (EMBRAPA, Empresa Brasileira de Pesquisa Agropecuária) has established a relatively new breeding program that emphasizes *AAAA* crosses with *AA* forms, but is also testing *AAAB* types (Shepherd et al., 1987).

In contrast to modern cultivars, local production at the smallholder or back garden scale, so widespread in tropical countries, is likely to change little unless selections are made for local use. Scope exists for developing high-yielding, disease-resistant bananas for the growing urban markets in the developing world. Already, more than two-thirds of Latin Americans live in towns and cities with more than 20,000 inhabitants. Urban populations in developing countries are growing especially rapidly in Latin America and Asia.

Multinational corporations, once heavily involved in banana breeding, are generally less involved in research and development. U.S. companies, for example, are now more concerned with processing and transportation than with land ownership and germplasm collections. Cooperatives and parastatal organizations in developing countries lack sufficient resources to collect, maintain, evaluate, and fully utilize banana and plantain germplasm. The onus for banana breeding is thus shifting to national agricultural research systems, which are often already stretched tackling other crops. A mechanism is clearly needed to catalyze banana research and development.

REALIZING THE POTENTIAL

In lieu of creating a new international center, the networking approach offers one of the more viable approaches to furthering research on banana, as well as most other tropical perennial crops. In response to the need to stimulate research on bananas and plantains, INIBAP was launched in 1984. INIBAP's purpose is to initiate and coordinate research, collect and exchange information, and to support training for researchers and technicians from developing countries.

With support from fourteen donors, INIBAP has established regional subnetworks in Western and Eastern Africa, Latin America and the Caribbean, and Asia and the Pacific. To help set priorities and chart future banana and plantain research, INIBAP organized an international workshop in 1986. This workshop drew specialists from around the world and identified needs and knowledge gaps. Research has been commissioned on chemotaxonomy of *Musa* following an INIBAP workshop on banana systematics in 1988.

To promote the safe exchange and widespread testing of banana germplasm, INIBAP helped arrange intermediate quarantine for *Musa* materials in Belgium and France. Such services are vitally important, particularly after the Royal Botanic Gardens at Kew phased out its quarantine operation for bananas in the late 1960s (A. G. Bailey, pers. comm.).

Banana materials move largely in the form of whole or pared-down rhizomes with at least stubs of the roots and leaf bases attached. Many pathogens, insect pests, and nematodes are thus spread at the local level. Washing in hot water or dipping in fungicides or pesticides is recommended to reduce the chances of spreading diseases and pests between countries or islands. In some cases, countries prohibit the importation of banana germplasm from areas with serious disease problems.

Some banana clones are now moving as tissue cultures, but research is still needed to devise reliable disease-indexing procedures for meristem cultures. In 1988, IBPGR, FAO, and INIBAP hosted a workshop on disease indexing with banana and recommended some currently available techniques for the attention of quarantine services. Tissue culture is used to produce banana germplasm free of cucumber mosaic virus, but no antisera or probes are available for bunchy top virus (Dale, 1988). Several research efforts are under way to devise a recombinant DNA probe and an antiserum for bunchy top virus.

Standard tissue culture techniques such as rapid multiplication are beginning to be used to mass produce "clean" banana clones. Morocco, for example, has 1,500 hectares of 'Valery' bananas under plastic greenhouses, all cloned by tissue culture (Janick, 1989). Tissue culture combined with thermotherapy and disease indexing could help improve and safeguard banana yields for many farmers in the tropics and subtropics.

BREADFRUIT

Breadfruit (*Artocarpus altilis*) is a tree synonymous with the exotic tropics. Its broad, fleshy leaves and generous dark green canopy provide a fitting frame for tropical scenes in films and for the canvas. Breadfruit was the centerpiece in the infamous mutiny of the HMS *Bounty*. In 1789, King George III dispatched Captain Bligh to Tahiti to secure breadfruit trees for British possessions in the Caribbean, where the trees would serve as an abundant and cheap source of food for slaves. But on the return voyage the crew became unhappy with their captain. Unrest grew among the deckhands in part because precious fresh water on board was being used to keep a thousand young breadfruit trees alive (Oster and Oster, 1985). One of the first acts the rebellious crew performed after taking control of the *Bounty* was to toss the breadfruit trees overboard. Bligh survived several weeks in a longboat after being cast adrift in the Pacific, and his second attempt to bring breadfruit to the New World succeeded (Barrow, 1831; O'Brian, 1987). In 1792, young breadfruit trees were gathered on Tahiti and Timor, and 333 of them were planted in the St. Vincent Botanic Garden, the earliest botanic garden established by colonial powers in the New

World. A further 347 young breadfruit trees were unloaded at Port Royal, Jamaica.

The extraordinary lengths to which the British went to secure breadfruit underscores the enormous food value of the handsome tree. The starchy fruits usually weigh about a kilogram, but some varieties produce fruits as heavy as 5 kilograms (Figure 6.6). Breadfruit is a compound fruit with thin, green skin that is peeled to reveal a pulpy yellow flesh. The cannon-ball-sized fruits are baked, roasted, boiled, steamed, or fried. Inhabitants of Tahiti, the Marquesas, eastern Samoa, and a few other Polynesian islands preserve surplus fruits by burying them in pits to ferment; the resulting paste can be stored for over a year and is baked before eating (Barrau, 1961:53; Ferdon, 1981). Breadfruit chips are now a common snack food in the Pacific. In parts of Southeast Asia, the fruits are stewed and pickled or candied. The carbohydrate-rich fruits are nutritionally superior to some other staples, such as rice (*Oryza sativa*) and potato (*Solanum tuberosum*), since they contain significant amounts of thiamine, riboflavin, nicotinamide, and vitamin C (IBPGR, 1986).

Although in terms of global food production breadfruit pales when compared with the major cereal crops, in some parts of the humid tropics it is one of the perennial staffs of life. In parts of Southeast Asia and much of the Pacific, particularly Polynesia, breadfruit is daily fare. Peoples of Polynesia rely exclusively on vegetatively propagated crops for the bulk of their carbohydrate needs, and breadfruit is a basic staple alongside sweet potato (*Ipomoea batatas*), yams (*Dioscorea* spp.), cassava (*Manihot esculenta*), various aroids (*Alocasia, Colocasia,* and *Cyrtosperma*), bananas (*Musa*), the pineapple-like fruits of screwpine (*Pandanus* spp.), and coconut (*Cocos nucifera*). Only coconut and screwpine are native to Polynesia; the rest, including breadfruit, were introduced across vast stretches of the Pacific in several waves. As people moved out from New Guinea and Southeast Asia to colonize the Pacific between 2000 B.C. and A.D. 800, breadfruit went with them.

Breadfruit's significance in the Pacific depends on location. It is a major crop on the volcanic islands and coral atolls of French Polynesia, particularly in the Society Islands and in the Marquesas (Pétard, 1986:141). One indication of the importance of breadfruit to the robust inhabitants of the Society Islands is the impressive number of times the tree or its fruit appear in the sensuous paintings of Paul Gauguin: *Bare-Breasted Tahitian Woman Holding a Breadfruit* (ca. 1892); *E hare se i hia* (Where are you going?; 1892); *Ia orana Maria* (I salute you, Maria; ca. 1891); *Te nave nave fenua* (The delightful land; 1892); and *Te raau rahi* (The big tree; 1891; Figure 6.7). Breadfruit is also featured on French Polynesia's 20-franc coin.

At least twenty-six varieties are cultivated in the Society Islands, another measure of its significance to the inhabitants of such fertile islands as

FIGURE 6.6 'Paea' variety of breadfruit (*Artocarpus altilis*), Bora Bora, Society Islands, August 1989.

Tahiti, Moorea, Huahine, Raiatea, and Bora Bora (Pétard, 1986:146). Tongans recognize at least nine breadfruit varieties (Yuncker, 1959:98). Breadfruit is also a major food in Micronesia, especially Truk, Pohnpei, and the atolls (D. Ragone, pers. comm.). In Melanesia it is less significant, except in Vanuatu and the Temotu Province of the Solomon Islands.

Javanese traders seeking clove and nutmeg in the Moluccas took bread-fruit to Java well before the Portuguese arrived in the Moluccas early in the sixteenth century. Colonial powers took breadfruit farther westward to Malaysia, Mauritius, India, and Sri Lanka between 1750 and 1800. Bread-fruit entered Africa in 1899 by way of the Camayenne botanic garden near Conakry in Guinea (Busson, 1965:102). All the breadfruit trees planted in West Africa apparently stem from that single introduction. Although not an important food overall in Africa, breadfruit is a particular favorite of Zaireans.

FIGURE 6.7 *Te raau rahi* (The big tree) by Paul Gauguin, 1891. The large-leafed tree on the right is breadfruit (*Artocarpus altilis*). (The Cleveland Museum of Art, gift from Barbara Ginn Griesinger 75.263.)

Breadfruit is primarily grown for its edible fruit, but the tall, evergreen tree provides other useful products. The male inflorescences are eaten with syrup, and the leaves are suitable for livestock feed (Kervégant, 1937). In the Society Islands, breadfruit is used for a wide variety of medicinal purposes, including treatments for coughs, bronchitis, asthma, and ear problems. In Polynesia and Micronesia, the latex is used for caulking boats, and in some parts of breadfruit's range the white sap is used in the preparation of sticky mixtures to trap birds. In New Guinea, breadfruit trees are included in rain rituals (Powell, 1976).

BREADFRUITS ORIGIN AND NEAR RELATIVES

The origins of breadfruit are unclear, but it probably arose by spontaneous hybridization among two or more wild species of *Artocarpus*, possibly in New Guinea. Species of *Artocarpus* are outcrossers, so people must have selected from spontaneous hybrids on many occasions. Breadfruit is found wild in swamp forests bordering rivers of New Guinea (Powell, 1976). Wild breadfruit trees in the forests of New Guinea are often individually owned and cared for by villagers. Spontaneous populations of breadfruit were recorded on Jamaica in the last century, but they could have arisen only if some seeded forms were introduced (Grisebach, 1864: 279).

Seeds of a near relative of breadfruit, *A. mariannensis*, are roasted and eaten on Guam, where it is native (Kajale, 1989; B. Stone, 1970). In various parts of Micronesia, *A. mariannensis* also introgresses spontaneously with breadfruit.

Several of breadfruit's near relatives have been domesticated or are sporadically cultivated for a variety of purposes. Jackfruit (*A. heterophyllus*) is indigenous to the western Ghats of India, where wild populations are threatened by deforestation (Arora and Nayar, 1984:21; Hawkins, 1986: 337). Like breadfruit, jackfruit is a backyard tree grown primarily for its fruits, which are produced from the trunk and main branches of the tree and can be even larger than breadfruit. In parts of South and Southeast Asia, a saffron-colored dye is obtained from the wood for coloring the loose robes of Buddhist monks. Champeden (*A. integer*) has similar fruits to jackfruit and was probably taken into cultivation in the Malayan Peninsula. Champeden is largely confined to Southeast Asia, and wild forms are still found in the remaining forests of Malesia. In the Philippines, *A. odoratissimus* is planted for its fruits, whereas *A. hirsutus* is grown for its fine timber.

Breadfruit's wild relatives are usually large and inhabit mature forest in the wetter parts of tropical South and Southeast Asia. Some of them have been gathered in the wild for a long time. On Sri Lanka, for example, fragments of *A. nobilis* fruits have been found at the Beli-Lena archaeological site dating to 10,500–8000 B.C. (Kajale, 1989). Sri Lankans still gather fruits of *A. nobilis* for their seeds, which are roasted or boiled.

The genus *Artocarpus* contains some fifty species and was revised in the 1950s, when two subgenera, *Artocarpus* and *Pseudojaca*, were created (Jarret, 1959). Breadfruit belongs to the Incisifolii series, along with five other species: *A. blancoi*, *A. treculianus*, *A. horridus*, *A. pinnatisectus*, and *A. multifidus*, all native to the Philippines and the Moluccas. This grouping suggests that breadfruit and its closest relatives arose in an area embracing the Moluccas, New Guinea, and the Philippines. Fosberg (1960) suggested that breadfruit's origins can be traced to a much wider area that includes much of Malesia and New Guinea. More distant relatives extend farther afield in the Asian tropics and subtropics. For example, *A. lanceolatus* is a dominant tree in the seasonal rain forests that blanket the hilly portions of southern Taiwan (Huen-pu et al., 1989).

TRADITIONAL SELECTIONS AND VARIATION

Breadfruit has been cultivated for thousands of years, and with its wide distribution and the rich tapestry of ethnic groups that tend the crop, it is not surprising that many local forms have arisen. Some varieties produce fruit for up to five months, whereas others produce fruit all year round. Over fifty years ago, a French agronomist called for the introduction of fresh cultivars to Martinique so that more fruits would be available throughout the year to satisfy the heavy demand for the fruit among the poorer segment of society (Kervégant, 1937).

Rural people and villagers have selected cultivars from seedling progenies, and such selections, adapted to the local climate, soils, and cultural tastes, have thrived to the present. Selection has been strong for seedless forms, and these are propagated by planting suckers cut from the extensive surface roots. Seedless cultivars originated in Polynesia by careful observation and propagation of sports from somatic mutation of root shoots (D. Ragone, pers. comm.).

Seeded forms are propagated from seed or can be cloned by planting suckers. Breadfruits with seeds contain less pulp than the more common seedless varieties, and the flesh may be stringy. Although seeded forms of breadfruit are less common than seedless kinds, they remain in the repertoire of varieties because the roasted seeds provide a nutritious snack (Coenan and Barrau, 1961).

The diversity of breadfruit cultivars is little understood. Some varieties may be sterile hybrids arising from natural crosses between wild diploids and tetraploids. Mutation has also played a role, with some interesting new forms "saved" for future generations by vegetative propagation.

Numerous breadfruit cultivars were recorded by C. Wilkes on explorations led by the United States in Fiji and Samoa in 1845. More recently, the South Pacific Commission systematically studied varieties of the fruit in the Pacific (Coenan and Barrau, 1961). Several hundred varieties of breadfruit are grown throughout the humid tropics, with the greatest diversity occurring in the Pacific.

Although modern cultivars are not being bred and therefore are not sweeping aside traditional varieties of breadfruit, the cultivated gene pool is nevertheless experiencing some genetic erosion. Competition from other crops and foodstuffs is prompting the decline of breadfruit in some parts of its range. In the past few decades, the importance of breadfruit in the Society Islands has declined as the custom of eating baguettes made from imported wheat has spread from Tahiti to neighboring islands. In Indonesia, breadfruit cultivation is also waning (S. Sastrapradja et al., 1981). As breadfruit loses favor in some regions, traditional selections are likely to be lost. As regions develop, people with rising incomes tend to abandon food items associated with the poor.

Conservation Efforts and Needs

Breadfruit produces recalcitrant seeds that cannot be dried and frozen without losing their viability. The only way to conserve cultivars of the crop is to maintain the genotypes in the field, either in farmers' garden plots and backyards or in field gene banks.

Several attempts have been made to assemble germplasm collections of breadfruit for study. Over 200 accessions of breadfruit from the islands of Tokelau, French Polynesia, Cook Islands, Tonga, Papua New Guinea, Rotema, Fiji, the Solomon Islands, and the New Hebrides have been introduced to Western Samoa for study, but few accessions remain conserved today. Only a few field gene banks have been established for breadfruit, and only one has more than 100 accessions.

The largest germplasm collection for breadfruit is maintained on Maui, Hawaii. Between 1985 and 1988, several hundred breadfruit accessions were collected from forty-five islands and sent to the Kahanu Gardens on Maui operated by the National Tropical Botanic Garden, which is headquartered on the island of Kauai (Ragone, 1987, 1988, 1989). The 47-hectare Kahanu Gardens currently maintains thirty-eight breadfruit varieties as live trees and eventually plans to have living examples of 116 varieties, making it the largest germplasm collection of breadfruit in the world. With support from the IBPGR, this initiative has identified twelve superior cultivars for possible wider dissemination. Among the promising selections are fleshy varieties from volcanic islands that could be grafted onto seedier plants adapted to the calcareous soils of certain atolls.

In spite of breadfruit's widely recognized food value, no breeding efforts have been initiated for the crop. Breadfruit's potential therefore remains only partially exploited. Much research would be needed to underpin any breeding; for example, the nature of seedless breadfruit is not fully understood and seedless breadfruit may be due to parthenocarpy or sterility. More work is needed on breadfruit diseases and possible sources of resistance to them. Pingelap disease has undercut breadfruit production in the Marianas and Marshall islands. This little-known disease causes wilting and dieback, but some cultivars apparently withstand attack (Zeiger and Zentmeyer, 1967). In India, soft rot caused by a fungus, *Rhizopus artocarpi*, damages breadfruit, and the disease is currently treated with copper-based sprays.

The potential value of breadfruit's near relatives for breeding purposes is unknown. Accordingly, it is difficult to identify particular relatives of breadfruit for high priority conservation. Because *A. mariannensis* may have contributed to the ancestry of breadfruit, that species warrants special attention for in situ conservation. Known locally as dugdug, populations of *A. mariannensis* on limestone outcrops on Guam and Palau are threatened by land clearing (Fosberg, 1960). Several of breadfruit's relatives, such as *A. camansi, A. elasticus,* and *A. rigidus,* have been used as rootstocks.

Breadfruit is particularly well suited to agroforestry and as a backyard tree for the urban poor. In Sri Lanka, for example, breadfruit is a common overstory tree in home gardens (MacDicken, 1990). Breadfruit provides generous shade, leaves for feeding pigs and other livestock, and a nutritious fruit. If a suite of high-yielding varieties could be developed that produce year-round, breadfruit could become an important weapon in combating malnutrition in the burgeoning cities of the lowland tropics.

PEACH PALM

Peach palm (*Bactris gasipaes*) is an important food for people in many parts of the humid lowlands and foothill zones of Central and South America. Known by many names, including *pejibaye* (Central America), *pupunha* (Brazil), *chontaduro* (Colombia), and *pijuayo* (Peru), peach palm provides two edible products: starchy fruits and celery-like heart-of-palm. The ovoid fruits are boiled in saltwater, and after the fibrous outer skin is torn off, the floury, orange flesh is relished from Amazonia to Mexico. In addition to being a good source of carbohydrates, peach palm fruits are frequently rich in vitamin A and contain more protein than most tropical tuber and root crops (Balick, 1984a; Zapata, 1971). Some peach palm fruits are also significant sources of oil (Johannessen, 1966a).

Peach palm is normally planted by seed and can live for over fifty years. Often clumps are formed as new stems emerge from basal suckers. Various weevils (species of *Derelomus* and *Phyllotrox*; Curculionidae) and possibly scarab beetles (*Cyclocephala* spp.) are the main pollinators of peach palm (Beach, 1984; Mora-Urpí, 1982; Mora-Urpí and Solis, 1980; Uhl and Dransfield, 1987:524). Stingless bees visit peach palm flowers, but are apparently little involved in pollination. The fruits, usually 50 to 200 per ⸱alk, are harvested with a sharp knife tied to a long pole. In some cases, ⸱⸱nches are cut down by climbers equipped with special devices to help them avoid the spines that arm the trunk in concentric rings (Balick, 1984b).

Peach palm provides a wide range of useful products in addition to food (Patiño, 1963b). The roots are reputed to dispel intestinal worms, while the finger-length spines serve as needles. The hard, woody stem of peach palm makes excellent bows, fishing poles, floors, walls, and mortars

(Spruce, 1908b:447). The tough leaves are fashioned into baskets, while some Indian groups flavor dishes with peach palm flowers.

Alcoholic drinks are also prepared from peach palm. In Costa Rica, *coquillo* is a palm wine prepared from sap tapped from the trunk (Johannessen, 1966a). A fermented drink is made from mashed fruits that have been boiled, sometimes with plantain, and left to ferment a few days; this beverage is used during ceremonial dances in the Colombian Amazon (P. H. Allen, 1967).

Indigenous groups of the Colombian Amazon celebrate the peach palm harvest by dancing in bark-cloth masks for three to four days (Schultes, 1988:23, 112). The versatile palm is so appreciated among the Jivaro of the Ecuadorian Amazon that a festival is held in its honor at the end of March (Cordero, 1950:173).

In season, street stalls selling peach palm are a common sight in many Latin American towns and cities, such as San José, Costa Rica, and Manaus and Belém in Brazil. In Pará, Brazil, the globular fruits ripen in the rainy season and compete with popcorn and candy for the attention of moviegoers and other pedestrians in the larger towns. In some regions, such as along the Pacific coast of Colombia, the widely appreciated fruit is an important source of calories and cash income (Angel, 1978).

ORIGIN AND SPREAD

The origins of peach palm are still obscure. Peach palm has been postulated as a spontaneous hybrid between wild species in southwestern Amazonia that was subsequently domesticated by indigenous groups. It is possible that various hybrid combinations have occurred at different places and times, and some of these hybrids were selected by local people. Among the several candidates involved in the possible hybrid origin of peach palm, *Guilielma microcarpa* stands out.

Peach palm may be derived directly from *G. microcarpa* (C. R. Clement, 1988). In a discriminant analysis, continuous variation has been observed from the smallest *G. microcarpa* to the largest *B. gasipaes*. In either case, extreme southwestern Amazonia in northwestern Bolivia and southeastern Peru has been postulated as the palm's center of origin (C. R. Clement, 1988).

In another interpretation, truly wild peach palm allegedly occurs from Santa Cruz Province in Bolivia to the Darien Gap in Panama, and possibly the Atlantic coast of Nicaragua (Corrales and Mora-Urpí, 1990). Wild peach palm is thought to occur on the Pacific side of Ecuador. Different wild subspecies and possibly valid taxonomic species may have been brought into cultivation by various indigenous groups. Thus the palm may have no single point of origin, akin to the story of avocado and some other tropical perennials.

The domestication process for peach palm, as for many tropical perennial crops, may have started when hunters and gatherers enriched certain campsites or trails with the species. Later, peach palm may have been planted in villages, and finally incorporated in swidden systems.

Slash-and-burn fields are especially propitious for selecting interesting new forms of peach palm. Swidden plots are relatively small, and when intercropped with peach palms, only a few individuals are involved. Genetic drift is thus promoted (C. R. Clement, 1988). Individuals are selected from a limited number of parents, and inbreeding helps lead to distinct varieties. Aboriginal fields are generally isolated from one another in Amazonia, thus opportunities for foreign genes to "contaminate" small, contained populations are limited. Finally, the relatively short duration of swidden fields curtails generation overlap.

This cultural shaping of peach palm evolution led to many different forms and continues to this day (Mora-Urpí et al., 1984). The distinct landraces suggest that the crop has been cultivated for a protracted period; at least eight landraces are recognized in Amazonia (C. R. Clement, 1988). The Ticuna of the Upper Amazon, for example, save seeds from especially large or flavorful fruit for planting (Kerr and Clement, 1980). Around Yurimaguas in the Peruvian Amazon, some virtually spineless peach palms have been selected by indigenous groups. Spineless specimens with large, floury fruits are favored in the Colombian Amazon near Leticia (C. R. Clement, 1988). In Costa Rica, the Boruca maintain both spineless and armed forms (D. Stone, 1949).

If peach palm originated in Amazonia, it was taken to Central America in precontact times, probably through a low pass in the Andes into the well-watered lowlands of western Ecuador and then up the Pacific coast of South America (Johannessen, 1966b; Schultes, 1984a). Today peach palm is cultivated from Honduras south to northern Bolivia and is especially common in Costa Rica, where it is cultivated in both urban and rural areas (Johannessen, 1966a). When Spaniards first penetrated the Sixaola Valley in Costa Rica near the border with Panama they found a single plantation with tens of thousands of peach palms (Brücher, 1989:133).

Near Relatives

The precise number of peach palm's near relatives is not known, mainly because of taxonomic confusion. The genus *Bactris* varies considerably in size and form and includes some 200 to 250 species, all in the New World tropics (Balick, 1984a; Blombery and Rodd, 1985:59; McCurrach, 1960: 22; Uhl and Dransfield, 1987:524). Some scientists place peach palm in the genus *Guilielma* (Samson, 1986:287; R. E. Schultes, pers. comm.). Generally, *G. gasipaes*, *G. speciosa*, *G. utilis*, and *G. chontaduro* are considered synonyms of *B. gasipaes*, and *Guilielma* is regarded as a subgenus of *Bactris* (C. R. Clement, 1988).

Most of peach palm's near relatives inhabit openings in lowland tropical forests. Some such species are *B. caribaea*, *B. macana*, and *B. granatensis* in Colombia and Venezuela; *B. microcarpa* in Brazil; *B. insignis* in Bolivia; *B. ciliata* in Peru; and *B. piritu* in Colombia. Others, such as *B. porschiana* in Costa Rica, are understory trees (Beach, 1984). Some disturbance, either natural or artificial, is thus important for the survival of the rich diversity of peach palm's cousins.

Little is known about the natural history or potentially useful attributes of peach palm's near relatives. At least some near relatives appear to introgress spontaneously with peach palm (C. R. Clement, 1988). Eventually, peach palm may be crossed with its near relatives to garner useful traits. Several near relatives are harvested by indigenous groups and could be used for breeding or domestication in their own right. The Bora Indians of the Peruvian Amazon gather the sweet fruits of *B. macroacantha*, which allegedly help people relax and feel drowsy (Balick, 1984a).

Germplasm Collections

Germplasm expeditions in search of peach palm began in the early 1960s, when Carl Johannessen and colleagues began scouring parts of Central America for diverse forms of the palm. Their finds were subsequently planted at the Center for Tropical Agricultural Research and Training (CATIE, Centro Agronómico Tropical de Investigación y Enseñanza) in Turrialba, Costa Rica. In the 1970s, Jorge Mora-Urpí started collecting germplasm of the palm in widely scattered locations in Latin America and has established a field gene bank under the auspices of the University of Costa Rica.

With support from the U.S. Agency for International Development, scientists at the National Institute for Amazonian Research (INPA, Instituto Nacional de Pesquisas da Amazonia) in Manaus and at Brazil's National Center for Genetic Resources (CENARGEN, Centro Nacional de Recursos Genéticos e Biotecnologia) in Brasília organized several peach palm germplasm collecting expeditions in Amazonia in 1983 and 1984. Samples obtained from these far-reaching expeditions greatly enriched some existing field gene banks for peach palm and helped establish new ones.

The largest collection of peach palm germplasm, with over 1,100 accessions, is maintained by the University of Costa Rica at Guapiles (J. Mora-Urpí, pers. comm.). INPA, Manaus, maintains 450 accessions, up from 283 in the early 1980s (C. R. Clement et al., 1982). CATIE holds 400 accessions, while three field gene banks in Colombia, which date from the mid-1970s, contain between 100 and 400 accessions each. Peach palm collections were started in the mid-1980s at Yurimaguas and Iquitos, Peru, and along the Rio Napo in Ecuador. Such collections, now with around 200 genotypes each, are useful sites for studies on comparative physiology and fruit characteristics, among other traits. However, only a small portion of the palm's gene pool is represented in field gene banks, and several institutions are encountering difficulties maintaining the collections (C. R. Clement, 1991).

No modern, high-yielding cultivars of peach palm have been released to displace traditional forms, but genetic erosion is locally serious. The cultural and ecological integrity of many indigenous groups are under assault by expanding national frontiers, and some unusual and potential valuable forms of peach palm are undoubtedly threatened with extinction. The Colombian government's recent decision to set aside several million hectares for Indian reserves in Amazonia will surely help preserve unique forms of the palm and other crops.

POTENTIAL FOR FURTHER DEVELOPMENT

Peach palm is already used for many purposes by indigenous populations and to a lesser extent by other rural and urban folk in Latin America. The heart-of-palm industry provides one of the most promising avenues for further development of peach palm, but other new or expanded uses include flour meal for human consumption, livestock feed, and oil. Ironically, scientists have been calling for more research on peach palm's potential for almost a century (F. Müller, 1895; W. Popenoe and Jiminez, 1921), but only recently has the palm excited much research attention.

Peach palm contains from 3 to 18 percent protein on a dry-weight basis and therefore would make an excellent flour for incorporation in other foodstuffs. Furthermore, the fruits contain most amino acids, including tryptophane (C. R. Clement, 1988). Peach palm thus has considerable potential for further improving the diet of both the urban and rural poor. In many cities in the humid tropics, bread is an important source of protein because wheat prices are subsidized. Other sources of vegetable protein that can be grown nearby, such as peach palm, would help provide local employment and reduce the outflow of foreign exchange.

The growing market for heart-of-palm, both within Latin America and abroad, has opened new possibilities for peach palm as a plantation crop. As wild stands of palms, particularly species of *Euterpe*, are increasingly destroyed for the heart-of-palm (palmito) canning industry and local consumption, prospects for raising palms to supply the industry brighten (Balick, 1976). At least 100 million palm trees were being destroyed annually in Brazil in the 1980s (Johnson, 1987).

Peach palm is a leading candidate for plantations to supply the palmito industry because it grows rapidly and can be harvested within a year. Peach palm sprouts readily after being cut, thereby reducing the cost of replanting. Furthermore, harvesting is constant, thus plantations provide year-round employment both for fieldworkers and canners.

The selection of spineless peach palms by indigenous people will facilitate the establishment of heart-of-palm plantations. Peach palms are generally armed with large, downward-pointing spines to thwart rodents and other predators from reaching the fruits. The spineless trait, an apparent mutation under domestication, assists the harvesting of peach palm for palmito. An estimated 0.5 percent of peach palms in Costa Rica, for example, have virtually no spines (Johannessen, 1966a).

Spineless palms may be a disadvantage in large plantations for fruit production because of rodent damage. Even in heart-of-palm plantations, gophers can be a nuisance. In Costa Rica, for example, spineless peach palms are more vulnerable to pocket gophers (*Orthogeomys cherriei*), known locally as *taltuzas*, which eat roots of spineless forms (J. Mora-Urpí, pers. comm.). Pocket gophers are a delicacy in some parts of Central America (Walker, 1975:736), so plantation managers could encourage locals to trap the rodents for food.

Several peach palm plantations for palmito were established in Costa Rica starting in the 1970s (Figure 6.8). By 1990, entrepreneurs in Costa Rica had laid out 2,500 hectares of peach palm for heart-of-palm exports. Average palmito yields on Costa Rican plantations are 1.3–1.8 metric tons per hectare (Mora-Urpí, 1989). Peach palms with spines are used because they are relatively high-yielding and reasonably disease-resistant. As more uniform plantings of peach palm appear on the landscape, however, diseases are likely to become more of a problem. At the moment, various fungi (*Phytophthora palmivora*, *Fusarium moniliforme*, *Dreschlera incurvata*, and species of *Colletotrichum*, *Pestalotiopsis*, and *Mycosphaerella*) and a bacterium, *Erwinia chrysanthemis*, cause generally minor damage to peach palm leaves in Costa Rica (Vargas, 1989a, b).

Scientists are pursuing research on the potential of spineless peach palm for palmito production at the University of Costa Rica and CATIE, Costa Rica; INPA, Manaus, Brazil; the Cacau Research Institute (CEPLAC, Commissão Executiva do Plano da Lavoura Cacaueira) in Brazil; and ICA (Instituto Colombiano Agropecuaria) in Cali, Colombia. Brazilians are interested in establishing peach palm plantations because of the widespread destruction of *Euterpe edulis* in the Atlantic forests of eastern and southern Brazil (Nodari and Guerra, 1986). In Pará, açaí (*E. oleracea*) sprouts again if cut down at the base, so natural stands can withstand greater harvesting pressure if properly managed. If açaí is pruned selectively for heart-of-palm, fruit production can be enhanced on the remaining stems (A. B. Anderson, 1990).

FIGURE 6.8 Harvesting heart-of-palm in a plantation of peach palm (*Bactris gasipaes*), Agro Palmito, S.A., Guapiles, Costa Rica, June 1990.

High-yielding peach palms for local consumption of fruits could also be profitable. As cities in Latin America continue to grow, backyard production in urban and rural areas may eventually be insufficient to meet demand. Peach palm could also be incorporated more into agroforestry systems to diversify income for farmers; it is occasionally used as a shade tree for coffee and cacao (C. R. Clement, 1989; Johannessen, 1966c). With periodic fertilization, fruit yields of 30 to 50 metric tons per hectare could probably be achieved on peach palm plantations (C. R. Clement, 1988).

With assistance from Canada's International Development Research Center, the native fruit project of Peru's agricultural research system is investigating the potential of increased food production in Amazonia using superior selections of peach palm. To boost consumption levels, however, breeders need to develop varieties that fruit at different times of the year, have less-perishable fruit, and are higher-yielding (Tracy, 1985).

Penetration of the animal feed market will require selections that produce drier fruits. At the moment, only inferior fruits are fed to pigs and chickens at the household level. The animal feed market is growing worldwide, and feedgrains are the most important item in international trade for feeding cattle, pigs, and poultry. Sorghum, triticale, or maize imports can be costly for a developing country. Maize is grown in tropical and temperate areas, but local production may be insufficient. Tests in Costa Rica demonstrate that peach palm meal can substitute for maize in chicken feed without any reduction in weight gain (Murillo, 1990).

After several breeding cycles, it may be possible to achieve reasonably uniform characteristics, such as spineless stems. A more reliable way to propagate valuable attributes is to clone superior selections. Peach palm can be reproduced from shoots that frequently sprout at the base of the tree (Johannessen, 1967), but the success rate is low, and subsequent growth is usually disappointing (C. R. Clement, pers. comm.). Labor costs of vegetative propagation are high, and a limited number of desirable clones can be produced with conventional cloning methods. Eventually, micropropagation techniques may be used to multiply large numbers of high-yielding peach palms.

SAGO PALM

The sago palm (*Metroxylon sagu*)[2] is a significant source of carbohydrates for rural people in various parts of Southeast Asia. Sago palm is wild in swamp forests of eastern Indonesia and New Guinea and may have originated in the Moluccas (Davis, 1986). In Papua New Guinea alone, wild sago palm covers an estimated one million hectares, with an additional 20,000 hectares under cultivation (Rauwerdink, 1986).

Between 100 and 400 kilograms of starch can be obtained from an individual palm. In a typical swamp, yields are in the 100- to 160-kilogram range. With a density as high as sixty-three palms per hectare, about 20 million calories per hectare can be obtained from swamp forests, assuming 35–40 percent water content of the sago starch (Barrau, 1958:39). Sago palm is thus an ideal plant for food production in a marginal environment.

Sago palm was taken to Malaysia, Indonesia, and some islands of Polynesia early (Yen, 1973). Introduced sago palms are well adapted to poorly drained parts of Southeast Asia, where they soon became "half-wild." Marco Polo noted sago palm cultivation on Sumatra in 1298.

Sago is harvested by cutting down the palm shortly before it flowers

2. A spiny-sheathed form of sago palm has sometimes been placed in a separate species, *Metroxylon rumphii*.

(Wallace, 1986:383). Sago palm takes from eight to fifteen years to mature and stores enormous quantities of starch in the trunk (stem) as flowering time approaches. The pith from the trunk is cut out and kneaded and washed to obtain the starch (Figure 6.9). Much of the sago starch is still extracted by hand in the field, although some small mills have been established to extract the starch. Sago starch is boiled with water and flavored with salt, lime, and chili pepper. Soft, easily digested sago bread is also baked in homemade ovens. Sago flour is prepared by drying raw sago in the sun, then pounding the desiccated pith into a powder for sieving. The residue left after the starch has been washed out is often fed to livestock.

Sago starch is also made into pellets or flakes, so that it can be transported more easily to local and regional markets. The development of "pearl" sago, apparently by the Chinese, opened up overseas markets for sago. Granular pearl sago is prepared by pressing the wet, starchy paste through a sieve and then drying the fine material on a hot surface. Malaysia and Indonesia are the chief exporters of sago products.

Sago alone is not a balanced food. With only a trace of protein and minerals, no fats, and deficient in vitamins A, B, and C, a diet of pure sago

FIGURE 6.9 Extracting sago starch on Ceram in the Moluccas in the nineteenth century. (From Wallace, 1986.)

would soon lead to malnutrition. People who consume sago obtain essential amino acids, minerals, and vitamins from other vegetable and animal food sources. In industrial countries, sago products are incorporated into gravy mixes, custard powders, confectionery, and the flavor enhancer monosodium glutamate.

Sago is an important emergency food when rice supplies are low. Many households in Southeast Asia hoard sago starch as insurance against hunger. If sago stocks are low, more can be prepared within a day by visiting a sago palm grove. In this regard, sago palm fulfills a function similar to that of cassava (*Manihot esculenta*) as a backup to protein-rich cereal foods in Latin America and Africa.

In addition to food, sago palm is employed for various industrial products, in folk medicine, construction, and for fuel. Adhesives containing sago starch are used to make chipboard. In the Moluccas, midribs of sago fronds have long been used to build houses (Wallace, 1986:333). Sago fronds last up to seven years as roof thatch (Whitmore, 1973). Gouged-out trunks of sago palm are sometimes converted to sawdust for fuel, flooring, or matting. Sago starch is used as a bulk filter in the pharmaceutical industry and for stiffening textiles (Ruddle et al., 1978). Sago palm is used locally to treat impotence and viral shingles (Gimlette, 1939).

Near Relatives

Five wild species of *Metroxylon* are found from Thailand to the Solomon Islands, New Hebrides, Samoa, Fiji, and the Carolines, but only *M. sagu* and *M. salomonense* are used to produce sago (Barrau, 1959; Rauwerdink, 1986; Soejarno, 1980; Uhl and Dransfield, 1987:246). On some Pacific isles, locals fashion palm "ivory" buttons from the hard seed of *M. salomonense*, but sago from this species is now used only for pig feed (Barrau, 1958:37). In the New Hebrides, *M. warburgii* was formerly a food source, but the fronds and trunk of are still used for buildings (Barrau, 1958:38).

Other Sources of Sago

A wide variety of palms from other genera are felled occasionally to produce sago in Melanesia, including several species of *Caryota*, toddy palm (*Borassus flabellifer*), *Phoenix farinifera*, and nipah palm (*Nypa fruticans*). Toddy palm and *Caryota* palms are tapped for their sugary sap, which is used directly as a syrup or is fermented to produce an inexpensive, lemonade-colored alcoholic drink. The nomadic Penan in Sarawak rely heavily on the wild Bornean sago palm (*Eugeissona utilis*), often associated with poor soils, for food (Uhl and Dransfield, 1987; Whitmore, 1990:156).

Some palms introduced to Melanesia from tropical America for ornamental purposes, such as species of *Roystonea* and *Mauritia*, are also employed to produce sago (Johnson, 1977). In the Caribbean and northern South America, the oil fruits of *Roystonea* palms are fed to pigs, while the vitamin C–rich fruits of *Mauritia* palms are harvested to make juices and ice cream. Starch is extracted from *Mauritia flexuosa* along the Orinoco River, particularly in the Delta region (Ruddle et al., 1978:8).

SELECTION AND BREEDING

Despite sago palm's penetration of international and regional markets, no breeding efforts have been launched to stretch the genetic potential of *Metroyxlon sagu*. Selection for high starch yields is feasible, but germplasm resources need to be screened and selected forms must be propagated. Sago palm is propagated vegetatively, so superior selections can be readily maintained.

Superior cultivated forms should be collected and tissue culture methods should be developed for rapid propagation of exceptional material. At the same time, a strategically planned genetic conservation program is needed to ensure that representatives of different genetic patterns of diversity are available for future use.

Sizable in situ reserves are needed in the core of sago palm's native area encompassing the Moluccas, Irian Jaya, and Papua New Guinea. Some semiwild groves in introduced areas might also warrant conservation in a system that allows harvesting of forest products and some planting by local people.

Indonesia is one of the few countries concerned with the conservation of sago palm germplasm. But much more fieldwork and analysis of genetic variation are needed to launch a manageable conservation plan for sago palm. Indigenous knowledge should be canvassed on variation in sago palm; the Marind distinguish at least eight different forms of sago palm on the basis of such characteristics as smooth or thorny fronds, size, color, growth, and yield (Barrau, 1958:39). Rural development projects in Southeast Asia, particularly on Sumatra, Kalimantan, and Irian Jaya, should incorporate funds for genetic resource conservation to improve the sustainability of their efforts, which would include extractive reserves.

Malaysia intensified research on sago palm in the 1980s with the establishment of the Talau Research Station and the Mukah Sago Workshop, both on Sarawak. The research station conducts agronomic trials, botanical studies, chemical and soil studies, and pest management. Work has begun on conventional and induced-mutation breeding of sago and micropropagation techniques. The Mukah Sago Workshop undertakes processing experiments to help improve the quality of the finished product at small-scale factories.

PROSPECTS

Sago will retain an important role locally in food production in Southeast Asia and portions of the Pacific because the crop thrives in marginal habitats. It grows well in swamps, an environment unsuitable for most crops. Sago palm is likely to become a more significant crop as some marginal areas of the tropics become crowded and as demand for quality starch continues to grow in international markets. Rather than attempt to clear and drain more swamps for rice production, people should use such areas for sago. Sago cultivation would disrupt the environment less than rice, particularly for fisheries.

Because of competition from other starch sources, such as tapioca derived from cassava, the volume of sago exports is likely to grow only if higher-yielding germplasm is more widely disseminated, cultivation methods improve (Floch, 1983), and processing is made more efficient. Increased investment in local processing plants coupled with possible mineral and vitamin enrichment of sago products would enhance the palm's local nutritional contribution. With increased selection efforts and agronomic and processing research, sago could generate more income as well as improve the diets of people living in sago-producing countries (D. Sastrapradja and Mogea, 1977).

Prospects for sago are particularly bright for some nontraditional uses. Several developing countries have expressed an interest in establishing industries for the bioconversion of sago starch to protein (Raimbault, 1980). Sago palm has recently been used by local industries to produce ethanol, which can be added to gasoline to stretch petroleum supplies, or, with appropriately designed engines, can be used directly as a transportation fuel (J. H. Baker, 1980; Holmes and Newcombe, 1980). As oil prices rise again in the 1990s, such alternative energy sources are likely to assume greater importance, as they did in the mid to late 1970s.

"Ivory" buttons, fashionable in developed countries in Victorian times, have recently come into style again, and species of *Metroxylon* may benefit from this "rainforest badge." In the lowland forests of Colombia and Ecuador, for example, ivory buttons made from species of *Phytelephas* and *Ammandra* are being exported to developed countries (R. Bernal, pers. comm.).

To help further research and development of sago palm, scientists can exchange ideas and findings during meetings. For instance, several international symposia on sago have been organized by the Food and Agriculture Organization (FAO); an informal lobbying unit and networking node has formed in Papua New Guinea; and the Sago Advancement Group Office, formerly known as the Sago Palm Research Network, operates from the Office of Economic Services, East Sepik Provincial Government, in Papua

New Guinea. The Sago Advancement Group Office publishes a newsletter on sago palm that is sent to approximately 100 individuals in several European and Southeast Asian countries. One of the aims of the Advancement Office since its establishment in 1965 is to secure funding from donors to establish a regional research institute dedicated to sago and possibly some other palms. Unfortunately, such fundraising appeals have been largely unheeded.

More recently, the International Center for Underutilized Crops has been asked to collaborate with the Tree Crop Program of the International Fund for Agricultural Research to develop a strategic plan for sago. A strategic plan, based on a synthesis of existing knowledge and the pooling of expertise on which directions to take, strengthens proposals sent to donors and development agencies for consideration.

CHAPTER 7

Fuelwood, Fodder, and Woody Grasses

One of the daily dramas for millions of women in the Third World is finding enough fuel to cook the day's meals (Figure 7.1). Most staple foods have to be cooked, but because few rural people in developing countries have access to electric or gas stoves, obtaining sufficient supplies of fuelwood or charcoal is critical. Indeed, fuelwood has become so scarce in some areas that households may eat only one cooked meal a day (Pimentel et al., 1986). Furthermore, as women spend more tedious hours combing the countryside for fuelwood and carrying a heavy load for several kilometers back home, home life suffers. Less time is available for attending to crafts and caring for children. And, because women perform much of the agricultural work in developing countries, the reduction of time spent on farm plots probably reduces crop yields (Eckholm, 1975; N. J. H. Smith, 1981b).

The dietary and health implications of the growing fuelwood scarcity in developing countries are far-reaching. Leftovers from the one cooked meal a day may spoil by dinner or breakfast the next day. Little wood is available to boil drinking water; most households in developing countries are not connected to piped, potable water. Consumption of foods that require prolonged cooking, such as protein-rich beans, may plummet.

Another worrisome dimension to the fuelwood crisis is that animal manure and waste from corrals and stockades are increasingly used to cook meals instead of being recycled back onto farmland. Stacks of dried cattle dung for sale are a common sight along roadsides in India and many parts of Africa. Millions of extra mouths could be fed if more fuelwood could be located.

By the year 2000, as many as 2.4 billion people may suffer fuelwood shortages (FAO, 1981; Postel and Heise, 1988). The scarcity is likely to

FIGURE 7.1 Farmer returning home with fuelwood gathered from tropical montane forest. Mulungu, Kivu, Zaire, April 1989.

affect as many as fifty developing countries. The problem is particularly
acute in Africa and India. Reforestation rates lag behind the demand for
fuelwood and other forest products (Spears, 1983). To help overcome the
growing shortage of fuelwood, many developing countries are promoting
reforestation and afforestation schemes (D. Anderson and Fishwick, 1984;
Eckholm, 1979; Postel and Heise, 1988; N. J. H. Smith, 1981c). Various
models of more efficient stoves, made largely from local materials, are also
being promoted.

Leucaena (*Leucaena leucocephala*), a fast-growing leguminous tree native to southern Mexico and some other parts of Central America, is one
of the most commonly planted trees for fuelwood production worldwide.
Although leucaena did not originate in closed-canopy tropical forests,
many of its near relatives are found in such environments. Leucaena's close
relatives are being tapped in efforts to upgrade its growth form as well as
its resistance to diseases and pests.

Leucaena is profiled here over many other species used for fuelwood
production, such as species of *Eucalyptus, Calliandra,* and *Acacia,* because
it is a multipurpose tree par excellence. It provides a copious supply of
firewood, and the leaves can be fed to livestock and used as green manure.
Leucaena also fixes nitrogen through symbiotic *Rhizobium* bacteria and is
thus particularly suitable for interplanting with food or cash crops because
it enriches the soil.

Construction materials are vital to provide shelter and piping for drinking and irrigation water. As forests retreat from urban areas and around
homesteads in the countryside, the cost of procuring wood for building or
repairing homes escalates. Bamboo has long been used for such purposes
and has been cut from tropical forests for millennia. Some species have
been domesticated and are a common sight in villages and around fields in
most parts of tropical and subtropical Asia. Light-loving bamboos benefit
from some human disturbance of forest ecosystems, but outright forest
destruction eliminates many bamboo species.

Bamboo could be more rationally managed from existing forests as well
as planted more systematically. In some tropical forests, such as those in
southwestern Amazonia and parts of Central America, native bamboos are
hardly exploited. Bamboos, now introduced to most tropical and subtropical countries, can also be used for food, paper production, and livestock
fodder. Quick-growing bamboo provides additional rural income as well
as employment in urban areas for truckers and retailers. Women, in particular, generate income from home-based crafts that employ bamboo.

LEUCAENA

Leucaena (*Leucaena leucocephala*) is a fast-growing, multipurpose tree native to Mexico and now cultivated in many tropical lands (Brewbaker,

1987a; Sorensson, 1989a). Approximately 2–5 million hectares are planted to leucaena worldwide, making it one of the most widely disseminated fuelwood and fodder crops (Brewbaker and Sorensson, 1990). Leucaena is also planted as a living fence for livestock and as a source of mulch (Figure 7.2). As a member of the legume family, leucaena also performs the valuable function of enriching the soil with nitrogen (Nair, 1990:14).

Livestock feed is one of the most important uses of leucaena. The pinnate-leafed tree is highly productive, yielding up to 20 metric tons of feed per hectare per year on a dry-weight basis (Lahane et al., 1987). An advantage of leucaena is that leaves remain green long into the dry season, when other forage species, such as grasses, have died back. Leucaena can be fed fresh to livestock or converted into dried pellets. While toxic levels of mimosine can be a problem, leucaena has a relatively high protein content in its beans and leaves. If leucaena is less than 30 percent of the feed, livestock usually do not suffer any ill effects from mimosine. Mimosine helps ward off insect attack, so the compound confers some advantages to the crop (Brewbaker and Kaye, 1981).

Leucaena has historically been more important for human food than for livestock feed (Zarate, 1984). As early as the first millennium A.D., the Mayans in the Yucatán Peninsula may have relied on *Leucaena* as a green manure and for food (Brewbaker, 1979). The name of a commonly used species, *L. esculenta*, was given in recognition of its importance as a food source for Indians of highland Mexico (Brewbaker, 1987a). Green beans from immature pods are eaten raw or cooked as an ingredient in several dishes. Mature seeds are also eaten after they are boiled. The dry seeds are shiny brown and are ideal for leis, bracelets, and other ornamental products (Brewbaker, 1987b).

The ability of leucaena to thrive in relatively poor soils has propelled the versatile tree into the forefront of species used in reforestation and erosion control in the tropics (NRC, 1984). Leucaena's role in rehabilitating degraded areas is increasing as development organizations and government agencies focus more attention on the need to bring mismanaged areas back into production, rather than destroy remaining wildlands.

Hundreds of varieties of leucaena are recognized, but they can be grouped into two main forms: common and giant. The common type is short and bushy, reaching an average height of 5 meters. The giant form has a single trunk soaring to 20 meters.

Because of its sensitivity to cold, leucaena is restricted to the bottomlands, coastal plains, and lower reaches of mountains in the tropics and subtropics. Two species from northern Mexico, *L. retusa* and *L. greggii*, display some tolerance to frost (Brewbaker, 1987b; Sorensson, 1989a). Leucaena is well adapted to seasonally dry areas and can survive areas with as little as 350 millimeters of annual rainfall. Under optimal growing conditions, however, leucaena can grow as high as 6 meters in one year

FIGURE 7.2 A Living fence of leucaena (*Leucaena leucocephala*), Luzon, Philippines, June 1986.

and can be coppiced regularly (Vietmeyer, 1986). Productivity of leucaena declines markedly in very acid or waterlogged soils.

CENTER OF ORIGIN AND SPREAD

Central America including Mexico is home for the fourteen species of *Leucaena*, with an apparent center of diversity in Guatemala and in Oaxaca and Chiapas, Mexico. Whereas *L. leucocephala* is native to rather dry areas in the lowlands, other species are found in moist and highland tropical forests. Indigenous peoples harvested several species for food and helped disperse them from Texas to Ecuador. Within this extensive distribution, both wild and spontaneous forms are found, a pattern also found with some other tropical perennial crops, such as avocado (*Persea americana*).

After the Spanish conquest of Mexico, leucaena crossed the Pacific to the Philippines in the 1600s on galleons carrying silver and gold to be bartered for spices and Chinese goods. Ships leaving Acapulco used leucaena for livestock fodder and bedding during the long sail across the Pacific.

Leucaena thus established a beachhead in the Philippines and Guam at the onset of the colonial period. Locals learned to use the novel plants for fodder and fuelwood. Leucaena was soon put to use as a shade tree for cash crops such as coffee and cacao and to support vanilla and black pepper. Leucaena poles are used to prop up banana plants in some areas. By the nineteenth century, leucaena had spread to Indonesia, Malaysia, Papua New Guinea, Hawaii, Fiji, northern Australia, India, parts of Africa, and the Caribbean (NRC, 1984). Leucaena has become a spontaneous plant in disturbed sites in many introduced areas, such as Samoa (Parham, 1972: 70).

FUELWOOD, PAPER, AND CONSTRUCTION

The heating value of some species of *Leucaena* reaches as high as 16,000 BTU/kg, which is about a third of the heat output of natural gas (AID, 1981). The common form of leucaena can yield 88 cubic meters of stacked wood per hectare per year, and productivity of the giant type can be twice as high. Leucaena also makes excellent charcoal, which is a particularly useful product for urban areas because it is lighter to transport than freshly cut wood (USDA, 1961). The ability of leucaena to coppice readily makes it ideal for fuelwood production. After a tree has been cut, numerous sprouts typically form and can reach a diameter of 4 centimeters within a year (Dutt and Jamwal, 1987).

When properly cured, leucaena wood can be fashioned into inexpensive furniture, parquet flooring, crates, poles, and rafters. The wood has short fibers, approximately 1 millimeter long, that produce a high yield of pulp for the manufacture of paper and rayon. Pulp is suitable for cardboard products, waffleboard, chipboard, and pressed board. The durable wood of *L. salvadorensis* has long been exploited as a source of weight-bearing posts in house construction.

RESEARCH HISTORY AND CURRENT NEEDS

The wide distribution of leucaena, combined with its many uses, has spurred research efforts along several fronts. Research on the soil-restoration properties of leucaena began in the early 1900s in Indonesia (Dijkman, 1950). A little later, scientists in Hawaii and Australia began research on fodder management and improvement of leucaena (Oakes, 1968; B. Pound and Martinez, 1983; Takahashi and Ripperton, 1949). Starting in the early 1980s, the International Institute of Tropical Agriculture (IITA) in Ibadan, Nigeria, began extensive trials intercropping leucaena with maize and other crops (Figure 7.3).

Research is currently under way on many aspects of leucaena. Improved varieties have been developed with higher yield potential for fodder and

FIGURE 7.3 Alley cropping trial with leucaena (*Leucaena leucocephala*) and maize (*Zea mays*) at the International Institute of Tropical Agriculture (IITA), Ibadan, Nigeria, May 1983.

fuelwood. The flexibility of leucaena in various agroforestry systems has been a major factor in the crop's high priority on the research agenda of agricultural and forestry research institutes. Several methods of intercropping leucaena are being field-tested in Africa, Indonesia, and India (Brewbaker, 1989).

A major challenge to leucaena production has recently emerged. A jumping plant louse (*Heteropsylla cubana*), which is native to the Caribbean, reached the Pacific and Southeast Asia by unknown means in the mid-1980s and began causing extensive damage. Areas reporting heavy infestation of this psyllid pest have experienced declines in fodder yield of up to 60 percent. Livestock productivity has declined in areas affected by the pest, and cottage and small-scale industries have also suffered

(Napompeth et al., 1987; Russel-Smith, 1986). Scientists at the University of Hawaii are screening for resistance to the pest, and efforts are also under way to locate a suitable biocontrol agent. Some genotypes of the giant form of leucaena exhibit varying degrees of tolerance to the pest, but high levels of resistance are found only in species other than *L. leucocephala*. Resistance has been validated in *L. pallida*, *L. esculenta*, and *L. collinsii*.

To help counteract threats to leucaena, scientists have assembled a number of germplasm collections. With support from the USDA, researchers at the University of Hawaii initiated collections of *Leucaena* in 1967. Other germplasm collections were started in 1978 by the Commonwealth Scientific and Industrial Research Organization (CSIRO) in Australia, and in 1984 by the Oxford Forestry Institute, United Kingdom. Germplasm of leucaena is maintained both as seed and in arboretum gene banks. The largest and most comprehensive collection is maintained by James L. Brewbaker at Waimanalo, Hawaii. This University of Hawaii collection comprises over 1,000 accessions. A collection of 700 *Leucaena* accessions is located in Queensland, Australia. Both collections contain mostly *L. leucocephala*. Recent collections of leucaena and related mimosoid genera under the direction of C. Hughes of the Oxford Forestry Institute have expanded interest in the genetic resources of *Leucaena*.

Materials from this and other collections are being used to develop varieties suitable for the varying ecological and cultural conditions where leucaena is grown. Interspecific hybridization has produced unique growth habits, wider ecological adaptation, better wood and fodder quality, and increased insect resistance (Brewbaker and Sorensson, 1990). Asia and the Pacific need better seed production, site adaptability, and pest control (IDRC, 1983). Scientists in Australia are exploiting leucaena's potential as a fodder crop, with a focus on selections with high yield and psyllid resistance. India and Indonesia are screening a range of germplasm to develop materials suitable for increased wood and fodder production.

Cold tolerance has been transferred from *L. retusa* to various interspecific hybrids. Leucaena may thus be able to contribute to the fuelwood and fodder needs of people living in the subtropics and highland tropics. Those areas often are densely settled and have particularly severe fuelwood crises.

Tolerance of acid soil has been noted in accessions of *L. diversifolia*, *L. lanceolata*, and *L. shannoni* (Sorensson, 1989b). Lines from a triploid hybrid developed from the diploid *L. diversifolia* and *L. leucocephala* have shown promise on acid soils with toxic levels of aluminum (Hutton, 1984, 1990).

Most species of *Leucaena* are diploid, while four species, including *L. leucocephala*, are polyploids. Seedless hybrids (triploids) are especially useful for wood production and increased ecological adaptability. Interspecific hybridization using *L. pallida* and other species has led to seedless leucaenas useful for plantations, intercropping, and for urban shade. Seedless forms require special care to avoid spreading pests and diseases because they must be propagated vegetatively.

Interspecific crosses have also been assessed for improved wood quality. One of the earliest hybrids, *L. pulverulenta* × *L. leucocephala* has shown promise because of its fast growth and tolerance to cool temperatures (Brewbaker and Sorensson, 1990). Unfortunately, early hybrids from these two species are susceptible to psyllid attack. More recently, hybrids of *L. leucocephala*, *L. pallida*, and *L. diversifolia* have shown some resistance to psyllid pests.

Ongoing research on leucaena, particularly the widecrossing efforts, are likely to continue to pay dividends. Leucaena is a valuable crop for subsistence and cash income in many parts of the tropics, and investments in collecting and maintaining germplasm collections are likely to prove sound. Consideration also needs to be given to conserving suitable habitats for wild species of leucaena so that future options will be open for improving and sustaining leucaena yields.

BAMBOOS

Bamboos occur throughout the tropics and subtropics, and in some cases, the sturdy, woody grasses thrust through the canopy of tropical forests. Bamboos may grow 4 centimeters a day and reach as high as 40 meters (Hanke, 1990). Many species of bamboo commonly reach 17 meters at maturity and remain in the shady understory. Some bamboos form impressive clumps (Figure 7.4), thereby facilitating the harvest.

Bamboos are an exceptionally diverse assemblage of species. About 1,250 species in seventy-five genera are included in the Bambusoideae, a grass subfamily. China and India have 400 and 130 species, respectively. Many species are harvested in the wild. Given the enormous diversity of bamboos and their wide distribution, it is not surprising that several species have been taken into cultivation.

Bamboos are a widespread plant resource in the tropics and subtropics, especially in Asia (Y. M. L. Sharma, 1987). Most bamboo is harvested in the wild or from village environments, and at least 2.5 billion people regularly use the sturdy grass (Manokaran, 1990). Wild bamboo (*Guadua angustifolia*) has been harvested for at least five thousand years along the Cauca Valley in Colombia and in northwestern Ecuador (Parsons, 1991).

Tropical Asia and the Neotropics are particularly rich in bamboos, but the majestic grasses are important economic plants in all warm, humid

FIGURE 7.4 A clump of bamboo in mountain gorilla habitat, with a view of Mount Sabinyo in the background. Virungas Range, Rwanda, March 1989.

regions, including the Pacific. In all tropical and subtropical countries, bamboos have been introduced to produce a mix of wild and semidomesticated or cultivated species. In the warmer parts of Asia and the Pacific, bamboos are the single most important product from tropical forests for local communities (Y. M. L. Sharma, 1985). In Indonesia, bamboo ranks with banana and coconut as one of the most important multipurpose "trees" for satisfying domestic needs (Yudodibroto, 1987).

ECONOMIC AND CULTURAL SIGNIFICANCE OF BAMBOO

The contribution of bamboos to local and regional economies is poorly documented. Sometimes bamboos are classified as minor products or are not inventoried at all. Accurate data on areas covered by particular species and the frequency with which they occur are rarely available. Bamboos can be difficult to survey from the air because many species are hidden in the understory and others are cultivated in villages. Nevertheless, bamboo numbers can reach staggering proportions: a survey of villages in Bangladesh revealed 190 million mature bamboo plants and 558 million younger ones (Master, 1981).

People throughout the tropics and subtropics have put bamboos to a wide variety of uses. Construction is the most widespread and important use of bamboos (Parsons, 1991). The sturdy grasses have been employed in house building, bridges, and rafts for a long time (McClure, 1973; Ubidia, 1985). Split bamboo makes durable floors and walls. The woody grasses are also fashioned into tool handles, fishing poles, and agricultural implements, including poles to support climbing vegetables. Other uses for bamboo include coarse mats, baskets, fish traps, musical instruments, blowguns, lances, knives, toys, and chopsticks. Hollow stems of the larger bamboos serve as containers and plant pots (McClure, 1966). Bamboo shoots are a staple in the Orient and are delicacies in Chinese restaurants around the world.

In China, bamboos have played a prominent role in human affairs for at least five millennia (Hsiung, 1987; Hsiung and Ren, 1983). Alongside the chrysanthemum, the plum, and the orchid, bamboo is considered one of the four noble plants in China. The graceful grasses represent serenity, gentleness, and modesty in Chinese culture and are incorporated in many landscape and wildlife paintings. Bamboo symbolizes harmony with nature in art inspired by Taoism.

Bamboo mats and baskets have been unearthed in the ruins of Hemudu and Shishen built between 4800 and 5200 B.P. in Zhejiang Province, China. Bamboo articles are recorded in the oldest Chinese characters inscribed on bone and tortoise shell in ruins of the Yin Dynasty, which held power in Anyang from 1600 to 1000 B.C. Fishing rods, writing strips, and musical instruments were made from bamboo during the Zhou Dynasty (1100–300 B.C.), and bamboo shoots were also eaten during Zhou's dynasty. Bamboo arrows were unleashed in ancient Chinese battles, and the origins of papermaking with bamboo can be traced to 1700 B.P. in the western Jin Dynasty. A manual on bamboo (*zhupu*) produced in the Jin Period (A.D. 317–420) describes sixty-one bamboo types and their uses (Wu and Ma, 1987).

TWO WIDELY CULTIVATED BAMBOOS

Of the many species of bamboo that have been domesticated, only a couple will be treated here, *Bambusa vulgaris* and *Phyllostachys pubescens*, to illustrate their far-reaching cultural and economic importance. Round-stemmed *B. vulgaris* is native to tropical Asia, but has been introduced widely in Africa, Latin America, and the Pacific. The strong stems are straight and easily worked, hence the species' popularity. The 20-meter-high species tolerates a broad range of ecological conditions. It is grown in community stands, and wild or semiwild stands are harvested.

People propagate *B. vulgaris* vegetatively. One cutting can produce ten new stems and up to fifty new shoots within three years (Hasanbahri, 1983). This moisture-loving bamboo is at home in backyards, along creeks, and as a hedgerow between fields. For construction purposes, villagers generally cut the bamboo after only three or four years of growth. Furthermore, the bamboo can tolerate many cutting cycles before replanting becomes necessary, so time and labor costs are saved.

Stems of the bamboo are air dried for building and furniture. Around Lake Victoria in east-central Africa, *B. vulgaris* is used for making water pipes (Clayton, 1979). Samoans fashion strong but flexible fishing poles from the species (Parham, 1972:92). For making baskets and mats, the stems are split lengthwise. Ply bamboo is prepared by gluing several woven bamboo sheets together and then pressing them under heat. For pulp production, the dry stems are chipped and treated by the Kraft process. Bamboo is sometimes mixed with wood for pulping, although pure bamboo is also used. Pulp made from *B. vulgaris* is suitable for wrapping paper and boxes. Other bamboo species are used for making better grades of paper, such as bond paper and light, onionskin paper for air mail.

Shoots cut from *B. vulgaris* are relished in many countries. The shoots are harvested from village groves, and potential exists to expand the canning industry. The young, tender shoots are eaten fresh, cooked as a vegetable, dried, or made into pickles. The vitamin A–rich shoots are daily fare in many Asian nations.

A number of diseases plague *B. vulgaris*, some of which could eventually be controlled by breeding for genetic resistance. Bamboo blight, caused by *Acremonium strictum*, is a serious problem in Bangladesh. Current control measures for this fungal pathogen, which also attacks fig, betel vine, sunflower, and maize, are restricted to the deployment of fungicides and steps to improve plantation hygiene (Rahman, 1987). Other locally important bamboo diseases include culm decay in India and Indonesia, black leaf spot and leaf rust in India, shoot rot in Thailand, and root rot in India, Pakistan, and the Philippines (Boa, 1987).

Moso bamboo (*Phyllostachys pubescens*) is China's major commercial bamboo. Production of this tropical to subtropical species has increased 80 percent since 1950 (Hsiung, 1987). Its primary gene center is in central China, especially Szechwan, with a secondary center of diversity of cultivated forms in Japan, where *P. pubescens* was introduced from China about 1714 (Oshima, 1982). Moso bamboo can attain 20 meters and is used for construction, walking canes, umbrella handles, and pulp.

In Japan, most groves of *P. pubescens* are harvested for edible shoots and woody stems. Japan exports canned shoots of moso bamboo to Europe and North America. Other countries where moso bamboo is intensively cultivated cannot satisfy local demand for the product. Interest is growing in enhancing this species through breeding, but few fertile seeds can be located because the species only flowers after sixty years, and then it dies after the seeds have fallen.

GENETIC RESOURCES AND BREEDING

The protracted generation time for tropical bamboos has discouraged breeding efforts. Most tropical bamboos take three decades to flower, while some take as long as 120 years (Hanke, 1990). The shortest period before a breeder can expect a tropical bamboo to flower under normal conditions is fifteen years; a rice breeder could observe progeny from as many as forty-five generations during the same period. Synchronous flowering at lengthy intervals among bamboos has evolved to escape heavy seed predation. Saturating the environment with seed at periods far exceeding the lifespans of seed-eaters assures survival of enough progeny. In some cases, 15 centimeters of seed accumulates on the floor of a bamboo grove, far exceeding the appetites of rats, birds, and other animals. Bamboos also propagate vegetatively; seed production allows recombination and the emergence of fresh genotypes. Little is known, though, about the genetic variation of bamboo populations.

While such a seed production strategy has helped keep tropical bamboos one step ahead of seed predators, it has prevented any selective breeding programs for the woody grasses. New tissue culture techniques, however, promise to dramatically shorten flowering time for bamboo. Indian scientists have managed to induce flowering in two species of bamboo, *B. arundinacea* and *Dendrocalamus brandisii*, after only three subcultures in vitro (Nadgauda et al., 1990). Explants from seedlings of the two species were cultured on a medium supplemented with a cytokinin and coconut milk. After some further refinement, this procedure promises to open up an array of improvement opportunities both within species and in the development of high-yielding interspecific hybrids.

Tissue culture may also permit rapid multiplication of desirable bamboo clones. When such procedures may become routine is hard to predict, but they will have an impact on the genetic diversity of cultivated bamboos. Experiments with production of somatic embryos and the culturing of nodal explants and multiple shoots from seeds are encouraging (Rao and Rao, 1990).

A genetic narrowing of backyard and plantation bamboos is likely if superior clones are mass produced in vitro for farmers and villagers. Breeders will need a wide array of genetic material on hand to replace clones subject to disease and pest epidemics, which are likely threats if plantings are genetically uniform. Slow-growth tissue cultures could be used as an adjunct to field gene banks and in situ conservation so that an array of materials are close at hand for breeders to work with.

As the door opens for breeding bamboo, taxonomic research will need to accelerate on bamboos, particularly experimental work to unravel evolutionary affinities. As would be expected with so many species occupying the same area, some bamboos appear to be hybrids. Bamboos fall into two morphological groups, and these categorizations have implications for cultivation. One group puts out long, thin underground stems or rhizomes, from which shoots emerge. Above ground, the plants are evenly spaced. This monopodial growth form typifies bamboos of the drier subtropics. In the sympodial growth habit characteristic of the tropics, stubby rhizomes produce dense clumps of bamboo.

To launch breeding programs for bamboo, breeders should probably target no more than a dozen or so species for immediate attention. In this manner, the research effort will not dissipate, and quick gains are more likely to generate support for research on lesser known species. Candidates for early attention from breeders include *B. vulgaris, Dendrocalamus asper, D. strictus, Gigantochloa apus,* and certain species of *Phyllostachys, Thyrostachys,* and *Arundinaria* in Asia; *Oxytenanthera abyssinica* in tropical Africa; and several species of *Guadua* in tropical America.

Some breeding work has already begun with moso bamboo. In China, interspecific hybrids have been achieved between *P. pubescens* and *B. perviridis, B. textilis,* and *B. sinospina.* Improvement efforts have been geared to papermaking and construction. For pulp production, selection criteria include fiber morphology and its formation rate. For construction purposes, mechanical strength and durability have been the major traits sought after by bamboo breeders (Zhang and Chen, 1985).

Several near relatives of moso bamboo, such as *P. bambusoides, P. hemonis,* and *P. viridis* also merit attention. These close relatives of moso bamboo are cultivated and may be useful parents in hybridization efforts. The genus *Phyllostachys* includes some forty species, which occur from South Asia to East Asia.

Studies on the botany and genetic potential of other bamboos would help breeding efforts. At least four of the seventy species of *Bambusa* are cultivated. The *Bambusa* gene pool stretches across three continents and represents enormous potential for improvement.

BAMBOO GERMPLASM COLLECTIONS

Interest in bamboo improvement is growing, particularly in Asia, and several introduction gardens have been established to observe and distribute promising material. Modest but growing "orchards" for bamboo have been set up by the Forestry Research Institute at Chittagong in Bangladesh; the College of Forestry at the University of the Philippines in Los Baños; the Ministry of Lands and Land Development, Sri Lanka; and the Royal Forest Department and Kasetsart University, Thailand. India has established a field gene bank containing forty-five accessions of bamboo at Basar in Andhra Pradesh (Paroda and Mal, 1989). The Basar gene bank contains accessions of eight bamboo species: *B. tulda, B. pallida, B. nutans, B. arundinaceae, B. balcooa, B. multiplex, Dendrocalamus sikkimensis,* and *D. giganteous.*

China has assembled the most bamboo collections. Bamboo germplasm is maintained at several locations, including the Guangdong and Guanxi Institutes of Forestry, the Amji Bamboo Garden of the Institute of Subtropical Forestry of the Academy of Forestry Sciences, Hangzhou Botanical Garden, Wangjiang Park, and the Bamboo Sample Garden of Nanjing Forestry University. The latter university established a Bamboo Research Institute in 1984. In 1987, the Institute of Botany of the Academia Sinica and the local government of Guang County, Sichuan, agreed to establish a botanic garden to help conserve rare and endangered plants, including *Bambusa,* in Sichuan Province (Xianpu and Yenfeng, 1987). For the most part these bamboo collections are for research and have not evolved into gene banks containing representative diversity.

In Costa Rica, where *B. vulgaris* is a common ornamental and is used for construction, more than twenty species of nonnative bamboos are maintained at CATIE (Centro Agronómico Tropical de Investigación y Enseñanza) near Turrialba. CATIE's collection is composed mostly of giant bamboos that form clumps.

In the United States, the USDA keeps information on bamboo introductions, and the American Bamboo Society has established a garden in Encinitas, California. Most of the bamboos introduced into the United States are ornamentals. All bamboos imported into the United States are held for up to eighteen months for postentry quarantine. Smut, caused by *Ustilago shivaianae*, is a major quarantine concern. Eventually, bamboos will be exchanged as tissue cultures. That practice, when combined with disease indexing procedures and thermotherapy, will greatly reduce the chances of spreading pathogens (Rao et al., 1989).

Tissue culture techniques may also help store bamboo germplasm. The Forest Research Institute in Chittagong, Bangladesh, the National University of Singapore, and several institutions in India have worked out procedures for tissue culturing certain species of *Schizostachyum, Thyrostachys, Bambusa*, and *Dendrocalamus* (Mahta et al., 1982; Nadgauda et al., 1990).

Bamboo seeds are probably orthodox, although special handling will be required to establish seed gene banks. Seed gene banks would enable institutions to keep many more genotypes of bamboo than is possible in outside plots. More research is needed on techniques for drying bamboo seed so that their viability is not impaired.

PLOTTING A COURSE FOR IMPROVEMENT

More surveys are needed to inventory bamboo resources and to understand current patterns of exploitation. Such information would help in drawing up plans for rational management of tropical and subtropical forests. Preparation is under way on monographs of bamboo species found in several countries, such as India, Sri Lanka, Indonesia, and China, but much remains to be done before existing monographs are finished. More such monographs will be needed for other tropical and subtropical areas.

In some tropical regions, increasing numbers of shifting cultivators are putting considerable pressure on wild stands of bamboo. The dynamics of swidden agriculture and the regeneration of bamboo warrant further study. Forest clearing and burning for fields favor some species of bamboo, such as *Melocanna bambusoides* and *Dendrocalamus hamiltonii* (W. J. Peters and Neuenschwander, 1988:46), but rampant deforestation will inevitably shrink or even eliminate many bamboo gene pools. Reserves are thus needed for conserving bamboo germplasm in forest environments.

Bamboo normally falls within the purview of national forestry departments. Forestry departments often regard bamboos as weeds, rather than as important resources in their own right. In many government structures, a fault line separates agriculture and forestry, when in reality they are closely interlinked, particularly in the humid tropics. The distinction between forestry species and backyard or field crops is often blurred in tropical forest regions. Cooperation between foresters and agricultural scientists would help further progress in the development of bamboo resources. In Colombia, various regional development agencies and the National Federation of Coffee Growers are promoting afforestation with bamboo (*Guadua angustifolia*), particularly as an intercrop for coffee, plantains, and cassava (Parsons, 1991).

A number of products from bamboo could be more fully utilized. Bamboo hay has four times the protein content of fodder grasses, a lesson apparently learned long ago by the giant panda (Hanke, 1990). Cattle ranchers in tropical countries could benefit from incorporating bamboo leaves into cattle feed. Bamboo paper is superior to newsprint, and bamboo cultivation could prove ecologically more suitable in some areas than growing exotics such as eucalyptus and Caribbean pine (*Pinus caribaea*) for pulp.

Fortunately, several collaborative efforts are under way in Asia to promote research on bamboo. Research networking focused on bamboo has been fostered for a decade in particular by Canada's International Development and Research Centre (IDRC). Ottawa-based IDRC has long emphasized networking to further agricultural research and development, and this approach is particularly well suited to the widely grown bamboos. Past and current research is currently being reviewed by the Program on Tropical Tree Crops of the International Fund for Agricultural Research with a view to developing a sharpened research strategy and a more solid funding base.

Successful networks rest on the drive and self-interest of participants (Plucknett and Smith, 1984; Plucknett et al., 1990). Many tropical countries, such as China, the Philippines, and India have accorded bamboo high priority for research and improvement. Any international strategic plan must emphasize the better harnessing of the genetic resources of bamboos. Such a plan should not only focus on the major commercial species, but should include some minor bamboos in backyards and those harvested in the wild. Also, synergies between agriculture, forestry, and the environment are needed because of the role of bamboos in stabilizing the environment, particularly in marginal areas, and in rehabilitating degraded lands.

Recent international interest in tropical deforestation and the need to enhance local employment opportunities have mobilized some international donors in 1989 to commission an in-depth study of research needs for bamboo and rattan (IFAR, 1991). The study underscored the need for resource assessment and conservation and better training and extension. New partnerships will have to be forged between nongovernmental organizations and governments to further research on bamboo and to work out management strategies to sustain the resource. Fresh vision and more creative working relationships are also needed for most of the other nontimber products from tropical forests (De Beer and McDermott, 1989; NRC, 1991).

CHAPTER 8

Spices and Natural Food Colorants

Spices and other plant products for flavoring and scents may not figure prominently in international trade statistics, but they are nevertheless important to many regions and developing countries. Considerable local marketing of spices in the tropics escapes official records, but one does not have to wander far in a market in a developing country before the spice section is encountered. Stalls specializing in a pungent and intriguing array of spices can be found in virtually all tropical countries. Sprawling Addis Ababa, bustling Delhi, and seemingly chaotic Bangkok, in particular, abound in spice markets. Spices are thus a significant source of income for millions of farmers, traders, and retailers. Some spices and flavorings, such as vanilla, cinnamon, and clove, also generate substantial export earnings.

Although not as essential as the staffs of life, spices are integral parts of many cultures. In India, for example, the skillful blending of spices in various curries has reached an art form. Virtually all societies flavor their food in some way, if only with a sprinkling of black pepper (*Piper nigrum*). In addition to enhancing the flavor of food, spices have long been used to help preserve meats and to mask culinary items on the verge of spoiling. Refrigeration technology was developed only in the middle of the last century, and refrigerators have become widely available only in the last half of this century. Many consumers in the Third World cannot afford refrigerators, so spices still have a role to play in preserving foods.

Little wonder, then, that widespread demand for spices stimulated ancient trade routes overland in Asia and across seas and oceans. From biblical times, when dhows crisscrossed the Arabian Sea, to the sixteenth century, when Spanish galleons plied the Pacific, spices have long been high on the shopping list of seafarers and explorers.

Early in the Neolithic, expanding cities became manufacturing centers and magnets for trade. Rare commodities, such as oriental spices, were avidly sought, and trade routes were soon pioneered by entrepreneurs from Arabia, India, and Southeast Asia. Many of those trade routes persisted after the decline and collapse of empires. As early as 2000 B.C. some trade routes stretched from Western Europe to remote tropical archipelagos halfway around the world.

By about 800 B.C., civilizations in the Mediterranean, Southwest Asia, South Asia, and China were active in the spice trade. Palestinians imported cinnamon from the Orient in Old Testament times, while Hellenic trade penetrated as far as India. Nomadic peoples, such as the Scythians and inhabitants of the steppes of central Asia, provided links between Europe and China. China soon emerged as an entrepôt for many spices, such as cloves and nutmeg, obtained from the Moluccas. Persia and China were closely linked by trade by A.D. 220. The famous silk road (Figure 8.1) was a conduit for a variety of products, including fine porcelain and spices.

The virtually insatiable demand for spices and other luxury goods by Imperial Rome and the expansion of Indian civilization into Southeast Asia provided a major boost to the spice trade. By the sixth century A.D., the volume of spice shipments had increased markedly. Pliny the Elder wrote about the trade in spices and its impact on Rome's balance of payments.

Spices are dried plant products, so they have always been easy to transport. In addition to products of woody perennials from the tropics that have long entered the spice trade, numerous herbaceous plants from warmer climates are used to flavor food, including ginger (*Zingiber officinale*), turmeric (*Curcuma domestica*), chilies (*Capsicum* spp.), and coriander (*Coriandrum sativum*). Chilies, originally from tropical America, are now cultivated throughout the tropics, in part because they can be adapted to different soils and climates.

Spices have thus been at the center stage of history for thousands of years. The spread of civilizations and major religions opened up new markets for spices and extended trading routes. Colonial powers jostled for control over the supply of many tropical spices, such as clove, nutmeg, and cinnamon, and brought some New World flavorings, particularly vanilla, into international commerce. Demand for spices still remains strong because synthetic concoctions often fail to capture the many nuances and distinctive flavors of natural spices.

Four spices are profiled in this chapter: clove, cinnamon, vanilla, and annatto. Clove and cinnamon are from tropical Asia, while vanilla and annatto are native to the Americas. Vanilla is a climbing orchid in forests of Central America; the others are woody perennials. All were domesticated to flavor food or beverages long before the colonial period. In our sample of perennial spices, we explore such issues as whether the species is under- or overexploited; the degree to which synthetic compounds can

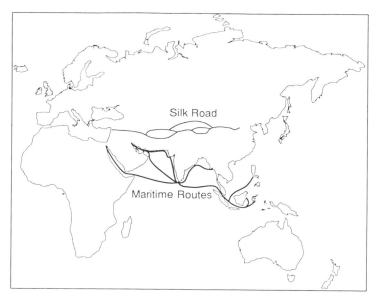

FIGURE 8.1 Major land and maritime trade routes between Asia and the eastern Mediterranean, 500 B.C. to A.D. 1000.

replace natural products; and the role of local production and processing of spice plants to provide value-added exports.

Clove, cinnamon, and vanilla are processed to obtain essential oils. These aromatic oils should not be confused with oils derived from fatty fruits, such as oil palm and peach palm. The latter are not volatile and are used for cooking and various industrial uses. Essential oils flavor food, tobacco, beverages, and are blended in perfumes. Clove is also used medicinally, for example, as a palliative for toothache. Extracts from some of these spices are used to synthesize vitamins and insecticides.

The genetic diversity of clove, vanilla, cinnamon, and annatto is hardly exploited and little information is available about variation in wild populations. More sampling, preservation of herbarium specimens, and screening are needed to determine the range and quality of chemical compounds in wild populations and to better understand taxonomic relationships. When the evolutionary affinities between species are better understood, breeders will be able to more readily tap primary and secondary gene pools.

Expensive and sophisticated breeding programs are not necessary to upgrade spice plantations. Much could be accomplished with improved selection of planting stock to boost yield and the quality of the harvest. The quantity and quality of several constituents in natural oils vary from plant to plant, so gas chromatography is increasingly used to identify superior genotypes for clonal propagation. Gas chromatography screening therefore helps ensure a more even supply of high-quality essential oils, thereby avoiding mixtures from inferior plants.

The degree to which investments should be made to bolster breeding efforts of spice plants and food colorants hinges to a large extent on the market for synthetics. Flavors of vanilla and cinnamon can be produced in the laboratory from other plant industrial wastes (Eiserle and Barreto, 1969). Industrial countries can often obtain food colorants more cheaply by synthesizing them than by importing plant products. Cost of production is thus a critical consideration. Market forces and processing trends occasionally force crops out of production.

All the spices considered here, however, all are expected to survive as viable crops, although current cropping patterns are likely to shift, new varieties will be deployed, and agronomic practices will be altered. One of the great advantages of natural essential oils is the range of chemicals present in addition to the major sought-after constituent. It is this subtle mix of oils and essences that is so attractive to the perfume industry, a mix that cannot be replicated so easily on the laboratory bench. For example, cinnamon bark oil has twenty-seven constituents, leaf oil, twenty-six, and root bark oil, twenty; only twenty ingredients are common to all three oils (Senanayake et al., 1976). Cinnamon exemplifies the complex nature of natural oils used in flavorings and perfumes.

Spices thus remain important plantation crops in many developing countries, although for each spice, a handful of countries usually dominates production. Innovations by the private sector have always been critical to the success of many plantation crops, and many new technologies and agronomic practices have been adopted by small as well as large growers (Corley, 1989). Spice plants lend themselves to mixed-cropping and agroforestry systems, thereby increasing their appeal to small farmers. Also, research and incentives could focus on the better integration of these cash crops with local processing and herbal drug companies. In this manner, growers of spices and food colorants would be less dependent on international markets for their income.

The expense of assembling and managing large gene banks and the tendency of a handful of countries to dominate production of spices and food colorants preclude extensive ex situ germplasm collections for spices and food flavorings. Scope exists, however, for some carefully planned gene banks to serve these important crops. Vanilla could be conserved in vitro or as seed, but the germplasm of cinnamon cannot be stored by such space-saving means. Cinnamon easily becomes naturalized in local vegetation, and these populations could be monitored with minimal expense. In some cases, nature reserves might be artificially enriched with cinnamon and then used as extractive reserves. The continued supply of cinnamon would come from a broad genetic base, and, because the reserves would generate immediate economic returns, they would be relatively safe from development pressures.

The major thrust for germplasm conservation of spices and food colorants, such as annatto, would best be channeled to safeguarding habitats where wild populations of the cultivated species and their near relatives are found. Existing or proposed reserves should be studied to determine patterns of variation, particularly in economically important plants. Protected areas should encompass a wide degree of intraspecific variation to help ensure the long-term survival of populations and to provide breeders with more promising options for crop improvement. This task is by no means simple: periodic field surveys are required, followed by checking the range of diversity present over time.

Decisions as to where parks and reserves should go are not easy to make. A range of political and economic issues have to be considered, in addition to scientific criteria. Often park planners have little hard data on species present or their variation, vital information for drawing up priorities for conservation. Given the limited resources available for conservation, tough decisions have to be made. In the case of India, for example, four near relatives of cinnamon occur in three distinct phytogeographical provinces: the northwest Himalayas, the eastern Himalayas, and western peninsular India (Arora and Nayar, 1984).

A concerted effort to protect natural habitats combined with some prudently assembled gene banks should provide breeders with sufficient material to improve spice plants. To be effective, however, such a strategy will require closer cooperation among a number of genetic resource programs, as well as support from industry. Spice plants are cash crops, and collaboration has historically been more difficult in germplasm issues when the commodity is geared entirely to international commerce.

No single government institute can devise and implement a genetic conservation and utilization strategy for any of the tropical spices and food colorants. The progenitor wild species often transcend national borders, and the main producing areas are sometimes far removed from the native range of the species. Whereas scientists are often willing to share germplasm resources, national policies sometimes restrict the flow of germplasm for cash crops because of fears that competitors will gain an advantage. Requests are made to exchange germplasm resources, for instance black pepper from India, or cloves from Indonesia, but in practice difficulties often arise.

To help overcome barriers to germplasm exchange, governments need to adopt a consortium approach in which concerns about equal sharing of benefits are clearly addressed. Unless a collaborative framework is established, conservation and breeding efforts will proceed on a piecemeal basis.

When research and conservation efforts are pursued in isolation, their effectiveness is usually diminished (Plucknett et al., 1990).

CLOVE

Cloves are dried flower buds of an evergreen tree (*Syzygium aromaticum*). Synonyms include *Caryophyllus aromaticus* L., *Eugenia aromatica* Kuntze, *E. caryophyllata* Thumb, and *E. caryophyllus* (Sprengel) Bullock and Harrison. Native to the Moluccas in Indonesia and first described scientifically by a German botanist in 1741 (Rumphius, 1741), cloves have long been one of the most sought-after spices on world markets, and they have provided an incentive for traders and explorers to penetrate Southeast Asia. Spanish and Portuguese galleons could finance a year's adventure on the high seas if they could secure a cargo of cloves for delivery to their sponsors in Iberia. One surviving ship from Magellan's trip around the world returned to Spain in 1522 with 26 metric tons of cloves, sufficient to pay for the entire cost of the expedition (Purseglove et al., 1981a:231)

Clove products can be separated into four categories: clove buds, clove stem oil, clove leaf oil, and mother-of-cloves. Clove buds are the most important product harvested from clove trees. Dried clove buds are used whole, shredded, or are employed to extract oleoresin. Clove buds are also distilled to obtain clove bud oil. The major use of clove buds is for flavoring cigarettes in Indonesia. Cloves are an essential ingredient in many curries and enhance the flavor of rice pudding, apple pie, and sugared ham. In Indonesia, hot tea is spiced with whole cloves, rather than taken with milk as in the United Kingdom and many other countries. Clove is an antispasmodic, and in parts of the Third World, such as Brazil, a clove bud is sometimes placed between the gum and cheek to relieve toothache.

Ground cloves are used in the manufacture of many pickles and sauces (Purseglove et al., 1981a). Food companies in Western Europe and North America employ oleoresin in several products, while clove bud oil is used to flavor food and in certain pharmaceutical products. Clove oil resembles the smell of fresh carnations and is used to scent soaps, bath salts, and certain perfumes.

Clove stem oil is distilled from dried flower stalks (peduncles) and is employed for purposes similar to those of clove bud oil. Although cheaper than clove bud oil, clove stem oil is an inferior substitute. Whole leaves and twigs are used to produce clove leaf oil, from which eugenol is extracted for use in perfumes, the pharmaceutical industry, and in dentistry.

P. Pool provided substantial input for this profile of clove.

In the past, eugenol has been used as a source of vanillin, but artificial vanillin is now synthesized in the laboratory from other sources. Finally, the fruits, known as mother-of-cloves, are occasionally used as an adulterant and for medicinal purposes in Indonesia (Burkill, 1935).

The largest producer of cloves, Indonesia, is also the largest market for the spice. Indonesians are fond of clove cigarettes, which contain tobacco mixed with 30–40 percent dried, shredded cloves. The *kretek* cigarette industry, as clove cigarettes are known in Indonesia, started in 1916 and now encompasses two large firms and close to 700 smaller ones (Purseglove, 1974; UNCTAD/GATT, 1982). Between 1980 and 1984, Indonesia's cigarette industry used between 45,000 and 50,000 metric tons of cloves every year. In contrast, the remainder of the world market is satisfied with an annual production of 6,000 metric tons of cloves. Although the smoking of clove cigarettes is largely confined to Southeast Asia, particularly Indonesia, imported *kretek* cigarettes enjoyed brief notoriety among some U.S. teenagers in the 1980s, until health officials became alarmed and prohibited their sale. In industrial countries, smoking is on the decline, but it is still increasing in developing nations. Demand for clove cigarettes is expected to continue to rise, at least in the near future (A. E. Dann, pers. comm.). Eventually, though, the use of cloves for cigarettes is likely to diminish as health concerns about smoking take root around the world.

From 1980 to 1984, world production of cloves averaged approximately 55,000 metric tons a year, with Indonesia accounting for about two-thirds of the annual production. Other major clove producers include the Malagasy Republic, the Tanzanian islands of Pemba and Zanzibar, the Comoros Islands, Sri Lanka, and more recently, Brazil (Figure 8.2). As recently introduced witches'-broom takes its toll of cacao plantations in Bahia, clove is likely to assume greater importance in that coastal state of Brazil. The global pattern of clove production matches fairly closely that of cinnamon. India and Malaysia are investing heavily in clove production (UNCTAD/GATT, 1982).

Clove is well suited to small-scale producers. The compact tree can be grown as a monocrop or in mixed stands, as in Sri Lanka where clove shares backyard gardens with coconut, betel palm, nutmeg, citrus, mango, breadfruit, jackfruit, banana, and rambutan (MacDicken, 1990). Clove picking and drying also provide local employment.

ORIGIN AND SPREAD

Clove is indigenous to five volcanic islands in the North Moluccas of Indonesia: Ternate, Tidore, Motir, Makian, and Bacan (Figure 8.3). The Moluccas, commonly referred to as the Spice Islands, are also home to

FIGURE 8.2 A small plantation of clove (*Syzygium aromaticum*) near Itabuna, Bahia, Brazil, January 1990.

nutmeg. The germplasm base for clove outside its native range is extremely narrow. The limited genetic variation of clove plantations outside the Moluccas can be understood if one traces the spread of the crop to other tropical lands.

The first known references to the use of clove are found in Chinese books of the Han dynasty, from 220 to 260 B.C. According to these ancient writings, courtiers were required to sweeten their breath with cloves before addressing the emperor (Crofton, 1936). In China and Persia, clove was considered a stimulant, an aphrodisiac, and a carminative. Such potent properties stimulated an early interest in developing trade routes to secure the spice.

The Chinese not only controlled the early trade in cloves, but were probably also responsible for encouraging the planting of clove trees farther south in the Moluccas on Ambon and Seram (Purseglove et al., 1981a). Clove is known as *cengkeh* in Indonesia, a name derived from Chinese (Burkill, 1935).

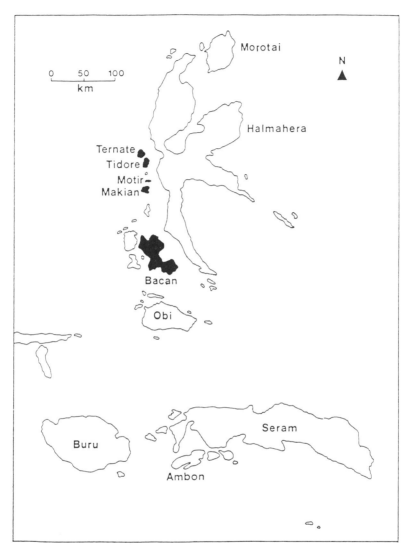

FIGURE 8.3 The Moluccas, including the islands where clove (*Syzygium aromaticum*) is native (shaded). (Adapted from Pool, 1988; Pool et al., 1986; Purseglove, 1974:402; Tidbury, 1949; Wit, 1976.)

In India's ancient Ayurvedic system of medicine, the chewing of cloves and cardamom (*Elettaria cardamomum*) wrapped in betel-nut (*Piper betle*) leaves is recommended to aid digestion (Rosengarten, 1969). Known as *pan*, certain concoctions, including clove, wrapped in betel-nut leaves are still common in India. *Pan* is taken by all classes in India, particularly after meals.

Cloves reached Alexandria by A.D. 176, and, from Roman writings, it is clear that cloves were well known around the Mediterranean by the fourth century (Parry, 1969). Arab traders brought cloves to the Mediterranean and kept their origin a close secret. From the eighth to tenth centuries, Radanite Jewish traders brought cloves and other spices to Spain, France, and other parts of Western Europe. Cloves reached Europe and the Mediterranean via entrepôts in Java, India, and Sri Lanka.

The high value of cloves in Europe prompted a search for their source. The Portuguese were the first European explorers to obtain cloves when Vasco da Gama rounded the Cape of Good Hope in 1498 and secured some of the spice at Calicut on the Malibar coast of southwest India, which in turn had obtained them from the Moluccas. In the scramble to control trade routes to the Spice Islands, the Portuguese established a string of outposts stretching from Moçambique in East Africa to Goa in India, Malacca on the Malayan Peninsula, Macao in southern China, and Timor in Indonesia.

In 1514, the Portuguese reached the Spice Islands and occupied Ternate and several other islands in the Moluccas. The Portuguese controlled the clove trade for close to a century until they were usurped by the Dutch in 1605 (Ridley, 1912). Holland then held a monopoly over the lucrative clove trade for almost 200 years. In 1651, the Dutch began eliminating clove trees on all islands in the Moluccas except Ambon, where the tree is introduced, in order to ensure their grip on the highly sought-after commerce in cloves. Under penalty of death, clove cultivation was prohibited on other islands. This drastic measure clashed with local custom, since a clove tree is planted for every child born in the Moluccas; if a clove tree is cut down, ill fortune may befall the child for whom it was planted.

Over the next thirty years, approximately three-quarters of the clove trees in the Moluccas were destroyed. This widespread destruction increased the profits of the Dutch East India Company by driving up the price of clove. But it also forced the species through a genetic bottleneck. At least some young seedlings probably escaped the death sentence, but the targeting of islands where the tree is native must have dramatically reduced the clove's wild gene pool in the seventeenth century.

By the late eighteenth century, the Dutch began to lose their grip on the Moluccas. In 1770, a French administrator on Mauritius (then Ile de France) managed to procure a few clove plants from Seram. A few of these clove seedlings were successfully established on Mauritius (Figure 8.4). Two years later, additional clove seedlings were obtained from the Moluccas and distributed to Réunion (then Bourbon) and the Seychelles, and around 1789, to Cayenne (Maistre, 1964; Sheffield, 1950). Only one tree is thought to have been the ancestor of the whole clove industry on Réunion. From these sources cloves were subsequently introduced to the Car-

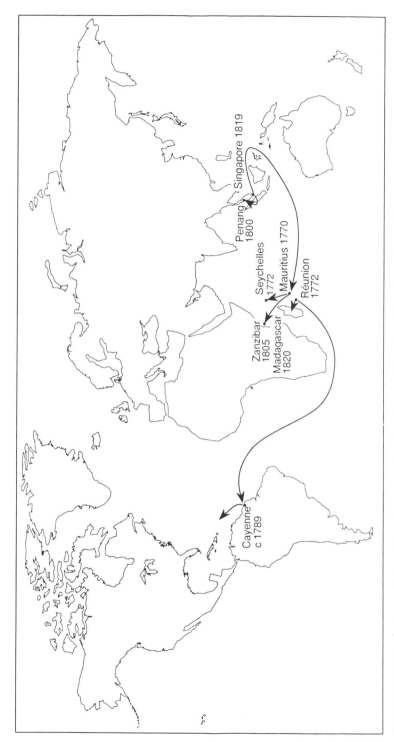

FIGURE 8.4 Introductions of clove (*Syzygium aromaticum*) seed or seedlings for cultivation. (Adapted from Maistre, 1964; Purseglove, 1974: 402; Purseglove et al., 1981a:232; Sheffield, 1950; Tidbury, 1949; Wit, 1976.)

Singapore 1819

Penang 1800

Seychelles 1772

Mauritius 1770

Réunion 1772

Zanzibar 1805

Madagascar 1820

Cayenne c 1789

ibbean islands of Dominica, Martinique, St. Kitts, St. Vincent, and Jamaica (Ridley, 1912; Tidbury, 1949).

Clove reached Madagascar in the 1820s, and the present clove industry in the Malagasy Republic is based on a single tree seedling introduced from Réunion (Wit, 1976). Clove seedlings reached Zanzibar in the early 1800s. Harameli bin Saleh, an Arab trader, may have brought clove planting material to Zanzibar from Mauritius in 1818 (Tidbury, 1949). Clove was soon thriving on Zanzibar, and the Sultan quickly realized the economic potential of the crop. The Sultan instructed landowners to cultivate clove on Zanzibar and Pemba, and extensive plantations were operating on Zanzibar by the mid-1800s (Moorehead, 1971). Even a devastating cyclone in 1872 did not deter Zanzibar and Pemba from becoming the world's leading producers of clove, a position they held until quite recently (Nutman and Roberts, 1971).

In 1800, clove from the Moluccas was established on Penang, an island off the northwestern shores of Malaysia. A few seedlings from this consignment were also dispatched to Madras and Calcutta in India, as well as the Royal Botanic Gardens, Kew, in England (Purseglove et al., 1981a). Sir Stamford Raffles attempted to introduce cloves to Singapore in 1819, but they did not fare well on the island. In 1823, extensive plantings of the spice were established by British settlers at Bencoolen (now Bengkulu) in Sumatra, but these plantings apparently perished when the Dutch took over (Ridley, 1912). A single consignment of clove seeds reached Sri Lanka in 1830, possibly from Penang.

During the nineteenth century, clove seeds or seedlings from the Moluccas were gradually introduced to other parts of Indonesia. Over 95 percent of present-day clove production in Indonesia comes from individual holdings of less than 10 hectares. Clove is thus well suited to small farmers. Clove is now grown throughout Indonesia, but the most important producing areas are Aceh, West Sumatra, Lampung, West and Central Java, North Sulawesi, and the Moluccas (Cut, 1977).

In an attempt to upgrade clove plantations in Indonesia, the Dutch brought a few seeds from Zanzibar to Java in 1932. The seeds were planted in West Java, Bengkulu, West Sumatra, and Ambon (Cut, 1977). The resulting trees, known as the Zanzibar type, were preferred by farmers because their yield was higher than that of traditional varieties. The greater commercial potential of the Zanzibar type soon prompted farmers to adopt the Zanzibar clove in other parts of Indonesia. Clove imported from Zanzibar now dominates Indonesian production.

It is ironic that Indonesia, the home of clove, has established a thriving clove industry that is based on a few germplasm introductions from a distant clove population. The progeny of just a few seeds imported from stock that had already gone through a genetic bottleneck has thus narrowed the variability of cultivated clove in Indonesia. Clove on Zanzibar and neighboring Pemba exhibits little variation (Tidbury, 1949).

The trend to genetic homogeneity of clove production in Indonesia is accelerating. In the 1960s and 1970s, Indonesia dramatically increased plantings of genetically uniform Zanzibar type clove in order to achieve self-sufficiency. This push to reduce reliance on imports was triggered by a decline in clove production in Zanzibar as a result of political changes in 1964, when clove plantations were nationalized and owners and skilled supervisors were evicted. Some clove plantations were broken up for small-scale farmers. Zanzibar's position as a major exporter of cloves thus suffered.

With the decline in production and quality of cloves in Zanzibar, Indonesia felt the need to increase plantings dramatically. This rapidly implemented campaign to expand clove production focused exclusively on the planting of Zanzibar type seedlings. By promoting genetic uniformity in plantations, the clove industry on Indonesia has become more vulnerable to disease and pest outbreaks. Diseases have become a more serious problem in clove plantations in Indonesia, accounting for an estimated 20 percent crop loss (A. E. Dann, pers. comm.). In Sulawesi, for example, clove leaf drop has recently damaged clove plantings, but the causal agent is unknown (Whitten et al., 1987:605).

PATTERNS OF VARIATION

Whereas little variation is found in clove planted outside Indonesia and in the more recent plantations established within the country, distinct morphological differences are noted in traditional planting material and wild cloves (Cut, 1977). (In fact, wild clove is sometimes considered a distinct species, *S. obtusifolium*.) Wild cloves growing in forests of North and Central Moluccas are rarely exploited by farmers. The flower buds are larger than domesticated forms, but they produce fewer inflorescences. Other unfavorable characteristics are that the flowers contain much less essential oil and they are especially low in eugenol. For these reasons, wild cloves do not enter commerce.

Wild cloves are nevertheless more vigorous than domesticated forms. Wild cloves branch lower down on the trunk and have rounder crowns than cultivated selections, attributes that would facilitate harvesting. In addition, the leaves of wild cloves are appreciably longer, broader, thicker, and less brittle than those of cultivated forms. Wild cloves flower at a different time than cultivated varieties. The vigorous trait of wild cloves could prove especially useful in improvement programs. Fertile hybrids between wild and cultivated cloves can be obtained easily (Wit, 1976).

Three main clove types are recognized outside the Moluccas (Cut,

1977). The Sikotok and Siputih types occur in West Sumatra, while the Zanzibar form is more widely disseminated (Brinkgrieve, 1933; Hadiwijaya, 1979). Neither the Sikotok nor the Siputih type has ever been formally described, so neither can be considered a botanical variety. Zanzibar clove branches low on the trunk, has dense, dark green foliage, and produces light green flower buds. Siputih trees branch farther up the trunk, sprout fewer and lighter colored leaves, and sport yellow-green flower buds. Unfortunately, such distinctions are seldom obvious in the field. Furthermore, all three clove types apparently originated from the same narrow genetic base (Jarvie and Koerniati, 1987).

In spite of the considerable "genetic thinning" instigated by the Dutch in the Moluccas during the colonial period, several indigenous populations of clove survive in the North and Central Moluccas (IPB, 1973; LPTI, 1974). The size, shape, and color of cloves in the Moluccas vary significantly (Sahertian, 1980). Indigenous clove populations in the Moluccas are more diverse than plantations in other regions and are particularly variable with regard to tree shape, branching habit, and the size, shape, and color of leaves and buds (Pool et al., 1986). A particularly interesting trait in one indigenous clove population is the propensity for harvestable flower buds to fall before opening. Because cloves are gathered by hand, a laborious task, this characteristic may prove valuable.

Morphological and isozyme analyses of a wide range of wild clove material in the Moluccas confirm a major difference between cultivated and wild cloves (Pool, 1988). In addition, these more precise measurements reinforce the impression that indigenous cultivated cloves within the center of origin exhibit more diversity than cloves grown outside the Moluccas. Of the cloves cultivated away from the center of origin, the Zanzibar type is the least diverse.

BREEDING OBJECTIVES

The low levels of genetic variation in cultivated clove outside the Moluccas will hamper pantropical breeding initiatives. Indonesia, then, is better placed to make tangible progress in clove breeding over the short term. Initial improvements in clove accomplished by the Dutch were negated during World War II, when much of the working material and many of the records were lost. Indonesia has not attempted to revive clove-breeding efforts.

The most pressing need for clove improvement is to boost yields. Three obstacles immediately confront those concerned with raising clove yields. First, little information is available on the yield potential of different clove types, or even of individual trees within a population. Second, how yield potential is inherited remains unknown. Third, a single generation in clove breeding requires six to ten years.

Breeding procedures need to be adopted that minimize the number of generations required before results can be obtained. Also, several simultaneous breeding approaches may be needed to avoid stalling an improvement program in the event that a single strategy proves fruitless. A multiple-strategy approach to clove improvement will inevitably cost more than one relying exclusively on a single tack.

A simple mass-selection program would be a logical beginning. Although advances could be expected over single generations, success of such a strategy hinges on high heritability for yield. Some controlled crossing experiments should be initiated parallel to a mass-selection effort so that progeny can be assessed carefully for yield potential and good parents can be identified for further breeding efforts. Crossing schemes should include indigenous populations of cultivated cloves from the Moluccas in order to broaden the genetic base of breeding pools. Both of the above strategies would operate initially through planting seed, with a later switch to clonal propagation if reliable techniques for vegetative propagation of clove are developed.

Some scope exists for improving clove yields by exploiting genetic material of cultivated cloves outside the Moluccas (Pool, 1988). Breeders working with other crops, however, often reach a roadblock when they discover insufficient variability in the source material for their crossing experiments. Given clove's limited number of introductions, it is safe to assume that the crop's genetic base needs considerable broadening if significant yield gains are to be achieved.

Breeders are more likely to obtain dramatic yield increases in clove if they tap the genetic reservoir of cultivated clove in the Moluccas. Clove cultivated in backyards and on smallholdings in the Moluccas is particularly well suited to this end because of its rich genetic variability and overall similarity to commercial types, such as Zanzibar, Sikotok, and Siputih.

It is impossible to tell at this point which wild clove populations should be used because of our lack of understanding of patterns of variation in traditional cloves planted by seed. Crosses should be conducted in such a way as to obtain estimates of general and specific combining ability in order to identify the most promising parents for future use.

The raising and sustaining of clove yields will require an ongoing effort, particularly in Indonesia, where domestic demand continues to grow. Although the area under clove cultivation is still expanding in Indonesia, increased yields would release land for other purposes, such as growing food crops or conservation.

Diseases and Pests

Any high-yielding clove varieties developed in the future will be of little practical use if they are susceptible to diseases and pest epidemics. Three principal diseases threaten clove plantations in Indonesia: Sumatra disease, *cacar daun*, and *gugur daun Manado*. Sumatra disease, first recognized in western Sumatra in 1961 (Purseglove et al., 1981a:251), is a relatively recent disease and underscores the highly dynamic nature of diseases and pests. It is also the most serious ailment of clove plantations in Indonesia, where it causes high mortality among clove plantations in Sumatra and West Java. Sumatra disease is caused by a xylem-inhabiting strain of *Pseudomonas solanacearum* (S. J. Eden-Green, pers. comm.) and is transmitted by two species of cercopoid leafhopper, *Hindola fulva* in Sumatra and *H. striata* in West Java (Eden-Green et al., 1985, 1986).

Cacar daun, caused by a fungal pathogen (*Phyllosticta* sp.), now affects most clove-growing areas in Indonesia (Kasim et al., 1980). The disease is spreading, but its epidemiology remains elusive. Although *cacar daun* is not as destructive as Sumatra disease, clove yields often decline steeply when the trees are attacked by the disease. Even less is known about the epidemiology and etiology of *gugur daun Manado*, a disease that has surfaced recently on cloves in North Sulawesi.

Because of the propensity for diseases to spread, breeding efforts need to locate resistance to the three major diseases of clove and incorporate the responsible genes into productive new varieties. Initial searches should concentrate on cultivated cloves, both within and outside the center of origin of the crop. Because of the generally low eugenol content of flowers from wild clove trees, wild cloves should be screened only if no resistance genes are discovered in cultivated forms.

Not all clove trees are affected to the same degree with *cacar daun*. This differential field response to the pathogen suggests that genetic resistance may eventually be found. Breeding for disease resistance against Sumatra disease is likely to be a more elusive goal, as indicated by an extensive, but discouraging search of cultivated and wild cloves (Pool, 1986). Further screening of clove, particularly in the Moluccas, may yet uncover resistance to the virulent disease. Sources of resistance to Sumatra disease are known in some of clove's near relatives, comprising some 500 species in the genus *Syzygium* (Jarvie and Koerniati, 1987). But widecrossing clove with its near relatives would be time-consuming and difficult, at least with existing techniques.

Sumatra disease illustrates the futility of expecting genetic resources to solve all crop-production problems. In the near term, Sumatra disease may be more easily controlled by restricting the spread of vector populations (C. Lomer, pers. comm.).

A range of other diseases and pests attack clove. In Zanzibar, for example, "sudden death," caused by a fungus, *Valsa eugeniae*, appeared in clove plantations in the mid-nineteenth century. Sudden death also attacks clove on Pemba. Dieback, caused by another fungus, *Cryptosporella eugeniae*, invades wounds created by wind damage or harvesting. By 1953, more than half the mature cloves on Zanzibar had died of sudden death or of dieback (Purseglove, 1974:407). No satisfactory control measures have been devised to combat these fungal diseases.

The choice of crop-improvement strategies depends on the breeding system, life cycle, and means of propagating the crop. Clove's breeding pattern conforms to a mixed-mating model in which cross-pollination dominates and a small amount of inbreeding is tolerated (Pool, 1988). Clove is presently propagated by seed, but some progress has recently been achieved with vegetative propagation (Lukmonohadi et al., 1983). Like production of many other commercially important perennial crops, production of clove in the future is likely to be dominated by clonally propagated varieties.

CONSERVATION STRATEGIES

If clove follows the pattern of other crops, the spread of clonally propagated cloves will tend to reduce further genetic variability in orchards and plantations. Consideration thus needs to be given to conservation needs now, before genetic erosion accelerates. Swift implementation of strategies for the conservation of clove germplasm in Indonesia is imperative.

Only three ex situ germplasm collections of clove have been assembled, two near Bogor in West Java, and the other at Solok in West Sumatra. These field gene banks do not have any duplicated material and thus are vulnerable to catastrophic loss. None of the gene banks is comprehensive because all contain very little wild clove and no cultivated clove from the Moluccas. Furthermore, all three gene banks are threatened by Sumatra disease. The lack of cultivated cloves from the Moluccas in gene banks is worrisome. Some indigenous populations of clove in the crop's center of origin are threatened by diseases and clearing. A dieback condition of unknown etiology has afflicted at least one population of cultivated clove in the Moluccas (Pool et al., 1986).

Clove germplasm collecting trips should begin immediately in Indonesia to gather representatives of the Zanzibar, Sikotok, and Siputih types, as well as all known indigenous populations of cultivated and wild cloves growing in the Moluccas. Some germplasm should also be obtained in other clove-producing countries and regions, such as the Malagasy Republic and Zanzibar. In addition, any populations that exhibit unique or unusual characteristics should be sampled. Until vegetative propagation techniques for clove are perfected, collectors will have to obtain open-pollinated seed. Therefore, collecting expeditions will be largely confined to a few months each year. The compilation of detailed passport data, conforming to standards established by the International Board for Plant Genetic Resources (IBPGR) in consultation with experts, will be a crucial adjunct to any germplasm collecting missions.

A concerted effort should also be undertaken to locate novel populations of clove. Clove is reputed to grow on the Moluccan islands of Saparua, Bisa, Kasirutu, Muari, Morotai, and Halmahera, but such reports need to be confirmed (Pool et al., 1986). Wild cloves are known to occur in Irian Jaya, but little is known about populations on the eastern fringe of clove's range (Cut, 1977).

A comprehensive gene bank for cloves is needed in the Moluccas, where most of the diversity for the crop is found. Another major advantage of the Moluccas for a conservation collection of clove is that the islands are free of Sumatra disease. A duplicate collection would be needed near Bogor, where most of the breeding work on clove is likely to be carried out in the future.

Development of in vitro culture techniques of shoot tips or embryos would permit the storage of clove germplasm under slow-growth conditions in relatively small facilities. If cryopreservation of clove tissue cultures becomes feasible, in vitro methods will permit the operation of both working and base collections. Tissue culture research on clove is in its infancy, however, because of the paucity of biotechnologists in developing countries and the need to work on crops with greater commercial or food value. At present it is virtually impossible to propagate clove asexually (Litz et al., 1985), but IBPGR is encouraging work on in vitro culture of clove in the hope of overcoming some of these obstacles.

In situ reserves would be a vital complement to field gene banks and in vitro collections. Parks and reserves are essential to the long-term viability of clove germplasm, since in situ preservation allows continued coevolution with pests and diseases and is ultimately cheaper and more effective than trying to capture all the genetic diversity in a field gene bank or test-tube collection.

In situ conservation in the Moluccas should target both wild and cultivated clove. Populations of planted and wild clove in the Moluccas vary from scattered individuals in the forest or backyards to groups of five to ten trees concentrated in clearings. Clove populations in the center of origin are widely dispersed, and insufficient data have been gathered to indicate any one area in the Moluccas for high-priority conservation. A number of small, scattered reserves might be needed to encompass sufficient variation, but administrative and management costs would be higher than for a single, large reserve. Whatever the mix of parks and reserves that are eventually devised for the Moluccas, rigorous protection will be required to prevent boundary encroachment.

One management tactic that warrants study is the artificial pollination of wild trees in certain locations with pollen from other areas, thereby increasing local diversity. Natural selection would then act on the resulting progeny, with results attainable only after a century or more. Such long-term experiments have never been attempted, but they could be a valuable management technique, particularly if populations shrink because habitat is destroyed and gene flow ceases.

Another strategy would be to identify an area in the Moluccas where environmental heterogeneity is greatest and plant numerous seedlings from all parts of clove's range. Such action is fraught with many technical difficulties, since the scientific community would be "managing" evolution. The implications of starting new evolutionary trajectories for clove will need to be considered over the long term. Quarantine precautions would have to be followed rigorously in order to avoid introducing new diseases or pathotypes of existing pathogens. One "rotten apple" in a germplasm consignment could wipe out a unique clove population.

Numerous other interventions for preserving and enriching clove's gene pool need to be considered. Wild cloves could be established as borders around cultivated forms in field gene banks and after outcrossing, seeds could collected and replanted in the wild. Field gene banks would thus be transformed into orchards of evolution, rather than museums where emphasis is on maintaining the integrity of individual accessions.

In situ conservation also needs to consider primitive cultivated forms. From the distant past to the present, clove has been sown in parts of the Moluccas and the buds periodically harvested. In some cases, individual cultivated trees are found in various plant communities and are exploited by local people. Although these indigenous cultivated types may be engulfed by forest in some cases, they remain distinct from wild forms. If people are excluded from the forest, some of the wild forms would be less vulnerable to felling, but some of the primitive cultivated types would probably disappear. In situ conservation should thus encompass areas where slash-and-burn farming continues as well as zones where no significant disturbance of the forest is allowed.

CINNAMON AND CASSIA

True cinnamon (*Cinnamomum zeylanicum*) is native to Sri Lanka, which still produces the finest cinnamon from wild and cultivated trees. Aromatic bark from cinnamon and cassia (*Cinnamomum* spp.) has been long appreciated in the Orient for flavoring food, for medicinal purposes, as an ingredient in incense, and as an aphrodisiac. In addition to the aromatic bark, the commercial products include oleoresins and an essential oil extracted from various parts of the trees. Essential oils are obtained by steam distillation of the bark, twigs, or leaves and are used as flavorings, as well as in perfumes, soaps, and pharmaceuticals.

Essential oils include cinnamon bark oil from Sri Lanka and cinnamon leaf oil from Sri Lanka, Seychelles, and Madagascar. The oils contain eugenol, which varies according to source; importers thus "rectify" the oil before sale. Cassia oil comes largely from China and is obtained from a mixture of twigs, bark, and leaves. Soft drink manufacturers in the United States are the largest market for cassia oil.

The etymology of cinnamon is derived from the Greek word for spice and the prefix "Chinese." The Greeks borrowed the word from the Phoenicians, indicating trade with the East from early times. Both cinnamon and cassia are recorded in Sanskrit, in the Old Testament, and in Greek medicinal works (Plucknett, 1979; Ridley, 1912:207). Cinnamon or cassia was employed by Egyptians for embalming purposes as early as 1485 B.C. The Arabs controlled and protected much of the trade in cinnamon and cassia to the West until the Middle Ages, when the Portuguese, and later the Dutch and English, penetrated tropical Asia.

India's Malabar Coast has figured prominently in the spice trade through the ages. Arabs purchased spices from such ports as Calicut and Mangalore for sale in Europe. The warm Malabar Coast of India was suitable for growing cinnamon and other spices and also served as a crossroad for maritime trade routes farther east.

Cassia, rather than the more famous cinnamon, was probably the first known spice of this group of species. Cassia has a higher oil content than cinnamon and has a stronger flavor. Cassia is harvested from India eastward to China, Japan, and the Pacific. Cassia is collected from wild trees and several cultivated forms. In the wooded mountains of central and southern Vietnam, for example, villagers collect bark from several wild species (Schmid, 1989). Saigon or Vietnamese cassia (*C. aromaticum*; syn. *loureirii*) occurs wild in Vietnam and is also cultivated to a limited extent in other parts of Southeast Asia, China, and Japan. Chinese cassia (*C. cassia*) is tended or cultivated in southeastern China where it grows naturally (Plucknett, 1979; Purseglove et al., 1981a:100). Indonesian or Padang cassia (*C. burmanii*) occurs wild in parts of South and Southeast Asia and was formerly cultivated in Malaysia. Indonesian cassia is now mostly grown in parts of Sumatra, Java, and Timor.

Until recently, considerable confusion has characterized the literature on types of cinnamon and cassia. This confusion stems from the plethora of wild and cultivated species that furnish cinnamon, or cinnamon-like flavors. True cinnamon has often been mixed with one or more of the cassias and labeled "cinnamon." In India, *C. tamala* is sometimes referred to as cassia cinnamon, although the wild and cultivated tree is a source of tejpat leaves, which are used as a spice in the northern part of the country (Purseglove et al., 1981a:104). Indonesian cassia was originally called Batavian cinnamon, after the capital of Dutch Indonesia, now called Jakarta.

Although only a handful of countries produce cinnamon or cassia, over fifteen countries re-export products of the trees. At least fifty countries regularly import cinnamon and cassia products (A. E. Smith, 1986). Fortunately, no tariffs inhibit trade in cinnamon or cassia.

Confusion has also stemmed from taxonomic uncertainty regarding *Cinnamomum*. Before the Linnaean system for classifying plants and animals was devised in the eighteenth century, common names and incomplete descriptions are generally all that is available for many representatives of the genus. Kostermans (1980) pointed out the difficult task of separating the species described in the *Hortus Indicus Malabaricus*, a twelve-volume opus published between 1678 and 1685. Even this landmark study of plant resources in Asia, written by Hendrik von Rheede, leaves us in doubt about the identity of certain spices. Taxonomic disputes continue to this day; one specialist assigns only fifty valid species to *Cinnamomum*, whereas another recognizes over 200 species in the genus (Purseglove et al., 1981a:104).

CASSIA

Cassia should not be confused with shrubs and trees of the genus *Cassia*, which belong to the legume family. Senna is produced from the leaves and pods of both wild and cultivated *Cassia senna* in the drier zones of Egypt, Sudan, and the Sahara, and from *C. angustifolia* in Somalia and Arabia. Other species of *Cassia* are used widely in India, Indochina, and Southeast Asia for fodder and green manure; some are cultivated in India for tanning materials and for folk remedies.

Wild cassias (*Cinnamomum* spp.) are generally found in lowland to hilly tropical forests with a monsoon climate from India to China. Several cassias have been domesticated and are propagated by seed and by cuttings. In addition to the aromatic bark, the unripe fruits of some cassias, known as "cassia buds," are dried and used as a flavoring. Cassia buds were exported to Europe in the fourteenth century (Ridley, 1912). Cassia is generally considered inferior to true cinnamon. China and Indonesia are the main exporters of cassia, and the United States is the principal market, followed by Japan, Saudi Arabia, India, Pakistan, Germany, and the Netherlands.

Cinnamon

Cinnamon has been known in southern Europe for well over a thousand years, but became more widely known in other parts of Europe during the Middle Ages when the Islamic world extended its influence to the Orient and east to Iberia starting in the eighth century A.D. Iraq and Iran were central nodes along caravan trading routes that stretched across the steppes of central Asia. Camel caravans plied these extensive trading routes laden with exotic spices, ivory, rare woods, and ceramics. Ships also brought luxury items to Baghdad and Iranian ports from the Malabar Coast and beyond.

Cinnamon and other aromatic plants were used to produce comfit boxes in Europe during the fifteenth century. Comfit was a home remedy kit composed of about a dozen different pastilles made from sugar, honey, saffron, and spices. In addition to cinnamon, lozenges of nutmeg, clove, ginger, anise, and fennel complemented the assortment in the comfit box. Cinnamon drops were regarded as a tonic, a sedative in childbirth, and a remedy for many common disorders. Cinnamon served as a breath sweetener in an age when oral hygiene was not what it is today. In medieval times, cinnamon was distilled to produce cordials, ostensibly to aid in digestion. In the Orient, cinnamon and its near relatives are still widely used for local remedies, particularly for gastrointestinal and respiratory disorders. In the Philippines and the Pacific, cinnamon is taken to relieve headaches. In the West, cinnamon is used mainly for flavoring food, as an ingredient in perfumes, and in the case of Mexico, to enhance the flavor of coffee. In Colombia, however, cinnamon sticks are chewed to speed parturition (Lang, 1988:87).

While Arabs dominated the trade in cinnamon and many other spices in the Middle Ages, several kingdoms in Europe aspired to wrest control of the lucrative trade in spices. The Portuguese were the first to establish a foothold in tropical Asia, when Vasco da Gama reached the Malabar Coast of India in 1498. The Portuguese soon established colonies in Goa on India's west coast and in Sri Lanka. Sri Lanka was a coveted prize because of its extensive wild stands of premium cinnamon, and the Portuguese took control of cinnamon production on that island by 1536.

The Portuguese held on to Sri Lanka for a century until they were dislodged by the Dutch, who were in turn ousted by the British in 1796. While Sri Lanka was under Dutch control, cinnamon was intensively cultivated on the island for the first time in 1770 (Ridley, 1912:207). The Dutch also started plantations of cinnamon on Java in 1825. Cinnamon has thus been truly domesticated since only the eighteenth century, much later than some of the cassias.

Cinnamon still occurs wild on Sri Lanka as well as in the Western Ghats in India and in parts of the Malay peninsula. The fast-disappearing forests of the Western Ghats also contain several of cinnamon's wild relatives (Pascal, 1988), as well as wild species of another tropical perennial crop, jackfruit (*Artocarpus heterophyllus*). Wild populations of cinnamon are threatened by widespread deforestation.

After introduction, spontaneous populations of cinnamon have become established in various parts of the tropics because fruits are eaten by birds, which disperse the seeds. For example, cinnamon became naturalized on Jamaica over a century ago (Grisebach, 1864:279). Spontaneous populations of cinnamon may prove useful sources for novel traits in the future, particularly if wild populations of cinnamon in its native range are severely reduced.

Like cassia, cinnamon is propagated by seed or cuttings. Growers cover the fruits to protect them from birds if they wish to harvest the seeds for later planting. Wild trees reach 20 meters in height, but on plantations, cinnamon is cut every two years. A flush of straight shoots is generated by coppicing, suitable for obtaining cinnamon quills. Cassia is coppiced in similar fashion. Stems are cut so that the bark can be peeled off in two longitudinal strips. The bark strips are packed together in heaps for a day or two, then the outer bark is scraped off. The inner bark then curls to form the cinnamon quill of commerce. Quills are air dried in the shade to prevent spoilage before sale. Quills are graded prior to export. Poor-quality quills are used for oil extraction. Smaller pieces from twigs and broken bark sections are mixed together as "quillings" and are ground up and steamed to extract the essential oil.

Cinnamon trees are coppiced for about a decade, then rootstocks are split up and replanted. Alternatively, seedlings or rooted cuttings are transplanted in the field. Cinnamon thrives when temperatures average 27°C and rainfall is plentiful, with over 2,000 millimeters of rain a year. Tolerant of a wide range of soils, cinnamon is usually grown in lowlands, particularly in southwestern Sri Lanka, which continues to dominate world production of the spice.

Sri Lanka maintains its lead in high-quality cinnamon. Why cinnamon grown in other countries fails to match the quality of the spice produced on Sri Lanka is not clear, but the sandy soils of southwestern Sri Lanka may be a factor. Other producers of cinnamon include southern India, Seychelles, Madagascar, and Brazil. Mexico, the United States, United Kingdom, Spain, Germany, Algeria, Brazil, Colombia, Peru, and Chile are the principal markets for cinnamon.

Little breeding of cinnamon or cassia has been attempted. A number of cultivars are recognized on Sri Lanka, mainly on the basis of qualitative descriptions of the bark, such as sweet, honey, or astringent. Cinnamon cultivars are simple selections from the progeny of wild, heterogeneous populations that have been maintained by vegetative propagation. No tissue culture methods have been devised for rapid propagation of selected materials.

Wild populations of cinnamon and its near relatives could contain genotypes with especially high levels of useful compounds (C. K. Allen, 1939; Tétényi, 1970). Very little screening of wild populations of cinnamon has been attempted, in spite of some potentially high levels of cinnamon oil and other compounds that could be found in forest gene pools of the species. Cinnamic aldehyde in true cinnamon varies between 30 and 70 percent, and it seems likely that higher-quality forms could be found in the forest. Care has to be exercised in any such screening because of variations between different parts of the plants.

Some of cinnamon's near relatives are rare or threatened by overzealous collecting and forest clearing. One cinnamon relative, *C. cappare-coronde*, was found in Sri Lanka 250 years after a specimen was collected there and subsequently deposited in a herbarium in Geneva (Kostermans, 1973). The bark of this rare species exudes a clovelike aroma and is employed in local folk remedies. Bark collectors are threatening this interesting species with extinction. A similar story probably prevails for some of cinnamon's other near relatives. More floristic inventories are needed, coupled with ethnobotanic studies to elucidate the distributions, status, and use of cinnamon's close relatives.

Cinnamon is relatively free of diseases, which partly accounts for the absence of any breeding efforts. A fungal pathogen (*Coticium javanicum*) attacks cinnamon on Java during the rainy season, but the disease is controlled with Bordeaux mixture. In high-rainfall areas of Hawaii, a fungus (*Phytophthora cinnamomi*) attacks the bark of *Cinnamomum* and *Cassia* trees. No germplasm collections of *Cinnamomum* have been assembled because of the lack of breeding programs.

How well cinnamon plantations will withstand future challenges to productivity from pests and diseases is unclear. A crop may be relatively free of pathogens and pests for a while and then succumb to a new pathogen or a virulent mutation of an existing disease. Complacency is risky in view of the dynamic nature of pest and pathogen populations. A wiser recourse to sustain cinnamon productivity in the future would be to safeguard wild cinnamon and its near relatives in the forests of tropical Asia and the Pacific and to establish a few, carefully planned gene banks.

As with many crops, scientific aspects of germplasm conservation and utilization are only part of the equation. The viability of cinnamon as a commercial crop hinges on a constellation of political, socioeconomic, and agronomic factors. In Sri Lanka, for example, domestic prices paid to growers for cinnamon have been very unstable, in spite of public subsidies (Moore, 1978). The size of plantings on the island has fluctuated in response to production shifts from other areas, such as the Seychelles and Madagascar, as well as the size stocks held by importers. Furthermore, cassia continues to provide vigorous export competition for cinnamon and therefore also warrants germplasm conservation.

For cinnamon and cassia products to remain competitive on world markets, their yield and quality will have to be raised. Alongside crop-improvement programs, better marketing efforts are needed to increase and sustain demand for cinnamon products (Lazaroff, 1989). Advertising campaigns are needed to promote the qualities of true cinnamon and other natural spices and food colorants. Inexpensive synthetic cinnamon flavors are available, but the spice is likely to continue to generate income for growers in developing countries and delight the palate and olfactory senses of consumers the world over for the foreseeable future.

VANILLA

Vanilla (*Vanilla fragrans*) belongs to the Orchidaceae, a botanical family better known for producing exotic tropical flowers. Most orchids are epiphytes, but vanilla is a climbing vine anchored to the soil. Native to the lowland forests of Central America and parts of northern South America, *V. fragrans* produces long, podlike capsules. The "beans" are cured to bring forth the distinctive aroma. Vanillin is extracted with alcohol and water. Finely ground vanilla beans or the extract are used to flavor ice cream, chocolate, various desserts and beverages, confections, and some perfumes. Some pipe tobaccos contain vanilla. In the past, vanilla has been employed as a nerve stimulant and as an aphrodisiac.

The unripe beans are harvested when the tip turns black. Several processes have evolved to cure vanilla, depending on the region where it is cultivated or collected. In French territories, or areas once under French control, the "Bourbon" process is used, which involves immersing the beans in hot water. In other areas, such as Central America, the beans are first exposed to the sun for several hours on mats or blankets. Heated beans are then wrapped in blankets and enclosed in boxes to retain moisture for one or two days. While in the warm, humid boxes, the beans ferment and vanillin is produced. After they have fermented, the beans are dried in the sun. This procedure may be repeated several times. Finally, the beans are sorted according to size and quality and are then ready for shipping. The curing and drying process can take from six weeks to several months.

Two other species of *Vanilla* are cultivated, but they do not rival the quality of vanilla from Central America. In the Society Islands, *V. tahitiensis* occurs wild in Tahiti and is cultivated to a limited extent in French

Polynesia. Some question the taxonomic validity of *V. tahitiensis*; Pétard (1986:123) suggested that it arose from a cross between *V. fragrans* and *V. pompona* on Tahiti in the last century. West Indian vanilla (*V. pompona*) occurs wild in southeastern Mexico, other parts of Central America, Trinidad, and northern South America and is cultivated to a limited extent in Guadeloupe, Martinique, and Dominica.

The fleshy perennial vine climbs by putting out adventitious roots that grasp trees of wet, lowland forests. Planted trees or trellis wires support cultivated vanilla. Vanilla plants are trained to shoulder height to facilitate hand-pollination and harvesting. Vanilla grows best in hot, moist climates with frequent rain. The optimum temperature range is from 21° to 32°C, with rainfall relatively evenly distributed throughout the year (Purseglove, 1975). Two months of reduced rainfall help arrest vegetative growth and promote flowering. Support trees provide partial shade and help prevent desiccation.

ORIGIN AND SPREAD

Although vanilla has been collected in the wild for a long time, like coffee, it became an important crop only during the colonial period. Indigenous groups in lowland Mexico may have cultivated vanilla on a limited scale, particularly as population densities increased and large tracts of forest were cleared. Because vanilla is a recent domesticate with little breeding effort, cultivated forms are little changed from the wild.

When the Spaniards arrived in Mexico, the Aztecs were using vanilla to flavor chocolate drinks. In 1520, Montezuma and other Aztec leaders consumed *chocolatl* drinks laced with vanilla, maize meal, and capsicum peppers in front of Hernán Cortez and his officers. The Aztecs called vanilla *tlilxochitl*, derived from *tlilli*, which means "black," and *xochitl*, which is interpreted as "pod" or "flower" (Correll, 1953; Hernandez, 1651). The Aztecs occupied parts of the dry, and seasonally cold, Valley of Mexico, an environment unsuitable for vanilla plants, so they traded for vanilla with lowland cultures along the Gulf Coast of Mexico.

In Mexico, Vera Cruz is still known for its vanilla production. The Totonoco people southwest from Tuxpan are known for their skill as vanilla farmers and curers. San José Acateno remains a major market town for dried vanilla beans and extract. San José Acateno, Papantla, and Gutierrez Zamora are the main curing centers. The versatile flavoring has thus been collected and traded by a succession of civilizations in Mexico.

Vanilla apparently reached Spain before Cortez set foot in Mexico. Vanilla was brought to Europe about 1510 for use as a perfume, presumably after explorers, including one of Columbus's four voyages, obtained some beans from Central America (Morren, 1839). By the second half of the sixteenth century factories in Spain were making chocolate drinks mixed with vanilla and sugar. Although Spain is not well known for its chocolatiers, Iberia was well ahead of Switzerland, the Netherlands, and Belgium in using both of these New World crops.

Vanilla was first taken to England in 1733, but was then apparently lost and reintroduced around 1800 (Purseglove et al., 1981b:645). In 1812, vanilla cuttings were sent from Charles Greville's garden in Paddington, London, to botanic gardens in Paris and Antwerp (Ridley, 1912:25). In 1819, the Antwerp botanic garden in turn sent two vanilla plants to the Bogor (then Buitenzorg) botanic garden on Java, but only one plant survived (Figure 8.5). The surviving plant flowered in 1825 in the newly opened botanic garden, but produced no beans.

In 1827, the Jardin des Plantes, Paris, dispatched vanilla cuttings to Réunion (then Bourbon) in the Indian Ocean. From Réunion, vanilla plants were sent to the Pamplemousses botanic garden on Mauritius (then Ile de France), which served as a major staging area for exotic plantation crops for the French Empire (A. E. Smith, 1986; N. J. H. Smith, 1986). Réunion also sent vanilla cuttings to Madagascar about 1840 and to Seychelles in 1866. Vanilla was introduced to Tahiti from Manila in 1848; Pétard (1986:123) claimed that Admiral Hamelin was responsible for bringing the plants to the botanic garden at Papeete, whereas Hermann et al. (1989:20) asserted that William Ellis, a missionary from the London Missionary Society, brought cuttings from the Philippines.

The spread of vanilla cultivation is closely tied to the discovery of the need to hand-pollinate the plant. Bees, such as *Meliponia beechii*, pollinate vanilla within its native range (Dodson, 1967). Without its tropical American pollinators, vanilla is barren. The turning point for vanilla cultivation came in 1836 when Charles Morren of Liege obtained fruits by hand-pollinating the flowers. Two years later, Neumann repeated Morren's achievement in Paris. In 1841, a former slave, Edmond Albius, devised a practical method for pollinating vanilla flowers with a slender stick, and this procedure is still used (Correll, 1953).

Once the fertilization problem had been overcome, plantations were established in tropical Asia and on islands in the Indian Ocean and the South Pacific. Introduced to Europe as a luxury item in commerce in the sixteenth century, vanilla only became a plantation crop three hundred years later. The French were largely responsible for the launching of vanilla plantations in the tropics.

Vanilla production is now concentrated in areas where it has been introduced, a not uncommon occurrence with tropical perennials. Madagascar is the main producer, and vanilla is an important source of foreign exchange for that poor country. By 1967, the Malagasy Republic was producing close to three-quarters of the world's natural vanilla. Over 70,000

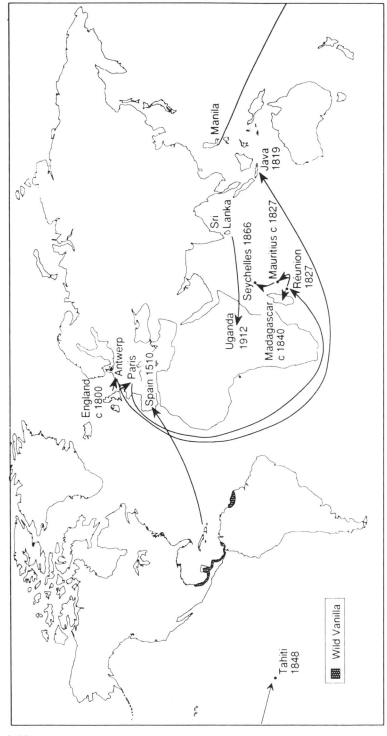

FIGURE 8.5 Introductions of vanilla (*Vanilla fragrans*) and its native area. (Adapted from Aublet, 1775:78; Hermann et al., 1989:20; Morren, 1839; Purseglove et al., 1981b; Ridley, 1912:25.)

small farmers grow the labor-intensive crop on Madagascar (RAFI, 1987). Réunion, the Comoro Islands, and Indonesia are the other main exporters of vanilla. Vanilla exports generate between $60 and $80 million in foreign exchange for producing countries.

THE PRECARIOUS GENETIC FOUNDATION

Global production of natural vanilla rests on a dangerously narrow genetic base. The vanilla plantations of Réunion, Mauritius, Seychelles, and the Malagasy Republic all derive from a single cutting introduced to Réunion from the Jardin des Plantes in Paris (Purseglove et al., 1981b:651). Superior material is cloned by stem cuttings, and commercial vanilla groves throughout the tropics are based on only a few genotypes.

Vanilla is ripe for a concerted breeding effort to broaden its genetic foundation. Despite its importance, virtually no breeding has been attempted with vanilla. Vanilla production would benefit from the release of clones with higher vanillin production and inherent resistance to a number of diseases and pests. Anthracnose (caused by *Calospora vanillae*) and root rot (caused by *Fusarium batatis*) are among the most serious and widespread diseases of vanilla (Declert, 1984). Anthracnose causes wilting and falling of fruits. Root rot has been a limiting factor in vanilla production in Puerto Rico, as well as some other countries. Some resistance to root rot has been obtained from one of vanilla's near relatives, *V. phaeantha* (Delassus, 1962).

Vanilla is also attacked by numerous other pests and diseases including such fungi as mildew (*Phytophthora jatrophae*) in Madagascar; *P. parasitica*, which causes fruit rot; *Glomerella vanillae*, which attacks roots; *Vermicularia vanillae* on Mauritius; *Gloeosporium vanillae* in Colombia, Mauritius, and Sri Lanka; *Uredo scabies* in Colombia; *Macrophoma vanillae* in Brazil; *Pestalozzia vanillae* in Brazil; *Physalospora vanillae* in Java; *Atichia vanillae* in Tahiti. Insect pests include *Trioza litsae*, which attacks buds and flowers, creating puncture spots that can become sites for secondary infection on Réunion; the pantropical emerald bug (*Nezara smaragdula*); two species of weevil (*Perissoderes oblongus* and *P. ruficollis*) that attack vanilla on Madagascar; *Cratopus punctum*, a weevil that bites holes in flowers and can destroy the column in Malagasy, Réunion, Mauritius; lamellicorn beetles (*Hoplia retusa* and *Enaria malanichtera*) that also damage flowers in Malagasy; and various species of moth caterpillars that cause widespread damage (Purseglove, 1981b).

Closer scrutiny of wild populations of vanilla and its near relatives would likely reveal variation in vanillin content, resistance to pests and diseases, as well as other useful traits. West Indian vanilla (*V. pompona*), *V. barbellata*, and *V. phaeantha* resist root rot and can be hybridized with

V. fragrans. A wild relative of vanilla, *V. planifolia* f. *gigantea*, in the rain forests of Mato Grosso in north-central Brazil has pods that exceed 28 centimeters (Mors and Rizzini, 1966:76). The beans of this wild climbing orchid are twice as long those of vanilla, and Brazilians aptly refer to the vanilla relative as *vanilão* (big vanilla). The rapid disappearance of forests in Latin America is closing off options for future improvement of vanilla by destroying populations of the crop and many of its near relatives.

Plant breeders have little to work with in vanilla germplasm collections because the collections are dominated by a few clones. In India, a regional station of the Central Plantation Crops Research Institute (CPCRI) maintains a vanilla collection, including accessions of *V. aphilla*, a wild relative of vanilla from southern India. Another small vanilla collection is maintained by CATIE (Centro Agronómico Tropical de Investigación y Enseñanza) in Costa Rica but has suffered losses from diseases, pests, and inclement weather (Jarret and Fernandez, 1984). Field gene banks for vanilla germplasm are highly vulnerable to virus infection (Zettler et al., 1990).

Future efforts to conserve vanilla germplasm should focus on collecting material threatened by deforestation in Central America. Tissue culture collections may ultimately reduce the cost of maintaining ex situ germplasm collections of vanilla, particularly after cryopreservation techniques are worked out. At present, vanilla tissue cultures can be stored for up to 18 months at 18°C before reculturing is necessary (Jarret et al., 1986). At least some wild populations of vanilla should be safeguarded in parks and reserves.

NATURAL VANILLA'S COMPETITION

Certain aromatic plants, such as tonka bean (*Dipteryx* spp.), vanilla-plant (*Trilisa odoratissima*), and little vanilla (*Selenipedium chica*), have been used as vanilla substitutes, but the emergence of synthetic vanillin has, until recently, dampened cultivation of vanilla.

German scientists were responsible for launching artificial vanillin in the marketplace in 1874. Tiemann devised a process of oxidizing coniferin, a glucoside obtained from the cambium of various coniferous trees, with acid (Katz, 1943). In 1891, a French chemist, De Taire, extracted vanillin from eugenol, a constituent of clove oil. Eugenol is now the main source of artificial vanillin. Synthetic vanilla is also derived from the waste sulfite liquor of paper mills and from coal-tar extracts. The low cost of artificial vanillin combined with the ability to manufacture it in the major consuming areas of North America, Europe, and Japan have led to its virtual domination of the market for most of the twentieth century.

Close to 95 percent of vanilla flavor used worldwide still comes from artificial vanillin, but after half a century of stagnation, demand for true vanilla has been rising since the early 1970s (Purseglove et al., 1981b:705). Demand for natural vanilla remains high, despite the lower cost of artificial vanillin, in part because artificial vanillin does not embody the flavor nuances of true vanilla. The fragrance and flavor of real vanilla are due to numerous aromatic compounds, including resins (Purseglove et al., 1981b: 644). Some 170 volatile constituents have been identified in 'Bourbon', a vanilla cultivar from Madagascar. Unlike some of the natural dye plants, synthetic compounds have not eliminated vanilla as a viable commercial crop.

In addition to its superior flavor, natural vanilla is staging a comeback because of increased demand for "natural" products, particularly in foodstuffs (Joshi et al., 1980:46). The packaging on premium ice cream emphasizes "all natural ingredients," and black flecks in the better-quality ice creams are a sure sign that vanilla beans are one of the ingredients. Natural vanilla is also in demand for up-market chocolates, puddings, and confections.

The U.S. Food and Drug Administration strengthened the position of the natural vanilla industry in 1965 when it required that frozen desserts indicate whether they contain natural or artificial vanilla (Purseglove et al., 1981b:644). Americans consume over half of the world's vanilla, and ice cream accounts for half of this amount. Developments in the U.S. ice cream industry thus have major impacts on vanilla production and prices. In France, a law passed in 1966 requires that if vanilla is used in any food products, it must be clearly labeled as artificial or natural.

Test-Tube Vanilla

The surge in demand for natural vanilla may not necessarily produce large dividends for growers in the Third World, however. Vanillin occurs in several parts of the vanilla plant, not just the beans, and scientists are exploring the potential of culturing cells in vitro to produce natural vanilla (Jarret and Fernandez, 1984; Konowicz and Janick, 1984). Tissue culture would retain the many auxiliary compounds present in the natural extract (CMR, 1988). A number of companies and research institutions are investigating the potential of tissue culturing vanilla, including David Michael & Co., a private company collaborating with scientists at the University of Delaware (RAFI, 1987).

If tissue culture proves a viable method of producing natural vanilla, the area devoted to the crop is likely to shrink. Alternate cash crops would need to be found for the affected farmers. Tissue culture may prove cheaper than growing vanilla because less labor is required, particularly for pollination and curing. It remains to be seen whether the flavor of vanilla extract produced by tissue culture rivals the natural flavor prepared by traditional means. Many growers of vanilla have survived the onslaught of artificial vanillin, and perhaps a market will still persist for field-grown and village-cured vanilla. Vanilla "as nature intended" may one day be relegated to a specialty market where consumers are willing to pay a premium.

Should tissue culture of vanilla become practical on a large scale, production would not necessarily shift completely to industrial countries. Many developing countries have advanced their skills in various biotechnologies within the last two decades and are likely to seize the opportunity to produce vanilla in the laboratory. Brazil, India, Mexico, the Philippines, and Thailand, for example, all have good facilities for a range of biotechnology work in both the public and private sector.

PROSPECTS

In spite of the tissue culture "cloud" looming over vanilla growers, several countries are stepping up efforts to increase production of natural vanilla. Prices for natural vanilla improved sufficiently in the 1980s to justify such investments. A vanilla plant bears fruit in two to three years and remains productive for about a decade. Because vanilla starts producing a crop within a few years, some growers are willing to take the risk that vanilla prices may dip again in the future. Governments in French Polynesia are providing incentives and advice on how to boost production of vanilla, particularly on Tahiti, Moorea, Raiatea, and Tahaa (Wisler et al., 1987a, b). Other parts of the South Pacific may also shortly take up or expand vanilla production (Menz and Fleming, 1989).

Quarantine precautions are needed for vanilla because it is propagated vegetatively, is attacked by numerous diseases, and several pathogenic viruses thrive on alternate hosts. In the Society Islands, for example, the increasing popularity of imported orchids to grace homes and offices threatens to increase the incidence of several viruses that can damage vanilla plants, such as cymbidium mosaic virus (CyMV), odontoglossum ringspot virus (ORSV), and a recently discovered mosaic-inducing potyvirus (Wisler et al., 1987a, b). Clean-up procedures for vanilla germplasm destined for export and for local planting would help eliminate viruses and other pathogens, but genetic resistance would be a surer recourse.

ANNATTO

Annatto (*Bixa orellana*) is an ancient crop of the tropical lowlands of Latin America, where it is extensively grown for the seeds that produce bixin, a bright orange-red dye. Known also as *achiote* in Spanish, *urucú* in Portuguese, and *rocouyer* in French, annatto was probably domesticated so that people would have a ready source of body paint. Such body paint was used for decorative purposes, to ward off evil, and to go to war (Stearman, 1989). Annatto paint is also reputed to repel insects (Standley, 1924). Several Indian groups in Amazonia, such as the Siona and Secoya of eastern Ecuador and the Waika of the Upper Rio Negro, still occasionally paint their bodies with the caratenoid dye (Mee, 1988:111; Vickers and Plowman, 1984). The use of annatto for dyeing cloth and coloring food came later. Rope is sometimes made from the fibrous bark.

Annatto is also employed in a wide range of folk remedies (Cordero, 1950; Standley, 1924). Various concoctions of annatto are used to treat such ailments as epilepsy, fever, dysentery, and venereal diseases. Aphrodisiac properties are attributed to annatto. Leaves are applied as a poultice to the forehead to relieve headaches, and the pulp is applied to burns. A decoction of annatto leaves is gargled to relieve sore throat. The ground-up flowers are used to treat itchy skin, while a potion is taken as a purgative and to relieve inflammation (Baer, 1976:177).

Annatto was a viable cash crop in Amazonia and other parts of tropical America until the advent of synthetic food colorants in the mid-twentieth century. During the nineteenth century, the Brazilian Amazon exported significant quantities of annatto powder to Europe (Santa-Anna Nery, 1885:100). In the 1920s, large quantities of the vermilion powder were exported from South America to Europe and the United States (Standley, 1924:835). After World War II, Red Dye No. 3 largely replaced annatto as a colorant for food and cosmetics. Since 1970, however, the artificial dye has been implicated as a carcinogen, and in early 1990, the U.S. Food and Drug Administration finally decided to ban outright its use. Red Dye No. 3 has long been defended by the food and pharmaceutical industry because of its low cost and steadfastness in fluids (Waldman, 1990). But the price to be paid for red cherries that do not bleed in fruit cocktails and pink antacids with stable color was ultimately thought to be too high. The ban on Red Dye No. 3 sparked renewed interest in annatto among growers, food processors, and cosmetic companies because the natural red dye is safe for consumption and for skin applications, including lipstick. Industrialized countries use bixin extensively to color butter, cheese, ointments, and chocolate. Bixin was first identified in 1825, but has not been synthesized (Baer, 1976:45).

Annatto is a bush that produces oval capsules with 5 to 100 seeds (Figure 8.6). The oval capsules are easily split open by hand to extract the pellet-sized seeds, normally around 40 to 60 per capsule. After the pulpy exterior of the seeds is dried, either with solar driers or on the ground, the resulting cake is exported if the bixin content is at least 2.5 percent, or the

FIGURE 8.6 Annatto (*Bixa orellana*) with ripe capsules intercropped with pasture (*Brachiaria* sp.). Near Igarapé-Açu, Bragantina Zone, Pará, Brazil, August 1988.

seeds are ground up into powder for local consumption. Market stalls with brilliant orange-red piles of annatto powder are common in many parts of Brazil, such as Bahia, where *urucú* is stirred into stews and rice dishes (Figure 8.7). Bixin is found in various parts of the plant, but is concentrated in the fleshy mesocarp. Seed yields can approach 2 metric tons per hectare, and there is growing interest in using the starchy seeds as a by-product, possibly for livestock feed.

Annatto is a common dooryard shrub in many lowland areas of tropical America, where housewives periodically pluck off seed capsules as needed. A cottage industry of drying annatto for local merchants has sprung up in many regions of Latin America, and prospects are bright for expanding this loosely organized, local trade. In several parts of Brazil, such as Pará and São Paulo, medium-sized plantations have sprung up in response to quickening demand for annatto on the world market. Some Costa Rican farmers are also planting sizable fields to the crop. Annatto plantations have also been established in nonnative areas, such as Guadeloupe, India, Zanzibar, Southeast Asia, and parts of the Pacific.

In the Bragantina zone of Pará, annatto fares well when planted in poor soils that have undergone many slash-and-burn cycles for food production.

FIGURE 8.7 Annatto (*Bixa orellana*) powder for sale in a market
stall, Itabuna, Bahia, Brazil, January 1990.

Some growers interplant black pepper (*Piper nigrum*) with annatto when
the former begins to suffer from *Fusarium* infection; in this manner, the
annatto can take advantage of any fertilizer that may accumulate in the
soil (Falesi, 1987). For continued high yields, however, annatto plants require periodic fertilization. In backyards, organic mulches and kitchen
waste provide needed nutrients, but on plantations, inorganic fertilizer is
occasionally needed to shore up yields.

Annatto is not grown extensively outside of tropical America. Red *Capsicum* peppers, dried and ground into a powder, appear to occupy the
place of annatto in markets of tropical Asia and in Ethiopia. Annatto has
been imported into several tropical regions more as an ornamental or for
folk remedies, rather than as a food colorant. In French-speaking equatorial Africa, a decoction of annatto leaves is used medicinally, and the
fruit pulp is ingested to treat fevers, dysentery, and kidney ailments (F. R.
Irvine, 1961:72). In Java, annatto is commonly planted as a hedge (Van

Hall, 1914:276). A physician took annatto to Tahiti as an ornamental in 1845 (Pétard, 1986:236).

Inhabitants of the Marquesas Islands were already familiar with annatto when the first missionaries arrived. Peoples of the volcanic Marquesas mixed annatto dye with coconut oil to smear on their bodies (Pétard, 1986:236). It is not clear how or when annatto reached the Marquesas. Perhaps the Portuguese brought annatto seeds to Macao, Timor, and the Moluccas in the sixteenth century, and traders took them farther afield. Like sweet potato, which was brought to the Pacific before Europeans arrived, annatto seeds may also have been on board Peruvian rafts or Polynesian longboats.

AN OCEAN OF GENES

Annatto is probably native to Acre in southwestern Amazonia, where it may have arisen from *B. excelsa*, a forest tree (Schultes, 1984a). Apparently unknown in a wild state, *B. orellana* reported in swamp and marsh forests along rivers in the Guianas (Van Roosmalen, 1985:54) may be remnants of abandoned plantings.

Annatto was probably domesticated for ceremonial and supernatural reasons and may have been among the earliest plants domesticated in tropical America. After domestication, probably during the Paleolithic, annatto was adopted widely and diffused to Central America and the Caribbean, where it has become feral. For example, annatto occurs along the shores of Barro Colorado Island in Panama and is an escapee in El Salvador, especially in the Lempa Valley, and in the Isthmus of Tehuantepec in Mexico (Coe and Diehl, 1980:88; Croat, 1978:618; Standley and Calderón, 1927:152). Annatto persists around abandoned homesites and in old secondary forests along the Pacific lowlands of Costa Rica (P. H. Allen, 1956:138; Bentley, 1983). Spontaneous annatto plants thrive on Puerto Rico, St. Croix, St. Thomas, and Trinidad and Tobago (Britton, 1918; Britton and Wilson, 1924:588; Marshall, 1934).

In Amazonia, "wild" or spontaneous annatto is found in a variety of habitats. The now-extinct Bacairi of the Batoví River, an affluent of the Xingu, used to irrigate wild annatto trees (Lévi-Strauss, 1963a). The annatto trees watered by the Bacairi were probably seedlings from older plantings. In eastern Amazonia, apparently wild annatto is found in swamp forests along the Gurupi and Turiaçu rivers (Balée, 1989b). On Ilha de Maracá in Roraima, shrubby annatto grows on the edges of flooded grassland and in riverside vegetation (W. Milliken, pers. comm.).

The large stature of *B. excelsa* would pose problems for annatto breeders, but this possible ancestor of annatto might contain other valuable properties, such as disease and pest resistance. Breeders should have an ample pool of traditional cultivated material to choose from. Annatto is mostly planted as seedlings, and the genotypic variation is enormous.

Two other near relatives of annatto are also forest trees in Amazonia: *B. platycarpa* and *B. arborea* (Baer, 1976:1). The former has flat fruits without spines. Another relative, *B. urucurana*, forms shrubs or thickets along water courses and is harvested in the wild by indigenous groups for body paint (Brücher, 1989:164).

Indigenous groups have a particularly rich assortment of local forms of *B. orellana*. The Siona and Secoya in the Ecuadorian Amazon cultivate three distinct types of annatto (Vickers and Plowman, 1984). Even the largely nomadic Maku of northwestern Amazonia plant annatto (Milton, 1984).

BREEDING AND SELECTION CHALLENGES

CATIE near Turrialba, Costa Rica, has initiated an annatto improvement program that is designed to tap more of the genetic potential of the crop. Otherwise, little breeding is under way with annatto, and most selections are from spontaneous seedlings. Desirable genotypes can be propagated vegetatively, thereby helping to achieve higher yields with consistent quality. Annatto can be propagated by rooted cuttings, grafting, or by culturing shoot tips. Currently, most commercial plantings of annatto are seedling orchards in which low-producing individuals are uprooted and replaced with potentially more useful seedlings. A shift to grafted material will likely become more common in the near future.

For the most part, no insect pests or pathogens cause serious damage to annatto. One reason that insects generally do not cause extensive damage to annatto is that extrafloral nectaries attract ants that patrol the shrubs, keeping intruders at bay. This relatively healthy state of affairs is also due to the sporadic and highly dispersed pattern of annatto plantings in backyards and gardens that has prevailed for most of the crop's history. Now that plantations are coming on-line, with a focus on a handful of cultivars, disease and pest pressure could intensify.

Some insect pests that are already causing a local nuisance in annatto fields include leaf cutter ants (*Atta* spp.); an unidentified hemipteran bug of the Coreideae family known as *chupão das cápsulas;* a beetle (*Capsus* sp.) that attacks annatto leaves; and a bruchid beetle that bores into the seeds (Falesi, 1987). Any of these insects might eventually reach epidemic proportions in large fields planted to annatto.

Two potentially worrisome diseases have been recorded on annatto in Amazonia. Anthracnose, probably caused by *Colletotrichum gloeosporioides*, attacks young leaves of some bushes, causing excessive formation of new twigs. Because anthracnose also attacks avocado fruits, annatto and avocado would not be suitable for intercropping. Witches'-broom, caused by a serious fungal pathogen (*Crinipellis perniciosa*) of cacao and its near relatives, has been observed on annatto in the vicinity of Tomé-Açu in Pará (Falesi, 1987). The infected annatto was close to cacao heavily infested with witches'-broom, so annatto and cacao would not be advisable candidates for intercropping in Amazonia.

Genetic solutions should be sought now for annatto's pests and diseases. Leaf cutter ants are a perennial problem with many crops in Amazonia; at the moment growers combat leaf cutter ants with insecticides, or abandon their fields if the seemingly ubiquitous ants become too numerous. Perhaps some annatto genotypes contain a compound in their leaves that inhibits the fungus upon which the leaf cutter ants feed.

Other desirable characteristics for commercial annatto production include low stature, a well-developed crown, high bixine content of seeds, and a large number of seeds per capsule. 'Wagner', selected in Rio de Janeiro and the most widely planted annatto variety in Brazil, has a relatively low bixin content.

PROSPECTS AND NEEDS

The future of annatto is bright, but more research is needed on its genetic potential. Consumers in industrial countries, and increasingly in developing nations, are health-conscious and have developed an aversion to artificial ingredients in their foodstuffs. Still, market fluctuations can discourage growers. After good prices for annatto on the world market in the mid-1980s spurred overproduction, prices fell by 1989. Producers with a better-quality product and efficient management are more likely to survive swings in commodity prices.

It is risky to stake most of one's income on a single crop. Annatto lends itself well to multiple cropping with pasture or fruit crops, and farmers who employ this technique may be less vulnerable to wild fluctuations of commodity prices. Annatto is highly suitable to intercropping with pasture because cattle eschew the leaves, which are rich in calcium oxalate crystals (Baer, 1976:80).

The U.S. government ban on Red Dye No. 3 should help some annatto growers saddled with excess production. Annatto exports are unlikely to make a big dent in the several hundred billion dollars owed by Latin American nations, but every crop that finds a new or expanded market helps the economy of developing countries and provides additional employment. Another benefit of increased annatto production is nutritional, since the seeds are an excellent source of vitamin A.

CHAPTER 9

Nuts

Nuts from tropical perennials are an important source of income for many smallholders in the tropics. Much of the world's cashew production, for example, comes from small-scale growers in India, East Africa, and Brazil. Although most cashews are eaten in industrial countries, cashew farming and processing in Third World countries provide jobs and many useful byproducts.

Three tropical nuts are profiled here: cashew, Brazil nut, and macadamia. Cashew (*Anacardium occidentale*) did not originate in tropical forests, but many of its close relatives are forest trees. Cashew's relatives have yet to be tapped for a host of useful qualities, including disease resistance and enhanced flavor of the nuts and fruit. Wild populations of cashew are found along the coasts of tropical forest areas, as well as inland on islands of scrub savanna. Brazil nut (*Bertholletia excelsa*) is a legitimate tropical rainforest tree, but wild stands are rapidly being cut down. Fortunately, efforts are under way in various parts of Amazonia and Southeast Asia to cultivate the forest giant. The ultimate success of this move to full-scale domestication will hinge on a better understanding of genetic resources of Brazil nut and the survival of sufficiently large parts of the wild gene pool. Macadamia (*Macadamia* spp.) is native to the tropical forests of northeastern Australia and is becoming an important orchard crop in various developing countries.

The three nut crops examined here represent only the tip of an iceberg. Numerous "minor" perennial plants that bear nuts are tended or cultivated in various parts of the tropics. Indeed much more needs to be learned about these obscure species tucked away in remote areas. Today's minor crop can blossom into tomorrow's hot commodity.

Of the many tropical nut crops of local importance, a few are worth mentioning here to illustrate the cross-cultural diversity of plant exploitation. In the Philippines, the pili nut (*Canarium ovatum*) is gathered in the wild and from sporadic plantings along roadsides and around fields. Pili nuts are avidly gathered and eaten fresh, or the oil is expressed for cooking and lighting. Pili nuts make a nutritious snack since they contain 11 percent protein and are high in calories because of the elevated oil content (Rosengarten, 1984:308).

Cola nut (*Cola acuminata*) is a sought-after masticatory in West Africa, where it is native to the rapidly shrinking forests of that region. The caffeine, theobromine, and kolanin in cola nuts act as stimulants, and the nuts are considered a tonic to the digestive system. Long a trade item along caravan routes in the Sahara, cola nut was domesticated in West Africa, where it retains an important place in the regional culture as a symbol of hospitality.

Another masticatory, betel nut palm (*Areca catechu*), is also cultivated. Native to Malesia, betel nuts are in great demand in India, Bangladesh, Burma, Sri Lanka, Indonesia, and southern China. The nut is chewed by hundreds of millions of people, and pieces of the nut are mixed with lime and other flavorings and rolled up in a leaf of betel pepper (*Piper betle*) to form a quid. Known as *pan* in India, quids containing betel nuts are placed in the mouths of Indians from all walks of life. The alkaloids in betel nuts endow it with powerful medicinal properties. Unripe fruits are taken as a laxative, while ripe nuts are eaten to purge intestinal worms. A multipurpose palm, areca nut is also used for orange dye in the batik industry, and other plant parts are used in weaving.

Tropical nut trees are ideal candidates for agroforestry. Cashew, for example, is commonly intercropped with other perennial crops in northeastern Brazil. Cashew is also a valuable backyard tree in urban areas. Some nut trees, such as tropical almond (*Terminalia catappa*), provide welcome shade as well as nutritious nuts. The pulpy mesocarp covering tropical almonds is relished in its own right throughout the tropics. Tropical nut trees thus offer a wide variety of products, both for cash income and domestic consumption. Nut production in the tropics is low in relation to their nutritive and cash value, and much could be done to bolster their local consumption and export for domestic and international markets.

CASHEW

Cashew (*Anacardium occidentale*) is a multipurpose tree well suited to poor lateritic or sandy soils. Because the tree provides a range of products and can be planted in degraded areas, it is a good candidate for rehabilitating areas degraded by poorly managed pastures or slash-and-burn farming in which fallow periods have become too short. Cashew tolerates high temperatures and drought, but cannot grow in cold areas or in poorly drained locations. It is a valuable component of some agroforestry systems; in Brazil's Northeast region, for example, cashew is intercropped on small farms with coconut, citrus, and banana (Johnson and Nair, 1985). Cashew is also intercropped with staple food crops, such as cassava and maize (J. Ascenso, pers. comm.). Cashew-processing factories provide employment in urban areas, particularly for women (Figure 9.1). India alone has some 550 cashew-processing factories that employ 280,000 people (Das, 1986).

Cashew provides a nutritious kernel for cash income and is a valuable source of foreign exchange for several developing countries. Cashew ranks third after almond and hazelnut in world trade of tree nuts (IBPGR, 1986:5). Most cashew nuts are eaten as snack foods in industrial countries, and are often marketed in cans of mixed nuts. Cashew nuts are expensive, so only a small segment of society in developing countries usually consume the nuts in any quantities. Broken cashew nuts are also used in confectionery and in baked products.

In addition to valuable nuts, cashew provides vitamin C–rich "apples" and several folk remedies. Cashew apples are swollen pedicels or stalks to which the nut is attached (Figure 9.2). Cashew apples are eaten fresh, made into preserves and candies, or squeezed to make fresh juice. Brazilians are particularly fond of cashew juice and about 30,000 metric tons of cashew juice are bottled in Brazil, all for the domestic market (Ascenso, 1986a).

Cashew juice is also used to make alcoholic drinks in several developing countries, including Brazil. In Goa, India, people prefer to make cashew brandy, known locally as *fenni*, from the apples rather than drink fresh juice. Cashew "wine" is consumed in a number of countries, such as Guatemala and Moçambique. The Krobos of the Gold Coast in West Africa ferment cashew juice to make an alcoholic drink (Dalziel, 1948:336).

Cashew byproducts are used to make several industrial products. Cashew nut shell liquid (CNSL) is extracted from cashew shells at factories where the nuts are processed and graded; the phenolic liquid is used for waterproofing and as a preservative (Purseglove, 1974:19). After distillation and further refinement, CNSL is employed in varnishes, lacquers, paints, and brake linings. In Kerala, India, cashew factories burn the pressed shells to distill CNSL, which is exported to Japan, Korea, and Australia. The shells contain a little over 20 percent CNSL, but only about half the CNSL is recovered from shells with present extraction techniques, and only a small fraction of the shells are processed for CNSL.

Cashew also has medicinal uses. The Desana who inhabit the Uaupés

FIGURE 9.1 Woman grading cashew nuts (*Anacardium occidentale*) in the Kamanth factory, Kasaragod, Kerala, India, December 1990.

and some of its affluents in northwestern Amazonia scrape cashew bark and mix it with water to treat wounds (Buchillet, 1988). CNSL is used to treat such ailments as scurvy, sores, warts, ringworm, and psoriasis (Rosengarten, 1984:44). In West Africa, an infusion of cashew bark and leaves is used to relieve toothache and sore gums and is taken internally for dysentery (Dalziel, 1948:337).

Cashew wood is used locally for boat building, light construction, and for fuel. The bark exudes a yellow gum that repels insects and is reportedly used in some countries as a bookbinding gum (Dalziel, 1948:337). In West Africa, the bark yields a yellow dye and is employed for tanning.

ORIGIN AND SPREAD

Native to the northeast coast of Brazil, cashew was domesticated long before the arrival of Europeans at the close of the fifteenth century. How far cashew spread before contact is disputed; it probably reached the West Indies before Europeans penetrated the region, but does not appear to be have been present, or at least common, when Cortez landed in Mexico.

Portuguese explorers soon recognized the value of cashew apples and

FIGURE 9.2 Highly variable cashew apples (*Anacardium occidentale*) collected from backyard trees on Careiro Island, Amazon River, Brazil, August 1972.

nuts and took the crop to their possessions in the Old World. Portuguese took cashew to Cochin on the Malabar coast of India during the sixteenth century. Cashew is recorded in Cochin by 1578 and in Goa by 1598. Cochin thus served as the staging point for the spread of cashew in India and possibly other parts of tropical Asia (Johnson, 1973a). Cashew may have been introduced to India for its medicinal properties and to make wine, rather than for its nuts. In sixteenth-century Brazil, cashew apples and juice were taken to treat fever, to sweeten breath, and to "conserve the stomach" (Sousa, 1971:187).

Cashew was taken from India to East Africa, where it soon became naturalized. Although cashew self-propagates in eastern Africa and parts of India, little secondary diversity has arisen (Nayar, 1983). A French businessman, Auguste Goupil, introduced cashew to Tahiti from Central America around 1875 (Hermann et al., 1989:20).

Cashew nuts started entering world commerce only at the beginning of this century. Today, India and Brazil are the world's leading cashew producers. Moçambique and Tanzania used to dominate global trade in cashew nuts, but output in those countries has declined because of political instability in the case of Moçambique and inappropriate agricultural policies in Tanzania. In 1975, Moçambique accounted for 27 percent of world cashew production, but by 1984, the insurrection-plagued country was responsible for only 5 percent of world production (Samson, 1986:282). Tanzania's share of world cashew production fell from 19 percent to 12 percent during the same period.

Brazil and India have taken up the slack caused by East Africa's declining output of cashew nuts. Brazil's cashew production increased 350 percent between 1975 and 1989 (Homma, 1989a; Samson, 1986:282). From 1970 to the mid-1980s, Brazilian growers planted approximately 200,000 hectares to cashew, mostly in the Northeast (Ascenso, 1986a). Almost three-quarters of Brazil's cashew production comes from the coastal areas of Ceará, Rio Grande do Norte, and Pernambuco (Johnson and Nair, 1985; Figure 9.3). Cashew is also an important cash crop in Benin, Guinea-Bissau, Indonesia, Kenya, Madagascar, Nigeria, Peru, Senegal,

FIGURE 9.3 A village scene in Pernambuco, Brazil, with cashew (center), coconut (background and right), and papaya (left). (Illustration by Barboza Leite, in *Tipos e Aspectos do Brasil*, Instituto Brasileiro de Geografia e Estatística, Rio de Janeiro, 1975, p. 200.)

Thailand, Togo, Uganda, Vietnam, and Venezuela. Cashew, then, is another New World crop that has made a major contribution to African and Asian agriculture.

VARIETAL DEVELOPMENT AND BREEDING CHALLENGES

Most of the cashews grown for local consumption and export are produced from trees planted by seed. Consequently, plantings are generally heterogeneous, which is good from the standpoint of disease resistance, but less desirable with respect to consistent yields and quality. The gap between yields on estate plantings, around 300 kilograms of nuts per hectare per year, and theoretical yields of 10 metric tons of nuts per hectare per year, indicate much scope exists for improving yield (Ascenso, 1986b).

Yield depends on a variety of factors, including soil type, rainfall patterns, crop management, and genotype. To help boost yield, breeders tried grafting techniques on cashew beginning in the early 1970s. Now routine procedures have been worked out to graft superior cashew clones in such countries as Brazil, India, and Moçambique. Indian farmers increasingly employ the apical wedge minigraft to propagate superior cashew clones. One farmer near Puttur in Karnataka state along India's Malabar coast grafts five selections to improve his cashew production and to sell to neighbors (Figures 9.4, 9.5).

Yields would also improve if clones were selected that have a higher proportion of perfect flowers. A major problem for many cashew growers is that fruits often fall before maturity. More studies are thus needed on floral biology (Northwood, 1964).

To reduce the dangers of catastrophic outbreaks of diseases and pests, growers need to interplant several high-quality clones, as has been done by the farmer in Karnataka. More genetic resources will have to be tapped to provide growers and breeders with sufficient material to work with.

Over 100 insects and mites attack cashew, in some cases causing significant losses. In India, for example, the tea mosquito (*Heliopeltis antonii*) is a major pest of cashew, resulting in 30 percent yield declines in some areas (NRCC, n.d.). Cashew is pollinated primarily by insects, particularly bees and flies in the genera *Ligyra* and *Helophilus*, although wind may also disperse pollen (IBPGR, 1986:6). Extensive use of insecticides in orchards is not only expensive but can reduce fruit set. Biocontrol measures and the deployment of genetically resistant cultivars would be a more enduring, and an environmentally safer, approach to pest management.

In India, the National Research Center for Cashew (NRCC) near Puttur, Karnataka, is screening germplasm for resistance to tea mosquito, among other objectives. Negligible damage from tea mosquito has been observed in five accessions in NRCC's field gene bank (NRCC, 1988:4). NRCC has

FIGURE 9.4   Farm worker grafting a superior cashew (*Anacardium occidentale*) selection onto a rootstock. Near Puttur, Karnataka, India, December 1990.

FIGURE 9.5 Farmer (left) and helper with grafted cashews ready for the farm and for sale. Near Puttur, Karnataka, India, December 1990.

also identified several dwarf accessions, such as M 44/3 and M 10/4, that are early-bearing and highly productive (NRCC, 1988:3). An advantage of dwarf clones is that plantations can be more compact, thereby boosting overall production of nuts.

WILD POPULATIONS

The original range of wild cashew has been obscured by human agency. Aboriginal populations along Brazil's coast were very dense, especially among the Tupinamba, when Europeans arrived, and cashew has been cultivated there for a long time.

Cashew quickly establishes spontaneous populations, and these occur from Mexico to Peru and southern Brazil (Purseglove, 1974:19). Spontaneous cashews grow on pockets of savanna in the Guianas and near Santarém, Pará, in the middle Amazon (Spruce, 1908a:66; Van Roosmalen, 1985:1). Wild cashews on the Rupununi grassland of Guyana are harvested (Baichoo, 1981). Stands of wild cashew dot the coastal plain of

French Guiana and are also spontaneous in the drier parts of Martinique (Aublet, 1775:392; Kervégant, 1937).

Spontaneous populations of cashew are found in introduced areas, such as Tanzania and Moçambique (Rosengarten, 1984:38). By the mid-nineteenth century, cashew was naturalized along the coasts of East and West Africa, tropical Asia, and the East Indies (Dalziel, 1948:336; J. D. Sauer, 1988). Adventitious cashew reproduces prolifically on the Ikoyi plains near Lagos, Nigeria (Kennedy, 1936:188).

Cashew is indigenous to coastal dunes of northeastern Brazil (J. D. Sauer, 1988:24). Wild cashew has been found at several sites anchoring sand dunes along the coast of Brazil from Ceará, at 3° S, to southern Bahia at 18° S (Johnson, 1973b; Mori, 1989; Morton, 1961; W. Popenoe, 1914:183). Wild cashew grows in areas experiencing a six-month drought, as in Ceará, to areas in Bahia with no appreciable dry season.

Cashew may have evolved in the scrub savannas (*cerrado*) of central Brazil and later colonized the tropical coast of Brazil (Mitchell and Mori, 1987). The *cerrado* populations of *A. occidentale* differ in several characteristics, such as size and flavor of the apple, from the wild populations along Brazil's coast. Cashew plants from coastal populations were taken into cultivation, whereas wild cashew on the *cerrado* may only have been gathered.

Wild cashew on savannas of Colombia, Venezuela, and the Guianas may also be indigenous, rather than escapees from cultivation (Mitchell and Mori, 1987). Truly wild cashew may have reached widely scattered grasslands during drier climatic cycles that may have occurred in Amazonia during the Pleistocene. During dry climatic cycles, avenues of open habitat may have been created, thereby providing dispersal routes for wild cashew. Electrophoresis, or more precise genetic fingerprinting, such as restriction fragment length polymorphism (RFLP), may help sort out the origins and affinity of disjunct cashew populations.

Cashew probably had a wide distribution even before people arrived in the Americas. Mature apples can float and may have been dispersed along seashores by ocean currents (Johnson, 1973a). Wild animals also spread cashew, particularly after people take the crop inland. The enlarged pedicel attracts dispersal agents. Large fruit-eating birds, such as toucans (Ramphastidae), and fruit bats are strong enough to carry off the apple and drop the nuts. In the middle of the last century, residents of Tefé (then Ega) along the Amazon River asserted that bats eat cashew apples and guava fruits (Bates, 1864:402).

Fruit bats have learned to feed on cashew in areas where the crop has been introduced. In East Africa, Wahlberg's fruit bat (*Epomophorus wahlbergi*) relishes the juice of cashew apples and discards the nuts after feeding (Mitchell and Mori, 1987). Along the Malabar coast of India, bats raid cashew orchards and are considered a nuisance.

Cattle can also wreak havoc in cashew groves since they are partial to the apples. In some cases, they may help spread cashew. In East Africa, elephants are fond of cashew apples and may be responsible for some of the spontaneous populations of the tree (Johnson, 1973a).

The nuts themselves are largely safe from mammalian and avian predators because the shells contain cardol, which severely blisters skin and mucous membranes (Mitchell, 1990). Some parrots, bats, fish, and tapirs are reportedly able to bite through the pericarp of the nut with impunity (Mitchell and Mori, 1987).

NEAR RELATIVES

The size of the genus *Anacardium* has been disputed for some time, but a recent revision suggests ten valid species (Mitchell and Mori, 1987). Four of the nine near relatives of cashew occur in the forests of Amazonia (Figure 9.6). Caju-açu (*A. giganteum*) can reach up to 50 meters and is found throughout Amazonia and the Guianas. This denizen of mature, upland forests could prove an important source of resistance to anthracnose, caused by *Colletotrichum gloeosporioides*, a disease that plagues cashew and several other tropical perennial crops (C. R. Clement et al., 1982; Weber, 1973). Caju-açu's apples are exceptionally juicy with a hint of strawberry flavor, and the nuts are the equal of cashew in flavor (Cavalcante, 1988:72). The Tiriyo of the Paru River collect large quantities of caju-açu apples in season and prepare a fermented drink from the swollen pedicels that they mix with cassava flour. Caju-açu apples also find their way to the bustling markets of Belém. The timber is locally important (Mitchell and Mori, 1987).

Cajurana (*A. microsepalum*), also known as *cajuí da várzea* and *cajutim*, is native to the seasonally inundated forests of the Rio Negro basin south to Rondônia, where rural people collect and roast the nuts. Probably dispersed by water, the nuts of *A. microsepalum* are eaten by fish (Mitchell and Mori, 1987). Another cashew relative in Amazonia's forests, cajuí (*A. spruceanum*), has a sour, but edible pedicel. Cajuí has a light wood suitable for construction and making boxes (Loureiro et al., 1979). The spectacular green and white foliage on cajuí's outer branches would lend this handsome tree to the ornamental trade. At home in floodplain forest and in the uplands, *A. parvifolium* occurs in a broad band on either side of the Amazon River from its mouth to the western part of the basin.

Adjacent to the Amazon basin, a shrubby relative of cashew, *A. fruticosum*, is known only from savannas of the upper Mazuruni river system in Guyana (Mitchell and Mori, 1987). Little is known about this near relative, and its fruits have not yet been collected. One of the largest trees in tropical America, *A. excelsum*, is confined to dwindling forests from southern Honduras to Ecuador and Venezuela (Mitchell and Mori, 1987).

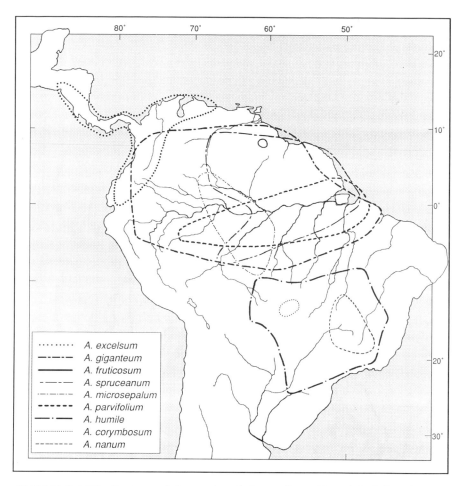

FIGURE 9.6 Distributions of the species of *Anacardium*. (From Mitchell and Mori, 1987.)

This species has edible apples and nuts, after they are roasted, and the wood is used in general carpentry. The macerated bark is used as a fish poison in Panama. Known by various names, including *marañon*, *A. excelsum* is cultivated in Cuba and Ecuador.

Of the four other near relatives of cashew, three occur mostly or entirely in central and southern Brazil: *A. humile*, *A. corymbosum*, and *A. nanum*. The fate of natural habitats in Brazil is thus critical to the biodiversity of cashew and its near relatives.

Ex Situ Conservation

In India, NRCC maintains a field gene bank of 213 accessions. This gene bank is composed mostly of clonal progenies, but has only three near relatives, represented by one accession each. Nevertheless, the gene bank has helped the center launch nineteen varieties adapted to local growing conditions. One reason that the gene bank is making a contribution to ongoing breeding efforts is that collection is well documented, including evaluation data for many of the accessions. A further 600 cashew accessions are held at five field gene banks in other parts of India (Gulick and Van Sloten, 1984).

Moçambique has a field gene bank of 530 cashew clones maintained at the Ricathla research station near Maputo. This sizable collection was enriched with accessions garnered by José Ascenso and co-workers between 1968 and 1973 in farmers' orchards in all provinces except Cabo Delgado. The status of the gene bank is unknown, but it is likely suffering from neglect due to the protracted civil war in the former Portuguese colony. Thailand maintains a collection of 744 cashew types gathered in the country. The Philippines has a collection of 1,300 cashew accessions at Los Baños, mostly spontaneous plants from the archipelago country.

In the center of origin for cashew, only one sizable collection of cashew germplasm has been assembled. The Ceará Agricultural Research Enterprise (EPACE, Empresa de Pesquisa Agropecuária do Ceará) looks after 130 accessions of cashew in an outside plot near Pacajus, Ceará, Brazil. This field gene bank contains mostly commercial cultivars; only three near relatives are represented, with one accession each (Gulick and Van Sloten, 1984). Northeastern Brazil is particularly rich in cultivated forms of cashew (Johnson, 1973a). In Pernambuco alone, over forty types are recognized. A Portuguese chronicler remarked on the great diversity of cashew in Bahia over four centuries ago (Sousa, 1971:187). More collecting of cultivated genotypes in the varied agroecological zones of northeastern Brazil is therefore warranted.

More prospecting for interesting and potentially useful forms of cashew is not enough. Materials need to be systematically screened and made available for breeders. Unfortunately, cashew germplasm is not readily exchanged because of national policies that restrict the export of materials that could help a competing country on the world market. As we have seen in the case of some other tropical crops, this situation is not unique to cashew. If a free-exchange policy is not possible, assigning value to swaps involving germplasm of another crop is difficult. Some developing countries are likely to retain restricted lists of crops for germplasm exchange purposes for the foreseeable future.

PROSPECTS

Demand for cashew nuts continues to grow, and prices have remained high enough to stimulate further plantings (Samson, 1986:282). In view of the importance of cashew in world trade and the value of the nuts, apples, and industrial byproducts of the tree, more research on genetic resources is needed. More degraded tropical lands will need to be restored, either to some semblance of their original habitat, or rehabilitated to accommodate people. Cashew trees thrive on poor soils in the lowland tropics with a pronounced dry season, areas particularly susceptible to ecological degradation.

The vast majority of the world's cashew production comes from small-holdings, and this pattern is likely to prevail. Given the importance of cashew to the economies of several African countries, Brazil, as well as some fast-growing Asian nations, a more strategically planned effort to sample genetic variation among wild and spontaneous populations is called for. Many of cashew's near relatives are confined to Neotropical forests and the central Brazilian *cerrado*, so these environments need to be safeguarded so that breeders will be able to access wider gene pools in the future.

BRAZIL NUT

Almost all Brazil nuts eaten today come from wild trees in the forests of Amazonia. North Americans and Europeans eat raw Brazil nuts mainly at Christmas time, but they also consume them at other times of the year in chocolates and in cocktail snacks mixed with other salted and roasted nuts. In Amazonia, Brazil nuts are eaten raw and are grated and mixed into gruels. In the Brazilian Amazon, shelled Brazil nuts are sometimes grated with the thorn-studded stilt roots of *Socratea* palms, producing a savory, white mash, known as *leite de castanha*, that is stirred into manioc flour.

A widespread wild tree with tasty, nutritious nuts is sure to be incorporated into the diets of many local peoples. Not surprisingly, Brazil nut has been an important food source for many indigenous groups in Amazonia. It is a staple for the Moré of the Bolivian Amazon during certain times of the year and, in the sixteenth century, was the most important source of food for Tacanan-speaking groups of the Beni and Madre de Dios drainage (Métraux, 1963b). Tacanan people even traded Brazil nuts with Andean groups. The Amanayé of eastern Pará add Brazil nuts to manioc flour to improve its taste (Nimuendajú and Métraux, 1963). Brazil nuts are also a significant part of the diet for the Yuruna of the Upper Xingu (Nimuendajú, 1963a), the Maué of the middle Amazon (Nimuendajú, 1963b), the Tupí-Cawahíb of the Machado River in Rondônia (Lévi-Strauss, 1963b), the Cayabí of the Paranatinga River, a tributary of the Upper Tapajós (Nimuendajú, 1963c).

Brazil nut occurs in varying concentrations throughout forested portions of the Amazon basin and parts of the Guianas. It is an important supplement to the diet of people in the largest tropical rain forest because the nuts are high in protein (14–17 percent) and oil (63–69 percent). With such an elevated oil content, Brazil nuts burn like miniature candles when lit. The oil can be expressed and used for cooking, lamps, soap, and livestock feed. Because of their high oil content, Brazil nuts are also high in calories, a concern for those on the cocktail party circuit but a valuable asset to the region's rural and urban folk.

Brazil nut trees (*Bertholletia excelsa*) have several other uses. Although it is illegal to fell a Brazil nut tree, large numbers are cut for its fine timber, particularly in southern Pará and northern Mato Grosso. In the municipality of Claudia in Mato Grosso, for example, sawmills have obtained permission to cut down Brazil nut trees to make way for "urban expansion" (R. Lima, pers. comm.). Such activities impoverish the region's long-term potential for food production and cash earnings. Brazil nut wood has long been sought by boat builders because it resists rotting (Santa-Anna Nery, 1885:183). Empty Brazil nut capsules (*ouriços*) are used to carry small, smoky fires to discourage attacks of black flies (*Simulium* sp.) while people work in their fields during the rainy season in the vicinity of agrovila Castelo Branco along the Marabá-Altamira stretch of the Transamazon Highway. Opened capsules are also sometimes used to collect latex from rubber (*Hevea brasiliensis*) trees.

Brazil nuts are gathered in the Amazon during the early months of the rainy season, which varies in different parts of the basin. In Brazil, groves of *B. excelsa* are known as *castanhais*, and Brazil nut gatherers are called *castanheiros* (Figure 9.7). Owners of Brazil nut groves grubstake *castanheiros* with food and other supplies so that they can concentrate on gathering the fallen nuts. *Castanheiros* construct makeshift huts in the forest, where after gathering fallen capsules, they sit and chop them open with machetes to release the nuts. Mules carry the nuts to the nearest stream or river, where they are loaded onto boats and taken to the owner for payment.

Castanheiros are paid by the hectoliter. One hectoliter is the equivalent of six cans of kerosene, 108 liters, or approximately 50 kilograms of nuts including shells. Once the owners have subtracted the cost of the gatherers' supplies, *castanheiros* typically leave with meager earnings. After the two- to three-month collecting season is over, *castanheiros* disperse to look for other employment, such as gold mining, tapping rubber, or cutting forest for large landowners. Before the Tucurui dam flooded substantial portions of the Tocantins River, many *castanheiros* spent the hot, dry season panning for diamonds along the sandy banks in the once-clear water.

The Tocantins valley accounts for more than half of the Brazil nuts produced in Amazonia (Figure 9.8). Marabá, at the confluence of the Tocan-

FIGURE 9.7 *Castanheiros* gathering nuts in a grove of Brazil nut (*Bertholletia excelsa*) trees. (Illustration by Percy Lau, in *Tipos e Aspectos do Brasil*, Instituto Brasileiro de Geografia e Estatística, Rio de Janeiro, 1975, p. 23.)

tins and Itacaiúnas, grew up from a few huts to a small town at the beginning of this century largely in response to the growing demand for Brazil nuts. Marabá is still in the center of the richest Brazil nut country (Dias, 1959; D. Silva, 1973). Most of the 24,634 metric tons of Brazil nuts exported from Pará in 1979 came from the Marabá area.

Given the importance of Brazil nuts to the regional economy, it is not surprising that the extractive product has had a major impact on rural and urban life, as well as transportation, in the Tocantins valley. Many medium-sized commercial houses in Marabá grew and prospered with the Brazil nut trade. A sizable proportion of buildings along the riverbank at Marabá are used for storing Brazil nuts. Until 1970, Brazil nuts from Marabá were taken downstream to Belém where large commercial houses export both unprocessed and shelled nuts.

In the 1920s, a railroad was built between Tucurui (then Alcobaça) and

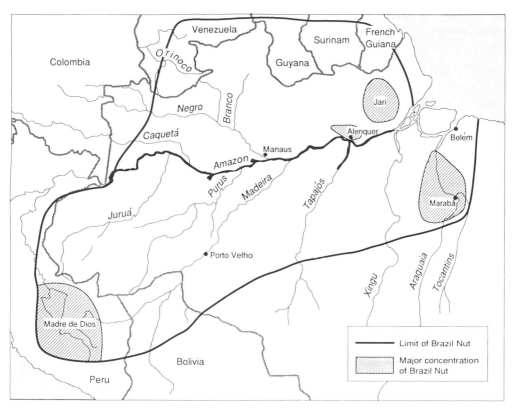

FIGURE 9.8 The main production areas for Brazil nut (*Bertholletia excelsa*).

Jatobal to circumvent rapids along the lower reaches of the Tocantins where many barges laden with Brazil nuts had foundered. The treacherous rapids of the lower Tocantins cost many lives and were the subject of much folklore about drowned spirits. Fueled by wood from the nearby forest, American-built locomotives plied the 117-kilometer railroad for over forty years, occasionally attacked by the Parakanã Indians.

The advent of pioneer highways in 1970 quickly made the Tucurui-Jatobal railroad redundant. Since the arrival of a road connecting Marabá to the Belém-Brasília Highway in 1970, and other all-weather roads to Belém that have crisscrossed Brazil nut country in the 1970s and 1980s, virtually all Brazil nuts now reach Belém by road. Much of the railroad and all of the rapids of the lower Tocantins now lie beneath the still waters of the Tucurui reservoir.

Once the Brazil nuts reach Belém, they are handled by several large exporting firms. The nuts are graded by size and sealed in 18-liter tins. More recently, shelled Brazil nuts are also being exported in 20- to 50-kilogram laminated bags that are vacuum sealed. Brazil nut factories in Belém stockpile nuts so that they can operate year-round. One factory in Belém employs 1,000 people, an important enterprise considering the widespread unemployment and underemployment in many Amazonian cities.

Although Pará dominates Brazil nut production, other areas of Amazonia also yield significant quantities of the nutritious nuts. In Amazonas, Manaus and Itacoatiara serve as entrepôts in the long series of steps before Brazil nuts reach the tables of consumers in industrial countries. In 1986, exports worth $5.7 million were shipped from Manaus (Associação dos Exportadores da Zona Franca de Manaus; C. Miller, pers. comm.). Normally, annual production of Brazil nuts is under 600,000 hectoliters, but in 1966 over 1 million hectoliters were exported, triggering a decline in the price of the product (Homma, 1989a:53).

Although some Brazil nuts were exported to Europe during the colonial period (Rosengarten, 1984), the main push to exploit Brazil nut trees for export began in this century, particularly after the collapse of the rubber boom in the 1920s. As more *castanheiros* fanned out into the forest, conflicts sometimes arose with indigenous groups, particularly in the Tocantins watershed (Laraia and Matta, 1978). *Castanheiros* have sometimes had violent confrontations with the Gaviões, Parakanã, and Akuawa-Assurini (a branch of the Parakanã). Disputes flared over the ownership of Brazil nut trees, but the trees were left intact. Now, however, many Brazil nut trees are succumbing to the axe, fire, and reservoirs.

NATURAL HISTORY

Brazil nut trees are among the forest giants in Amazonia and have been described as perhaps the noblest trees in the region (Spruce, 1908b:357). A mature Brazil nut tree can reach 50 meters and is easily distinguished by its thick, straight trunk covered by fissured gray bark and a spreading, cauliflower-like crown. Flowers of towering Brazil nut trees start to form in the drier months and drop soon after opening during the early part of the wet season. Unfertilized flowers immediately fall to the ground. Brazil nut trees put out a flush of shiny new, copper-colored leaves during the late dry season.

Brazil nut trees are self-incompatible, and the hooded flowers cannot be pollinated by wind (Figure 9.9). Flowers are pollinated by euglossine, anthophorid, and apine bees in the genera *Xylocopa, Bombus, Centris, Epicharis,* and *Eulaema.* These large bees are strong enough to pry open the curved hoods of Brazil nut flowers and crawl in for their nectar reward (Nelson et al., 1985). The males of at least some of the bee species responsible for pollinating Brazil nut depend on orchid flowers to attract their mates. Male bees gather odors from certain orchids and form a lek to

FIGURE 9.9 Flowers of Brazil nut (*Bertholletia excelsa*). Jari, Pará, January 1990.

attract females for mating (Prance, 1984). Without a source of appropriate orchids, mating might not occur.

During a study of the ecological effects of forest fragmentation near Manaus, several euglossine bee species became locally extinct as forest patches diminished (Lovejoy et al., 1986). Some euglossine bees benefit from clearings, such as *Euglossa prasina*, *E. augaspis*, and *Eulaema mocsaryi*, but it is not known if they are significant pollinators of Brazil nut.

When the marble-sized Brazil nut flowers fall to the ground, several months after fertilization, they are avidly eaten by certain game, such as brocket deer (*Mazama americana*) and the nocturnal paca (*Agouti paca*). Along the Marabá-Itaituba stretch of the Transamazon, hunters wait at night in hammocks or platforms near the base of Brazil nut trees when the cream-colored flowers litter the forest floor (N. J. H. Smith, 1976).

Mature nuts are produced approximately fifteen months after fertilization. A pronounced dry season is essential for good fruit set. The dry season was unusually wet in Pará in 1989, and the Brazil nut harvest was correspondingly poor in 1991. Conversely, the Brazil nut harvest was good in 1990 in Pará because little rain fell during the 1988 dry season. *Castanheiros* sometimes set small fires at the base of Brazil nut trees during the dry season so that the capsules can be more easily collected later; heat or smoke from the fire might stimulate flowering, and thereby increase yields.

Amazonia's climate shifts from one part of the basin to another because of the forest's enormous girth as well as the varied topography; this climatic heterogeneity affects the phenology of Brazil nut trees. The onset and duration of the rainy season differ markedly as one travels from south to north and from east to west. In southern Pará, for example, the rainy season typically begins in November and ends in April. Around Marabá, the harvest of Brazil nuts extends from December through January, while at Jari, Brazil nut *ouriços* fall from the end of January until the end of March. Some parts of the region receive on the order of 1,500 millimeters of rain a year, while other areas, such as around Belém, have nearly twice as much rainfall. The differing rainfall patterns in Amazonia have surely influenced the emergence of genetically distinct populations of Brazil nut trees.

Because of its influence on temperature, topography also influences the onset of flowering of Brazil nuts. At Serra dos Carajás, for example, 3 Brazil nut trees at the top of a 675-meter plateau (N5) all had capsules, whereas of 310 Brazil nut trees sampled lower down between 100 and 200 meters near the base of the Carajás range, only 1 had capsules.[1]

The hard capsules of Brazil nut trees protect the ten to twenty-five enclosed nuts from macaws (*Ara* spp.) and certain terrestrial seed predators, such as peccary. Nevertheless, macaws manage to gnaw through some Brazil nut capsules while they are still green and partially consume the immature nuts. Macaw-damaged *ouriços* are commonly encountered by Brazil nut gatherers (Figure 9.10).

Brazil nuts on the ground, are dispersed mainly by squirrels (*Sciureus* spp.) and agoutis (*Dasyprocta* spp.). Although these rodents eat the nuts, they also bury some for later consumption. Occasionally, squirrels and agoutis forget the location of their caches or are eaten by predators themselves, including humans in the case of agoutis. The observation that agoutis are involved in Brazil nut dispersal was recorded in the early 1900s by a Yorkshire botanist (Spruce, 1908a:44), although Indians have probably been aware of the role of agoutis and squirrels in absconding Brazil nuts for a long time. If conditions are right, some of the buried nuts may germinate.

Under what conditions the nuts germinate is unclear. Brazil nut seedlings are rare or apparently absent in some forest with mature Brazil nut trees, such as at Jari (C. Miller, 1990). The seed coat needs to be broken

1. Trees were sampled by counting all mature Brazil nut trees within 200 meters on either side of roads or rivers. The lower-elevation sampling areas, all within 40 kilometers of the Carajás range, included the 32-kilometer road from the N1 plateau to the Itacaiúnas River (22 kilometers of which contained 236 Brazil nut trees); a 1.5-kilometer stretch of the Itacaiúnas from the ferry crossing along the N1 road to the Carreira Comprida rapids, which contained 16 Brazil nut trees on either side of the river; 28 Brazil nut trees were seen along the 8-kilometer road from the railroad head to the sediment trap reservoir at Barragem Esteríl Norte; and 10 kilometers of road east of Parauapebas, which contained 30 Brazil nut trees.

FIGURE 9.10 Brazil nut capsules damaged by macaws, kilometer 19 of the Altamira-Marabá stretch of the Transamazon Highway, Brazil, 1974.

down before it can germinate, and light is probably important for further growth (S. Mori, pers. comm.). Forest gaps created by tree falls would be one propitious environment for Brazil nut seedlings. Humans also create clearings in the forest, and Brazil nuts may be part of the successional pattern in some abandoned fields. Agoutis thrive in second-growth areas and might carry nuts from surrounding forest.

Brazil nut trees also sprout from roots when the "mother" tree is blown or cut down, in a manner similar to groves of sequoia (*Sequoia sempervirens*) along the northwest coast of the United States. Along the road leading from the Serra dos Carajás to the Itacaiúnas River, Brazil nut trees are regenerating from roots in areas bulldozed during road construction (M. F. F. Silva and Rosa, 1986). Similarly, at least some Brazil nut trees at Jari appear as sprouts from old root systems (C. Miller, 1990). Brazil nut trees thus appear to depend on natural or artificial disturbance to reproduce. A better understanding of the ecology of patch dynamics in Amazonian forests is needed.

Many Brazil nut groves were originally planted by indigenous groups. This practice probably extends back to the time when hunters and gatherers penetrated the region tens of thousands of years ago. The distribution, density, and propagation of Brazil nuts may thus be a cultural imprint on the forest.

DISTRIBUTION AND DENSITIES

Brazil nut occurs widely in Amazonia, though its density varies considerably. Trees are found from southwestern Amazonia in the Madre de Dios of Peru, across southern Amazonia as far as 11°45′ S in the area of the Chácobo in the Bolivian Amazon, to the eastern fringe of the Amazon forest in Maranhão (Boom, 1989; Mori et al., 1990:63). North of the Amazon, Brazil nut trees occur south of the Rio Branco and east of the Negro to Amapá.

Northwest Amazonia contains large stretches of sandy soil, which may partially explain why Brazil nut trees are not found there. But many other soil types also occur in that part of the basin, such as heavy yellow oxisols. Indigenous peoples have surely tried to establish groves of the tree along the right-bank affluents of the Negro. The lack of a well-defined dry season is probably the main reason why so few Brazil nut trees are found in northwestern Amazonia. At Barcelos along the middle Rio Negro, for example, at least 100 millimeters of rain normally falls during the driest month, while at Marabá, in the center of Brazil nut country, at least six months of the year have less than 100 millimeters of rain (Diniz and Bastos, 1974).[2] In July and August, typically less than 10 millimeters of rain falls in the vicinity of Marabá.

Rainfall differences cannot explain all anomalous distributions of *B. excelsa*. The upper Juruá has few Brazil nut trees, while the Rio Acre, some 400 kilometers to the east, is rich in the valuable trees, which provide income and food for local inhabitants (Hecht and Cockburn, 1989:21).

People have had a lot to do with the densities of Brazil nut stands. Particularly large concentrations of Brazil nut trees are found in the Tocantins Valley, especially around Marabá, Pará; in the upland forest between Alenquer and the Jari River north of the Amazon; and in the Madre de Dios in Peru. Locally large concentrations of Brazil nut trees have been noted along the Curuphy River, an affluent of the upper Anapú, and along elevated portions of the south shore of the Rio Branco in Roraima (Coudreau, 1899:127; Spruce, 1908a:494). Castanhal, a boom town in the Bragantina zone east of Belém, was once famous for its Brazil nut trees, but few trees remain standing near Castanhal today.

The density of Brazil nut trees in the forest may range from a low of 1 tree per 10 hectares to as many as 29 trees per hectare (Mori and Prance, 1990; C. Miller, 1990). Along a 22-kilometer stretch of road between

2. Rainfall data for Barcelos and Marabá are for the 1931–60 and 1952–58 periods, respectively (Metereological Service, Brazil).

Serra Norte and the Itacaiúnas River in Pará, 236 mature Brazil nut trees were counted along a 100-meter strip on either side of the road (440 hectares), an average population density of 0.5 trees per hectare. In a survey of 36 hectares of forest in the Tapajós National Forest, the average density of Brazil nut trees was 7 trees per hectare (Lopes et al., 1986). In four 1-hectare plots at Jari, densities of the tree ranged from 10 to 29 trees per hectare (C. Miller, 1990).

The link between people and Brazil nut trees has been suspected for some time, but only recently has the evidence become clearer. Brazil nut groves are often found associated with archaeological sites, such as along the Itacaiúnas River, which drains the northern slopes of the Carajás range in Pará (Balée, 1989a). The Kayapó who inhabit interfluvial forests and woody savanna between the Tocantins and Xingu raise Brazil nut seedlings and claim that their ancestors planted the dense stands of Brazil nut trees on parts of their land (Hecht and Cockburn, 1989:153).

Another hint that human interactions in the forest are responsible for Brazil nut tree groves is that the tree may need sunlight to germinate and grow. Natural tree falls provide light gaps for one or two trees, but they cannot explain dense stands of Brazil nut. The Brazil nut seeds sprout in second growth, provided that forest is nearby to provide a seed source. Also, when farmers clear forest for their temporary fields they spare Brazil nut trees; many of these giant trees survive the subsequent burn. Some of their progeny may find favorable conditions in fields as they are gradually abandoned to second growth.

Cross sections of some Brazil nut trees provide evidence of stress, possibly provoked by fire or insect attack. Several incomplete dark rings are sometimes seen in cross sections near the base. The rings are caused by traumatic resin or gum canals (P. Gasson, pers. comm.). Dark gum may accumulate in noncrystalliferous parenchyma cells (Mori et al., 1990:38). Eucalyptus trees also develop resin or gum rings in response to insect attack or fire (M. R. Jacobs, 1937).

Intermittent black rings have been noted in downed Brazil nut trees in the Jari watershed and in other parts of Amazonia (P. Alvim, pers. comm.; C. Miller, 1990). A cross section of a mature Brazil nut tree that had been blown down by a storm near the Esteril Norte reservoir at Serra dos Carajás revealed eight dark rings. Several of these rings were closely spaced; if they were provoked by fire, indigenous populations had probably become dense in the past, forcing a reduction in fallow periods. Brazil nut trees may reach several hundred years of age, so it appears that parts of the Carajás range were densely settled in the past.

It is unlikely that resinous rings in cross sections of Brazil nut trees could have been caused by natural forest conflagrations. When clearings are burned, flames rarely leap far into the surrounding forest. The moist microclimate of Amazonia's forests usually inhibits fire. Fires set by Brazil nut gatherers to clear space for collecting nuts, or more likely field burns, may cause dark rings in the trees.

THREATS TO BRAZIL NUT POPULATIONS

The greatest concentration of Brazil nut, in the Tocantins valley, is also the most threatened. A host of development activities, ranging from the Transamazon Highway, which was open for traffic between Marabá and Altamira in 1971, to the Tucurui reservoir, which began to fill in 1984, are drastically shrinking the wild gene pool.

Although aboriginal groups have occupied the Tocantins valley for a long time, they appear to have enriched the forest with Brazil nuts, rather than destroy them. In the 1970s, a wave of pioneer settlers along the Transamazon and a "grass rush" of new pasture development, particularly around Paragominas along the Belém-Brasília highway, ushered in an era of rapid forest retreat. Of 15 million hectares of artificial pasture created in the Brazilian Amazon since 1970, more than half was on cleared forest in Pará. When Paragominas was linked to Marabá in 1970 by the PA 332 highway (then PA 70), rich Brazil nut groves, once exploited by the Gavião Indians, succumbed to fields cleared by new colonists, sawmill operators, and cattle ranches. South of Marabá, more pioneer highways soon crisscrossed formerly mature forest studded by the characteristic rounded domes of emerging Brazil nut trees. Gold strikes in the shadow of the Carajás range, especially at Serra Pelada, fueled an even greater influx of settlers. Serra Pelada is now a town of 10,000, and Parauapebas, at the foot of the Carajás range, was first settled in 1980 and became a bustling town of at least 20,000 within a decade. In 1970, Marabá was the largest town along the Tocantins, with 10,000 inhabitants; by 1990, the sprawling town had more than 50,000 dwellers.

Within a 47,000-square-kilometer area southwest of Marabá, the deforested area rose from approximately 300 square kilometers in 1972 to 8,200 square kilometers by 1985 (Figure 9.11). Over 40 percent of the Brazil nut–rich forests surrounding Marabá had been devastated by ranchers, farmers, and loggers by the late 1980s (Homma, 1989a, b). A flight between Parauapebas and Marabá in February 1990 revealed that at least half the forest has been cleared and is now mostly pasture.

Another threat to Brazil nut groves in the Tocantins watershed is the Tucurui dam. The Tucurui reservoir has flooded 2,000 square kilometers of forest, and tens of thousands of Brazil nut trees have drowned (Figure 9.12). Brazil nut and other valuable trees were not even used for timber because the company that was contracted to log the area to be occupied by the reservoir did not perform. Smaller hydroelectric dams are planned farther up the Tocantins and some of its tributaries, such as the Itacaiúnas. Hydroelectric resources are needed for urban communities and industry, but the long-term costs of such projects in terms of lost genetic resources are hard to gauge.

A more recent threat to Brazil nut and other forest resources has emerged with plans to establish a string of pig-iron smelters along the Carajás-Itaqui railroad. This 890-kilometer railroad, opened in the mid-1980s, takes iron and manganese ore from the mineral-rich Carajás range to a deepwater port near São Luis, Maranhão. Understandably, Brazilians would like to process some of the iron ore to create jobs and generate more export earnings.

Estimates vary on the quantity of forest that would have to be cleared to supply the pig-iron smelters, but all underscore that vast areas of forest would have to be cleared to fuel blast furnaces and to supply carbon for the reduction process in smelting. If the projected pig-iron smelters are established, some 1.2 million metric tons of charcoal would be required a year (Mahar, 1989:44). As much as 200,000 hectares of forest would have to be cut each year to supply the necessary charcoal. The economic and ecological dimensions of the ambitious pig-iron program have profound implications for some of the richest remaining stands of Brazil nut.

Other concentrations of Brazil nut are in jeopardy, although not on the scale of changes under way in eastern Pará. A planned highway to open access to Pacific ports from Acre in Brazil will surely endanger Brazil nut groves in the Madre de Dios region of Peru. Multilateral development banks have shied away from funding such a road from Cruzeiro do Sul in Brazil to Pucallpa, Peru, because of environmental concerns, and the Japanese government has apparently cooled to the idea of helping build such a road following pressure from the U.S. government (Swinbanks and Anderson, 1989). Even without external help, though, Brazilians and Peruvians will eventually collaborate in building the road, just as Brazil and Paraguay worked together to build one of the largest hydroelectric dams in the world at Itaqui.

GENETIC VARIABILITY AND CONSERVATION

Virtually no studies of the genetic variation of Brazil nut have been carried out. One study compared two populations from widely separated areas, one in Acre at 10°45' S, 68°10' W; the other in Amazonas at 3°23' S, 59°50' W (Buckley et al., 1988). The study showed little difference between the populations, suggesting that the genetic diversity of *B. excelsa* may be preserved within one or two large populations. Much more research is needed before such conclusions can be drawn.

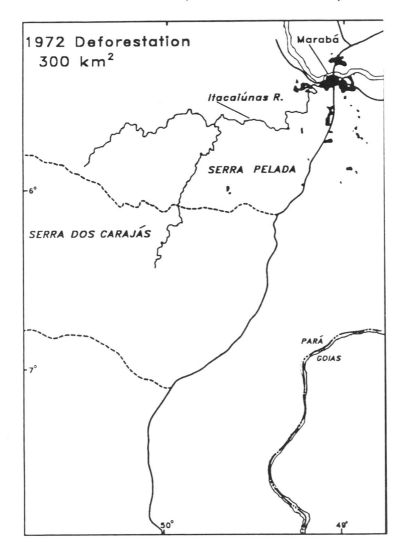

Genetically distinct populations of Brazil nut must have evolved in different parts of Amazonia since they occur on a variety of soils, and some populations may vary genetically in response to edaphic conditions. Brazil nut trees normally occupy well-drained sites, but some populations occur on seasonally flooded alluvial soils along the Amazon near Alenquer, Pará (I. Falesi, pers. comm.).

Fortunately, some of the largest remaining groves of Brazil nut are on biological preserves, Indian and extractive reserves, or within the confines of large corporate holdings. The 1.5-million hectare Manu Biosphere Reserve in the Peruvian Amazon contains populations of Brazil nut, and thus far at least, is well protected. Brazil nuts are a significant source of cash

FIGURE 9.11 Deforestation within a 47,000-square-kilometer area southwest of Marabá, Pará, Brazil, between 1972 and 1985. (Adapted from Mahar, 1989:14.)

income for several extractive reserves established recently in Acre. An average of half of the income for inhabitants of these extractive reserves is currently derived from Brazil nuts, an encouraging sign for the survival of the trees (Schwartzman, 1989). In the Brazilian Amazon, Brazil nut groves are also relatively safe on large corporate properties, at least for the short term. Indian reserves, national forests, and biological reserves are highly vulnerable to illegal logging, squatters, and gold miners. It is too early to tell how well the integrity of extractive reserves in Acre will be protected. Corporate properties are better patrolled, and trespassers are promptly evicted.

Two notable areas rich in Brazil nut are safeguarded by Jari and the

FIGURE 9.12 Brazil nut tree (*Bertholletia excelsa*) flooded by the Tucurui reservoir on the Tocantins River, Pará, Brazil, August 1988.

Companhia Vale do Rio Doce (CVRD). Only 10 percent of Jari's 1.6 million hectares is to be cleared for silviculture for pulp production; the remainder is slated to remain in forest. At Carajás, CVRD, which is a joint venture between private industry and the government, has a 411,000-hectare concession to exploit minerals.

While CVRD's area is relatively small in Amazonian terms, the company's influence on the protection of surrounding areas is significant. Adjacent to CVRD's area, the Xikrin Indian reserve occupies 439,000 hectares. Straddling part of the northern borders of CVRD's area and the Xikrin reserve is the 190,000–hectare Tapirapé-Aquiri national forest. Adjacent to this national forest, which has been set up for sustainable forestry, is the 103,000-hectare Tapirapé biological reserve. Finally, a buffer zone of forest has been designated along the northern margin of CVRD's area. This buffer zone, known as an environmental protection area, encompasses 21,600 hectares and is rich in Brazil nut trees, among other valuable forest resources. The Xikrin reserve is under protection of the Indian service (FUNAI, Fundação Nacional do Indio), and the border with CVRD's area will afford it some protection. The environmental protection area, the Tapirapé biological reserve, and the Tapirapé-Aquiri national reserve are under the jurisdiction of the national agency for the environment and renewable natural resources (IBAMA, Instituto Brasileiro de Meio Ambiente e dos Recursos Naturais Renováveis).

IBAMA lacks the personnel to oversee these reserves, so CVRD has been given the authority by the federal government (decretos 97.718, 97.719, and 97.720) to administer the properties jointly. CVRD's armed guards, backed by helicopters, speedboats, and vehicles, are capable of detecting and expelling intruders. The total area thus protected is 1,164,600 hectares.

Nonetheless, substantial areas of forest in other parts of the Amazon basin should be conserved to safeguard the Brazil nut gene pool. The Carajás, Jari, and Madre de Dios areas have been designated as the highest priority for conservation by biologists (Prance, 1990; Rylands, 1990), and these portions of the basin are particularly important for the long-term survival of Brazil nut's gene pool. Field gene banks are also needed to backstop breeding and selection efforts. Brazil nut seeds are recalcitrant, so they cannot be dried and frozen. With current technologies, live trees are the only way to maintain germplasm of the species. Some research is under way at the Seed Bank of the Royal Botanic Gardens, Kew; preliminary results suggest that protocols might well be developed for storing seeds. If such protocols work out, ex situ conservation of the genetic resources of Brazil nut would be much simpler.

Very little of the genetic variability of Brazil nut is represented in field gene banks. The Belém station of Brazil's agricultural research system

(EMBRAPA, Empresa Brasileira de Pesquisa Agropecuária) maintains forty-five accessions, but these are all clones from materials collected in the vicinity of Alenquer. EMBRAPA also has minor collections of grafted trees at two other sites in the Amazon. Finally, the Rio Branco station of EMBRAPA has a collection of 200 trees, mostly obtained from local groves and sixteen clones from the Alenquer area (C. Miller, 1990). EMBRAPA-Belém plans to establish a more comprehensive field gene bank for Brazil nut on 1,000 hectares south of Belém along the Moju River. Attempts to set up an ex situ collection of Brazil nut trees on an island (Base 1) in the Tucurui reservoir, based on collections of nuts from trees that were ultimately flooded, was largely unsuccessful because of poor germination rates.

THE MOVE TO DOMESTICATION

Several entrepreneurs have already started planting Brazil nuts on a small and large scale. The 10,000-hectare Fazenda Aruanã at kilometer 215 of the Manaus-Itacoatiara road has the largest Brazil nut plantation, with 4,000 hectares of grafted trees. The trees were established on a degraded pasture, so some are doing poorly. The Fazenda Aruanã plantation was started in 1982 by Sérgio Vergueiro with fiscal incentives of SUDAM (Superintendência do Desenvolvimento da Amazônia), a regional development agency. Grafting of superior selections is done at the Aruanã ranch, and with the help of fiscal incentives of the government of Amazonas, some 400,000 seedlings were ready for grafting and distribution to interested growers in November 1989.

The initial failure of a Brazil nut orchard in the Aleixo district of northern Manaus and a few other experimental plantings led to speculation that plantations would probably not be viable unless substantial tracts of forest were protected near such orchards (Mori and Prance, 1987). After a cacao intercrop in a Brazil nut orchard near Manaus was fertilized, however, the Brazil nut trees produced sizable harvests. Extreme soil infertility, rather than the absence of suitable pollinators, may have inhibited nut production (P. Alvim, pers. comm.). Jesuits planted Brazil nut at Tauaú along the Acará River well over a century ago, and it would be interesting to learn how those trees are doing (Spruce, 1908a:14). A seedling introduced to the Peradeniya Botanical Garden in Sri Lanka in 1880 fruited in 1906, and by the 1930s was producing nuts every year (Macmillan, 1935:230). Malaysia has had some success establishing Brazil nut, but attempts elsewhere have been largely unsuccessful.

Another potential drawback for the development of Brazil nut as a plantation crop is that excessive smoke may discourage pollinators. Reduced yields of Brazil nuts in the Marabá region have been attributed to smoke drifting into the forest from fires ignited to clear space for fields and ranches (Kitamura and Müller, 1984). It is true that forest burning in the Marabá region has increased dramatically during the last two decades, but it is not clear whether increased smoke during the dry season is interfering with pollination of Brazil nuts.

Other factors may account for annual variations in yield, such as the length and severity of dry seasons. Also, the drop in Brazil nut production from the Tocantins valley in Pará during the 1980s is more likely linked to a mass exodus of rural people to pan for gold. Large numbers of Brazil nut gatherers found the lure of the numerous gold strikes in Pará and elsewhere in Amazonia more enticing than collecting nuts (Cleary, 1990:171). At the Serra Pelada strike, which is near Marabá and is an important Brazil nut producing area, 100,000 miners were at work in 1983.

Although Brazil nut seedlings take about ten years to start producing, grafted material may begin to bear nuts within four years (C. H. Müller, 1981). Well-managed plantations could greatly exceed the productivity of Brazil nut groves in the forest. If several selections are cloned and grafted onto stocks, cross-fertilization will be more common and yields should be higher than if a single clone is used.

Even with the deployment of several clones in plantations, insect and disease pressures are sure to increase. In the wild, some Brazil nut trees "bleed" a sticky red resin, apparently in response to boring insects. If extensive plantations of a single, precocious, and high-yielding clone are established, pests could significantly undercut production.

The diseases and insect pests of Brazil nut are poorly understood. Breeders will need much more information in order to screen for resistant selections. Also, soils in Amazonia are highly variable, so genotypes adapted to different soil textures and fertility levels will be needed. If sufficient numbers of Brazil nut trees are left in forests in various parts of Amazonia, breeders are more likely to be able to sample them and surmount challenges to raising and sustaining production on plantations.

FUTURE PROSPECTS

As forests in Amazonia retreat, Brazil nut production will increasingly come from plantations. By the middle of the next century, perhaps half or more of Brazil nuts will come from plantations of high-yielding clones. Already, several entrepreneurs in the Brazilian Amazon are experimenting with Brazil nut plantations, and other tropical countries, such as India and Malaysia, are exploring the potential of Brazil nuts to diversify their agricultural economies (Chadha and Pareek, 1989). With support from Canada's International Development Research Center (IDRC), the Peruvian agricultural research system is exploring the potential of Brazil nut tree plantations in the Amazon region. To ensure adequate pollination, plantations will need a mix of clones; fruit set is likely to be poor with a single variety (Mori et al., 1990:118).

The ultimate success of plantations hinges on a better understanding of the ecology of wild Brazil nut populations and ecogeographic variation of the species. The sooner such studies begin the better, or it may be too late. Every year, countless Brazil nut trees perish in flames, in rising waters of reservoirs, or to the power saws of logging crews.

If diseases and pests become too severe in Brazil nut plantations in Amazonia, other regions, such as Southeast Asia, may become the center of Brazil nut production in the next century. Malaysia may do to Brazil nut what it has done to rubber and oil palm: turn a relatively recent domesticate into a successful plantation crop. Although, such a turn of events may bring dismay to Brazilian development planners and entrepreneurs, farmers in Amazonia can try their hand at many other underutilized endemic or exotic perennials. Plantation agriculture has been based on swapping species between continents and islands.

Disease and pest pressure in the native region of Brazil nut might be reduced if the trees are incorporated into agroforestry. Brazil nut trees are an excellent candidate for an overstory species in mixed cropping systems (Nair, 1980). Japanese-Brazilian farmers are interplanting cacao and Brazil nuts in the vicinity of Tomé-Açu, Pará, in eastern Amazonia.

The market for Brazil nut remains strong, and new and expanded markets are on the horizon. For example, one premium ice cream maker in the United States is now selling a crunchy ice cream with blended Brazil nuts. Also, domestic demand for Brazil nuts, particularly in large urban areas, is likely to increase.

New uses for Brazil nut and its byproducts may also be found. For example, the Rio Doce ranch at kilometer 46 of the Belém-Brasília highway mixes Brazil nut shells with manure to fertilize *Brachiaria humidicola* pasture (Figure 9.13). The hard, woody shells are obtained from Brazil nut processing plants in Belém, and using them as fertilizer is a good way of recycling at least some of the nutrients extracted from the region.

MACADAMIA

Savored fresh and after roasting, and mixed in ice cream, chocolate, and various candies, macadamia nuts are probably the most expensive nuts in the retail business. Although macadamia is a minor crop in terms of the world nut trade, it is finding increasing favor among farmers in the low- to mid-altitude tropics and subtropics. Hawaii dominates world production of macadamia, but other tropical areas are fast entering the picture. Brazil, Costa Rica, Guatemala, Jamaica, Kenya, Malawi, Paraguay, Samoa, South

FIGURE 9.13 Shells of Brazil nut (*Bertholletia excelsa*) composted with cattle manure for fertilizing pasture. Fazenda Rio Doce, kilometer 46 BR 316, Pará, Brazil. May 1990.

Africa, Thailand, and Zimbabwe have all established commercial plantations in recent years (Duke, 1989; IBPGR, 1986). Near Thika in Kenya, macadamia is intercropped with coffee, and the nuts are exported to Japan and Switzerland.

Macadamia is the only commercial food crop that originated in Australia. It is a long-lived evergreen tree that belongs to the Proteaceae, a predominantly Australian family that includes other species used as ornamentals. Macadamia itself is often used as an ornamental tree because of its handsome stature and its attractive, holly-like leaves.

The genus *Macadamia* contains ten species: five in eastern Australia, three in New Caledonia, one in Sulawesi, and another in Madagascar. The Australian members of the genus were confused for a long time, since several species were aggregated into *M. ternifolia*. European settlers found *Macadamia* in the forests of eastern Australia, but the nuts were thought to be inedible and poisonous for many years largely because *M. ternifolia* was collected. Then further taxonomic work recognized the smooth-shelled macadamia (*M. integrifolia*) and the rough-shelled macadamia (*M. tetraphylla*). Once the latter species were described, the negative view of macadamia by European immigrants changed. Macadamia nut trees of both cultivated species can produce commercial crops even after a century of growth.

The smooth-shelled and rough-shelled species are the basis of the macadamia nut industry. Nuts of other species of *Macadamia* are distasteful because of a cyanogenic glycoside (Rosengarten, 1984). Piles of macadamia nut shells suggest that aborigines collected macadamia nuts for food in the forest and brought them back to campsites for consumption (McConachie, 1980). In addition, some aboriginal groups partially extracted oil from the nuts. The nuts have a high fat content, largely oleate with some palmitoleate and a little palmitate.

Gympie, or maroochy nut (*M. ternifolia*), is distinguished from the two cultivated species by smaller, lanceolate leaves and a small, bitter nut. The smooth-shelled and the rough-shelled macadamias differ in such characteristics as leaf shape, flower color, seed texture, and sugar and oil content (Storey, 1979). Differences in sugar and oil content led to a misconception that *M. integrifolia* produced better nuts than *M. tetraphylla*. When *M. tetraphylla* nuts are roasted at the high temperatures suitable for *M. integrifolia* nuts, they become brown and unattractive. Thus the two commercial species must be processed differently. Otherwise, the quality and flavor of the nuts is roughly comparable. The smooth-shelled macadamia is grown in Hawaii, whereas the rough-shelled species is cultivated in Australia and California.

ECOLOGY

Both cultivated species of *Macadamia* originated in the tropical and subtropical rainforests along the coasts of Queensland and New South Wales. The cultivated macadamias are adapted to a relatively wide range of climatic conditions (Duke, 1989; Ohler, 1969). However, *M. integrifolia* is native to more tropical areas than *M. tetraphylla*. Macadamia reportedly tolerates rainfall as low as 700 millimeters and as high as 2,600 millimeters, and since the trees also grow within a relatively broad mean annual temperature range of 15° to 25°C, the crop is adapted to both the lowland and highland tropics (Duke, 1989).

Although young trees are sensitive to frost, mature trees can withstand moderate freezes. However, macadamia has few lateral roots and is thus easily uprooted by strong winds. The trees grow on a wide range of soils, from strongly acidic (pH 4.5) to alkaline (pH 8), and do particularly well on deep, well-drained loams and sandy loams. Coastal sands, heavy clays, or gravelly ridges are unsuited for macadamia plantings (Duke, 1989). Fertilization is necessary to maintain high production levels.

Cultivated macadamia is largely self-incompatible and needs to be cross-pollinated by bees (Storey and Saleeb, 1970; Urata, 1954). Nuts mature in six to eight months and fall to the ground when fully developed. On macadamia plantations near Turrialba in Costa Rica, schoolboys earn some pocket money by gathering the fallen nuts after school and on weekends. Some smooth-shelled macadamias produce nuts all year, whereas the rough-shelled species is more of a seasonal bearer (Ohler, 1969).

Most trees bear six to fifteen years after planting. Grafted trees generally produce nuts earlier than those planted from seed. Annual yield levels vary according to such factors as soil and fertilization rates, but 30 kilograms of nuts and their shells per tree is average. Under exceptional circumstances, trees can produce as many as 91 kilograms of nuts per year.

Large yield gaps between harvests gathered by farmers and those recorded under ideal experiment station conditions are common with all crops that have undergone scientific breeding. In the case of macadamia, the yield potential appears to be about three times the levels typically achieved on commercial plantations. While it is unlikely that all macadamia plantings will eventually realize theoretical yield limits, the difference between productivity under experimental conditions and the field indicates that plenty of scope exists for improving genetic stock and crop management.

HISTORICAL PERSPECTIVES

Allan Cunningham was the first person to describe macadamia, after he discovered the trees between the Logan River and the base of Mt. Dunsinane in 1828 (Anonymous, 1988). Cunningham referred to macadamia as "Moreton Bay chestnut" and regarded the tree primarily as an ornamental, although he noted the abundance of nuts, which deserved attention by farmers.

Ludwig Leichhardt first collected a botanical specimen of macadamia in 1843, but it remained unnamed in the Melbourne National Herbarium until 1857, when specimens collected by Ferdinand von Mueller and Walter Hill were named *Macadamia* (Rosengarten, 1984:120). The genus is named after John Macadam, the secretary of the Philosophic Institute of Victoria, later to become the Royal Society of Victoria. The species collected was *M. ternifolia*. Macadam may never have seen the tree named after him or eaten any macadamia nuts (McConachie, 1980).

Walter Hill was the first superintendent of the Brisbane Botanic Garden where he planted *M. ternifolia* and later *M. integrifolia*. An assistant tasted the nuts and to Hill's surprise, pronounced them delicious. Hill had assumed that all macadamia nuts were poisonous. When Hill became aware that nuts from the smooth-shelled macadamia were edible, he realized the tree's economic potential.

Toward the end of the nineteenth century, macadamia was introduced to various parts of the world, such as the College of Agriculture at Berkeley, California, in 1870; Kukuihaele on Hawaii (1881); and the Singapore Botanic Garden (1882). The Singapore Botanic Garden, which has played such a prominent role in the movement of tropical perennial crops, such as rubber (*Hevea brasiliensis*), was the staging point for introducing macadamia to the Malayan Peninsula and to Java. In the early part of the twentieth century, macadamia was planted in Chile, China, Vietnam, Papua New Guinea, and several Pacific islands. Specimen trees reportedly survived the harsh winter of 1890–91 in southern France, and, by 1920, macadamia was planted in the Mediterranean, although not on a large scale.

All these early introductions were based on seed, and plantings were therefore heterogeneous with varying quality and yield. The unpredictable nature of macadamia orchards as an investment discouraged large-scale cultivation of the crop. Two countries nevertheless pursued the economic potential of macadamia by investing in research: Australia and the United States (California and Hawaii).

Macadamia Cultivation in Its Native Region

Macadamia was first planted in orchards at Rou's Mill, near Lismore, New South Wales, in 1888 (Rosengarten, 1984:122). The original trees, all rough-shelled macadamia, were planted on 3 acres and are still productive. As local demand for the creamy nuts increased, other orchards were established in New South Wales.

J. B. Waldron, a bantamweight boxer, was an early promoter of macadamia in the late 1800s and the first half of the twentieth century. Waldron came across macadamia in the Brisbane Botanic Garden where he trained. Intrigued by the trees, he purchased a 79-hectare property, where he planted a thin-shelled form. After the harvest, he cracked the shells by hand, a practice he continued until his death in the 1960s.

In the early days of the macadamia industry, whole kernels were roasted in a drum that was turned manually over a fire. Broken nuts were mixed with honey, and the oil was extracted with a hand-operated press. Kernels scorched in the roasting process were ground and mixed with coffee to produce "almond coffee."

Waldron tried to improve the thin-shelled form by repeatedly planting selected seed, but the trees did not breed true to type. In the 1930s, three brothers named Angus purchased macadamia kernels from Waldron and started commercial production. But the lack of genetic improvement stymied the crop's progress. By the 1940s and 1950s, macadamia acreage had declined because of seedling variability, insect pests, and insufficient knowledge about appropriate agronomic practices. Less than 200 hectares of macadamia were planted in Australia in the early 1950s, and all were seedling progeny (Storey, 1954). More efficient processing was also needed; a breakthrough in this area was possible only when it was realized that the moisture content of the kernels had to be reduced to 1 or 2 percent before the shells could be machine-cracked effectively.

In 1962, the Australian Macadamia Industry pinpointed four major constraints to improved production (McConachie, 1980). First, better grafting techniques were needed to improve survival and growth of clones. Second, insufficient information was available about varietal requirements for Australia. Third, insect pests were making serious inroads on macadamia production in many areas. Finally, data on orchard yields were unreliable.

HAWAII ENTERS PRODUCTION

The highly successful development of macadamia in Hawaii can be traced in part to the introduction of a wide genetic base. William Purvis was responsible for the first recorded introduction of macadamia to Hawaii in 1881. Purvis obtained seeds of the smooth-shelled macadamia from the area surrounding Mount Bauple, north of Gympie, Queensland, and planted them at Kukuihaele on Hawaii (Rosengarten, 1984:123). One of the trees planted by Purvis is still producing nuts (Hamilton et al., 1983). Captain R. A. Jordan visited friends near Brisbane in 1892 and also returned to Hawaii with seeds of the smooth-shelled macadamia. Jordan planted at least six macadamia trees in his garden in Honolulu.

Macadamia was planted in reforestation programs near Honolulu on land that was later set aside for the Hawaii Agricultural Experiment Station (HAES). HAES was established in 1900, and for sixty years scientists have periodically imported macadamia germplasm from Queensland and New South Wales. Macadamia trials were conducted to identify superior performers, and by 1918, 1,000 trees had been distributed to growers. A concerted effort by such scientists as G. T. Pope, H. J. Beaumont, William Storey, and Richard Hamilton have furthered our knowledge about the genetic resources of macadamia and appropriate crop management strategies.

Ernest Sheldon Van Tassel stimulated the commercialization of macadamia on Hawaii. Van Tassel came to Hawaii in 1910 to convalesce and soon developed a passion for macadamia nuts. In 1920, he returned to Hawaii to plant the smooth-shelled macadamia. Although early planting attempts failed, Van Tassel founded the Hawaiian Macadamia Nut Company at Keauhau on Hawaii. To further the industry, he urged the territorial legislature to pass an act exempting all macadamia acreage from taxation until 1932. In 1931, he established a processing factory.

In addition to reforestation areas, macadamia was planted on land formerly occupied by sugarcane. The Honokaa Sugar Company, for example, became the largest macadamia producer for over thirty years and used seed from the original Purvis introduction to establish plantations (Rosengarten, 1984:123). The trend of replacing sugarcane with macadamia continues in Hawaii to this day, particularly on the big island of Hawaii. In 1918, the coffee industry also began to diversify into macadamia production. By 1932, 432 acres were in macadamia, but yields were low because of the considerable variability among the trees.

Cloning and grafting superior selections are the most effective ways of raising yield and quality on macadamia plantations, as has been proved by macadamia production in Hawaii. The first successful attempt at grafting occurred in 1926, when W. T. Pope of HAES brought in a branch that had been snapped by wind three weeks earlier but had remained attached to the tree. The detached limb waited two days before Ralph Moltzau, a high school student employed during vacations, tried to graft it. After several weeks, it was apparent the grafts succeeded. Further research by W. W. Jones in 1936 revealed that scion wood of macadamia needed to be girdled before grafting onto rootstock (W. W. Jones and Beaumont, 1937; Storey, 1959).

The fact that all orchards planted on Hawaii since 1938 have employed grafted material has been a decisive factor in Hawaii's ascendency in global macadamia production (Storey, 1954). Between 1965 and 1981 alone, the macadamia acreage in Hawaii almost tripled (Heinz, 1983).

HAES has thus had an extensive and highly successful program for selecting varieties (Pope, 1929). Selection was based on fourteen criteria: high yield, vigorous growth, numerous branches, round or conical crowns, crown neither too open nor excessively dense, nuts uniform in shape and size, resistance to diseases and insect pests, high proportion of kernel, kernel round and full, good color, high oil content, superior taste, thin shells, and satisfactory cracking by machines. By 1948, HAES had launched five varieties that met most of the above criteria: 'Pahau', 'Keauhou', 'Nuianu', 'Kohula', and 'Kakea'. By the early 1980s, thirteen clones had been selected in Hawaii, several of which have been adopted in other countries (Hamilton et al., 1983). In Costa Rica, for example, a mixed farming operation near Turrialba has 700 hectares of macadamia and was grafting 'Keeau' (selection number 660), 'Kakea' (508), and 'Kau' (344) onto *M. tetraphylla* rootstock in June 1990 (Figure 9.14).

THE CALIFORNIA EXPERIENCE

An agricultural scientist, C. W. Dwinelle, first brought macadamia to California in 1879 (Rosengarten, 1984:134). Dwinelle acquired seeds of

FIGURE 9.14 Worker grafting 'Keeau' (660) clone of smooth-shelled macadamia (*Macadamia integrifolia*) on seedling rootstock of rough-shelled macadamia (*M. tetraphylla*). Hacienda Atiro, near Turrialba, Costa Rica, June 1990.

409

the smooth-fronted macadamia from Australia and planted them in a garden for economic plants alongside Strawberry Creek at the University of California, Berkeley.

Macadamia did not immediately become popular in California. But after World War II, many U.S. servicemen returning from the Pacific theater brought back with them an appetite for macadamia nuts, which they had acquired after tasting the nuts in Hawaii. California's interest in macadamia cultivation was thus rekindled. In the early 1940s, only about 300 rough-shelled macadamia trees were planted in the booming state, mainly as a novelty. A decade later, the number of macadamia trees in California had tripled, mostly in gardens (W. W. Miller, 1955). The rough-shelled macadamia is better adapted to California's mediterranean climate than the smooth-shelled species and is grown mostly in the southern part of the state.

The growing consumer demand for macadamia after World War II was welcome because avocado production was suffering heavy losses due to root rot, and growers were looking for an alternative crop. Under the leadership of C. A. Schroeder, the University of California at Los Angeles initiated research on existing seedling trees and on the crop's potential. The first commercial orchard of rough-shelled macadamia was established at South Oceanside in 1946, and more soon followed. The California Macadamia Society formed in 1953 to disseminate information among growers and to promote the crop among the general public (Storey, 1959).

Breeding Efforts

Macadamia breeding started in Queensland, Australia, in 1968, and several selections have since been released. Improved cultivars have also been identified in Costa Rica, Brazil, Zimbabwe, and South Africa (Anonymous, 1974; Anonymous, 1978; Hobson, 1972; Ojima et al., 1976). The overall emphasis is on clonal production to increase yields and to limit the wide variation inherent in seedlings.

Selection has also generally focused on thin-shelled forms to facilitate mechanical cracking. Care has been taken to avoid material with shells that are too thin. In Hawaii, for example, excessively fragile shells are not desirable because mechanical crackers damage a higher percentage of the kernels. In Australia, very thin shelled forms do not protect the kernels adequately from insect pests, such as ants and a kernel-spotting bug.

The susceptibility of macadamia to insect pests and disease has alerted breeders to the need to screen for resistance to pathogens and damaging insects. The major disease problem in Australia, Hawaii, and California is a trunk and branch canker caused by *Phytophthora cinnamomi*. Current control measures include avoidance of poorly drained sites and damage to tree trunks and branches, orchard hygiene, and spraying with fungicides. In Hawaii, the canker is more frequent on seedlings than in established orchards.

Another problem in Hawaii is anthracnose, apparently caused by species of *Colletotrichum*. Symptoms of anthracnose on macadamia nuts include the darkening and necrosis of the husks, which causes mature nuts to remain on the tree instead of falling to the ground. Stick-tight nuts, as the disease is also known, is a serious problem because the affected nuts are difficult to gather and are inferior. Anthracnose can be especially serious in areas with high rainfall. Fortunately, some resistant varieties have been developed to combat the disease, such as 'Keauhou', 'Wailua', 'Kakea', and 'Ikaika' (Storey, 1954).

No spray program or genetic resistance has yet been devised to counter flower raceme rot, caused by several fungi (species of *Botrytis* and *Phytophthora*). Another disease, macadamia quick decline, has recently begun attacking plantations in Hawaii. At the onset of the disease, the canopy turns yellow then brown, and the tree dies (Nagao et al., 1990). Although the etiology of the disease is uncertain, one or more species of fungi may be involved. Cultivar 333 is highly susceptible, and growers are urged to replant with other materials in infected areas.

In Australia, a root rot caused by *Kretzschmaria clavus* is a significant disease problem. This fungal pathogen invades the roots and covertly progresses to the trunk. Few symptoms are evident until the infection has become extensive; at that point, trees often lose their leaves and perish (Ko et al., 1977). A rapid screening method has been devised to detect the pathogen (Ko and Kunimoto, 1986), but genetic resistance is urgently needed since the pathogen has spread to other areas. The root rot is becoming more of a problem in Hawaii, particularly in view of the discovery that two hosts for the pathogen have been found in Hawaiian forests (Ko et al., 1986).

WILD GENE POOLS AND GENETIC RESOURCES

The two cultivated species of macadamia grow naturally in rain forests on the eastern slopes of the Great Dividing Range in Queensland and New South Wales. Wild populations of *M. integrifolia* are found between 23° and 28°S, while *M. tetraphylla* occurs in a narrower band between 28° and 29°S. Wild stands of the cultivated macadamias are heterogeneous and are largely self-incompatible. The latter trait helps promote heterozygosity and genetic variation. In some valleys, the two species overlap, and spontaneous hybrids result.

Chance hybrids have also arisen under cultivation. In California, for example, a seedling orchard inspected in 1956 contained ninety-six trees with characteristics intermediate between the smooth-shelled and rough-shelled macadamias (Storey and Saleeb, 1970). The seed source for the orchard was an *M. tetraphylla* tree less than 10 meters from a smooth-shelled macadamia; branches of the two trees were interlaced. The rough-shelled macadamia had been selected for its large nuts. One of the leading macadamia cultivars in California, 'Beaumont', contains genes from *M. integrifolia* and *M. tetraphylla* (Rosengarten, 1984:135).

The widecrossing ability of macadamia opens up greater possibilities for genetic improvement. For example, *M. ternifolia*, has the same number of chromosomes as the two cultivated species and tolerates frost better. But the use of wild species for breeding has hardly been explored. If the problem of scion overgrowth can be solved, at least some of macadamia's wild relatives might prove useful as rootstocks. The nuts of *M. whelani*, a native of eastern Australia, are edible, and the species might provide additional flavors or textures. In Indonesia, *M. hildebrandii* has inedible nuts, but the species might broaden the adaptability of the cultivated macadamias to the wet tropics. Much remains to be learned about the potentially useful traits of other species of *Macadamia* and their evolutionary relationships (Storey, 1965).

Australia has a comprehensive plan to conserve natural ecosystems, so sufficient germplasm is likely to be conserved in situ. Indeed, of all the perennial crops of the tropics, macadamia's wild populations appear to be the safest, largely because Australia is a developed country with the resources to protect reserves. And more important, it has a well-fed population that does not need to invade forest areas to eke out a living.

Gene Banks

Macadamia nuts cannot be dried and frozen without destroying their viability, so field gene banks are the only recourse for ex situ collections. By far the largest ex situ germplasm collections of macadamia are in Australia, but even there the number of accessions is small: twenty-five samples of *M. integrifolia* and six of *M. tetraphylla*. About twenty-five accessions of the two cultivated macadamias are held in Israel. The U.S. clonal repository in Hilo, Hawaii, holds twenty-two accessions of the two cultivated species, as well as one sample each of *M. hildebrandii*, *M. praealta*, and *M. ternifolia*. The breeders' collection at the University of Hawaii contains over fifty accessions of the cultivated species. Only a handful of macadamia accessions are held in other countries, such as Costa Rica and Malawi. The closest relative to the cultivated macadamias, *M. ternifolia*, is hardly represented in field gene banks.

PROSPECTS

The United States is still the largest exporter of macadamia nuts, but there is room for other players. In addition to Europe, Canada, and Japan, the United States exports macadamia nuts to China, the Dominican Republic, French Polynesia, Malaysia, Mexico, Panama, Republic of Korea, and Uruguay. Macadamia production in Hawaii varies according to the weather and does not satisfy demands of the domestic market (Rowley and Nakamura, 1988).

More developing countries could produce macadamia nuts profitably for local consumption and foreign markets. Costa Rica has developed a thriving domestic confectionery industry, and macadamia nuts are used in chocolate and nougat bars. Other countries could follow Costa Rica's example. At the turn of this century, for example, Cuba's potential to produce macadamia was highlighted (Earle and Popenoe, 1915). The Caribbean and East Africa are well situated to intensify macadamia cultivation, particularly for export.

CHAPTER 10

A New Cornucopia

Tropical forests are rich hunting grounds for new plants to domesticate. Nowhere else on earth is such a bewildering array of plant species spread out for us to choose from. Tropical forests offer us a variety of plant forms, ranging from tall trees to vines and understory shrubs that could fit backyards or open fields if only uses could be found for them. Ironically, as human civilization seems poised for a new wave of plant domestication, the richest terrestrial biome on earth is rapidly being destroyed.

A balance must always be struck between pursuing new crops and maintaining yield gains already achieved with existing cultivated plants. It is often pointed out that only a handful of crops, such as wheat, rice, maize, and potato, feed most of humanity (Harlan, 1975; Heiser, 1990:61; Prescott-Allen and Prescott-Allen, 1990). Only ten to twenty staple crops provide the bulk of the calories consumed by the world's population, a tiny proportion of the rich array of plant species that could be deployed to better feed, clothe, and house the earth's growing urban and rural population.

More efforts should be focused on upgrading "minor" food crops or exploring the potential of new plants for agriculture (NAS, 1975; Plotkin, 1988). A pantropical push to diversify agriculture in order to improve sustainability and protect farmers from wild swings in commodity prices is under way and often entails the promotion of new or little-known crops (Hamilton, 1987). The knowledge base to support attempts at diversifying agriculture with minor crops is tenuous, and nowhere are poor peoples' crops more neglected than in the tropics (Kochhar and Singh, 1989).

Yet the world cannot afford a major shift of funds and support from the main food crops to lesser known species. The risks are too high. Reduce the funding for research on rice and wheat, for example, and yields of those important cereals are sure to decline. In a world in which population levels are expected to at least double before stabilizing, the staffs of life deserve continued high levels of support from the agricultural research community and donors.

Enthusiasm for a possible new wave of plant domestication must be tempered by the realization that crop improvement is a protracted affair. Most of our crop plants were "handed over" to modern science in relatively good shape. Traditional farmers did a superb job of selecting hardy genotypes adapted to a variety of ecosystems. Indeed, modern breeding has yet to make an impact on many of the tropical perennial crops. The payoff period for new perennial crops will be much longer than for novel annuals, such as Andean lupins.

Still, not all the top ten or twenty crop plants in terms of area grown or value and volume of annual production are suited to all cultures and environments. The ability to "stretch" the genetic potential of major crops to fit all growing conditions and cultural needs is limited. Minor crops, including many of the species profiled in this volume, fill niches left by the major cereals, pulses, vegetables, and root crops. Some might argue that we already have a full complement of crop plants and that there is no more "room" for additional cultivated species. Concern about slicing off more wedges from the already small research-funding pie undoubtedly account for some of those feelings.

In view of the difficulty in predicting the needs of humanity for the next century, it would be a wise investment to explore at least some options for further plant domestication. The ornamental trade is busily scouring tropical forests for new palms, aroids, orchids, and other showy plants to adorn homes and offices. Surely, some crop candidates in the remaining tropical forests could offer products for industry, food production, and beverages. Tropical forests are a virtual pharmacopoeia, with many plant species that could be used more extensively in medicine (Schultes and Raffauf, 1990). A single tribe in Amazonia may use more than a hundred plant species for medicinal purposes (Plotkin, 1988). It is premature to suggest that we have finished with the job of plant domestication.

THE PLANT DOMESTICATION PROCESS

Before we explore some potential candidates for cultivation in plantations, backyards, or fields, it will be useful to review briefly some concepts with respect to plant domestication. Terminology and definitions regarding the incorporation of plants into human affairs vary considerably. Scientists from diverse disciplinary backgrounds, ranging from botany to anthropology, archaeology, geography, and the agricultural sciences, have all offered suggestions about how the process of plant domestication might be viewed.

Sometimes the plant domestication process is portrayed as a relatively sudden event. At the close of the Pleistocene, it is argued, people in various parts of the world took up agriculture (Minc and Vandermeer, 1990). Population pressure apparently prompted people to adopt more intensive forms of food production (Cohen, 1977:11; Evans, 1980). According to this view, hunters and gatherers shifted from relatively passive roles to large-scale manipulators of the environment.

In the humid tropics at least, a more leisurely experimentation with plants took place, with tending and enrichment of natural and artificial habitats long before the appearance of open fields (C. O. Sauer, 1969). Some transplanting and protection of plants, particularly perennial species that furnish fruits, nuts, and medicinal uses, have probably occurred for tens of thousands of years in the lowland tropics. Antecedents to the Neolithic agricultural "revolution" in the New World may extend back at least 40,000 years (Lathrap, 1977).

Evolutionary and ecological dimensions to plant domestication have long been emphasized (Harris and Hillman, 1989). Not all crops pass through clear-cut stages, starting with harvesting of untended wild plants to propagation in cleared plots and later selection of varieties. Some disagreements exist on the taxonomy of plant domestication, but our concern here is with the starting point: when plants are perceived as valuable enough that they are tended to in some way.

The very earliest stage of plant domestication has been referred to as incidental domestication (Rindos, 1984:152). This category includes plants that benefit from human activity, such as artificial enrichment of stands and creation of open sites around settlements for "camp followers." Some plant species have populations that can be classified as incidental domesticates as well as highly developed genotypes in fields and gardens. Oil palm is tended wild or semiwild in West Africa and is also a plantation crop in Southeast Asia and Latin America. Wild breadfruit trees are looked after in the forests of New Guinea and are grown as a backyard crop throughout the tropical Pacific.

In another treatment of the plant domestication process, energy-input thresholds differentiate stages of plant care and tillage (Harris, 1989). Minimal energy inputs characterize the first stage, since plant products are simply gathered in the wild or plants are given limited protection by weeding. Vegetation may be burned, thus favoring some useful plants. In the second stage, some tillage may be practiced, and plants are replaced by vegetative propagation or sowing. Soils may be modified by drainage or irrigation. Crop candidates from tropical forests span the first two stages.

In the third stage, land is cleared for fields, and crop varieties are selected and propagated.

In reality, it is hard to separate categories. Plant domestication is a continuum, and progress through stages is not inevitable. No sharp dividing lines can always be drawn between exploitation of wild plant resources, management of wild plants, planting and care of selected species, and intensive agriculture (Hope et al., 1983). Some plants linger as essentially wild plants with minimal care for millennia, whereas others may be incorporated in large-scale food production systems relatively rapidly. Crops can become feral or revert to a wild state because they are no longer needed or the cultures that kept them going have disappeared.

A STARTING POINT FOR THE SEARCH

Given the great species diversity of tropical forests, the list of plants that could eventually be cultivated is vast. Funds for research on new plants is limited, so efforts will need to be highly focused. One of the most fruitful places to look for new crop plants is among indigenous peoples in tropical forests because of their long and intimate association with wild and domesticated plants.

Aboriginal groups have already identified many plants in the forest that contain products for a variety of human needs. Some of these wild plants are protected by various means. Seeds may be planted along trails, in grasslands, or around settlements. Seedlings may also be transplanted. The Kayapó of the Brazilian Amazon, for example, create "resource islands" of useful plants near hunting camps and along forest trails; over fifty wild plant species are thus manipulated (Posey, 1983, 1984, 1985). Ethnobotanical surveys among other indigenous groups in Amazonia have uncovered a wealth of species that are tended in the wild and in some cases artificially enriched by planting within the forest and in second-growth communities (Balée, 1986, 1987, 1988, 1989a, b; Frikel, 1978; N. J. H. Smith, in press). Forest patches with many economically important tree species, such as *Annona purpurea, Spondias* sp., and *Leucaena leucocephala*, were also established and protected by the ancient Maya in Yucatán (Gómez-Pompa, 1985; Gómez-Pompa et al., 1987).

Nontribal peoples in rural areas are also knowledgeable about plant resources in the forest. Even recent settlers can be helpful in identifying potentially useful plants if they have hunted or gathered forest products with old-time residents (Moran, 1981). In the Brazilian Amazon, an apparently planted uxi tree (*Endopleura uchi*) was found in the backyard of a smallholder near Castanhal, Pará, in May 1990. Oily uxi fruits are normally collected in the wild and sold in markets, such as in Belém. Creamy uxi fruits are used to flavor ice cream, and the oil approaches the quality of olive oil (Cavalcante, 1988:229). The tall backyard uxi tree may have been a remnant from the dense forest that once covered the area, but the smallholder's site has been heavily farmed for over a century and little of the original tree cover remains.

In West Africa, people often try out a new plant in a dooryard setting before committing to more extensive plantings in fields (Vermeer, 1979). Several authorities on economically important plants, including Nikolai Vavilov, have remarked on the relative paucity of crops domesticated in Africa compared with other tropical regions (Brücher, 1989:1). Almost all the fruits grown in tropical Africa are introduced from the Neotropics or from Southeast Asia (Owen, 1973:76). The African tropics generally have fewer plant species than comparable regions in Asia and the Neotropics, and this fact may partly explain the fewer number of plants domesticated by Africans. On the other hand, further research may well reveal that Africans cultivate or have semidomesticated a far broader range of plant species than was hitherto suspected (Ogbe, 1990).

SOME CROP CANDIDATES

Only a few of the many interesting and potentially useful plants from tropical forests are reviewed here. The species selected are not necessarily the most promising as crops; they merely reflect the ones for which at least some information is available. Talk to any field botanist in a developing country and he or she will soon come up with a shopping list of interesting plants that warrant investigation for their economic potential. Candidates for domestication reviewed here abound in the tropical forests of the Americas, Africa, Southeast Asia, and the Pacific and were chosen to illustrate the intricate range of products that tropical forest trees can offer.

Edible Oils

Patauá Palm. The patauá palm (*Jessenia bataua*) inhabits the banks of forest streams in Amazonia, the Upper Orinoco, Guyana, lowland Colombia, and western Panama. Local people climb its smooth, slender trunk to gather the dark fruits hanging from horse-tail stalks during the dry season. Flesh from the plum-sized fruits is made into a refreshing drink. Oil extracted from the fruit pulp is virtually identical to olive oil, one of the most expensive vegetable oils (Pesce, 1985:69; Plotkin, 1988). Patauá oil contains 80 percent oleic acid and 20 percent linoleic acid (Brücher, 1989:138).

The Guahibo Indians, who obtain patauá palm in gallery forests in the eastern Llanos of Colombia and Venezuela, use the oil to treat tuberculosis and respiratory problems such as asthma and cough (Balick, 1980). In Peru, the oil is mixed with almond oil and taken as a purgative, and the Bora Indians take an infusion of patauá seedlings as a snakebite antidote (Plotkin and Balick, 1984).

In addition to its oil- and protein-rich pulp, patauá has been exploited for other purposes. Long, black spines protruding from persistent leaf bases of younger specimens were once in demand for making blowgun darts (Wallace, 1853:31). The Chácobo Indians of northeastern Bolivia lash together the stout petioles of *J. bataua* to make doors (Boom, 1986).

The virtues of patauá oil have been known for well over a century, but supplies from scattered, wild palms were often unreliable and processing techniques primitive. In the mid-1800s, merchants in Belém at the mouth of the Amazon used to blend refined patauá oil in equal portions with imported olive oil (Seemann, 1856:271). The oils are so similar in taste and appearance that this ruse went largely unnoticed. During the two world wars, Brazil exported oil obtained from wild patauá stands (Brücher, 1989:138; Pesce, 1985:70). In the latter part of this century, other vegetable oils, from soybean and maize, for example, have reduced the quantities of patauá oil on the market.

Interest in commercial production of patauá oil is growing. Patauá thrives in poorly drained sites (Kahn and Castro, 1985), areas unsuited for most crops. Some genotypes are adapted to upland locations (Balick, 1984a). The solitary palm is thus a versatile plant for indigenous production of high-quality vegetable oil up to 1,000 meters in the humid tropics. *Milpesos*, as *J. bataua* is known in Colombia, is considered a promising crop for agroforestry schemes in the Chocó region (Budowski, 1989).

To help launch selection efforts in Amazonia, the Brazilian agricultural research system has established a field gene bank for patauá at its Belém station (Figure 10.1). The recently assembled germplasm collection has 169 accessions of the palm. The Las Gaviotas research station in the Colombian Llanos is also investigating the potential of *J. bataua*.

A major constraint that will have to be overcome early in any effort to improve patauá is the palm's slow growth. Field gene banks can help identify precocious genotypes. Other desirable attributes for plantation varieties would be high oil content and regular bearing. Production of patauá fruits varies considerably from year to year, a drawback for commercialization (Pesce, 1985:70). Patauá reaches up to 25 meters, a height that might be a disadvantage in plantations. Shorter genotypes would facilitate harvesting. The Guahibo cut down the palm to access the fruits, a practice that could deplete wild stands if oil production was geared to external markets. At least the Guahibo return a couple of months later to snack on the protein-rich grubs of palm weevil (*Rhynchophora palmarum*) that are pried out of the fallen patauá trunks (Balick, 1980).

FIGURE 10.1 Recently established field gene bank for patauá (*Jessenia bataua*), EMBRAPA, Belém, Brazil, August 1988.

Patauá is widely dispersed, but little is known about its genetic variation. Field surveys will be needed to establish whether much variability exists within and between populations. The Guahibo in Colombia and Venezuela recognize three botanical varieties of *J. bataua*; two are harvested for eating and the third is cut to make bows (Balick, 1980). One particular promising collecting ground for patauá germplasm would be the area surrounding San Carlos del Rio Negro in southern Venezuela, where stands of the palm are so thick that they appear to have been sown (Spruce, 1908a:479). Given the high esteem with which patauá juice mixed with cassava flour is held along the Rio Negro and Upper Orinoco, it would not be surprising if aboriginal groups had artificially enriched the forest with *J. bataua*.

Macaúba.   Another wild palm native to Amazonia that produces an edible oil is macaúba (*Acrocomia sclerocarpa*). Macaúba occurs widely in tropical parts of Brazil and Paraguay and is deliberately spared by slash-and-burn farmers, particularly in eastern Pará, where the species is relatively common. The oily fruits were esteemed by people living in the vicinity of Belém in the last century (Wallace, 1853:97), but the palm is little used today. A clear, edible oil can also be obtained from the kernel; oil from the seed is especially suitable for soap production (Cavalcante and Johnson, 1977). The pulp contains 33 percent oil, and half of the kernel is oil (Pesce, 1985:28).

The relatively high oil content of the fruit and kernels of macaúba make it an attractive prospect for domestication. As in the case of pataúa, uncertain supplies have discouraged the vegetable oil industry from taking on macaúba. The palm is well adapted to second growth and degraded soils in the Bragantina zone and thus would be ideally suited to rehabilitating degraded areas in the humid tropics.

Brazil's genetic resource program (CENARGEN, Centro Nacional de Recursos Genéticos e Biotecnologia) has been systematically collecting germplasm of macaúba with a view to domesticating the species (Coradin and Lleras, 1986). A field gene bank of 192 accessions of macaúba has been established in the Municipality of Coração de Jesus in Minas Gerais. This *Acrocomia* collection contains samples from fifty-nine populations in various parts of Brazil and Paraguay. Considerable variability is noted both within and between macaúba populations (Coradin and Lleras, 1986).

Such genetic variation will permit breeders to select appropriate genotypes for different growing conditions (Lleras and Coradin, 1986). Some genotypes are productive enough to suggest that under plantation conditions, oil yields could reach 10 metric tons per hectare. To obtain high yields, however, plantations would have to be established on better soils or be heavily fertilized.

NUTS

Many tropical forest trees produce potentially valuable nuts, but only a few will be discussed briefly here: sapucaia (*Lecythis* spp.), native to northern South America; and Polynesian chestnut (*Inocarpus fagifer*), a leguminous tree in forests of Malaysia and the Pacific. At least one species of *Lecythis* and Polynesian chestnut have been planted on a limited scale in the past, but they remain essentially wild species.

Sapucaia. Sapucaia belongs to the Lecythidaceae, the same family as Brazil nut (*Bertholletia excelsa*). Like Brazil nut, sapucaia trees can be forest giants. Of the twenty-six species of *Lecythis* distributed from Nicaragua to the Tropic of Capricorn, many occur in Amazonia (Mori et al., 1990). Two species in particular, *L. usitata* and *L. pisonis*, produce nuts that are equal to or better than Brazil nuts. The nuts of *L. pisonis* contain 50 percent edible oil and thus are high in calories as well as protein (Cavalcante, 1988:210). Furthermore, the wood of many species of *Lecythis* is hard and durable and much desired for furniture and construction (Mors and Rizzini, 1966:130).

Sapucaia pods sometimes fall to the ground during storms, and people gather them as they are encountered on trips after game or other forest products. The fruit cases of some *Lecythis* are large enough to serve as plant pots.

The species of *Lecythis* that produce desirable nuts have one major drawback. The lid to the capsule that contains the nuts falls off to allow bats to access the fleshy arils; the nuts are thus dispersed far from the tree. The lid to Brazil nut capsules, on the other hand, is fused shut, and capsules (*ouriços*) of Brazil nut drop to the ground, where they are opened by rodents, which then disperse the nuts.

Perhaps sapucaia trees could be treated chemically to induce them to drop their capsules before the lids fall off. An attempt to establish a plantation of *L. zabucajo* on Trinidad failed because greater spear-nosed bats (*Phyllostomus hastatus*) carried off the nuts before they could be harvested (Greenhall, 1965). Alternatively, genotypes might be located that fail to release the lid—a deleterious trait in the wild, but a valuable characteristic under cultivation.

Several species of *Lecythis* have been introduced experimentally to tropical Asia (Brücher, 1989:122). Tropical Asia also has its complement of fruit bats, and nowhere has sapucaia become a viable crop at present. Sapucaia seeds can accumulate toxic levels of selenium, so if plantations become a commercial possibility in the future, care should be taken to avoid establishing the trees on selenium-rich soils. Attempts to set up plantations of *L. pisonis* will need to make provisions for pollinators, such as the carpenter bee *Xylocopa frontalis*, as well as careful selection for desirable genotypes, since the fruits are highly variable and distinct populations or even subspecies can be recognized (Mori, 1990; Mori et al. 1990:298).

Polynesian Chestnut. Polynesian chestnut (*Inocarpus fagifer*) is a tall tree often found in dense stands in Polynesia, Melanesia, Micronesia, and Indonesia, where it is a locally important food. Its origins are unclear, but the majestic tree may have been introduced to Southeast Asia from the tropical Pacific (Douglas and Hart, 1985:96). The buttress-rooted tree was used extensively by ancient Polynesian cultures (Pétard, 1986:177). The flat, oval nuts approach the size of a child's clenched fist and are boiled or roasted before eating; they taste like chestnuts, but are denser. In Polynesia and some Micronesian islands, surplus pods are buried for later consumption (Barrau, 1961:55). Inhabitants of the rugged Marquesas make a puree from the cooked seeds, mixed with coconut cream. Known as *mape* in the Society Islands, the tree produces a sap that is collected by slashing the bark or immature nuts and applied to swellings. *Mape* leaves are an ingredient in a remedy for dysentery (Pétard, 1986:178).

On Bora Bora and Moorea, Polynesian chestnut is found along creeks that have been spared by farmers. On Moorea, the tree often forms pure stands on archaeological sites (Figure 10), reminiscent of ramón (*Brosimum alicastrum*) and other economically significant trees around Mayan ruins (Gómez-Pompa, 1987). Polynesian chestnut was probably planted on a large scale around ceremonial centers and villages by ancient Polynesians, and these plantings have reverted to the wild. On Tonga, Polynesian chestnut is found in a wild state and occasionally cultivated (Yuncker, 1959:143).

As in the case of Fe'i bananas, use of Polynesian chestnut appears to be on the decline, at least in the Society Islands. Western foods and changes in farming practices, such as cultivation of export crops such as vanilla and coffee, have left little room for the multipurpose tree. The wood of Polynesian chestnut is useful for general construction, drums, and tool handles. With a modest selection program, genotypes that produce abundant nuts and grow rapidly to provide timber might help Polynesian chestnut make a comeback.

FRUITS

Tropical forests are veritable fruit baskets, with tasty and sometimes astringent fruits of all shapes and sizes, designed to attract an array of seed dispersal agents. People have already domesticated a good number of them, ranging from mango to avocado and African plum, but many other fruit-bearing species linger on the threshold of domestication. In Latin America, for example, several species of *Inga* are cultivated, while others are only collected in the wild. Here we look at four tropical fruit species: Camu camu, ucuquí, frutão, and néré.

Camu Camu.   Camu camu (*Myrciaria dubia*) inhabits lake margins in various parts of Amazonia, particularly along the Negro and Xingu rivers (Prance, 1989). The cherry-sized fruits of camu camu are harvested from canoes because the fruits mature at high water. The fruits are popular in Iquitos, Peru, where they are made into drinks and ice cream. Although camu camu is abundant near Altamira, Pará, the fruits are not marketed there (W. Balée, pers. comm.). If more information was available about this promising fruit, it might be gathered more widely and eventually cultivated.

Camu camu deserves particular attention by breeders because the fruits contain the highest known concentration of vitamin C in a plant product. Camu camu contains twice as much vitamin C as Barbados cherry (*Mal-*

FIGURE 10.2 Pure stand of Polynesian chestnut (*Inocarpus fagifer*) on an old temple site, Marae Titiroa, Punohu Valley, Moorea, Society Islands, September 1989.

pighia punicifolia), which was domesticated somewhere in northern South America, Central America, or the West Indies, and thirty times as much vitamin C as citrus (Prance, 1989). Nevertheless, enthusiasm for the camu camu has been somewhat dampened by low yields and indifferent flavor, at least for some tastes (Arkcoll and Clement, 1989). Clearly, screening of considerable germplasm will be needed to start the transformation of this wild plant into a crop. The Peruvian agricultural research system is investigating the potential of camu camu from one of its stations in Iquitos, but the budget for the entire national agricultural research system is only $5 million.

Ucuquí.   Ucuquí (*Pouteria ucuqui*) is a forest giant of the upper Rio Negro watershed of Amazonia. Its large, juicy fruits, measuring up to 13 centimeters long, are greatly appreciated by indigenous groups and peasants in its native range (Cavalcante, 1988:225; J. Murça Pires, pers. comm.). The fruits are cooked and mixed with cassava flour to make a thick porridge. No information is available on the nutritional value of the fruits or how the tree might respond to cultivation.

Frutão.   A near relative of ucuquí, frutão (*Pouteria pariry*), is also a canopy emergent and produces a heavy fruit, weighing close to a kilogram.

Frutão occurs south of the Amazon from the Juruá River in the west to the Tocantins River in the east (Cavalcante, 1988:187). Frutão drops its soccer-ball-sized fruits in the rainy season, and the fragrance is pervasive enough to attract hunters for a welcome snack. Along the Transamazon Highway near Altamira, Pará, the soft fruits are eaten fresh, mixed with sugar for a dessert, or stirred with water for a refreshing drink (N. J. H. Smith, 1982:46).

Although frutão appears to bear fruit with some regularity, it is reputed to take as long as fifty years to start producing fruit (Cavalcante, 1988:187). With further investigation, at least some genotypes might be found that are capable of producing fruit much sooner. Locating precocious frutão would be a top priority in any selection program.

Néré. In West Africa, néré, or farobier (*Parkia biglobosa*), occurs wild in the forest and is semidomesticated, similar in that regard to oil palm (*Elaeis guineensis*). Néré's sweet pulp is eaten fresh, prepared in a drink, and made into flour in its native range (Busson, 1965:276). The seeds are mashed and fermented; after it is dried, the hard, fermented meal is cut into cylinders and scraped for use as a condiment or is stirred into sauces. Néré is abundant in Guinea-Bissau, where the fruits are also fed to cattle (Gomes e Sousa, 1929). Néré typically attains 10 to 20 meters in height and could be useful in some agroforestry systems.

Other wild species of *Parkia* merit close scrutiny. Indian scientists, for example, are keen to explore the economic potential of *P. javanica* (Paroda and Mal, 1989). Indonesians plant *P. speciosa* for its seeds, which are used as a vegetable. Some primitive cultivars have been selected, but breeding is needed to develop tasty and early-maturing varieties.

PROSPECTS FOR ADOPTION

Plant domestication is an ongoing process, and more ethnobotanical work is needed to identify species being brought in from the wild. In particular, we need to know what motivates people to go to the trouble of picking up another cultivated species; after all, even tending a semiwild plant requires time and labor. A better understanding of the criteria used by people in different areas in selecting plants and crop varieties would help steer efforts to domesticate new plants more systematically.

At the same time, genetic resources need to be collected, studied, and documented. Without adequate germplasm at the outset, decades of breeding effort can be wasted.

In some cases, a new crop might do better far removed from its native area. In this manner, it would escape some of the pests and diseases that afflict the species in its native range. Farmers are just as open to adopting exotics as they are to taking on a familiar plant from the nearby forest. Except for the ornamental industry, new plant introductions have declined precipitously since the golden age of tropical botanic gardens and plant explorers during the colonial period, and the activities of the U.S. Foreign Seed and Plant Introduction Service at the beginning of this century (Whittle, 1988). Lack of space and funds has undoubtedly constrained the number of plants that scientists can evaluate for economic potential.

The ultimate success of crop candidates will hinge largely on the market. Farmers need to be in tune with the needs of the food industry for reliable, high-quality supplies (T. Thomas, 1989). If farmers are convinced that a new plant will generate cash income, they are more likely to experiment with it. New crops will have to face the harsh realities of the marketplace. A potential crop needs to be scrutinized at the farmer, industry, and consumer levels. Farmers will want to know if the crop pays, can be easily inserted into prevailing cropping patterns, and can be accommodated with existing machinery and farm implements (Meadley, 1989). The agricultural processing industry will be more willing to take on a new crop if it is cheaper than existing sources of the product, requires no new machinery designs, or fits a profitable, unfilled niche.

Of course, people will also adopt a new crop if it supplies domestic needs in ways not fully satisfied by existing cultivated species. But research on commercialization of a new crop is just as important as research on its potential uses and genetic resources (Muchaili, 1989; Wallis et al., 1989). Newcomers to the market face many obstacles. Costly promotion and advertising needed for a new crop geared for the export market or large urban centers are not likely to be financed unless the plant has outstanding prospects. Agricultural lobbies in industrial countries may rise up against a new crop from tropical countries if they perceive it as a threat to their interests.

Another hurdle for new tropical crops that aspire to penetrate markets in industrial countries is the regulatory maze of food and drug administrations. Heightened concerns about health hazards of foodstuffs require that new food products be subject to prolonged testing before they are certified as safe. The cost of satisfying such regulations can be high. A sustained commitment to a new crop is needed to navigate successfully through often tempestuous political and economic waters.

Only a small fraction of widely circulated crop candidates are likely to gain a foothold in introduced countries. Akee (*Blighia sapida*), native to the tropical forests of West Africa, is a popular fruit in its native range but has never caught on elsewhere. It is true that unripe akee fruits are poisonous, but mature and opened fruits are safe to eat. An introduced plant may linger as a botanical curiosity for decades or even centuries before becoming an important crop. Witness the centuries lag before the potato became widely cultivated for food in Europe and the sudden emergence of carambola (*Averrhoa carambola*) as an up-and-coming crop in southern Florida after the crop had been relegated to a few backyards for decades in the warmer parts of the state.

Another stumbling block to more rapid dissemination of crop candidates is the overtaxed global quarantine service. Even with support from international agencies such as the Food and Agriculture Organization, national quarantine services are generally ill-equipped to handle current workloads. As has been pointed out for many of the perennial crops examined in previous chapters, increased support for quarantine services would help streamline the international exchange of plant germplasm. For new species, quarantine services are likely to be especially conservative. When in doubt, quarantine officers are likely to destroy incoming shipments. Usually, little information is available on diseases and pests of wild plants.

Several countries in tropical forest regions plan to launch efforts to domesticate more plants. A provisional list of priorities for plant domestication in African rain forests was drawn up at a meeting of germplasm scientists at the International Institute of Tropical Agriculture (IITA), Nigeria, in 1990. Under the leadership of F. Ogbe of the Department of Botany, University of Benin, and N. Q. Ng of the genetic resources unit at IITA, priorities were established in three main product areas: fruits (species of *Irwingia*, *Treculia*, and *Dacryodes*); rattans (*Calamus decratus* and *Oncocalamus*); and fibers (*Raphia*). The All India Coordinated Programme and the National Bureau of Plant Genetic Resources in New Delhi are exploring the potential of many underutilized native plants, including potential seed and nut crops, vegetables, roots and tubers, fruits, and spices (Paroda and Mal, 1989).

In Mexico City, the botanic garden of the National Autonomous University created a unit for research on genetic resources during the early 1980s. The main task of this multidisciplinary unit is to identify and evaluate native plants with economic potential. A guiding philosophy of the genetic resource team is the importance of wedding ethnobotanical knowledge and modern science. More such ventures are needed in tropical countries.

CHAPTER 11

Conservation Strategies

A strategy needs to be developed for each of the perennial crops to save representative parts of its gene pool. Such a strategy would necessarily be based on careful field surveys and examination of herbarium records and the use of laboratory techniques, such as isozyme analysis to assess differences among populations. Ecogeographic surveys are needed to identify which populations are found within park or reserve boundaries and how safe they are. For most perennial crops, including avocado, the underpinnings for such a strategy have barely begun.

Crop genetic resources are conserved by two main means: in off-site (ex situ) collections, using frozen seed, whole plants growing in field gene banks, and, less frequently, tissue culture collections; or in their natural habitats (in situ). For crops with orthodox seeds, such as rice, wheat, or beans, which can be dried and frozen, seed gene banks can capture a relatively large segment of the gene pool (Plucknett et al., 1987). For crops with recalcitrant seeds, such as many perennial tree crops in the tropics, which cannot be dried and frozen without killing the embryos, seed gene banks are not an option. Field gene banks can safeguard some, but by no means all the genetic variation of tropical perennial crops.

Strategic plans for conserving genetic resources are needed for all perennial crops. The task of devising such plans for collecting, maintaining, and evaluating the genetic resources of tropical perennial crops is truly daunting. Germplasm collecting missions will have to be organized over vast, often difficult terrain. Variation in plant gene pools will need to be mapped, and priority areas for conservation identified. Then various options for in situ conservation will need to be explored; compromise will usually be necessary between an ideal area for conservation and what may be left, or allowed, for preservation. The locations and scope of field gene banks will also need to be discussed by the interested parties.

Several models or hypothetical cases illustrate the complexity of devising adequate conservation strategies for the numerous tropical perennial crops and their near relatives.

Model 1. A gene pool naturally distributed over a vast area, such as 150 million hectares, in which the patterns of diversity are poorly understood.

Species that can be conserved as seed. Samples will need to be collected across a wide range of environments and stored in ex situ gene banks. In situ conservation should be carried out as a complementary strategy.

Species with unorthodox seed. (1) Species for which no in vitro techniques exist will require collecting and storage in field gene banks as well as analyses to designate in situ reserves. (2) species for which in vitro techniques are available will require collecting and a three-pronged conservation strategy that encompasses germplasm storage in tissue cultures, field gene banks, and in situ reserves.

Model 2. A gene pool naturally distributed over very restricted areas. Various scenarios are possible, as in Model 1, but targets can be set more easily and action taken more quickly. Gene pools may be endangered by encroaching human activities, and delays in formulating conservation strategies could be costly. On the other hand, species with restricted gene pools in tropical forests may be "minor" crops and therefore classified as low priority.

Model 3. Species for which the gene pool is largely cultivated, and the wild relatives are only distantly related. This model requires different parameters for collecting, especially when the cultivars are primitive but exhibit wide adaptation to disturbed sites. Much of the germplasm will be collected using a coarse grid sampling technique.

EX SITU CONSERVATION

For the most part, gene banks for tropical perennial crops contain materials of current interest to users and some obsolete cultivars. A few contain a scattering of wild material and some accessions of near relatives. Field gene banks for perennial crops are generally more working collections than attempts to conserve the broad range of genetic variation of the crop and its close relatives. For many tropical perennial crops, particularly the minor fruits, no attempts have been made to set up gene banks.

Field Gene Banks

To be effective, field gene banks need to be tied more securely to ongoing crop improvement efforts. A number of gene banks currently linger in forgotten corners of research stations because the principal staff member who once took an interest in the collection has left. Unless field gene banks are being used, they can become victims of space constraints at field stations.

Given space limitations, gene banks are unlikely to represent the entire variation of a crop's gene pool. But as in the case of some oil palm and cacao collections, they can include at least a range of the crop's variability. Ideally, field gene banks contain accessions from various parts of the crop's range in the wild as well as under cultivation. The samples should be taken from a variety of environments whenever possible, including different soils, drainage conditions, altitudes, and climates. Such materials could prove useful to breeders wishing to adapt crops to new areas or to changing conditions.

Few genetic resource collections exist for tropical perennial crops. For those crops that do have field or in vitro germplasm collections, some could be broadened to include more of the genetic variation in domestic and wild gene pools. More funds will be needed, however, to establish genetic resource collections or upgrade working collections. A typical genetic resource collection might take $2 million to set up and an annual budget of $1 million to maintain. These figures do not cover the fieldwork needed to devise initial strategic plans, obtain germplasm samples, train personnel, evaluate the materials, or conduct germplasm-enhancement (prebreeding) trials.

Large quantities of clones of superior selections of many of the tree crops will be needed to quickly reach a significant number of farmers. Field gene banks thus need to strike a balance between heterozygous genotypes that might be appropriate for breeding in the future and potential "winners" that could prove suitable for cloning in the near term.

Considerable research is still needed to orient germplasm collecting efforts. Information on genetic variation of most perennial crops is sketchy. More ecogeographic surveys are needed to map variation and to guide germplasm collectors.

Space limitations and operating costs will prevent field gene banks from becoming the main conservation strategy for most perennial crops in the tropics and subtropics. Those maintained by agricultural research stations and certain botanic gardens are essential for breeding efforts and can be cost-effective (N. J. H. Smith, 1986). Field gene banks are thus an important complement to breeding and conservation efforts. By underscoring the usefulness of novel germplasm, field gene banks can build support for larger efforts to conserve plant genetic resources.

TISSUE CULTURE COLLECTIONS

Germplasm collections held as tissue cultures can overcome some of the constraints of field gene banks, such as space limitations and vulnerability to storms and disease epidemics, but in vitro conservation poses a new set of problems. First, tissue culture procedures have not yet been tailored to conserve most perennial crops. Second, somaclonal variation can occur in test tubes and flasks, thereby altering the qualities of the accession. Such in vitro genetic changes can be useful to breeders seeking novel traits, but they undercut the task of germplasm curators. Third, as in the case of field gene banks, tissue cultures are removed from natural selection pressures and are thus no longer integral parts of ecosystems. When living materials are removed from tropical forests, one of the world's most active theaters of evolution, evolutionary trajectories are deflected.

IN SITU CONSERVATION

The importance of in situ conservation of the genetic diversity of tropical perennial crops and their near relatives is now widely recognized. In situ conservation can provide a staging ground for the selection of tree species or genotypes for multiplication and further testing, and if warranted, duplication in ex situ collections (Stern and Roche, 1974). But how germplasm can be best maintained in place raises a number of theoretical and practical issues. Several strategies can be pursued to conserve plant germplasm under natural conditions, but each has its limitations. Ultimately, a suite of tactics is needed that maintains crop gene pools under pristine and managed conditions and at the same time accommodates the understandable desire of nations to develop their tropical forests.

Wild populations of crops and their near relatives are found in two broad categories of in situ reserves: those designed to maintain conditions as pristine as possible and those that permit extraction or even clearing for agricultural and other purposes. National parks, biological reserves, and specialized gene parks fall into the first category, whereas national forests, Indian reserves, and extractive reserves allow for a range of economic activities such as harvesting of forest products. Another possibility is to employ farmers as custodians of traditional varieties and selections in their fields and backyards.

NATIONAL PARKS AND BIOLOGICAL RESERVES

An impressive number of national parks have been created in countries with tropical forests within the last two decades. In some developing countries, the proportion of national territory covered by parks exceeds that of many industrial countries. Close to one-tenth of Ecuador, Costa Rica, and Sri Lanka, and a little over one-quarter of Belize are in parklands (Ledec and Goodland, 1988:78; Sun, 1988; Tangley, 1988). National parks are designed to minimize disturbance of wildlands, an important advantage for wild populations and near relatives of many tropical perennial crops.

Biological reserves are generally set up to safeguard critical ecosystems and to provide opportunities for ecological research. Such reserves tend to be smaller, although reserves within the Man and the Biosphere Program can be substantial. National parks and biological reserves play an important role in maintaining the integrity of some tropical ecosystems and in protecting water catchments for urban and rural people.

In spite of the impressive and widespread effort to create national parks and biological reserves in developing countries, such parks and reserves cover only a small fraction of tropical forest environments. A little over 42 million hectares of tropical forests are nominally protected worldwide. Although this area is equivalent to the size of California, it is far from adequate (McNeely et al., 1987). In the case of Amazonia, parks and reserves account for less than 4 percent of the basin (Eden, 1990). If forests in the Amazon basin were reduced to only those areas within parks and biological reserves, many plants and animals would become extinct: some 69 percent of bird species would likely disappear (Simberloff, 1986), and some 66 percent of plant species in the New World tropics would vanish if the tropical forests shrink to existing parks and reserves.

More parks and reserves are on the drawing board or are being proposed, but they are unlikely to cover a substantial proportion of Amazonia or other tropical forest regions. In January 1990, specialists from many scientific fields attended a workshop in Manaus, Brazil, and identified priority areas for conservation in the Amazon basin (Prance, 1990). With species diversity as the main criterion, workshop participants suggested that 60 percent of the region should be set aside for conservation. In addition, the scientists cautioned that areas outside the zones recommended for conservation should be studied in detail in the event that they contain endemic species or valuable gene pools. The workshop proceedings are providing valuable information for park and reserve planning in Amazonia, but it is unlikely that over half of the basin will ever be designated as parks and biological reserves.

Another drawback to relying on national forests or biological reserves alone to safeguard the gene pools of crops and their near relatives is that they are frequently violated by loggers, gold miners, cattle ranchers, and farmers. In some cases, setting up a park is an invitation to squatters, land speculators, and loggers to come in and establish a homestead or extract resources. Nationalization of forests in Thailand, Nepal, and Niger, for example, has accelerated environmental degradation (Berkes et al., 1989).

In Amazonia, virtually all parks and biological reserves of any size have been compromised in varying degrees by illegal activities, particularly logging and gold mining.

Much greater provision needs to be made for safeguarding and properly managing parks and reserves. Most nominally protected areas are woefully understaffed. Unfortunately, few strong constituencies are in place in developing countries to adequately defend parks (Schwartzman, 1989). In the late 1980s, the Brazilian forestry and environment agency had only 800 wardens to patrol and manage twenty-eight national parks and fifteen biological reserves. Budgets for park maintenance need to be substantially increased, a difficult task in view of the economic problems and already tight governmental budgets of most nations with tropical forests.

While many tropical forests and biological reserves contain wild populations or near relatives of perennial crops, the locations of such nominally protected areas are usually decided on grounds other than conserving parts of the gene pools of cultivated plants. In most cases, botanical inventories have not been made, or only partial surveys have been conducted. Specialized gene parks also need to be created that have as their primary function the protection of important plant genetic resources.

Gene Parks

In rare instances, parks or reserves have been set up specifically to safeguard near relatives of certain crops, such as the citrus sanctuary in the Garo Hills in Meghalaya, northeastern India. This 10,000-acre reserve protects populations of wild orange (*Citrus indica*). A small, but valuable population of perennial teosinte (*Zea diploperennis*), a near relative of maize, was a major factor in the Mexican government's decision to create the Sierra de Manantlán Biosphere Reserve in 1987. Other efforts to set aside areas for endangered populations of crops and their near relatives are warranted.

Gene parks will need to "piggyback" on other motives if substantial forest areas containing wild populations of crops and their near relatives are to be maintained. Given pressures to open up land to settlement and other development activities, it may be hard to set aside sizable areas just for wild populations of crops and their close cousins. There are few "pandas" in the plant world to ignite widespread public and governmental support for saving areas from development. If, on the other hand, wild gene pools of economic plants can be saved alongside rare monkey or bird populations, for example, multiple conservation objectives can be obtained.

Crop genetic resources need to be cast as important assets for agricultural development. In that regard, the economic or utilitarian card can be played more strongly when proposing gene parks. If the short- as well as long-term benefits of maintaining crop genetic resources in the wild are better explained to people around the world, then governments are likely to be more convinced of the need to set up gene sanctuaries.

Gene parks need curators and associated researchers. Curators and collaborating scientists who work with germplasm that is still an integral part of an ecosystem will have an opportunity to study phenological cycles, breeding systems, and the role of pollinators and dispersal agents, and to make observations on apparent resistance to diseases and pests. Conservation teams will need to tap a wide range of disciplines to adequately monitor germplasm reserves and better understand ecological relationships (Brush, 1989). Gene parks could thus fulfil some of the functions of a biological or ecological reserve.

Gene parks should be as large as possible to increase the likelihood that they will contain "keystone" species that perform vital functions upon which the fabric of the community depends. If genetic reserves are too small, pollinators or dispersal agents may not survive because of insufficient food or other resources (Hubbell and Foster, 1986). Euglossine bees, important pollinators for a number of tropical forest trees in Amazonia, were adversely affected when isolated in forest islands following clearing near Manaus (Lovejoy et al., 1986). Obligate frugivores, some of which are involved in seed dispersal in tropical forests of Central and South America, Malaysia, and Borneo migrate in response to periodic food shortages (Terborgh, 1986). In montane forests, such migration is usually up and down mountainsides; in lowland tropical forests, migration patterns are poorly understood. If a gene park is sufficiently large, it is more likely to compensate for the patchiness of many food resources of pollinators and plant dispersal agents.

Evidence is accumulating of the dangers of losing species if forest shrinks to small patches. On 17-square-kilometer Barro Colorado Island surrounded by the artificially created Gatun Lake, for example, several forest birds have disappeared. A similar pattern has been observed in 1- and 10-hectare patches of forest north of Manaus in the Brazilian Amazon (Simberloff, 1986). The more plant and animal species are lost, the greater the chance that the long-term integrity of the reserve will be undermined.

If wild populations of a crop are found over extensive areas, several sizable, widely separated reserves rather than a single megapark would be preferable. The equilibrium theory of island biogeography argues that the larger the protected area, the more likely it is to contain more species and to retain them. But size alone is not the only consideration (Simberloff, 1986). Several forested areas that contain diverse habitats may be more preferable than a larger, more uniform nature reserve. A constellation of gene parks with diverse terrain and plant communities is likely to contain more interesting variants than a large but ecologically simpler reserve.

The degree of ecological heterogeneity within a reserve is thus an important consideration as well as the size and number of reserves. Ideally, a reserve should encompass a range of habitats and plant communities in order to increase the chances of variation within a species and the number of near relatives of a crop (J. T. Williams, 1988). For example, four species of passionfruit relatives in the *Passiflora vitifolia* complex occupy distinct sites ranging from white-sand forest to alluvial areas all within close proximity to Iquitos in the Peruvian Amazon (Gentry, 1989).

Another advantage of more than one sizable reserve for preserving the gene pool of a plant or animal is that fragmented populations are less prone to catastrophic destruction from such threats as disease epidemics or fire (Pimm and Gilpin, 1989). A single reserve can accomplish the same objective if it is large enough and contains some geographical barriers, such as mountain passes or major rivers. Several reserves would be preferable for gene pools targeted for protection as a hedge against any one protected area suffering from some unforeseen disaster.

The danger of opting for several smaller reserves, rather than a single one, is that they end up being too small. Given the inadequacy of the knowledge base, reserves should always be as large as possible. In the case of species that are pollinated and dispersed by wind, smaller reserves may not be as critical as for plants that require animals for fertilization and seed dissemination.

The optimum sizes and number of genetic reserves would depend on the amount of variation present in the species in question and how many of its near relatives should be included. For some crops with many near relatives, such as avocado and passionfruit, how many near relatives should be protected is a difficult question. Very little scientific basis exists to adopt any triage for the near relatives of perennial crops. Emphasis could be placed on near relatives that can be most easily crossed with the crop, but biotechnologies are constantly enlarging the gene pool. It is virtually impossible to predict which wild relatives of a crop might contain some useful genes in the future.

As formerly contiguous tropical rain forests are broken up into insular patches, some of them better protected than others, gene flow will be interrupted between populations. As forest islands become more isolated, new species may eventually arise. Although the immediate concern is safeguarding interesting variation in present populations, such variation will likely change over time, especially in parks and reserves. Such changes, resulting from genetic drift and mutations, may take millennia to manifest themselves, but one will not be preserving the status quo in many parks and reserves.

Minimum population size is an interlinked issue with dimensions of preserved areas. The notion that 500 individuals is enough to ensure the long-term survival of a species may not apply to all plants and animals (Lande, 1988). If a population dips below a critical density or size, it may be doomed. For most situations, at least 1,000 individuals are needed to safeguard a population against environmental uncertainty such as inclement weather, changes in predator populations, or catastrophic events such as fire or severe drought (Shaffer, 1987). For some species, tens of thousands of individuals may need to be managed in a reserve to prevent loss of viability and to reduce vulnerability to sudden changes in the environment.

Ideally, extensive research should be conducted to lay the groundwork for assessing the size and management plans for a gene park. In practice, though, such parks may have to be set up hastily, before development forces close in. A haphazard approach is often taken by default, but can restrict the types of habitats available for conservation. By the time conservation measures are taken, trees on more fertile, bottomlands may have been logged and cleared and forest relegated to steep slopes. Germplasm cleared from more optimal sites for agriculture and urban growth may contain genes of more immediate interest for silviculture and crop improvement programs (Ledig, 1988).

Gene parks, biological reserves, and national parks and other conservation units designed to keep out human influences as much as possible will play a useful role in saving genetic resources of plants and animals. But it would be risky to entrust the genetic resources of crops and livestock to such nature islands alone. Such nominally protected areas are too small and scattered in many cases to encompass sufficient variation in wild populations and to prevent extinctions. The majority of nature reserves worldwide are under 10,000 square kilometers, too small by an order of magnitude for many of the larger species, such as top mammal predators (Shaffer, 1987; Terborgh, 1986). Furthermore, the long-term fate of parks and reserves is by no means assured. If local populations do not appreciate the value of such protected areas, or are denied any benefits that might accrue from safeguarding wild areas, then they may invade them, as they have already done in numerous cases.

Conservation of crop genetic resources is thus embedded in the larger issue of regional development and distribution of benefits among the population (Ashton, 1988; McNeely et al., 1987). Local education efforts combined with recycling in tangible ways some of the payoffs from nature reserves would help their long-term survival. Parks and reserves are on a more secure footing if local inhabitants obtain rewards from conservation and participate in planning and managing such protected areas (Leader-Williams and Albon, 1988). More productive and sustainable agricultural systems would relieve the pressure to clear more forest and would help raise local living conditions. What happens outside parks and biological reserves will thus largely determine their fate.

What types of agricultural, silvicultural, and forest management systems evolve in the interstices between scattered parks will be crucial to the viability of gene pools of many tropical perennial crops. A detailed discussion of farming and pasture recuperation that can relieve pressure on Amazonian forests can be found in Alvim 1977, Eden 1990, and Serrão 1986. Here we focus on the role of national forests, Indian reserves, and extractive reserves in conserving gene pools of perennial crops and their near relatives.

NATIONAL FORESTS

For the most part, logging in tropical forests is from unmanaged, natural stands. Since the early 1980s, timber extraction from tropical forests has increased dramatically, particularly in Southeast Asia. Between 1975 and 1984, log production nearly quadrupled in the Brazilian Amazon (A. B. Anderson, 1987). Logging rates in most tropical forest areas are unsustainable, and much attention is currently being focused on silviculture and management of tropical forests for timber production.

Only a few national forests have been established in developing countries. In the Brazilian Amazon, only three such forestry reserves have been established: Tapajós (600,000 hectares), Jari (227,126 hectares), and Caxiuanã (200,000 hectares). These national forests are in Pará and cover only a minute fraction of Amazonia. Some cattle ranches in the Brazilian Amazon are close to the size of the Tapajós National Forest.

National forests in tropical forest regions are still largely experimental. Much research remains to be done to tailor cutting cycles to different forest environments. Yields from managed forests in the tropics are low and generally do not compete with the "creaming" activities of predatory loggers. If markets are developed for some of the lesser-known tropical hardwoods and techniques are worked out for managing forests on a sustainable basis, more national forests are likely to be established.

INDIAN RESERVES

The lands of indigenous peoples, if their boundaries are respected, may prove one of the most important areas for conserving the germplasm of perennial crops and their close relatives. Indigenous peoples have long harvested a wide array of products from the forest, ranging from building materials to medicinal plants, so they recognize the importance of maintaining substantial forested areas. Aboriginal groups are also especially knowledgeable about useful plants and animals; cultural diversity and ecological heterogeneity are thus both important in maintaining biodiversity (Ledec and Goodland, 1988:77).

Colombia is at the forefront in establishing sizable Indian reserves in tropical forests. A total of fifty indigenous groups have been given legal rights to 18 million hectares in the Colombian Amazon (Bunyard, 1989). At present, some 70,000 Indians control 41 percent of the Colombian Amazon, a tribute to lobbying efforts of groups promoting indigenous rights and nature conservation in Colombia. Northwestern Amazonia contains gene pools of *Hevea* and many locally important fruit trees.

Colombia's bold moves to protect indigenous lands is commendable, but is not likely to be emulated elsewhere, at least not on that enormous scale. Pressures to develop the Amazon in Colombia are not nearly as great as those in Brazil or Peru, for example. Apart from the 9-million-hectare Yanomamo reserve, created in November 1991, Indian reserves in the Brazilian Amazon are widely scattered and generally in the order of a few hundred thousand hectares. Already, rumblings can be heard from some quarters in Brazil that Indians are the largest landholders in the Amazon. While it can certainly be argued that indigenous peoples have every right to substantial tracts of land to maintain their livelihoods, growing streams of disenfranchised peasants and entrepreneurs are closing in on some Indian reserves. As the frontier begins to close in Amazonia, the plight of landless families both within Amazonia and in other regions may constrain the size of future indigenous reserves and may further threaten the integrity of existing ones.

Several indigenous reserves in the Brazilian Amazon contain important genetic reservoirs for crop improvement. The 439,000-hectare Xikrin Indian Reserve near the Serra dos Carajás, for example, contains numerous groves of Brazil nut. To the southwest, the even larger Kayapó reserve straddling the Xingu River, and to the north, the Parakanã reserve, also contain large sections of Brazil nut's gene pool. In southwestern Amazonia, several indigenous reserves in Rondônia, such as Lourdes and Guaporé, undoubtedly contain populations of rubber and cacao, as well as near relatives of those important cash crops.

The setting aside of land for indigenous peoples is a sensitive issue with many governments because of economic pressures and sovereignty issues. In Sabah, for example, indigenous groups have opposed some logging concessions. In the Kivu region of Zaire, tribal peoples sometimes contest the jurisdiction of the national government in issuing land permits. In spite of the delicate and often difficult issue of sorting out land titles, indigenous reserves are likely to grow in number and provide an important reservoir for plant and animal genetic resources.

EXTRACTIVE RESERVES

In the late 1980s, several extractive reserves were established in Acre, Brazil, to secure a livelihood for rubber tappers. If indigenous groups have the right to large blocks of forest to manage for their own ends, it is argued, then the much more numerous communities of peasants also deserve to have secure title to their lands.

Extractive reserves have elicited considerable interest both within Brazil and abroad. The main idea behind extractive reserves is that local communities own and control the harvesting of forest products. Chico Mendes galvanized the rubber tappers and initiated the drive to set up extractive reserves. Mendes was attempting to organize rubber tappers to defend the forest against encroaching development, particularly cattle ranches, until he was killed in December 1988.

Environmental and labor groups coalesced around the charismatic figure of Chico Mendes. On the basis of domestic and international pressure, the Brazilian government has created several extractive reserves and several more are in the planning stage. The Chico Mendes Reserve is located at Xapuri, Acre, and covers 900,000 hectares. The Reserva Extrativista do Alto Juruá, created on 14 March 1990 during President Sarney's last day in office, embraces 506,186 hectares and contains some 500 families. The vast size of extractive reserves means that they could prove most helpful in conserving plant and animal genetic resources.

The rubber tappers' movement was envisaged as a dual-purpose cause to improve regional living conditions and to preserve the forest. Although Mendes has been portrayed as a savior of Amazonian nature, it should be remembered that his struggle was primarily for the social rights of a poor and relatively disenfranchised group of people.

While extractive reserves have some appealing aspects, their ability to wrest rubber tappers from poverty and to safeguard the forest from wanton destruction is by no means assured (Homma 1989a, b). Dependence on gathering forest products alone is unlikely to raise and sustain living conditions for inhabitants of the humid tropics (Lavelle, 1987). If extractive reserves fail to improve the livelihoods of those living within their boundaries, then the impetus to set up more such reserves will slacken, and existing extractive reserves will be increasingly vulnerable to the axe.

Despite claims that extractive reserves are currently economically competitive with other land uses in Amazonia (Gradwohl and Greenberg, 1988:150), the economics of extractive reserves needs further work. The much-discussed finding that one hectare of forest near the village of Mishana on the Nanay River in the Peruvian Amazon can generate almost $650 a year from the sale of fruits, nuts, and latex is promising, but should be regarded with caution (Peters et al., 1989). The Mishana study suggests that sustainable exploitation of nontimber products from rain forests can produce three times the income derived from converting the forest to other uses, such as agriculture. But it is not clear how representative the Mishana study is of forest areas in Amazonia; the ecological heterogeneity of the region makes it difficult to extrapolate findings from one area to another. Also, it is doubtful that most patches of Amazonian rain forest can generate three times as much income as alternate uses; if that were so, then forests would surely be less threatened. The potential economic values of any area slated for an extractive reserve need to be calculated in advance in view of variations in species composition and diversity (Prance, 1989).

In some areas, extractive reserves may generate reasonably high income levels. In Ecuador, nuts from the forest tagua palm (*Phytelephas aequatorialis*) are to adorn sportswear in the United States. Under a licensing agreement with Washington-based Conservation International, fees will be used to support conservation and sustainable development of Ecuador's rain forests. In the Amazon estuary, extractive products from the floodplain forest, such as fruits from açaí (*Euterpe oleracea*) and buriti (*Mauritia flexuosa*), can provide appreciable income for local residents. In the region of Iquitos in the Peruvian Amazon, at least 120 fruit species are harvested from the forest (Vasquez and Gentry, 1989).

Still, extractive reserves might better be regarded as valuable supplements to the diet and income of people living in them. To significantly raise living standards, communities will have to undertake other activities, such as farming. Cultivation of crops will mean clearing parts of extractive reserves. To prohibit any forest clearing will condemn rubber tappers to poverty, at least for the foreseeable future.

In extractive reserves in Acre, the main economic product appears to be rubber, which is hardly likely to elevate communities to new levels of prosperity. Three-quarters of Brazil's rubber consumption is met by imports from Southeast Asia. Rubber tapped from forests cannot compete in the marketplace with rubber derived from plantations in Southeast Asia, a lesson learned in the 1920s. Natural rubber produced in Brazil has been subsidized at a rate three times the world price (Homma, 1989a). As the Brazilian government relaxes duties on a range of imported items, including rubber from Malaysia, the economic viability of tapping rubber from wild or plantation trees in Brazil will likely diminish.

The relative contribution of forest products to Amazonia's economy peaked in 1910 during the rubber boom, but has since been on a long historical decline. By 1965, the value of agricultural production surpassed forest products in the Brazilian Amazon (Homma, 1989b). Rubber tapping is unlikely to return as a mainstay to the regional economy.

Another concern about extractive reserves is their alleged role in forest conservation. Given hard economic choices, communities living in extractive reserves may opt to cut down substantial tracts of forest for pasture, plantations, or field crops. Discussions on how to manage extractive reserves could heighten tensions within communities. It is unlikely that most rubber tappers would want to spend the rest of their lives collecting latex.

Certainly, most of the younger generation will rather engage in other activities, particularly in towns and cities.

One way to strengthen the contribution of extractive reserves to the income of local people is to develop new markets for forest products. Cultural Survival, a nonprofit human rights organization based in Cambridge, Massachusetts, has established a nonprofit trading company that will help find markets for Amazonian forest products that have been collected in a sustainable manner. Through the efforts of Cultural Survival, one ice cream manufacturer in the United States, Community Products, is selling Rainforest Crunch, a blend of Brazil and cashew nuts. Community Products has agreed to repatriate 40 percent of their profits from Rainforest Crunch to Brazil nut gatherers. Body Shop, a British cosmetics firm, is testing oils and essences from Amazonian forests for its impressive line of creams and lotions (Pearce, 1990). While such efforts and developments are commendable, the dollar volume of this business remains miniscule. In 1989, for example, the production capacity of Rainforest Crunch was only 680 kilograms a day (Anonymous, 1989). To make a larger impact, rainforest products need to conquer shelf space in major supermarkets, restaurants, and fast-food chains.

Continued efforts to find expanded or new markets for rainforest products is warranted, but reliance on extractive activities alone to provide for the material needs of rainforest dwellers and to safeguard the forest is a shaky proposition. Moves to circumvent middlemen in marketing products from extractive reserves in order to increase profits at the local level could backfire. Finally, any plant product that becomes important is usually either domesticated or its relevant constituents are synthesized in a laboratory. Plants are domesticated to improve yields and other qualities, and forests would have to be cleared to cultivate them, a process that has occurred with the rubber tree and is under way with Brazil nut. Innovative agroforestry practices are needed to minimize clearing on extractive reserves.

If extractive reserves are cast as multipurpose resources, including the freedom to farm, set up plantations, and even to rear livestock, then they are more likely to contribute to the region's welfare. Each extractive reserve will have a different mix and density of economic plant species. Medicinal plants, potential new crop plants, and genes from wild populations of existing crops, may ultimately prove more valuable than forest conversion. Market prices for forest products fluctuate, as in the case of farming, silviculture, and ranching, thereby making it imperative to develop markets for a range of forest products as well as farm activities. Sustained harvesting of forest products will need more botanical and faunal inventories coupled with natural history studies. Such studies should combine scientific expertise and folk knowledge and involve locals in the research effort.

If they are properly managed and if provision is made for agriculture, silviculture, and some pasture development, extractive reserves have the potential to spare sizable forest stands. In spite of some opposition to extractive reserves, it will be easier to sell the idea of setting aside more extractive reserves than locking up vast areas of forest for nature preserves. If extractive reserves foster a range of economic activities, not just collecting forest products, they can help defuse social tensions and improve regional living conditions. If material standards of living rise for a broad segment of the population, the future of tropical forests will be more secure.

Managed forest environments, because they allow economic activities, are likely to be far more common than pristine reserves. From the conservation standpoint, though, national forests, Indian reserves, and extractive reserves pose some formidable challenges. The degree of forest alteration may be difficult to control, and ecological disruption may ultimately become so pronounced that the viability of wild populations of perennial crops, as well as other plant species, suffers.

National forests, Indian reserves, and extractive reserves are also vulnerable to encroachment by outsiders. Many Indian reserves have been violated by loggers and gold miners. Even though such activities are more likely to be spotted and reported than they would in national parks, owners of extractive or indigenous reserves may be virtually powerless to evict intruders. In some cases, members of indigenous groups have welcomed loggers as a means to obtain cash income, and at least some Yanomamo are keen to learn from gold miners how to seek the precious metal.

Another concern with managed forest environments with regard to conservation of plant genetic resources is that some activities may be detrimental to dispersal agents. Hunters in Indian reserves and extractive reserves can drastically reduce birds and mammals that feed on fruits and later defecate some of the seeds. In tropical America, large game birds such as curassows and guans (Cracidae) ingest some seeds and pass them unharmed, while much-procured agoutis (*Dasyprocta* spp.) bury nuts, such as Brazil nut, some of which later sprout. Over time, heavy hunting pressure will likely alter the composition of forest communities in favor of wind-dispersed species (Redford, 1992).

On the other hand, forest disturbance will help secondary growth species to flourish. Such communities often contain spontaneous populations of perennial crops, such as guava, papaya, and African oil palm, as well as near relatives of banana and passionfruit. Thus extractive reserves and Indian reserves should not necessarily be viewed as unnatural. People have been interacting with tropical forests for millennia and have altered the composition of plant communities in many areas. Indian reserves and extractive reserves may provide an opportunity to provide continuity to such patchy and ephemeral disturbances in the forest.

DOORYARD GARDENS AND FIELDS

Traditional varieties and highly variable seedlings of perennial crops are found in backyards and orchards of farmers throughout the tropics. Farmer-selected varieties and seedlings contain a rich gene pool of potential value for breeders and agricultural development (Oldfield and Alcorn, 1987). Genetic erosion is not yet a major problem for many perennial crops in the tropics, since most crops do not have breeding programs that have launched modern, high-yielding varieties to displace the traditional material. Nevertheless, genetic resources are being lost in some cases, particularly with the disappearance of tribal cultures and shifts in farming patterns.

For certain crops, genetic erosion of dooryard gardens and groves is likely to accelerate. Traditional avocados in some parts of Mexico, for example, will continue to be topworked to 'Hass' and other commercially desirable selections. Florida selections of mango, such as 'Keitt' and 'Haden', will continue spreading in such countries as Brazil and Peru. For most tropical perennial crops, though, improvement work has barely begun, and highly variable genotypes can still be found in gardens throughout the tropics.

Several scientists from a variety of disciplines have urged that traditional varieties be conserved in situ by custodial farmers (Altieri and Merrick, 1988). Wilkes (1989) urged the setting aside of evolutionary gardens, where crops hybridize spontaneously with wild populations and their near relatives. If a community wishes to voluntarily maintain old traditions and eschew modern varieties, they should certainly be allowed to do so. But few farming communities hermetically seal themselves from agricultural change and the chance to adopt new technologies.

A number of practical problems will impede the broad-scale application of these in situ strategies for conserving cultivated genotypes. In many developing countries, increasing production of food and cash crops is of far greater concern than holding on to obsolete planting materials. Proposals for in situ conservation of crop varieties include setting aside zones where farmers can plant only primitive cultivars. Such a tactic might increase the gulf between poor and more prosperous farmers.

Subsidies are proposed to compensate rural communities for not adopting modern varieties. Agricultural valleys zoned for museum farms might become tourist attractions, thereby helping to defray their cost (Wilkes, 1989). A number of difficult issues would have to be worked out including which parts of the countryside would be designated for in situ conservation, how subsidies would be calculated for different crops, and the logistics of administering the subsidies. Governments would have to find the money to pay farmers not to adopt modern varieties. Finally, how committed would farmers be to maintaining varieties they no longer deem useful?

A major task for in situ conservation of traditional planting material would be deciding where such areas should be designated. Some farmers might object to being held back for the sake of conservation; presumably they might be asked to relocate. Agricultural systems are highly dynamic and are constantly evolving. Many traditional varieties were abandoned long before scientific plant breeding and modern nurseries arose.

Preventing the release of modern varieties in certain areas is not the way to safeguard older materials. If a farmer wishes to topwork a traditional selection with a higher-yielding, more commercially acceptable variety then he or she should be free to do so. Farmers should not be forced to retain obsolete materials, nor should they be be made to adopt newer ones. Decisions on varietal mixes maintained by farmers will be determined largely by cultural needs and market forces.

Even in areas benefiting from the release of modern cultivars, the erosion of primitive forms is not always pronounced. Modern varieties are often seen as an adjunct to traditional farming practices in developing countries (Brush, 1991). Farmers often adopt new, high-yielding varieties for cash income and continue to cultivate traditional varieties for home consumption. Few hard data are available on the adoption rates of modern varieties of perennial crops in the tropics and their impact on the mix of traditional cultivars.

The continued loss of tribal cultures represents a serious loss of knowledge about plant resources and traditional selections. The process of assimilation with national society often results in depopulation and disappearance of villages and their fields, particularly in tropical America. Even when tribal numbers remain relatively stable after contact, some of their cultivated plants and varieties may be abandoned. Very little work is being done to rescue such carefully selected crops before they disappear. It is imperative, therefore, that the integrity of Indian reserves be respected so that indigenous societies can evolve at their own pace and share mutually beneficial knowledge and technologies with the national society.

CORPORATE HOLDINGS

At first glance it might seem that private corporations or parastatal enterprises would be inimical to the survival of forest. After all, companies and other large, market-oriented ventures remove forest to make space for plantations or to gain access to subterranean minerals. But companies are motivated to keep intruders out of their lands or concessions and have the means to enforce their property rights.

In the case of the Brazilian Amazon, large concessions operated by corporations contain some of the best-protected stretches of forest. Only 10 percent of Jari's 1.6 million hectares is slated for plantations of exotic eucalyptus, Caribbean pine, and Gmelina to produce pulp. The consortium of companies that operates Jari does not permit mining, hunting, or logging on their property. Regular jaguar sightings at Jari are one indication that the forest is left relatively intact.

The Jari property straddling the boundary between Pará and Amapá contains rich stands of Brazil nut as well as many timber species. The company has established a network of five genetic reserves totaling 7,600 hectares on its property, where phenological studies are under way on native forest species. The Jari operation is clearly concerned about environmental conservation and sees the genetic reserves as an investment in potential new economic plant species.

At Carajás, the Companhia Vale do Rio Doce (CVRD), a joint venture between private industry and the Brazilian government, has a 411,000-hectare concession to exploit iron ore and manganese, among other minerals. CVRD not only regularly patrols its undulating concession, which involves minimal forest disturbance, but has been given authority to help protect several contiguous biological and forestry reserves. The Xikrin Indian reserve shares a boundary with CVRD's concession, and will thus benefit from the company's strict policy of expelling intruders. The total area thus protected by CVRD's presence at Carajás is 1,164,600 hectares.

Ultimately, governments should effectively manage and protect reserves and parks. How long it will take for developing countries to obtain the necessary resources to fully implement conservation plans is difficult to tell. Progress will be spotty; in some regions large tracts of forests will be saved, while in others, forest destruction will likely continue unabated.

DEBT-FOR-NATURE SWAPS

One way to reinforce conservation efforts in tropical countries is to negotiate the reduction of part of their external debts in exchange for the establishment of parks and reserves. Proposed in the early 1980s by the Fund for Private Assistance in International Development (PAID) and Lovejoy (1984), the novel idea was first implemented in 1987. The notion that debt-for-nature swaps can help save vast areas of tropical forest has subsequently attracted a lot of attention in the press and scientific literature (Fuller, 1989; Neumann and Machlis, 1989).

A small portion of the external debts of Bolivia, Costa Rica, the Dominican Republic, Ecuador, the Malagasy Republic, the Philippines, and Zambia has been purchased at a steep discount to acquire tropical forest reserves. As of August 1990, $97 million of debt had been purchased at a cost of $16 million. The external debt of Latin America alone stands at over $300 billion.

A total of 1.5 million hectares of forest and grassland has been set aside in the Beni region of Bolivia with help of the San Francisco–based Frank Weeden Foundation (Walsh, 1987). Washington-based Conservation International was instrumental in packaging this deal, which entailed the purchase of $650,000 of debt for $100,000 (Gradwohl and Greenberg, 1988:68). The World Wildlife Fund in Washington, D.C., consummated the debt-for-nature agreement with Ecuador, which involved raising $354,500 to purchase $1 million of debt (Palca, 1988). Four debt-for-nature swaps have been concluded in Costa Rica since 1987; the $70 million dollars worth of debt involved represents 5 percent of the country's external obligations (Reed, 1989).

With leverage funds from the U.S. Agency for International Development, the U.S. branch of the World Wildlife Fund brokered the debt-for-nature swap on Madagascar which liberated $3 million for the island's parks (*Economist*, 19 August 1989, p. 31). Madagascar's parks are famous for their unique and endangered lemurs, but they also contain wild relatives of coffee. World Wildlife Fund was also instrumental in arranging the purchase of $2 million of the Philippines' debt at a 50 percent discount. The released funds will be managed by the Haribon Foundation and the Department of Environment and Natural Resources to improve management of protected areas and their buffer zones and to train personnel. Debt-for-nature swaps have provided opportunities for government agencies and private, nonprofit organizations, both at the local and regional level, to work together for conservation.

Debt-for-nature swaps are worth pursuing, but even with discounts as low as 15 cents on the dollar, only a fraction of the billions of dollars of Third World debt is likely to be bought for conservation purposes (N. J. H. Smith and Schultes, 1990). Hundreds of millions of dollars will have to be raised to buy down enough Third World debt to set aside significant portions of the world's tropical rain forests.

Also, provision needs to be made for management of any debt-relief parks. Otherwise, they may suffer the same fate of existing parks and reserves, most of which have inadequate protection. Debt-for-nature swaps may hold more promise if they are part of a larger package that includes rural development in areas surrounding proposed parks.

Another tactic to reinforce conservation efforts in debt-for-nature deals would be to encourage citizens of developing countries to "purchase" some of their government's domestic debt for conservation purposes. A number of nonprofit groups have sprung up in many developing countries, particularly Latin America, to promote conservation efforts. Such groups could help raise funds to purchase some internal government debt at a discount. In this manner, debt-for-nature swaps would not be perceived as alien interference with national affairs and would be backed up by grassroots support.

Given the daunting task of raising enough cash to buy down some of the Third World debt for conservation, debt reduction in exchange for conservation integrated with development needs may be a more promising avenue to pursue. Debt-for-development is likely to secure more forest from rampant destruction than debt-for-nature swaps alone. By addressing legitimate concerns about raising living standards as well as saving forest, debt-for-development is more appealing to governments. Debt-for-development swaps are under way or planned in such countries as Brazil, Nigeria, and Peru (*Economist*, 6 August 1988, pp. 62–63). Debt-for-development deals are also less likely to stir up sensitive sovereignty issues.

The conservation of wild relatives of crops can be difficult to sell, but the stakes are high. It is difficult for developing countries, besieged with economic woes and rapidly growing populations, to set aside vast tracts of forest because they contain crop genetic resources of unknown potential. Should Malaysia compensate Brazil for safeguarding Brazil nut and rubber germplasm and Nigeria for safeguarding oil palm? Should Brazil help Ethiopia, Sudan, and Kenya protect forests containing wild arabica coffee?

A consortium of donors, including Third World governments, multilateral development banks, commercial banks, private foundations, commodity groups, and bilateral aid agencies, is needed to help conserve and better utilize crop genetic resources in the dwindling tropical forests. Fortunately, a consortium of donors has begun to address some of the issues related to conserving tropical forests and their resources (McNeely et al., 1990). It is hoped that discussion will soon translate into action, while there is still time.

CHAPTER 12

Realizing the Potential

Over the past two centuries, the demand for natural resources has risen steeply. As the world's population has grown, so too has production of major commodities upon which the well-being of many people depend. But the capacity of the world's natural and managed ecosystems to meet the growing appetite for natural resources cannot continue indefinitely. Rampant forest destruction in the tropics is undercutting our ability to improve the qualities and yields of existing crops as well as the potential to domesticate new plants.

Tropical deforestation and other forms of environmental degradation have two major impacts, both of which undermine the capacity of societies to respond to future challenges. First, habitat destruction extinguishes species that may play key roles in the environment. Species loss has now reached alarming levels and is foreclosing our options for the future. Second, the range of genetic diversity of remaining species is often depleted, thereby reducing their potential for survival in the wild and restricting the maneuvering room for crop breeders.

It would be risky to assume that only those plants exploited widely at this time will satisfy the global community's future needs. Even for the major staple foods, such an assumption is questionable; for the myriad other plant products that are vital to livelihoods on virtually every continent, the assumption is clearly wrong.

The risks involved in the demise of tropical forests and the loss of their products and environmental services have been recognized for several decades. More recently, a wider appreciation has begun to emerge of the complex nature of driving forces of environmental change and the need to examine conservation in a wider context of socioeconomic issues. Yet strategies to ameliorate threats to tropical forests are still generally characterized by a fragmented, piecemeal approach.

The mounting scientific evidence dictates more concerted action to preserve the plant genetic resources of tropical forests. We have shown that the past and current use of plants from tropical forests represents but the tip of an iceberg of potential for improving existing crops and developing new ones. If the genetic cornucopia of tropical forests is wasted, one of the world's greatest treasures will have been squandered.

CONSERVATION AND SUSTAINABLE DEVELOPMENT

The rapid destruction of forest habitats is the greatest overall threat to the perennial crops discussed in this book. But we are not certain of the dimensions of even this overriding threat. Estimates on rates of deforestation and areas of remaining forest are fraught with assumptions and varying definitions of forest (J. C. Allen and Barnes, 1983; WRI, 1985a, b, c).

Disagreements or confusion with regard to definition of forest types or assumptions about deforestation rates should not be used as an excuse for inaction. While there is scope for discussion about the taxonomy of forest environments and whether certain areas will be essentially denuded within a decade or two, most would agree that tropical deforestation is a widespread problem that requires remedial attention. Indeed, discrepancies with regard to deforestation rates and areas of remaining forest underscore the need to monitor changes in forest cover more systematically.

In view of the multiple threats to tropical forests, a concerted effort to conserve tropical forest ecosystems is clearly needed. The Educational, Scientific and Cultural Organization of the United Nations (UNESCO) has played an early role in attacking the problem of conservation of tropical forests and other biomes from a global perspective. UNESCO spearheaded the innovative Man and the Biosphere Program, which is designed to conserve areas of all the world's major biotic provinces (UNESCO, 1985). The International Union for the Conservation of Nature and Natural Resources (IUCN) and the United Nations Environment Program (UNEP) have also been instrumental in helping to establish reserves. In addition, IUCN maintains a valuable database on endangered plant and animal species.

Before 1980, most conservation efforts worldwide focused either on saving individual species, particularly glamorous animals, or on framing relatively small vignettes of pristine nature. As the Brundtland report pointed out, little effort has been directed at a conservation approach that integrates economic and resource preservation concerns (WCED, 1987). The

World Conservation Strategy, commissioned by UNEP through IUCN, was a far-sighted attempt to bridge this gap (IUCN, 1980). Issued in 1980 with endorsements from FAO and the Worldwide Fund for Nature (WWF), the World Conservation Strategy focused global attention on the need for more conservation of resources and ecosystems and helped influence policymaking in some bilateral aid agencies, multilateral development banks, and private foundations.

During the 1980s, a concern for linking sustainable development and conservation surfaced at some international meetings and resulted in several initiatives. The specific need to conserve genetic resources for agriculture, championed since the 1960s, gained strong support in the 1980s. The World Commission on Environment and Development and the FAO Commission on Plant Genetic Resources, for example, both underscored the need to conserve plant genetic resources for the common good of humanity. In 1984, two important propositions emerged from an Inter-Parliamentary Union meeting and a Global Possible Conference organized by the World Resources Institute. Laudable as they are, however, international discussions alone achieve little; what is needed is action underpinned by education, training, communication, and scientific research.

To address the need for conserving biodiversity and critical ecosystems, several organizations have categorized a range of protected areas in the tropics.[1] A major point that has emerged from the analyses of these categories is that the scientific basis for choosing and managing protected areas is precarious.

Another worrisome aspect of conservation efforts thus far is that parks and nature reserves can conserve only a few populations of useful species. Moreover, designation of the existing reserves is not based on any knowledge of the patterns of variation within a particular economic plant species. Nor do data exist on which useful species are found in most reserves, let alone estimates of their variability. To be useful for plant improvement, economically useful species need to be monitored and tended by field curators.

Leadership for safeguarding and monitoring reserves in tropical forest areas needs to be nurtured among nationals of developing countries. One way to fortify research efforts in developing countries is to foster research networks among scientists investigating the conservation and sustainable development of tropical forests. Investments in training more research sci-

---

1. Categories established by the International Union for Conservation of Nature and Natural Resources and other conservation organizations include scientific and intact nature reserves, national and provincial parks, natural monuments and natural landmarks, managed nature reserves and wildlife sanctuaries, protected landscapes, resource reserves, multiple-use management areas, biosphere reserves, and world heritage sites. See Chapter 11 for a discussion of the advantages and disadvantages of the various categories of protected areas.

entists in the Third World need to be made with some urgency, considering that we have only thirty to fifty years before substantial areas of remaining tropical forests are gone.

Ideally, many parks and reserves should overlap critically important wild populations of our crop plants and at least some of their near relatives. They rarely do. Most parks and reserves were established to suit other purposes; one can only hope that significant crop gene pools will eventually be found in some of them. To what extent conservation of genetic resources can be married to broader aims of conservation remains to be seen. Considerable joint planning and scientific analyses will be required to coordinate such conservation efforts. Discussion is under way on this important issue among several research institutions and international bodies, including FAO, IUCN, and the International Board for Plant Genetic Resources (IBPGR).

The inadequacy of funds, both from international and national sources, is clearly hampering efforts to establish sufficient genetic reserves for useful species. The funding deadlock and insufficiency of research raise pressing issues for conservation efforts (Namkoong, 1986). How can we incorporate the conservation of plant genetic resources into development strategies more effectively? How can poor countries justify the upkeep of reserves and cover the costs of scientific monitoring when many benefits accrue to other countries, particularly industrial nations?

The history and use of genetic resources of some tropical perennial crops show that most of them are inextricably linked to the social fabric of rural areas. Poor people depend on such crops for food, cash income, and jobs. Conservation efforts may be hindered unless it can be shown that preservation and evaluation of plant genetic resources continue to benefit people who live in tropical forests, and just as important, future generations who will grow up in the humid tropics.

While international organizations grapple with such issues, immediate steps can still be taken to help promote both the conservation of genetic resources and the social welfare. Integrated landscape management that seeks to increase rural incomes and improve social equity while addressing conservation needs will help until a more comprehensive global plan for conservation emerges (Ingram and Williams, 1984). Unfortunately, few successful models of integrated sustainable development and conservation have been forged to guide policymakers (Fontaine, 1986).

In spite of the formidable difficulties related to inadequate funding levels, insufficient research, lack of political will, and a host of other issues competing for the attention of policymakers, broad-scale efforts to conserve genetic resources in tropical forests are under way. In 1985, an international task force convened by the World Resources Institute, the World Bank, and the United Nations Development Program (UNDP) drew up the far-reaching Tropical Forest Action Plan. This ambitious plan reviews the extent of deforestation, highlights the need for better conservation and utilization of forest resources, and boldly calls for concrete steps to save and better manage tropical forests (WRI, 1985a, b, c). The Tropical Forest Action Plan called for an investment of $8 billion over five years in order to make a real dent on tropical deforestation.

Only a fraction of that amount has been raised, and the plan has been criticized because it allegedly accelerated logging rates in some areas (Lohmann and Colchester, 1990; Westoby, 1989:165). Nevertheless, awareness of the importance of tropical forests has been raised by the plan. Extensive coverage of the Tropical Forest Action Plan in leading newspapers[2] not only brought the issue of tropical deforestation to millions of homes in North America and elsewhere, but helped alert politicians to the need to address concerns about the disappearing forests in developing countries.

SECURE RESOURCE BASES

Breeders have long been at work raising yields and improving other qualities, particularly for the major food crops. But it has often taken a major international effort to throw back the barriers to yield improvement in developing countries. The Green Revolution in wheat and rice took the combined efforts of the Rockefeller and Ford foundations as well as major players in agricultural research and development, such as FAO, UNDP, and the World Bank.

The international agricultural research centers (IARCs) under the umbrella of the Consultative Group on International Agricultural Research (CGIAR) emerged to promote research on a wide variety of staple crops in the tropics and subtropics (Hanson, 1986:206; Plucknett and Smith, 1982). The collection, storage, and evaluation of germplasm has been a cornerstone of many of the IARCs both within and outside the CGIAR system. A decade or more is usually required before germplasm collecting and evaluating efforts deliver tangible payoffs.

Research on the genetic resources of most tropical perennial crops lags far behind that of staple food crops. Breeding efforts have been largely limited to a few commercially important plantation species, such as coffee and rubber, but such efforts encompass only a small fraction of economically valuable species. It is time to perceive tropical forests as valuable

2. In 1985, coverage of the Tropical Forest Action Plan appeared in such newspapers as the *Washington Post*, Washington, D.C., 10 November; the *New York Times*, 29 July, 29 October, and 27 November; the *International Herald Tribune*, 11 November; the *Boston Sunday Globe*, 27 October; the *Christian Science Monitor*, 23 October; the *Los Angeles Times*, 11 August; and the *San Francisco Chronicle*, 23 October.

resources in their own right, full of plants that can be used to upgrade our crops, provide new medicines, or be domesticated, rather than as areas to be plundered and then cleared to make room for other activities.

A major reason that breeding efforts for perennial crops lag behind those for annual food crops is the time frame involved: some perennial species, such as bamboo and nutmeg, take decades to mature. Despite such obstacles, progress is being made in improving several perennial crops, and the pace is likely to quicken with increased use of biotechnologies. The breeding of trees rarely involves large crossing programs, with extensive backcrossing, so the provenance stands of foresters have served as their working collections. For this reason, foresters have been slow to initiate seed gene banks in which a wide sample of diversity can be conveniently stored, even for species with orthodox seeds.

A substantial number of useful trees can be conserved in seed or field gene banks. Also, several perennial species are amenable to tissue culture, and some of their germplasm can be stored in vitro. Tissue culture will become a more common method for storing germplasm of perennial species. Although such ex situ collections can maintain only a small fraction of the gene pool, they need to be developed further to assist breeders. Ex situ collections can offer a sample of the genetic variation of a species and provide breeders with clues as to the possibilities for upgrading crops.

Several biotechnologies are being increasingly deployed to collect and evaluate plant materials, eliminate pathogens, and transfer desirable traits. In vitro techniques such as organogenesis, somatic embryogenesis, embryo rescue, protoplast culture, and the raising of haploid plants from anther cultures are being applied to such crops as citrus, mango, papaya, passionfruit, pineapple, oil palm, bananas and plantains, and peach palm (Litz et al., 1985; Withers, 1989). Relatively simple in vitro techniques for collecting woody perennials are being used with some species, such as cacao and several palms (Withers, 1987). Inexpensive in vitro field kits for obtaining germplasm samples are often preferable to conventional methods, such as taking budwood or cuttings, which often lead to high mortality and can spread diseases and pests.

As biotechnologies improve and it becomes easier to culture perennial species in test tubes and flasks and to accomplish hitherto difficult crosses, tree breeders will seek more diverse germplasm sources to work with. Unless more comprehensive strategies for conserving genetic resources are in place, however, they may come up empty handed. Unlike the staple food crops that have been developed over millennia, first by farmers and then more recently by scientists, some of the minor fruit and nut crops are little different from wild forms, and much of the variation may be lost before attention is focused on them.

To help prevent such a tragedy, researchers should carry out intensive studies to develop strategic plans for the collection, conservation, evaluation, and use of a large selection of tropical perennial species. Under the sponsorship of IBPGR, some initial efforts on collection and conservation have been taken with oil palm, rubber, and citrus, but such endeavors now need to be refined and more detailed work plans drawn up. For most of the other perennial crop species, little if any work has begun on assessing the genetic potential and conservation needs of the hundreds of cultivated tree and shrub species that provide sustenance, shelter, and other products for millions of people.

REAPING THE HARVEST

Any discussion of perennial tree crop improvement raises an important political consideration. Who will benefit from the enhancement? Clearly, value judgments are involved. Many species are grown on both a commercial and subsistence scale. The goals and interests of plantation growers and small-scale farmers are not always the same. A plantation owner, for example, may be primarily interested in yield and the capacity of a fruit to withstand long-distance transportation. Small-scale growers, on the other hand, may be more interested in disease or pest resistance because they cannot afford pesticides. Local tastes may dictate selection for different fruit color and texture than varieties geared for export markets.

Genetic conservation efforts should not be biased by the current investment climate, which tends to funnel the limited funds available to germplasm collection and enhancement of major food staples or export-oriented perennial crops. While such efforts are needed to sustain gains in food production, jobs, and income, the needs of the rural poor also need to be addressed more directly with respect to neglected woody species. Thus more "minor" crops need attention for conservation and improvement purposes. Local people should be canvassed about the kinds of varieties they can use, and this information should be fed systematically to conservation efforts and agricultural and forestry research institutes. For example, international and national financial institutions can be encouraged to pay greater attention to helping countries design tree-planting schemes that are attractive to the rural poor as well as satisfy the need to generate foreign exchange (Gregersen and McGaughey, 1985; Murray, 1987a, b).

RESEARCH PRIORITIES FOR
MARGINAL LANDS

In the more marginal areas of the tropics, the vocation of the land is forest or at least tree-based production systems. Silviculture, plantations, agroforestry, and fruit and tree orchards protect the soil better and can be sustainable if properly managed and in tune with ecological and market realities. Tree crops are clearly well adapted to the tropics and can serve as an important vehicle for rural development. Considerable success has already been achieved in certain areas with such woody crops as oil palm, rubber, cacao, and coffee. While such major plantation crops will continue to play a valuable role in the economies of some Third World countries, more perennial tree crops need to be moulded to the diverse environments of tropical forest regions.

Areas that have gone through numerous slash-and-burn farming cycles and no longer support forest because fallow periods are too short are particularly appropriate for more intensive development with tree crops. Silviculture, plantations, and orchards on smallholdings in secondary forests or on degraded farmlands could provide a catalyst for regional development while ameliorating the environment.

Agroforestry holds great potential for furthering rural development (Mongi and Huxley, 1979; Nair, 1991). Perennial and annual crops have often been grown together or in relay fashion in the humid tropics for a long time (Plucknett and Smith, 1986). Tree crop production can thus expand without necessarily cutting into food supplies.

The successful marriage of appropriate tree crops and varieties with local culture and biophysical conditions rests on solid field information and screening of germplasm. Much of this research is fine-grained and cannot be adequately carried out at a single research station. More on-farm testing of protovarieties would help identify suitable varieties and candidates for agroforestry and orchards. Many planters have suffered in the past because varieties of nutmeg, clove, and coffee were released that proved inappropriate for the prevailing soil and climatic conditions.

Only rarely are development plans for raising yields on existing farmlands and for adopting more environmentally sound agricultural and silvicultural systems on marginal lands undergirded by careful analysis of biophysical and cultural resources. Geographic information systems are needed that contain baseline data on climate, soils, agricultural potential, cultural groups, and market conditions. Donor agencies thus far have been largely unwilling to fund such preproject studies; time is of the essence, it is argued, and projects need to be implemented quickly. Too often, however, agricultural development projects founder in the tropics precisely because the natural and social history of the region is poorly understood.

THE QUARANTINE BOTTLENECK

The task of improving tropical perennial crops needs to be attacked along a broad front, and quarantine is a vital step in this process. After a wide range of germplasm is collected, trials containing a diverse package of genetic materials need to be shipped for multilocation testing, similar to the pan-tropical nursery networks for rice and wheat (Plucknett et al., 1990). Plant materials can be fed into international trials only if they have a clean bill of health.

International nurseries allow breeders to obtain a reading on performance across a range of environments and can more easily identify broadly adapted material. Specialized nurseries can also be arranged for disease "hot spots" to readily identify resistant material. Given the variability of many disease pathogens, however, several disease-screening nurseries will have to be set up in the midst of different races of the pathogen. In this manner, germplasm resistant to a range of pathotypes can be identified, or selections that resist different strains of the pathogen can be made.

International nurseries for perennial crops are virtually nonexistent, primarily because of quarantine concerns. Most of the germplasm of perennial crops destined for variety trials would have to be shipped as budwood, cuttings, or preferably in vitro. Such materials are particularly hazardous in terms of spreading diseases and pests. Quarantine services therefore often prohibit the entry of live plant material. Intermediate quarantine services are needed away from where the perennial crops are grown to check and clean up material if necessary.

Because of quarantine concerns and because budwood and cuttings can be highly perishable, tissue cultures are being used to exchange germplasm of a few perennial crops, such as oil palm and banana, but only on a limited scale. When combined with disease-indexing procedures and thermotherapy, tissue culture is likely to be used more commonly in the future to transfer germplasm of perennial crops around the world.

PERSONNEL REQUIREMENTS

After many developing countries gained their independence in the 1960s and 1970s, tropical tree crops received much less attention. Plantation crops have suffered from a colonial image and from the perception that they are largely export-oriented. Donors and nascent national agricultural research systems focused most of their effort on food crops. The neglect of plantation crops in Third World countries has led to acute personnel shortages for most of the tree crops important for generating cash income and other needs. Lesser-developed countries have especially weak research capacity in such important crops as bananas and plantains, coconuts, tropical fruits, and other species suitable for agroforestry (Oram, 1988:30).

Much of the research on genetic resources of tropical perennial crops is still being undertaken by expatriate staff. While outside help is always useful to fill critical knowledge or expertise gaps, national program scientists are needed to help provide greater continuity to research efforts. Breeding cycles for many perennial crops are extensive, so indigenous scientific capacity is needed to accompany trials and to study the phenology of wild populations.

The need for more training, in situ conservation, and improved linkages between the preservation of genetic resources and crop improvement in Africa was explored at a Pan African Workshop on African Plant Genetic Resources. Organized by UNEP, IBPGR, and the Nigeria-based International Institute of Tropical Agriculture (IITA), the workshop was held in 1987 and attended by representatives from over forty countries. One delegate underscored the need for surveys of vegetation, identification of useful plant gene pools, and more action on collecting, conservation, documentation, and use of genetic resources. He also pointed out that his country has only one resident botanist. Scientists from the limited number of African countries with crop genetic resource programs stressed the need for rapid training in conservation biology covering a range of subjects from taxonomy and ecology to methodologies of preserving germplasm.

Tailormade, modular courses are needed for postgraduates to prepare them for the multiple tasks involved in surveying genetic resources, conserving germplasm both in natural environments and in ex situ collections, and evaluating materials of potential interest. Courses are needed on such subjects as ecogeographic surveying of gene pools, taxonomy, and management of reserves for genetic resources. At present, no such courses are offered anywhere in the world. A few universities offer courses or programs on genetic resources, which could be expanded with further funding. The renowned international training course on plant genetic resources conservation launched by J. G. Hawkes at the University of Birmingham, United Kingdom, has trained germplasm specialists for two decades. A consortium approach, linking several universities and partners in developing countries, would be a cost-effective way to further training of specialists in the conservation and maintenance of plant genetic resources in the tropics.

A crash program in training and survey work is needed because of the speed of human-induced ecological change in the tropics. A repeat performance emulating the leadership of the Ford and Rockefeller foundations three decades ago is needed. Over $3 million would be needed to train just 100 scientists from developing countries; the bill for a more comprehensive effort to survey, safeguard, and evaluate genetic resources of perennial crops would be much higher. Through its Commission on Plant Genetic Resources, FAO is considering the need for training in plant breeding, seed production, and the conservation of plant genetic resources. It is hoped that this initiative will soon spur governments to allocate more funds for training of germplasm specialists. The future success of efforts to conserve and utilize genetic resources of perennial plants hinges on sufficiently trained personnel. Considering the time lag in agricultural research between the onset of a project and tangible results, typically a decade or more, it is imperative that efforts to train scientists from developing countries be redoubled.

Responsibility for furthering training in conservation biology and genetic resources in the tropics rests largely with the international donor community. Governments in the Third World increasingly recognize the need to do more to safeguard and utilize genetic resources, but often lack sufficient personnel and funds to meet the challenge. Developing countries are treasure troves of genes, but ironically have the least capacity to respond to calls for action. Nor can we expect developing countries to rapidly change policies and investment priorities until scientific work in those nations shows potential payoffs.

FINDING A WAY FORWARD

Over the past decade, the monetary value of genetic resources has stirred numerous debates, particularly with regard to agribusiness profits and plant breeders' rights. Monetary assessments of plant genetic resources are oversimplified, but they do illustrate that the deployment of plant genetic resources generates income and profits.

Some of these profits are so large that a percentage should perhaps be channeled to conservation. Large companies and parastatal organizations involved with plantations of rubber, cacao, coffee, bananas, pineapple, and sugarcane invest some resources in breeding and the assembling of limited ex situ collections. But the longer-term need to conserve genetic resources in situ seems to fall between cracks. The private sector generally takes the position that conservation collections and nature preserves are the domain of government institutions. In industrial countries, well-heeled governments can take up this burden, but in developing countries, funds are scarce and other priorities sideline in situ conservation, at least on the scale that is needed.

Money alone is not the problem, however. Even if massive amounts of funds are mobilized for training and conservation, their impact can quickly dissipate if strategic plans have not been drawn up and vested interests block such measures. Governments in developing countries need to realign some of their policies so that growers are encouraged to plant crops for market as well as home consumption. National agricultural research systems and conservation organizations need to be fortified with better equipment, more consistent leadership, and incentives for increasing their productivity. Creative ways need to be devised to entice the private sector into this important venture.

The case that the future well-being of humanity hinges to a large extent on the conservation and enhanced use of plant genetic resources has been made for several decades. But, except for some of the major food plants, little funding has been make available to one of the most important resource issues facing the world. Petroleum, ozone depletion, and global warming command more headlines, but petroleum will be essentially used up a century from now, ozone depletion can be tackled by changes in refrigerant gases and insulating materials, and global warming may turn out to be one of the biggest nonissues of the late twentieth century. All the while, the genetic resources of the earth continue to hemorrhage.

Cries for more funds need to be made, but just waiting for large handouts to arrive can lead to paralysis. It is unlikely that a system of IARCs will evolve to handle germplasm and breeding needs of perennial crops. Other mechanisms are available, however, to promote the conservation and use of crop plant genetic resources that may require only modest lubricating funds. One such mechanism is the International Program for Tropical Tree Crops Conservation and Development, which is organized under the umbrella of the International Fund for Agricultural Research (IFAR) in Washington, D.C. Started in 1990, this program is cosponsored by IBPGR in Rome and is designed to draw up specific action plans for perennial crops and to mobilize resources to support research.

For woody crop species, partnerships can be forged between scientists in developing and industrial countries to tackle specific problems in genetic resource conservation. A networking approach uses existing facilities and staff and helps eliminate unnecessary duplication of effort. Some external funds are needed to develop strategic plans, finance international travel, hold workshops, and catalyze short training courses related to the network's objectives.

Collaborators in such networks should include a mix of government agencies, private voluntary organizations (PVOs), and businesses. PVOs often have a better feel for local conditions, while national agricultural research systems have some capacity for research and for developing germplasm and other technologies. PVOs would also become more effective in conservation work if they were equipped with carefully prepared manuals that outline local plant and animal resources and identify management strategies (J. T. Williams, 1991). Diplomacy is required to forge such partnerships; donors can play a helpful but discrete role in helping overcome turf and other potentially contentious issues among network participants. Networks to promote conservation of genetic resources should be grounded on scientific research and cultural sensitivity, rather than a political agenda, otherwise the corporations and the international donor community are likely to pull back from such important endeavors.

The threats to plant resources continue to mount, and action to salvage forest ecosystems and indigenous knowledge about plants and their environment can no longer await much-needed changes in policies or a funding windfall. The immediate welfare of hundreds of millions of peoples in the humid tropics and the economic future of over forty developing nations rest with progress in conserving and better utilizing plant genetic resources. Better mobilization of existing scientific capacity can be achieved with minimal funding to make dramatic inroads on some of the seemingly intractable problems facing organizations and governments as they grapple with ways to conserve the environment. Some early successes will whet the appetite of donors and should goad the global community into action.

Some domesticated perennial species, exclusive of strictly medicinal or ornamental plants, with wild populations in tropical forests, region where wild populations occur, source of information on the distribution of wild populations, and principal uses

Family and scientific name	Common name	Location of wild populations[1]	Main use(s)
Anacardiaceae			
Anacardium excelsum	Marañon	S. Honduras to Ecuador; Venezuela	Nuts, fruit
Source: Mitchell and Mori, 1987:36			
A. giganteum	Cajuí	Amazonia; Guianas	Fruit, nut
Source: Cavalcante, 1976:9			
A. occidentale	Cashew	Brazil; Venezuela; Guianas	Fruit, nut
Source: Mitchell and Mori, 1987:40			
Bouea macrophylla	Gandaria	(Malaysia; Indonesia)	Fruit
Mangifera caesia	Mango	Malaysia; Philippines	Fruit
Source: Morton, 1987:237			
M. foetida	Horse mango	(Indochina; Malaysia)	Fruit
M. indica	Mango	India	Fruit
Source: Hawkins, 1986:365			
M. odorata	Kuini	(Malaysia)	Fruit
Spondias dulcis	Golden apple	(S. Pacific)	Fruit
S. mombim	Yellow mombin	Amazonia; Barro Colorado Island, Panama; Costa Rica; S. Mexico	Fruit

Sources: Cavalcante, 1976:21; Croat, 1978:541; Holdridge et al., 1971:72, 155, 263, 373; Morton, 1987:245; Standley, 1924:656, 1927:22; Vickers and Plowman, 1984

S. purpurea	Red mombin	T. America	Fruit

Sources: Alcorn, 1984; Holdrige et al., 1971:72, 155; Morton, 1987:242; Standley, 1924:657

Annonaceae			
Annona cherimola	Cherimoya	Ecuador; Peru; Bolivia; Honduras	Fruit
Sources: Morton, 1987:65; Standley, 1930a			

APPENDIX 1—*cont.*

Family and scientific name	Common name	Location of wild populations[1]	Main use(s)
A. diversifolia	Ilama	Guatemala; El Salvador; S.W. Mexico	Fruit
Source: Morton, 1987:83			
A. glabra	Pond apple	W. Indies; Honduras; Venezuela	Fruit
Source: Standley, 1930a			
A. montana	Mountain soursop	W. Indies; Tropical S. America	Fruit
Source: Morton, 1987:87			
A. muricata	Soursop	W. Indies; Barro Colorado Island, Panama	Fruit, drink, ice cream
Sources: Croat, 1978:398; Morton, 1987:76			
A. purpurea	Soncoya	Costa Rica; El Salvador; Honduras	Fruit
Sources: Hartshorn and Poveda, 1983; Standley, 1930a; Standley and Calderón, 1927:83			
A. reticulata	Custard apple	Costa Rica	Fruit
Source: Hartshorn and Poveda, 1983			
A. scleroderma	Posh te	S. Mexico; Guatemala	Fruit
A. squamosa	Sweetsop	(T. America)	Fruit
Cananga odorata	Ylang-ylang	Malaysia	Essential oils
Source: Purseglove, 1974:626			
Duguetia stenantha	Jaboti	(Upper Amazon)	Fruit
Rollinia mucosa	Biribá	T. America; W. Indies; W. Amazonia	Fruit
Sources: Cavalcante, 1988:58; Morton, 1987:88			
Apocynaceae			
Carissa congesta	Karanda	(India; Sri Lanka; Burma; Malacca)	Fruit
Couma utilis	Sorvinha	Amazonas, Brazil	Chewing gum, caulk
Source: Cavalcante, 1988:211			
Funtumia elastica	Lagos silk-rubber tree	W. Africa	Latex
Source: Dalziel, 1948			

APPENDIX 1—*cont.*

Family and scientific name	Common name	Location of wild populations[1]	Main use(s)
Macoueba uitotorum Source: Schultes, 1987b		(Colombian Amazon)	Fruit
Bignoniaceae			
Crescentia alata Sources: Janzen, 1983; Standley, 1930a	Gourd tree	Costa Rica; Honduras	Gourd
C. cujete Source: Janzen, 1983	Calabash	T. America	Gourd
Bixaceae			
Bixa orellana Sources: I. Falesi, pers. comm.; Holdridge et al., 1971:128; Standley and Calderón, 1927:152	Annatto	W. Costa Rica; El Salvador; Amazonia	Food colorant, body paint
Bombacaceae			
Ceiba pentandra	Kapok tree	T. America	Mattress filler
Durio malaccensis Source: Whitmore, 1990:156	Wild durian	Malaysia	Fruit
D. zibethinus Sources: Coronel, 1983:164; Morton, 1987:287	Durian	Malaysia; Borneo; N.W. Sumatra; Sulawesi; Moluccas; Mindanao; Burma	Fruit
Quararibea cordata Source: Morton, 1987:291	Chupa-chupa	W. Amazonia; Cauca and Magdalena valleys, Colombia	Fruit
Burseraceae			
Canarium ovatum Source: Coronel, 1983:327	Pili nut	S. Luzon, Visayas, Mindanao, Philippines	Nut
Caricaceae			
Carica papaya	Papaya	T. America[2]	Fruit, meat tenderizer

APPENDIX 1—*cont.*

Family and scientific name	Common name	Location of wild populations[1]	Main use(s)
Caryocaraceae			
Caryocar nucifera	Sawari	Guianas	Fruit
Source: Van Roosmalen, 1985:72			
Chrysobalanaceae			
Coupeia bracteosa	Pajurá	C. Amazonia	Fruit
Sources: Cavalcante, 1976:38; Ducke, 1946			
C. longipendula	Castanha de galinha	C. Amazonia	Nut
Source: Cavalcante, 1976:40			
C. subcordata	Umarirana	C. Amazonia	Fruit, shade
Source: Cavalcante, 1988:226			
Combretaceae			
Terminalia catappa	Tropical almond	(Malaysia; Andaman Islands)	Fruit
T. kaernbachii	Okari nut	(New Guinea; Solomon Islands)	Nut
Ebenaceae			
Diospyros blancoi	Mabolo	Philippines	Fruit
Sources: Coronel, 1983:494; Morton, 1987:419			
D. digyna	Black sapote	C. America	Fruit
Sources: León, 1987:216; Morton, 1987:417			
Elaeocarpaceae			
Muntingia calabura	Jamaica cherry	T. America; W. Indies	Fruit
Source: NAS, 1980:58			
Erythroxylaceae			
Erythroxylum coca	Coca	Catatumbo R., Colombia	Stimulant
Source: Cortés, 1919:99			
E. novogratense	Coca	(W. Amazonia)	Stimulant

APPENDIX 1—cont.

Family and scientific name	Common name	Location of wild populations[1]	Main use(s)
Eurphorbiaceae			
Aleurites fordii	Tung	(S. China)	Oil
Antidesma bunius	Bignay	India; Sri Lanka; S.E. Asia, N. Australia[3]	Fruit
Sources: Coronel, 1983:478; Morton, 1987:210			
Baccaurea dulcis	Tjoepa	S. Sumatra	Fruit
Source: Morton, 1987:220			
B. motleyana	Rambai	Borneo; Malaysia	Fruit
Source: Morton, 1987:220			
B. racemosa	Kapoendoeng	Java	Fruit
Source: Morton, 1987:220			
B. sapida	Burmese grape	S. China; Thailand; Cambodia; Malaysia	Fruit
Source: Morton, 1987:220			
Hevea brasiliensis	Rubber	Amazonia	Latex
Source: Schultes, 1984b			
Manihot glaziovii	Ceará rubber	(Brazil)	Latex
Phyllanthus acidus	Otaheite gooseberry	(Madagascar)	Fruit
P. emblica	Emblic	T. Asia	Fruit
Source: Morton, 1987:213			
Flacourtiaceae			
Dovyalis hebecarpa	Ketembilla	(Sri Lanka)	Juice, jelly
Flacourtia inermis	Louvi	(T. Asia)	Fruit
F. jangomas	Paniala	(India)	Fruit
F. ramontchi	Ramontchi	(T. Africa; Madagascar; T. Asia)	Fruit
F. rukam	Rukam	(India; S.E. Asia; Oceania)	Fruit
Hydnocarpus wightiana	Maroti	W. Ghats, India	Oil
Source: Bringi, 1987			

Family and scientific name	Common name	Location of wild populations[1]	Main use(s)
Gnetaceae			
Gnetum gnemon	Bago	Philippines	Fruit
Source: Coronel, 1983:475			
Guttiferae			
Calophyllum antillanum	Galba	(W. Indies)	Fruit
Garcinia atroviridis		Malaysia	Fruit
Source: Whitmore, 1990:156			
G. kola	Bitter kola	W. Africa	Nut
Source: Dalziel, 1948:91			
G. mangostana	Mangosteen	Malaysia; Burma; Thailand; Cambodia; Vietnam; Sunda Islands	Fruit
Sources: IBPGR, 1986; Morton, 1987:301			
Mammea americana	Mammey apple	Puerto Rico; St. Croix; Tortola	Fruit
Source: Britton and Wilson, 1924:583			
Platonia insignis	Bacuri	Amazonia; Guianas; Paraguay	Fruit, ice cream
Sources: Ducke, 1946; Morton, 1987:308			
Rheedia acuminata	Bacurizinho	Amazonia	Fruit
Source: Cavalcante, 1988:54			
R. benthamiana	Bacuripari selvagem	Amazonia	Fruit
Source: Cavalcante, 1976:49			
R. brasiliensis	Bacuripari liso	Amazonas, E. Maranhão, Brazil; French Guiana	Fruit
Sources: Balée, 1986; Cavalcante, 1988:51			
R. macrophylla	Bacuripari	Amazonia	Fruit
Source: Cavalcante, 1976:49			

APPENDIX 1—*cont.*

Family and scientific name	Common name	Location of wild populations[1]	Main use(s)
Humiriaceae			
Endopleura uchi	Uxi	Amazonia	Fruit
Sources: Cavalcante, 1976:50; Ducke, 1946; N. J. H. Smith, 1982:41			
Icacinaceae			
Poraqueiba paraensis	Umari	Pará, Brazil	Fruit
Sources: Cavalcante, 1976:55; Ducke, 1946			
P. sericea	Mari	Amazonas, Brazil	Fruit
Source: Cavalcante, 1988:226			
Lauraceae			
Cinnamomum burmani	Cassia	S.E. Asia	Spice
C. camphora	Camphor	(S.E. Asia)	Essential oil
C. loureirii	Saigon cassia	Vietnam	Spice
C. zeylanicum	Cinnamon	Sri Lanka; S.W. India	Spice
Persea americana	Avocado	Mexico; Guatemala; Honduras; Costa Rica; Panama; N.W. Colombia	Fruit
Sources: Bergh, 1975, 1976; Croat, 1978:414; Holdridge et al., 1971:235, 360, 434, 443, 484, 507; W. Popenoe, 1920:14; Standley, 1931; Storey et al., 1987			
P. schiedeana	Chinene	Mexico to Panama	Fruit
Sources: Herbarium records at UNAM (Universidad Nacional Autonómo de México), Mexico City; Schieber and Zentmyer, 1981b; Schieber et al., 1984; Schroeder, 1977			
Lecythidaceae			
Bertholletia excelsa	Brazil nut	Amazonia; Upper Orinoco	Nut
Lecythis usitata	Sapucaia	Amazonia; French Guiana	Nut
Sources: Aublet, 1775:721; Silva et al. 1977			
Leguminosae			
Adenanthera pavonina	Circasian bean	(India)	Fruit

APPENDIX 1—cont.

Family and scientific name	Common name	Location of wild populations[1]	Main use(s)
Calliandra calothyrsus Source: Standley and Calderón, 1927:97	Calliandra	Santa Ana, El Salvador	Fuelwood, erosion control
Cassia fruticosa Source: Van Roosmalen, 1985:178	Lokonanjo, jorkatiki	Guianas	Fruit
C. grandis Source: Van Roosmalen, 1985:178	Marimari	Guianas	Fruit
C. leiandra Sources: Cavalcante, 1988:164; Ducke, 1946	Marimari	Amazonas, Pará, Brazil	Fruit
Casuarina equisetifolia Source: NAS, 1980:38	Casuarina	N. Australia; S. and S.E. Asia; Pacific	Fuelwood, erosion control
Copaifera copallifera Source: Dalziel, 1948:184	Gum copal	W. Africa	Resin
Cynometra cauliflora	Nam-nam	(Malaysia)	Fruit
Gliricidia sepium	Madre de cacao	(Tropical S. America)	Shade tree, fuelwood
Inga cinnamomea Sources: Cavalcante, 1976:69; Ducke, 1946	Ingá-açu	Amazonia; Guianas	Fruit
I. edulis Sources: Cavalcante, 1988:125; Standley, 1927:19, 1930a	Ingá-cipó	Barro Colorado Island, Panama; Honduras	Fruit
I. fagifolia Sources: Cavalcante, 1988:127; Ducke, 1946	Ingá cururu	Brazil	Fruit
I. macrophylla Source: Ducke, 1946	Ingapeua	Amazonia	Fruit
I. ruiziana Source: Hartshorn and Poveda, 1983	Ingá peus	Costa Rica	Fruit
I. velutina Source: Cavalcante, 1976:77	Ingá de fogo	E. Amazonia	Fruit

APPENDIX 1—*cont.*

Family and scientific name	Common name	Location of wild populations[1]	Main use(s)
I. vera	Guaba	(T. America; Caribbean)	Fruit
Leucaena leucocephala	Leucaena	Chiapas, Yucatán, Mexico	Fuelwood, fodder
Parkia biglobosa	Nere	W. Africa	Fruit
P. speciosa		Malaysia	Fruit
Source: Whitmore, 1990:156			
Pithecellobium dulce	Manila tamarind	Honduras; Colombia; Venezuela; Mexico	Fruit, fuelwood, construction
Sources: NAS, 1980:144; Standley, 1924:393, 1930a			
P. jiringa		Malaysia	Fruit
Source: Whitmore, 1990:156			
Sesbania grandiflora	Sesbania	(S. and S.E. Asia)	Fuelwood, fodder, vegetable
S. sesban	Sesbania	W. Africa	Mulch, fodder
Tamarindus indica	Tamarind	Sudan	Juice, vegetable, sauces, candy
Source: Morton, 1987:115			
Malpighiaceae			
Bunchosia glandulosa	Ciruela	Amazonia	Fruit
Source: Ducke, 1946			
Byrsonima crassifolia	Nance	Honduras; El Salvador; Amazonia	Fruit
Sources: Cavalcante, 1976:79; Morton, 1987:208; Standley, 1930a; Standley and Calderón, 1927:128			
Malpighia emarginata	Acerola	Yucatán	Fruit
Source: IBPGR, 1986			
M. punicifolia	Barbados cherry	Costa Rica; Lesser Antilles	Fruit
Source: Hartshorn and Poveda, 1983			
Meliaceae			
Lansium domesticum	Langsat	S. Malayan Peninsula	Fruit
Source: Coronel, 1983:277			
Sandoricum koetjape	Santol, sentul	Malaysia; Philippines[3]	Fruit
Sources: Coronel, 1983:399; Whitmore, 1990:156			

469

Family and scientific name	Common name	Location of wild populations[1]	Main use(s)
Moraceae			
Artocarpus altilis	Breadfruit	(New Guinea; W. Micronesia; Indo-Malaya)	Fruit
A. heterophyllus	Jackfruit	W. Ghats, India; Sri Lanka; Philippines[4]	
Sources: Hawkins, 1986:337; Parker, 1986:25; W. Popenoe, 1920:416			
A. integer	Champedak	Malaysia; India; Sri Lanka	Fruit
Sources: Chin and Yong, 1985:27; Morton, 1987:63; Whitmore, 1990:156			
A. odoratissima	Marang	Mindoro, Mindanao, Basilan, Sulu, Philippines	Fruit
Source: Coronel, 1983:500			
Castilla elastica	Balata	Honduras; Nicaragua	Latex
Sources: Belt, 1888; Standley, 1930b			
Parartocarpus venenosa		(New Guinea)	Fruit
Pourouma cecropiaefolia	Puruma	W. Amazonia	Fruit
Source: Cavalcante, 1988:157			
Treculia africana	African breadfruit	T. Africa	Fruit
Musaceae			
Ensete ventricosum	Ensete	(Ethiopia)	Starchy pseudostem
Musa acuminata	Diploid AA bananas	S.E. Asia	Fruit
M. textilis	Manila hemp	Philippines	Fiber
Source: Stover and Simmonds, 1987			
Myristicaceae			
Myristica fragrans	Nutmeg	E. Moluccas	Spice
Source: Purseglove et al., 1981a:174			

Family and scientific name	Common name	Location of wild populations[1]	Main use(s)
Myrtaceae			
Camponanesia lineatifolia Source: Cavalcante, 1988:113	Guabiraba	W. Amazonia	Fruit
Eucalyptus deglupta	Eucalyptus	New Guinea	Fuelwood
Eugenia stipitata Source: Cavalcante, 1976:96	Araçá-boi	W. Amazonia	Fruit
E. uniflora Source: Morton, 1987:386	Surinam cherry	Pilcomayo River, Paraguay	Fruit
Myrciaria floribunda Source: Morton, 1987:389	Rumberry	W. Indies; C. America; Northern S. America; E. Brazil	Fruit
Pimenta dioica Sources: F. Cardenas, pers. comm.; Purseglove et al., 1981a:286	Allspice	Jamaica; Cuba; Tabasco, Chiapas, Mexico; Costa Rica; Haiti	Spice
Psidium acutangulum Source: Cavalcante, 1976:96; Morton, 1987:366	Araçá pera	Amazonia; Guianas; Venezuela	Fruit
P. cattleianum	Cattley guava	(E. Brazil)	Fruit
P. friedrichsthalianum Source: Standley, 1930a	Costa Rican guava	Honduras; Colombia	Fruit
P. guajava Sources: Belt, 1888:203; Standley, 1930a	Guava	Nicaragua; Pacific slope, Honduras[2]	Fruit
P. guineense Sources: Cavalcante, 1988:36; Morton, 1987:366	Brazilian guava	T. America; W. Indies; Amazonia	Fruit
Syzygium aqueum	Water apple	(India; S.E. Asia)	Fruit

471

Family and scientific name	Common name	Location of wild populations[1]	Main use(s)
S. aromaticum Source: Jarvie and Koerniati, 1987	Clove	Moluccas	Spice
S. cumini Source: Coronel, 1983:147	Jambolan	Philippines[5]	Fruit
S. jambos	Rose apple	(Indo-Malaysia)	Fruit
S. malaccensis	Malay apple	(Malaysia)	Fruit
S. samarangense Source: Morton, 1987:381	Java apple	Malaysia; Nicobar and Andaman Islands	Fruit
Orchidaceae *Vanilla fragrans*	Vanilla	Vera Cruz, Mexico; Nicaragua; Barro Colorado Island, Panama; French Guiana	Spice
Sources: Aublet, 1775:78; Belt, 1888:57; Croat, 1978:307; Price, 1917			
V. pompona Source: Purseglove et al., 1981b:647	West Indian vanilla	S.E. Mexico; C. America; Trinidad; Northern S. America	Spice
Oxalidaceae *Averrhoa bilimbi* Source: Morton, 1987	Bilimbi	(Tropical Asia)	Fruit
Palmae *Areca catechu*	Betel nut	(Malaysia)	Nut
Arenga saccharifera	Sugar palm	(T. Asia)	Sap
Astrocaryum chambira Source: Ducke, 1946	Tucum	N.W. Amazonia	Fruit, fiber
Bactris gasipaes Source: C. R. Clement, 1988	Peach palm	W. Amazonia	Fruit, palmito
Cocos nucifera	Coconut	Pantropical	Fruit

APPENDIX 1—*cont.*

Family and scientific name	Common name	Location of wild populations[1]	Main use(s)
Copernicia cerifera	Carnauba	N.E. Brazil	Wax
Elaeis guineensis	Oil palm	W. and C. Africa	Oil
Sources: Blombery and Rodd, 1985:98; Busson, 1965; N. J. H. Smith, field notes			
Euterpe oleracea	Açaí	Amazonia	Fruit, palmito
Metroxylon sagu	Sago palm	Moluccas; W. New Guinea	Starchy pith
Source: Blombery and Rodd, 1985:124			
Salacca edulis	Salak	(S.E. Asia)	Fruit
Passifloraceae			
Passiflora edulis	Passionfruit	(S. Brazil)	Juice, sherbet
P. laurifolia	Water lemon	Amazonia; S. Venezuela; Guianas	Juice
Source: Morton, 1987:332			
P. ligularis	Sweet granadilla	(T. America)	Juice
P. maliformis	Sweet calabash	Cuba; Puerto Rico; Dominican Republic; Jamaica; Barbados; Venezuela; Colombia; N. Ecuador	Fruit
Source: Morton, 1987:335			
P. mollissima	Banana passionfruit	Andes from Venezuela to Bolivia	Fruit
Source: Morton, 1987:332			
P. quadrangularis	Giant granadilla	Nicaragua	Juice
Source: Belt, 1888:69			
Pinaceae			
Pinus caribaea	Caribbean pine	Honduras; Nicaragua	Pulp
Source: Standley, 1930a			
Proteaceae			
Grevillea robusta	Silk oak	Queensland, New South Wales, Australia	Fuelwood
Macadamia integrifolia	Smooth-shelled macadamia	E. Australia	Nut
Source: IBPGR, 1986			

Family and scientific name	Common name	Location of wild populations[1]	Main use(s)
M. tetraphylla	Rough-shelled macadamia	E. Australia	Nut
Source: IBPGR, 1986			
Rosaceae			
Prunus salicifolia	Capulin	Central valley, Mexico	Fruit
Source: Morton, 1987:108			
Rubiaceae			
Coffea arabica	Arabica coffee	S.W. Ethiopia; Boma Plateau, Sudan	Beverage
Source: Wrigley, 1988:1			
C. bengalensis	Coffee	Bengal; Sumatra	Beverage
Source: Wrigley, 1988			
C. canephora	Robusta coffee	W. and C. Africa	Beverage
Source: Wrigley, 1988:57			
C. eugenioides	Coffee	Kivu, Zaire; W. Uganda; W. Tanzania	Beverage
Source: Wrigley, 1988			
C. liberica	Liberian coffee	Côte d'Ivoire	Beverage
Source: Dalziel, 1948:395			
C. stenophylla	Narrow-leaved coffee	Guinea; Sierra Leone; Côte d'Ivoire	Beverage
Source: Dalziel, 1948:395			
Genipa americana	Jenipapo	W. Indies; T. America; Amazonia	Liquor, body paint, fruit
Sources: Cavalcante, 1988:147; Croat, 1978:806; Morton, 1987:441; R. O. Williams and Williams, 1951:173			
Rutaceae			
Casimiroa edulis	White sapote	C. Mexico	Fruit
Source: Morton, 1987:192			

APPENDIX 1—*cont.*

Family and scientific name	Common name	Location of wild populations[1]	Main use(s)
C. sapota	Matasano	S. Mexico; Nicaragua	Fruit
Source: Morton, 1987:193			
Citrus aurantifolia	Lime	N. India; E. Indies	Juice, flavoring
Source: Purseglove, 1974:498			
C. aurantium	Sour orange	(S.E. Asia)	Marmalade flavoring
C. grandis	Pummelo	(Thailand; Malaysia)	Fruit, juice
C. limon	Lemon	(S.E. Asia)	Juice, flavoring
C. reticulata	Tangerine	(Cochin China)	Fruit, juice
C. sinensis	Sweet orange	(S. China, Cochin China)	Fruit, juice flavoring
Clausena lansium	Wampee	(S. China; Vietnam)	Fruit
Feronia limonia	Wood apple	India; Sri Lanka	Fruit
Source: Morton, 1987:190			
Santalaceae			
Santalum album	Sandalwood	Timor	Essential oil, wood
Source: Hawkins, 1986:484			
Sapindaceae			
Blighia sapida	Akee	Côte d'Ivoire; Ghana	Fruit
Source: Dalziel, 1948:332			
Melicoccus bijugatus	Mamoncillo	(Colombia; Venezuela; Guianas)	Fruit
Nephelium lappaceum	Rambutan	Malaysia	Fruit
Source: Whitmore, 1990:156			
N. mutabile	Pulasan	Malaysia	Fruit
Sources: Chin and Yong, 1985:8; Morton, 1987:266; Whitmore, 1990:156			
Paullinia cupana	Guaraná	(Amazonia)	Fruit
Talisia esculenta	Pitomba	Brazil; Bolivia; Paraguay	Fruit
Source: Cavalcante, 1976:123			

APPENDIX 1—*cont.*

Family and scientific name	Common name	Location of wild populations[1]	Main use(s)
Sapotaceae			
Calocarpum sapota	Sapote	S. Mexico; N. Guatemala; Belize; El Salvador; N. Nicaragua; Guanacaste, Costa Rica	Fruit
Sources: Holdridge et al., 1971:86, 196, 213, 341, 348, 434; Morton, 1987:398; W. Popenoe, 1936; Standley, 1931			
Chrysophyllum africanum	African star apple	Côte d'Ivoire; Ghana	Fruit
Source: Dalziel, 1948:354			
C. albidum	White star apple	W. Africa	Fruit
C. cainito	Caimito	W. Indies; Barro Colorado Island, Panama	Fruit
Sources: Morton, 1987:409; Standley, 1927:27			
Manilkara achras	Sapodilla	N.E. and S. Mexico; Guatemala; Honduras; Belize; Venezuela; Antilles	Fruit
Sources: Alcorn, 1984; Morton, 1987:393; Standley, 1924:1119			
Pouteria caimito	Abiu	Amazonia in Peru, Brazil, Ecuador; S.W. Venezuela	Fruit
Sources: Cavalcante, 1976:126; Ducke, 1946; Morton, 1987:406			
P. campechiana	Canistel	S. Mexico; Belize; Guatemala; El Salvador	
Source: Morton, 1987:403			
P. lucuma	Lucmo	W. Chile; Peru; S.E. Ecuador	Fruit
Source: Morton, 1987:406			
P. macrophylla	Cutite	Pará, Amapa, E. Amazonas, Brazil	Fruit
Source: Cavalcante, 1976:128			
Solanaceae			
Solanum sessiliflorum	Tupiro	(N.W. Amazonia)	Fruit
Sterculiaceae			
Cola acuminata	Cola nut	S. Nigeria, C. Africa	Nut
C. nitida	Cola nut	Guinea; Sierra Leone; Liberia; Côte d'Ivoire; Nigeria	Nut
Theobroma angustifolium	Cacao silvestre	C. America	Beverage

Family and scientific name	Common name	Location of wild populations[1]	Main use(s)
T. bicolor	Pataste	C. America; Amazonia	Beverage, fruit
T. cacao	Cacao	W. and C. Amazonia; Guianas	Beverage, candy, juice
Sources: J. B. Allen, 1987; J. B. Allen and Lass, 1983; J. G. Myers, 1930			
T. grandiflorum	Cupuaçu	Pará, W. Maranhão, Brazil	Juice, ice cream, puddings
Sources: Balée, 1986, 1987; Cavalcante, 1976:131; N. J. H. Smith, field notes			
T. speciosum	Cacauí	Northern S. America; Southern C. America	Fruit
Sources: Cavalcante, 1988:64; Santa-Anna Nery, 1885:90			
T. subincanum	Cupuí	Amazonia; French Guiana; Venezuela	Fruit
Sources: Cavalcante, 1988; Santa-Anna Nery, 1885			
Tiliaceae			
Grewia asiatica	Phalsa	(India; S.E. Asia)	Fruit
Verbenaceae			
Gmelina arborea	Gmelina	S. and S.E. Asia; S. China	Pulp
Source: NAS, 1980:46			
Tectona grandis	Teak	S.E. Asia	Timber
Vitex cienkowskii	Black plum	W. Africa	Fruit
Source: Dalziel, 1948:456			
Zingiberaceae			
Costus afer	Bush cane	Nigeria	
Source: Dalziel, 1948:472			

Note: Sources given refer to all of the species that immediately precede them.

[1] C. America = Central America, including Mexico; T. America = Tropical America; T. Asia = Tropical Asia. Parentheses indicate that, although not documented, wild populations are suspected to occur here.

[2] Indigenous and naturalized populations; seeds from cultivated forms and escapes dispersed by birds and/or mammals.

[3] Introduced to Philippines, where it is cultivated and grows wild in second growth.

[4] Native to India, where wild populations are found in some mountainous areas; has become naturalized in the Philippines and Sri Lanka (Morton, 1987:59; W. Popenoe, 1920:416).

[5] Grows wild in Philippines, although introduced, possibly from Malaysia.

Abbreviations of some institutions involved in collecting, maintaining, and/or breeding tropical perennial crops

ARI	Agricultural Research Institute
ARO	Agricultural Research Organization
BALITAN	Malang Research Institute for Food Crops
BARI	Bangladesh Agricultural Research Institute
BPI	Bureau of Plant Industry
CATIE	Centro Agronómico Tropical de Investigación y Enseñanza
CEMSA	Centro de Mejoramiento de Semillas Agamicas
CENARGEN	Centro Nacional de Recursos Genéticos e Biotecnologia
CENIAP	Centro Nacional de Investigaciones Agropecuarias
CENRADERU	Centre National de la Recherche Appliquée au Développement Rural
CEPLAC	Comissão Executiva do Plano da Lavoura Cacaueira
CIAB	Centro de Investigaciones Agrícolas de El Bajío
CNPMF	Centro Nacional de Pesquisa de Mandioca e Fruticultura
CPAC	Centro de Pesquisa Agropecuária dos Cerrados
CSIRO	Commonwealth Scientific and Industrial Research Organization
DENPASA	Dendê do Pará, S.A.
DICOF	Dirección de Investigaciones de Citricos y Otros Frutales
DPI	Department of Plant Industry
EMBRAPA	Empresa Brasileira de Pesquisa Agropecuária
EMPASC	Empresa Catarinense de Pesquisa Agropecuária
FCAV	Faculdade de Ciências Agrárias e Veterinárias
FHIA	Fundación Hondureña de Investigación Agrícola
FONAIAP	Fondo Nacional de Investigaciones Agropecuarias
IAC	Instituto Agronômico de Campinas
IBPGR	International Board For Plant Genetic Resources
ICA	Instituto Colombiano Agropecuario
ICG	International Cocoa Genebank
IFAS	Institute of Food and Agricultural Sciences
IFCC	Institut Français du Café et du Cacao
IIHR	Indian Institute of Horticultural Research
IITA	International Institute of Tropical Agriculture
INERA	Institut National pour l'Étude et la Recherche Agronomique
INIAA	Instituto Nacional de Investigación Agraria y Agroindustrial
INIAP	Instituto Nacional de Investigaciones Agropecuarias
INRA	Institut National de la Recherche Agronomique
IPB	Institute of Plant Breeding
IRA	Institut de la Recherche Agronomique
IRFA	Institut de Recherches sur les Fruits et Agrumes
IRRDB	International Rubber Research and Development Board
LBN	Lembaga Biologi Nasional
MARDI	Malaysian Agricultural Research and Development Institute
MIDINRA	Ministerio de Desarrollo Agropecuario y Reforma Agraria
NIHORT	National Horticultural Research Institute
NRCC	National Research Centre for Cashew
TARI	Taiwan Agricultural Research Institute
TFRS	Tropical Fruit Research Station
UC	University of California
UH	University of Hawaii
UNAM	Universidad Nacional Autonómo de México
USDA	United States Department of Agriculture
WINBAN	Windward Islands Banana Growers' Association

APPENDIX 3

Common names and distribution of avocado's relatives

Persea species	Common name(s)	Distribution[1]
alba	Abacatinho	S. Brazil
alpigena		Jamaica (UNAM 225432)
anomala		S. Cuba; Haiti
aurata		C. Brazil
benthamiana		Brazil (Amazonas)
bernardii		Colombia (Merida)
bilocularis		Bolivia
boldufolia		N. Peru
borbonia	Carolina bay	Texas and S.E. USA
buchtienii		W.C. Bolivia
bullata		C. Ecuador
brenesii		Costa Rica
brevipes		Ecuador (Loja)
caerulea	Aguacate de monte, aguacatillo, aguacate cimarron	Costa Rica (nr. Alajuela); Honduras (nr. Tegucigalpa, Valle de los Angeles); Venezuela; Colombia (Sierra Nevada de Santa Marta; UNAM 332693); Nicaragua (Nueva Segovia; UNAM 391644); Panama (Ancon Hill); Bolivia; Peru
caesia		S. Brazil
campii		Ecuador (Azuay)
chamissonis		Mexico (Hidalgo, Puebla, Oaxaca, S. Luis Potosí, Veracruz)
chrysophylla	Aguacatillo	C. Colombia
cinerascens		Mexico[2] (Tacambaro, Michoacan, Zacuapan, Vera Cruz, Oaxaca)
conferta		S.C. Ecuador
corymbosa		N.W. Peru
costata		C. Colombia
cuatrecasasii		Andes (Colombia)
cuneata		E. Colombia
donnell-smithii	Sacsi	Guatemala (Volcán Quezaltepeque, Tactic, Alta and Baja Verapáz); Honduras (Mt. Uyuca, nr. Tegucigalpa); Mexico (Chiapas: UNAM 80687, 13575, 80674, 295220, 80673, 320587; Veracruz: UNAM 391798)
fastigiata		Venezuela; Colombia
ferruginea		S.W. Ecuador
floccosa[3,4]	Aguacate cimaron	Mexico (Volcán Orizaba,

APPENDIX 3—*cont.*

Persea species	Common name(s)	Distribution[1]
		Veracruz; Chinantla, Puebla; Chiapas: UNAM 66890, 80209, 418454, 403022, 80270, 283668)
fuliginosa		São Paulo, Brazil
fulva		C. and S.E. Brazil
fusca		Brazil (Goias)
grandiflora		S.W. Venezuela
haenkeana		S.E. Peru; W. Bolivia
hexanthera	Laurel	Venezuela (Merida); Colombia (E. Cordillera); C. Peru
hintonii	Ahuacate cimaron	Mexico (Nayarit: UNAM 115028, 192190, 13411, 13387, 13410; Veracruz: UNAM 207447; Michoacan: UNAM 343827, 343758, 316657, 336361, 119934, 30421, 92085)
hirta		Peru (Andes)
hypoleuca		Cuba
indica		Azores; Madeira; Canary Islands
jenmani	Canau-yek	Guyana; Venezuela
krugii	Canel	Hispaniola; Puerto Rico
liebmanni		Mexico (Sinaloa: UNAM 329455; Puebla: UNAM 146103; Oaxaca: UNAM 433223; Mt. Ovando, Chiapas: UNAM 80691)
lingue	Lingue	W.C. Chile
longipes	Aguatillo de anis	Mexico (Veracruz; Chiapas: UNAM 72789); Belize
maguirei		Venezuela (Amazonas)
major	Abacate do mato	E.C. Brazil
meridensis	Laurel negro	N.W. Venezuela
meyeniana		W.C. Chile
microphylla		Guanabara, Brazil
mutisii	Canelon, curo de paramo	Colombia (E. cordillera); Venezuela (Andes); C. Ecuador
nivea		Guyana; Surinam; French Guiana
nubigena[3,4]	Aguacate cimarrón (Mexico)	Guatemala (Nuca, Chichoy, Chuchumatanes); Honduras (La Tigra); Mexico (Volcán Orizaba, Veracruz; Chiapas); Costa Rica; Nicaragua (Matagalpa)

APPENDIX 3—*cont.*

Persea species	Common name(s)	Distribution[1]
oblongifolia		Santo Domingo
obovata		East C. Brazil
obtusifolia		Panama (Cerro Copete)
pallida		Costa Rica
palustris	Sweet bay	S. and E. USA; Bahamas
parvifolia[4]	Aguacate cimarrón, aguacatillo	Mexico (Veracruz)
pedunculosa		Minas Gerais, Brazil
peruviana		C. Peru; W.C. Bolivia
podadenia		Mexico (S. Luis Potosí: UNAM 410670; Chihuahua: UNAM 13389; Nuevo Leon: UNAM 80212; Chiapas: 401954)
primatogena[4]	Guaslipe	Nicaragua
pseudofasciculata		W.C. Bolivia; Brazil
punctata		S. Brazil
purpusii		Mexico (S. Luis Potosí)
pyrifolia		E.C. Brazil
raimondii		N. Peru
rigens	Pizarra	Costa Rica; Panama; Ecuador; Guatemala
rigida		São Paulo, Brazil
rufotomentosa		C. Brazil (e.g., Serra do Espinhaço, Minas Gerais: UNAM 263509)
ruizii		Peru (Andes)
schiedeana[4,5]	Coyo, yas, chinene	S. and S.E. Mexico; Guatemala (Verapáz, Huehuetenango, Chiquimula); Costa Rica (Volcán Barba, nr. Cartago, Turrialba); Panama (El Boquete, Chiriqui); Honduras (Lancetilla)
sericea		C. Colombia; S. Ecuador
sessilis		Guatemala
splendens		C. Brazil
standleyi	Aguacatillo (Mexico)	Guatemala; Mexico (Veracruz: UNAM 156874; Puebla: UNAM 191509, 115984; Chiapas: UNAM 80223)
steyermarkii[4]	Tepe-aguacate, aguacate de montaña	Guatemala (San Marcos, nr. Quezaltenango, Jalapa, Solola/Totonicapan, Vuelta de Quetzal); El Salvador; Mexico (Chiapas: UNAM 300730)
stricta		Peru (Andes)
subcordata	Laurelon	Andes Mountains

APPENDIX 3—*cont.*

Persea species	Common name(s)	Distribution[1]
theobromifolia		Rio Palenque, Ecuador
trollii		C. Bolivia
urbaniana	Sweetwood	Jamaica; Puerto Rico; Lesser Antilles
venosa		S.C. Brazil (e.g., Serra do Espinhaço, Minas Gerais: UNAM 263515)
veraguasensis	Aguacatillo	Panama (Chiriqui: UNAM 360749, 202477); Costa Rica; Guatemala (Tajumulco Volcano, Vuelta de Quetzal); Honduras (San Juancito Mts., La Tigra); Mexico (Chiapas: UNAM 80688, 80269)
vesticula		Mexico (Chiapas); Guatemala; Honduras
weberbaueri		S.E. Peru

Sources: Bell and Taylor, 1982:63; Blake, 1920; Coronel, 1983:24; Gentry, 1972; D. Giacometti, pers. comm.; Holdridge et al., 1971:184, 341, 496, 507, 519; IBPGR, 1986:49; Kopp, 1966; Morton, 1987:102; W. Popenoe, 1920, 1927; Pérez and Salán, 1986:216; Schieber and Zentmyer, 1976, 1978a, b, 1979, 1980, 1983; Schieber et al., 1984; Schroeder, 1977; Standley, 1924:291, 1928:183, 1930a; Standley and Steyermark, 1949; L. O. Williams, 1950, 1977; Woodson and Schery, 1948; Zentmyer, 1954, 1957; Zentmyer and Schieber, 1987.

[1]UNAM = Universidad Nacional Autonómo de Mexico, Instituto de Biologia, Herbario Nacional, Mexico City. C = central.

[2]Localities based on accession records of the avocado germplasm collection, Centro de Investigaciones Agrícolas de El Bajío, Guanajuato, Mexico.

[3]Possibly botanical varieties of *P. americana* (Storey et al., 1987).

[4]Closest relatives of avocado in the subgenus *Persea.*

[5]Cultivated on a small scale in Veracruz, Mexico, and in parts of Guatemala (Bergh and Ellstrand, 1987; Morton, 1987:102; W. Popenoe, 1935).

APPENDIX 4

Avocado accessions in germplasm collections

Cultivar	Origin and year selected[1]	Botanical variety[2]	Gene bank(s)[3]
'Agnes'	Miami, Fla.	G × M	1
'Ajax'	Homestead, Fla.	G × W	4
'Alboyce'	Riverside Calif., 1965	Hass sdlg.	1
'Alpha'	Mexico	M	3
'Amatengo'			6
'Amatlan'			6
'Anaheim'	Anaheim, Calif., 1910	G	2, 6
'Andy's Dandy'	Homestead, Fla.	W	1
'Antigua Market'	Antigua, Guatemala	G	1, 6
'Apakia'			6
'Apakia 2'			6
'Aquila 1'			6
'Aquila 2'			6
'Ardith'	Riverside, Calif., 1970	M × G	2, 6
'Arguello'	Managua, Nicaragua	W	1
'Arue'	Society Islands	W	1
'Atlisco 1'	Mexico	M	1
'Avila'	Quebradillar, Puerto Rico	W	4
'Avis'	San Diego, Calif.	M × G	2
'Avocatosa 2'			6
'Avocatosa 3'			6
'Avon'	Avon Park, Fla.	G × W	4
'Aycock Red 3'	Lancetilla, Honduras	W	1
'Aycock Red 19'	Lancetilla, Honduras	W	1
'Azul 1'			6
'Bacon'	Buena Park, Calif., 1951	M × G	1, 2, 4
'Balboa'	San Diego, Calif.	M × G	2
'Barker'	Bradenton, Fla.	W	5
'Basaldua'	Michoacan, Mexica	M	3
'Bassage'	S. Dade Co., Fla.	W × G	1
'Beebe'	Galeto, Calif.	M × G	2
'Belize'	Lancetilla, Honduras	W	1
'Benik'			6
'Bernecker'	South Fla.	W	1
'Beta'	Homestead, Fla., 1966		1
'Big Gladys'	S. Fla.	G × W	1
'Biscayne'	USDA, Miami, Fla.	W	1
'Bitte'	Redland, Fla.	W	4
'Black Prince'	Homestead, Fla.	W	4
'Bonita'	Homestead, Fla.	G × W	4
'Booth 1'	Homestead, Fla.	G × W	4
'Booth 2'	Homestead, Fla.	G × W	4
'Booth 3'	Homestead, Fla.	G × W	4
'Booth 4'	Homestead, Fla.	G × W	4

APPENDIX 4—*cont.*

Cultivar	Origin and year selected[1]	Botanical variety[2]	Gene bank(s)[3]
'Booth 6'	Homestead, Fla.	G × W	4
'Booth 7'	Homestead, Fla.	G × W	1, 4, 5
'Booth 8'	Homestead, Fla.	G × W	1, 4, 5
'Booth 9'	Homestead, Fla.		4
'Booth 10'	Homestead, Fla.		4
'Borchard'[4]	Ventura, Calif.		2
'Borrego'	Havana, Cuba	G × W	1
'Brogdon'	Homestead, Fla.	M	1, 4
'Brooks Late'	Homestead, Fla.	G	1, 4
'Brooksville'	Brooksville, Fla.	M	1
'Brooksville Sdlg.'	USDA, Miami, Fla.	M	1
'Buccaneer'	Fla.		4
'Butler'	Miami, Fla., 1909	W	1
'C-1'	Celaya, Mexico	M	3
'C-3'	Celaya, Mexico	M	3
'C-15'	Celaya, Mexico	M	3
'C-18'	Celaya, Mexico	M	3
'C-19'	Celaya, Mexico	M	3
'C-20'	Celaya, Mexico	M	3
'C-21'	Celaya, Mexico	M	3
'C-22'	Celaya, Mexico	M	3
'C-23'	Celaya, Mexico	M	3
'C-24'	Celaya, Mexico	M	3
'C-25'	Celaya, Mexico	M	1, 3
'C-26'	Celaya, Mexico	M	3
'C-28'	Celaya, Mexico	M	3
'C-29'	Celaya, Mexico	M	3
'C-30'	Celaya, Mexico	M	3
'C-31'	Celaya, Mexico	M	3
'C-32'	Celaya, Mexico	M	3
'C-37'	Celaya, Mexico	M	3
'C-39'	Celaya, Mexico	M	3
'C-40'	Celaya, Mexico	M	3
'C-41'	Celaya, Mexico	M	3
'C-42'	Celaya, Mexico	M	3
'C-43'	Celaya, Mexico	M	3
'C-44'	Celaya, Mexico	M	3
'C-47'	Celaya, Mexico	M	3
'C-48'	Celaya, Mexico	M	3
'C-79'	Celaya, Mexico	M	3
'C-81'	Celaya, Mexico	M	3
'C-118'	Celaya, Mexico	M	3
'C-120'	Celaya, Mexico	M	3
'Camp'			4
'Camulas'			4

APPENDIX 4—*cont.*

Cultivar	Origin and year selected[1]	Botanical variety[2]	Gene bank(s)[3]
'Canaries'	Canary Islands	W	1
'Capac'	Chota Valley, Ecuador	M	1, 4
'Capri'	Weslaco, Tex.		1
'Carchi'	Ibarra, Ecuador	M	1
'Case'	Hawaii	G × W	1
'Catalina'	Havana, Cuba	W	1, 4
'Catalina cardoso'			5
'Cellon's Hawaii'	Miami, Fla.	W	1
'Cellon's Hawaii Sdlg.'	USDA, Miami, Fla.		1
'Cerezo'			6
'CH 4'	Hawaii, 1970	M × W	4
'CH 5'	Hawaii, 1970	M × W	4
'Chandler'			4
'Chapultepec Park'	Chapultepec, Mexico, D. F.	M	1
'Charol'	Nuevo Leon, Mexico	M	3
'Chica'			4
'Chinini'			6
'Choquette'	Homestead, Fla.	G × W	1, 4, 5
'Chota'	Ibarra, Ecuador, 1921	M	1
'Clifton'	Glendora, Calif., 1949	(M × G) × M	2
'Colin V33'	Mexico	M × G	2
'Collinred Sdlg. B'	USDA, Miami, Fla.	G × W	1
'Collins'	Guatemala	G	1, 4
'Collinson'	Miami, Fla., 1920	G × W	1, 4, 5
'COM-49'	Comonfort, Gto., Mexico	M	3
'COM-50'	Comonfort, Gto., Mexico	M	3
'COM-51'	Comonfort, Gto., Mexico	M	3
'COM-52'	Comonfort, Gto., Mexico	M	3
'COM-53'	Comonfort, Gto., Mexico	M	3
'COM-54'	Comonfort, Gto., Mexico	M	3
'COM-55'	Comonfort, Gto., Mexico	M	3
'COM-56'	Comonfort, Gto., Mexico	M	3
'COM-57'	Comonfort, Gto., Mexico	M	3
'COM-58'	Comonfort, Gto., Mexico	M	3
'COM-59'	Comonfort, Gto., Mexico	M	3
'COM-60'	Comonfort, Gto., Mexico	M	3
'COM-61'	Comonfort, Gto., Mexico	M	3
'COM-62'	Comonfort, Gto., Mexico	M	3
'COM-63'	Comonfort, Gto., Mexico	M	3
'COM-64'	Comonfort, Gto., Mexico	M	3
'COM-65'	Comonfort, Gto., Mexico	M	3

APPENDIX 4—*cont.*

Cultivar	Origin and year selected[1]	Botanical variety[2]	Gene bank(s)[3]
'COM-66'	Comonfort, Gto., Mexico	M	3
'COM-67'	Comonfort, Gto., Mexico	M	3
'COM-70'	Comonfort, Gto., Mexico	M	3
'COM-74'	Comonfort, Gto., Mexico	M	3
'COM-75'	Comonfort, Gto., Mexico	M	3
'COM-77'	Comonfort, Gto., Mexico	M	3
'COM-78'	Comonfort, Gto., Mexico	M	3
'COM-83'	Comonfort, Gto., Mexico	M	3
'COM-85'	Comonfort, Gto., Mexico	M	3
'COM-86'	Comonfort, Gto., Mexico	M	3
'COM-87'	Comonfort, Gto., Mexico	M	3
'COM-88'	Comonfort, Gto., Mexico	M	3
'COM-89'	Comonfort, Gto., Mexico	M	3
'COM-90'	Comonfort, Gto., Mexico	M	3
'COM-91'	Comonfort, Gto., Mexico	M	3
'COM-93'	Comonfort, Gto., Mexico	M	3
'COM-94'	Comonfort, Gto., Mexico	M	3
'COM-95'	Comonfort, Gto., Mexico	M	3
'COM-96'	Comonfort, Gto., Mexico	M	3
'COM-107'	Comonfort, Gto., Mexico	M	3
'Comitan 1'			6
'Comitan 3'			6
'Compadre guerendiro'	Michoacan, Mexico	M	3
'Connor'			4
'Copete Colorado'	Michoacan, Mexico	M	3
'Courtright'	Lakeland, Fla.	M	4
'CPK 5'			6
'CRC 14-11'	Riverside, Calif., 1960	(M × G) × M	4
'CRC 14-16'	Riverside, Calif. 1960	(M × G) × M	4
'CRC 23-18'	Riverside, Calif., 1960	M × G	4
'CRC 151-2'			
'CRC 1751'	Riverside, Calif., 1960	(M × G) × M	4
'Creelman'	Ventura, Calif., 1965	'Fuerte' × Hass	1, 4
'Cristina'		W	4
'Cuerno morado'	Nuevo Leon, Mexico	M	3
'Cuevas'			6
'Cuevas desconocido No. 5'	Michoacan, Mexico	M	3
'D6'[4]	Riverside, Calif., 1960	M	2
'Dade'	Miami, Fla., 1920	W	1, 6
'Day'	Homestead, Fla.		1, 6
'DD17'	Riverside, Calif., 1965	M × G	2
'De-Bart'			6
'DeBeldts'			4

APPENDIX 4—*cont.*

Cultivar	Origin and year selected[1]	Botanical variety[2]	Gene bank(s)[3]
'Del Oro 1'	Clearwater, Fla.	M	1
'DF'			2, 5
'DF 3'			5
'DF 4'			5
'DF 6'			5
'DF 7'			5
'Dickinson'	Altadena, Calif., 1899	G	1, 2
'Donaldson'	Lancetilla, Honduras	W	1
'Dora'	Lancetilla, Honduras	W	1
'Dourado'	Londrina, Paraná, Brazil		1
'Dr. Dupuis 2'	Hialeah, Fla.	W	1
'Duke'	Oroville, Calif., 1912	M	1, 4, 5
'Dunedin'	Miami, Fla.	G × W	1, 4
'Dwarf Herman'			4
'D.W.I. Bank'	Danish West Indian Bank, St. Croix, Virgin Islands	W	1
'Edmonds'	Nr. Homestead, Fla.	G × W	4
'Edranol'	Vista, Calif., 1932	M × G	2
'Egas'	Chota Valley, Ecuador, 1921	M	1, 4
'Ein-Shemer'			6
'Ein-Vered'	Bet-Dagan, Israel	M × G	1, 6
'El Venado'			6
'Epigmento'	Michoacan, Mexico	M	3, 6
'Esther'	Riverside, Calif.	Nabal sdlg. × Thille sdlg.	2
'Ettinger'	Tel Aviv, Israel, 1965	M × G	1, 2, 4, 6
'Extra'			4
'Fairchild'	Miami, Fla.	Collinred sdlg.	4
'Family'	Miami, Fla.	W	1, 4
'Floreno'	Nuevo Leon, Mexico	M	3
'Fortuna'			5
'Frazer'	New Zealand	G	2
'Fuca'	Morocco	G × M	1
'Fuchsia'	Homestead, Fla., 1916	W	1, 4
'Fuerte'	Atlixco, Mexico, 1911	G × M	1, 2, 4, 5, 6
'Fuerte Popenoe'	Ecuador, 1937	G × M	1
'G6'[4]	Riverside, Calif.	M	2
'G755C'[4]	Riverside, Calif.	*P. schiedeana* G	2
'Gainesville'	Gainesville, Fla.	M	1, 4
'Galvan'			6
'Garcia 1'	Duarte, Calif.	G	4
'General Bureau'	Morocco	W	1
'General Escolar'			5
'General	Quito, Ecuador	M	1

APPENDIX 4—*cont.*

Cultivar	Origin and year selected[1]	Botanical variety[2]	Gene bank(s)[3]
Francisco Robles'			
'Girradin'	Fort Myers, Fla.	G × W	1
'Goering'		Fuerte sdlg.?	4
'Gordo'	Riverside, Calif., 1966	G	2
'Gottfried'	Miami, Fla., 1914	M	1, 4
'Gottfried Sdlg.'			1
'Grandão'			5
'Greengold'	Oahu, Hawaii	G × M	1, 5
'Green Hass'			6
'Greenstem'			4
'Gripina 5'	St. Croix, Virgin Islands	G × W	1
'Gripina 12'	Fortuna, Puerto Rico	G × W	1
'Gripina 13'			4
'Guat 1'			6
'Guayabamba'	Chapingo, Mexico	M	1, 6
'Guzman'			6
'Gwen'	Riverside, Calif., 1968	(G × M) × G	2
'H × 48'	Riverside, Calif., 1975	(G × M) × G	2
'H287'	Riverside, Calif., 1970	Hass sdlg.	2
'H670'	Riverside, Calif., 1970		2
'H696'	Riverside, Calif.	Hass sdlg.	2
'Haas'	Miami, Fla.		1
'Haile'			4
'Hall'	Miami, Fla., 1938	G × W	4
'HAL-R27-T8'	Mexico	M	3
'Hardee'		W	4
'Hashimoto'	Hawaii	G	2
'Hass'	La Habra Heights, Calif., 1936	(G × M) × G	2, 4, 5, 6
'Hass F2'			6
'Hass 599'			6
'Hawaii'	Hawaii	mostly W	1, 4
'Hayes'	Hawaii	Hass sdlg.	1, 2
'H. de Toro'	Nuevo Leon, Mexico	M	3
'H. de Toro A'	Nuevo Leon, Mexico	M	3
'Henry Select'	Escondido, Calif.		4
'Herculano'			5
'Hermann'	Fla.		4
'Hiwassee'	Orlando, Fla.	G × W	1
'Horshim'	Israel, 1971	M × G	1, 2
'HT 124 NE, NW'	Goulds, Fla.		1
'I392'	Riverside, Calif., 1982	G	2
'Ile de France'	Morocco	G × W	1
'Irving'	Carlsbad, Calif., 1935	M × G	2
'Irwing 34'			4

APPENDIX 4—*cont.*

Cultivar	Origin and year selected[1]	Botanical variety[2]	Gene bank(s)[3]
'Irwing 59'	Boynton Beach, Fla.	Complex hybrid	1, 4
'Irwing 65'	Boynton Beach, Fla.	Complex hybrid	1, 4
'Irwing 78'	Boynton Beach, Fla.	Complex hybrid	1, 4
'Irwing 96'	Boynton Beach, Fla.	Complex hybrid	1, 4
'Irwing 120'	Boynton Beach, Fla.	Complex hybrid	1, 4
'Irwing 134'	Boynton Beach, Fla.	Complex hybrid	1, 4
'Ishim'	Antigua, Guatemala	G	1
'Ishkal'	Guatemala City, Guatemala, 1916	G	1
'Itzamna'	Santa Maria de Jesus, Guatemala, 1916	G	1, 4
'Itzours'	USDA, Miami, Fla.	G × W	1
'Izidora'			5
'J241'	Riverside, Calif.	G	2
'Jan Boyce'	Riverside, Calif., 1965	M × G	2
'Jim'			6
'JJ9'			6
'JJ12'	Riverside, Calif., 1965	G	2
'JP Young'			4
'Jose Antonio'	Havana, Cuba	W	1
'JR-114'	Juventino Rosas, Gto., Mexico	M	3
'Julia'	Riverside, Calif.	G	2
'Julia N-526'			6
'Kahaluu 30-15'		W × G	4
'Kalusa'			4
'Kalusa Sdlg.'			4
'Kampong'			4
'Kanan'	San Lorenzo El Cubo, Guatemala, 1917	G	1
'Kanan Sdlg.'	Miami, Fla.	G × W	1
'Kashlan'	San Cristobal, Verapáz, Guatemala, 1917	G	4
'Kayab'	San Cristobal, Verapáz, Guatemala, 1917	G	1
'Key Largo'	Homestead, Fla.	G	1
'Kilo 4 Ceiba'	Tela, Honduras	W	1
'KL'			4
'KM 145'			6
'Kosel'	Miami, Fla.?	M × W?	1
'L28'	Riverside, Calif., 1970	Linda sdlg.	2
'L137'	Riverside, Calif., 1970	Linda sdlg.	2
'L141'	Riverside, Calif., 1979	Linda sdlg.	2
'La Cruz'			6
'Lamat'	Amatitlan, Guatemala, 1916	G	1
'Lana 1'			5
'La pinera IV'	Michoacan, Mexico	M	3

Appendix 4—*cont.*

Cultivar	Origin and year selected[1]	Botanical variety[2]	Gene bank(s)[3]
'La pinera V'	Michoacan, Mexico	M	3
'La pinera VI'	Michoacan, Mexico	M	3
'Larralde'	Nuevo Leon, Mexico	M	3
'Las campanas'	Nuevo Leon, Mexico	M	1
'La sierra'			6
'Lavshan'			4
'Leal'	Mexico	M	3
'Lima Late'	Honduras	G	1, 5
'Linda'	Antigua, Guatemala, 1914	G	1, 4, 5
'LL13'	Riverside, Calif., 1970	M × G	2
'Lula'	Miami, Fla.,1919	G × W	1, 5
'Lula 1'			4
'Lyon'	Hollywood, Calif., 1911	(M × G) × G	2
'Lyon 172'			6
'Lypps'	San Diego, Calif.		2
'MacDonald'	Honolulu,Hawaii, 1895	G	1, 4
'Major Schaff'			4
'Manik'	Antigua, Guatemala, 1917	G	4
'Marcianeco'			6
'Marcus Pumpkin'	Boynton Beach, Fla.	G × W	1
'Marfield'	Nr. Homestead, Fla., ca. 1930		4
'Margarida'	Londrina, Paraná, Brazil		1
'Marguerite'	Nr. Homestead, Fla., ca. 1930	G × W	4
'Marshelline'	Boynton Beach, Fla.	Complex hybrid	1, 4, 6
'Maskaria 1'			6
'Maskaria 2'			6
'Maxima'	Homestead, Fla.	W	1
'Mayo'	Covina, Calif., 1947	M	2
'Mayo 133'			6
'McGill'	Homestead, Fla.		4
'McKay Twin'	Lake Alfred, Fla.	M	1
'Melendez 2'	St. Croix, Virgin Islands	W × G	1
'Mesa'	Carpinteria, Calif., 1950	(M × G) × G	4
'Mexicola'	Pasadena, Calif., 1912	M	2, 4
'Mex Sdlg.'			4
'Monroe'	Homestead, Fla., 1937	G × W	1
'Montgomery Late'			4
'Moreno'			6
'Morocco 42'	Rabat, Morocco	M	1
'Morocco 43'	Rabat, Morocco	M	1
'Mrs. Lowe'			4
'MS9'	Riverside, Calif., 1970	G	2
'Murashige'	Hawaii	G hybrid	1

APPENDIX 4—*cont.*

Cultivar	Origin and year selected[1]	Botanical variety[2]	Gene bank(s)[3]
'N-7'	Riverside, Calif.	G	4
'N 12'			6
'N-34'	Riverside, Calif.	G	4
'N 66'			6
'NA37'	Riverside, Calif., 1970	Nabal sdlg.	2
'NA251'	Riverside, Calif., 1970	Nabal sdlg.	2
'Nabal'	Antigua, Guatemala, 1917	G	1, 2, 4, 6
'Nabal Sdlg.'	Boynton Beach, Fla.	G	1
'Nabal Sdlg.'			4
'Nadin'	Aladin City, Fla., 1943		4
'Naranja'			4
'Naranjo 1'	Oahu, Hawaii	M	1
'Naranjo 2'	Mexico	M	1
'Naranjo 3'	Mexico	M	1
'NB86'	Riverside, Calif., 1972	Nabal sdlg.	2
'Negra pera'			6
'Nena'			4
'Nesbitt'	Homestead, Fla.	G × W	1, 4
'Netaim'	Israel, 1971	G	1, 2
'Nezahualcoyotl'			4
'Nimlioh'	Antigua, Guatemala, 1917	G	2, 4
'Nir'			6
'Nirody'	Rockdale, Fla., 1922	G × W	4
'NN10'	Riverside, Calif., 1972	G × M	2
'NN39'			6
'NN63'	Riverside, Calif., 1972	G × M	2
'Noga'	San Francisco, Calif., 1980	G × M	2
'Nordshtein'	Israel	G	1
'Norman'			4
'Northrop'	Santa Ana, Calif., 1911	M	2
'Norwood'	Miami, Fla.	G × W	1
'Novillero'			6
'Nowels'	Huntington Park, Calif., 1940	(G × M) × M	2
'NT4'	Calif.		4
'Orco'	Orange County, Calif.		2
'Organal'	Mexico	M	3
'Orit'			6
'Orizaba'			3, 6
'P78'	Riverside, Calif.	Pinkerton sdlg.	2
'Pag-7'	Mexico	M	3
'Pancho'	Mexico	M	1
'Peterson'	Homestead, Fla., 1917	W	1, 4
'Pic 9615'			6
'Pinera-VII'	Michoacan, Mexico	M	3
'Pink'			6
'Pinkerton'	Saticoy, Calif., 1979	G × M	1, 2, 6

APPENDIX 4—*cont.*

Cultivar	Origin and year selected[1]	Botanical variety[2]	Gene bank(s)[3]
'Pollock'	Miami, Fla., 1896	W	1, 4, 5
'Pope'			4
'Progresso Late'	Tela, Honduras	G	1
'Proteco'			1
'Pumpkin'	Homestead, Fla.		4
'PT37'	Riverside, Calif.	G	2
'Queen'	Nr. Antigua, Guatemala, 1914	G	1
'Queen 8'	Boynton Beach, Fla.	G	1, 4
'Queen 9'	Boynton Beach, Fla.	G	1, 4
'Queretaro 1'	Queretaro, Mexico	M	3
'Queretaro 2'	Queretaro, Mexico	M	3
'Queretaro 3'	Queretaro, Mexico	M	3
'Queretaro 4'	Queretaro, Mexico	M	3
'Queretaro 5'	Queretaro, Mexico	M	3
'Queretaro 6'	Queretaro, Mexico	M	3
'Queretaro 7'	Queretaro, Mexico	M	3
'Queretaro 8'	Queretaro, Mexico	M	3
'Queretaro 9'	Queretaro, Mexico	M	3
'Queretaro 10'	Queretaro, Mexico	M	3
'Queretaro 17'	Queretaro, Mexico	M	3
'Quintal'	São Paulo, Brazil	W × G	5
'R16-T11'	Mexico	M	3
'Raul Arango'	Tela, Honduras	W	1
'Reed'	Carlsbad, Calif., 1960	G	1, 2, 6
'Reina Victoria'	Quito, Ecuador	M	1
'Reinecke 1'	San Diego, Calif.	M × G	4
'Reinecke 12'	San Diego, Calif.	M × G	4
'Reyna'	Michoacan, Mexico	M	3
'Rincon'	Carlsbad, Calif., 1948	M × G	2
'Rivera 3'	Mexico	M	3
'Roberts Sdlg.'		M × ?	4
'Rodriguez'	Nuevo Leon, Mexico	M	3
'Romain 1'	Rabat, Morocco	M	1
'Rosita'	Nuevo Leon, Mexico	M	3
'Ruehle'	Homestead, Fla., 1923	W	1, 4
'Ryan'	Whittier, Calif., 1927	(G × M) × G	5
'San Miguel'	San Miguel Allende, Mexico	M	3
'San Miguel Green'	Mexico	M	1, 3
'Santana'			6
'Santini'			4
'Scotland'			4
'Seedless Avocado'	Costa Rica		1
'Seedless Mexican'	Orange County, Calif.	M	1, 4

APPENDIX 4—*cont.*

Cultivar	Origin and year selected[1]	Botanical variety[2]	Gene bank(s)[3]
'Seleção ceasa'	Brazil		5
'Sem caroco'			5
'Semil'			6
'Semil 34'	Fortuna, Puerto Rico	W	1
'Semil 43'	Fortuna, Puerto Rico	W × G	1
'Semil 44'	Homestead, Fla.	W × G	1
'Sharwil'	Redland Bay, Australia	M × G	1, 2, 6
'Sholola'			6
'Shomrat'	Israel	G	1, 6
'Shiller-1'			6
'Simmonds'	Miami, Fla., 1913	W (Pollock sdlg.)	1, 4, 5
'Simmonds Sdlg.'			1
'Simon'	Homestead, Fla.	G × W	1
'SMA-97'	San Miguel Allende, Gto., Mexico	M	3
'SMA-98'	San Miguel Allende, Gto., Mexico	M	3
'SMA-99'	San Miguel Allende, Gto., Mexico	M	3
'SS23'	Riverside, Calif., 1975	M × G	2
'Steffani'			4
'Stewart'	Calif., 1956	(M × G) × M	2
'Streamliner'			4
'Stuart'			6
'Suguiura'			5
'T2'	Riverside, Calif., 1970	G (Thille sdlg.)	2, 6
'T32'	Riverside, Calif., 1970	G (Thille sdlg.)	2
'T142'	Riverside, Calif., 1970	G (Thille sdlg.)	2
'T181'	Riverside, Calif., 1970	G (Thille sdlg.)	2
'T205'	Riverside, Calif., 1970	G (Thille sdlg.)	2
'T362'	Riverside, Calif., 1970	G (Thille sdlg.)	2
'Tacamba Rey'			6
'Tacho'			4
'Tamayo'	Chota Valley, Ecuador, 1921	M	1, 4
'Tanachuri I'	Michoacan, Mexico	M	3
'Tanachuri II'	Michoacan, Mexico	M	3
'Tanachuri III'	Michoacan, Mexico	M	3
'Tano'	Nuevo Leon, Mexico	M	3
'Tappen'			4
'Tardia'	Havana, Cuba	G × W	1
'Taylor'	Miami, Fla., 1913	G	1, 4
'TC-04-74-8'	Nuevo Leon, Mexico	M	3
'TC-05-74-8'	Nuevo Leon, Mexico	M	3
'TC-06-74-8'	Nuevo Leon, Mexico	M	3

APPENDIX 4—*cont.*

Cultivar	Origin and year selected[1]	Botanical variety[2]	Gene bank(s)[3]
'TC-07-74-5'	Nuevo Leon, Mexico	M	3
'TC-17-74-4'	Nuevo Leon, Mexico	M	3
'Teague'	Riverside, Calif., 1968	Fuerte × Duke	1, 2, 6
'Tela'			4
'Tenango 8'	Tenango, Gto., Mexico	M	3
'Tenango 10'	Tenango, Gto., Mexico	M	3
'Tenango II'	Tenango, Gto., Mexico	M	3
'Tenerife'	Canary Islands	G × M	1
'Tensen'	Tenerife, Canary Islands	G × M	1
'Tertoh'	Mexico; Guatemala, 1917	G	1
'Tezivtalan'			6
'Thille'	Ventura, Calif., 1959	Hass sdlg.	2
'Thomas'[4]	Escondido, Calif.	M	2
'Tito perla'	Tela, Honduras	W	1
'Toltec'	Puebla, Mexico	M	1
'Tonnage'	Homestead, Fla., 1930	G × W	1, 5
'Topa Topa'	Ojai, Calif., 1907	M	2, 4, 6
'Tova'	Israel, 1971		1, 4, 6
'Tower 2'	South Dade, Fla.	W (Waldin sdlg.)	1
'Trapp'	Coconut Grove, Fla., 1984	W	1
'Trappson'	USDA, Miami, Fla., 1921	(GxW) × W	1
'Trinidad'			4
'Tucuata'			6
'Turner'			4
'TV-1'	Puebla, Mexico	M	3
'TV-2'	Puebla, Mexico	M	3
'TV-3'	Puebla, Mexico	M	3
'TV-4'	Puebla, Mexico	M	3
'TV-5'	Puebla, Mexico	M	3
'TX531'	Riverside, Calif., 1975	G × M	2
'TX753'	Riverside, Calif., 1978	G (sdlg. of Thille sdlg.)	2
'TX756'	Riverside, Calif., 1978	G (sdlg. of Thille sdlg.)	2
'USDA 21'	Riverside, Calif.		4
'Utuado'	St. Croix, Virgin Islands	G × W	1
'Vaca'			4
'Valdin'			5
'V-B-2'	Mexico	M	3
'Ver-1'	Veracruz, Mexico	M	3
'Ver-3'	Veracruz, Mexico	M	3
'Ver-5'	Veracruz, Mexico	M	3
'Ver-6'	Veracruz, Mexico	M	3
'Ver-9'	Veracruz, Mexico	M	3
'Ver-10'	Veracruz, Mexico	M	3
'Ver-12'	Veracruz, Mexico	M	3

APPENDIX 4—*cont.*

Cultivar	Origin and year selected[1]	Botanical variety[2]	Gene bank(s)[3]
'Ver-15'	Veracruz, Mexico	M	3
'Ver-16'	Veracruz, Mexico	M	3
'Ver-17'	Veracruz, Mexico	M	3
'Ver-18'	Veracruz, Mexico	M	3
'Ver-22'	Veracruz, Mexico	M	3
'Ver-23'	Veracruz, Mexico	M	3
'Ver-24'	Veracruz, Mexico	M	3
'Ver-27'	Veracruz, Mexico	M	3
'Ver-30'	Veracruz, Mexico	M	3
'Ver-33'	Veracruz, Mexico	M	3
'Ver-34'	Veracruz, Mexico	M	3
'Ver-35'	Veracruz, Mexico	M	3
'Ver-36'	Veracruz, Mexico	M	3
'Ver-37'	Veracruz, Mexico	M	3
'Ver-38'	Veracruz, Mexico	M	3
'Ver-39'	Veracruz, Mexico	M	3
'Ver-40'	Veracruz, Mexico	M	3
'Ver-41'	Veracruz, Mexico	M	3
'Ver-43'	Veracruz, Mexico	M	3
'Ver-44'	Veracruz, Mexico	M	3
'Ver-45'	Veracruz, Mexico	M	3
'Ver-46'	Veracruz, Mexico	M	3
'Ver-47'	Veracruz, Mexico	M	3
'Ver-48'	Veracruz, Mexico	M	3
'Ver-49'	Veracruz, Mexico	M	3
'Ver-50'	Veracruz, Mexico	M	3
'Vero'			4
'Vero Beach'	Vero Beach, Fla.	M hybrid	1
'Vero Beach 1'	Vero Beach, Fla.	M × G?	1
'Vero Beach 1–B'	Vero Beach, Fla.	M hybrid	1
'Vero Beach 5'	Vero Beach, Fla.	M × W hybrid	1
'Vero Beach 15'	Vero Beach, Fla.	M hybrid × W	1
'Vero Beach S.E. 2'	Vero Beach, Fla.	M hybrid × W	1
'Vitoria'			5
'Wagner'	Hollywood, Calif., 1914	G	4, 5
'Waldin'	Homestead, Fla., 1913	W	1
'Waldo'			4
'Wanda'			5
'Ward'	Sierra Madre, Calif., 1914	G × M	1
'Wertz'	Delray Beach, Fla.		1
'Wester'	Coconut Grove, Fla., 1871	W	1
'Whitsell'	Riverside, Calif., 1966	G	2
'Wilson Popenoe'	Cuba	W	1
'Winslow'	Miami, Fla., 1904	G	4

APPENDIX 4—*cont.*

Cultivar	Origin and year selected[1]	Botanical variety[2]	Gene bank(s)[3]
'Winslowson'	Miami, Fla., 1917	G × W	4, 5
'Winslowson Sdlg.'	Fla.	G × W	1
'Winter Late'			4
'Winter Mexican'	Palm Beach, Fla.	G × M	1, 4
'Wurtz'	Encinitas, Calif., 1940	G	2, 6
'WW5'			6
'Y-3'	Boynton Beach, Fla.	Complex hybrid	1, 4
'Y-381'	Boynton Beach, Fla.	Complex hybrid	1
'Y-423'	Boynton Beach, Fla.		
'Yama 175'	Riverside, Calif., 1970	M	4
'Yama 381'	Riverside, Calif., 1970	M	4
'Ygnacio'	Saratoga, Calif.	M	4
'Yon'	Homestead, Fla.	G × W	1, 4
'Young 1'	Gainesville, Fla.		1
'Young 2'	Gainesville, Fla.		1
'Young Slipskin'	Mulberry, Fla.	M	1
'Young Special'	Mulberry, Fla.	M	1
'Younghans'	Miami, Fla.	M	1
'XX3'	Riverside, Calif., 1970	G (Murrieta Green sdlg.)	2
'Zutano'	Fallbrook, Calif., 1941	(M × G) × M	2
'3/1/1'	Riverside, Calif., 1972	M (Mexicola sdlg.)	2
'23 Sdlg.'	Riverside, Calif., 1968	Hass sdlg.	2
'23-19'			4
'27-24'			4
'86'	Riverside, Calif., 1975	(G × M) × G	2
'102'	Riverside, Calif., 1972	Whitsell sdlg.	2
'6836'	Boynton Beach, Fla.		1, 4
'7315'			4
'9670'			4
'9741'			4
'9744'			4
'9746'			4
'9756'			4
'9759'			4
'9760'			4
'9775'			4
'9776'			4

Sources: Cultivar lists published periodically in the *Yearbook of the California Avocado Society*; O. Atkins, pers. comm.; B. O. Bergh, pers. comm.; M. Coffey, pers. comm.; W. Krome, pers. comm.; E. Michelson, pers. comm.; N. J. H. Smith, field notes; Toy, 1931; Wolfe et al., 1934.

[1]For some cultivars, the origin is unknown or the year selelcted uncertain.

[2]G = Guatemalan; M = Mexican; W = West Indian. Blank = unknown.

APPENDIX 4—*cont.*

[3]1 = USDA Subtropical Horticulture Research Station, Miami, Florida. 2 = Citrus Experiment Station, University of California, Riverside, and the South Coast Field Station, Irvine, California. 3 = Centro de Investigaciones Agrícolas de El Bajío, Celaya, Guanajuato, Mexico. 4 = Tropical Research and Education Center, Institute of Food and Agricultural Sciences, University of Florida, Homestead, Florida. 5 = Centro de Pesquisa Agropecuária dos Cerrados, Empresa Brasileira de Pesquisa Agropecuária, Planaltina, D.F., Brazil. 6 = Germplasm Bank of Avocado, Volcani Center, Bet-Dagan, Israel.

[4]Rootstock.

References

Addison, G.C., and M. Tavares. 1952. Hybridization and grafting in species of *Theobroma* which occur in Amazonia. *Evolution* 6:380–386.

Adegbola, M.O.K. 1981. Cocoa diseases of West Africa. In: *Proceedings of the 7th International Cocoa Research Conference, Douala, Cameroon*, pp. 230–243.

Adu-Ampomah, Y., F. Novak, R. Afzar, and M. van Durren. 1987. Determination of methodology to obtain shoot tip culture of cocoa. In: *Proceedings of the 10th International Cocoa Research Conference, Santo Domingo, Dominican Republic, 17–23 May*, pp. 137–142.

Afonso, F.M.A. 1979. *O cacau na Amazônia.* Comissão Executiva do Plano da Lavoura Cacaueira (CEPLAC), Itabuna, Brazil, Boletim Técnico 66.

AID. 1981. Leucaena leucocephala: *A Tree That "Defies the Woodcutter."* Agency for International Development, Office of Agriculture, Development Support Bureau, Washington, D.C.

Akihama, T., T. Schichijo, S.A. Tsuchiya, and T. Ueno. 1985. Evaluation and utilization of *Citrus* germplasm—the case of IBPGR *Citrus* project. In: *Proceedings of the International Symposium on South East Asian Plant Genetic Resources*, ed. K.L. Melira and S. Sastrapradja, pp. 104–111. LIPI, Bogor, Indonesia.

Akiyama, T., and R.C. Duncan. 1982. *Analysis of the World Coffee Market.* World Bank, Staff Commodity Working Paper 7, Washington, D.C.

Alcorn, J.B. 1984. Development policy, forests, and peasant farms: reflections on Huastec-managed forests' contribution to commercial production and resource conservation. *Economic Botany* 38:389–406.

——. 1989. An economic analysis of Huastec Mayan forest management. In: *Fragile Lands of Latin America: Strategies for Sustainable Development*, ed. J.O. Browder, pp. 182–203. Westview Press, Boulder, Colo.

Allen, C.K. 1939. Studies in the Lauraceae III: some critical and new species of *Cinnamomum* and *Neocinnamomum. Journal of the Arnold Arboretum* 20:44–63.

Allen, J.B. 1981. Collecting wild cacao in its centre of diversity. In: *Proceedings of the 8th International Cocoa Research Conference, Cartagena, Colombia*, pp. 655–662.

——. 1987. London cocoa trade Amazon project: final report phase 2. *Cocoa Growers' Bulletin* 39:1–94.

Allen, J.B., and R.A. Lass. 1983. London cocoa trade amazon Project: final report phase 1. *Cocoa Growers' Bulletin* 34:1–71.

Allen, J.C., and D.F. Barnes. 1983. *Deforestation, Wood Energy, and Development*. Resources for the Future, Washington, D.C.

Allen, J.C., and D.F. Barnes. 1985. The causes of deforestation in developing countries. *Annals of the Association of American Geographers* 75:163–184.

Allen, P.H. 1956. *The Rain Forests of Golfo Dulce*. University of Florida Press, Gainesville.

——. 1967. Indians of southeastern Colombia. *Geographical Review* 37:567–582.

Allen, P.W., and K.P. Jones. 1988. A historical perspective of the rubber industry. In: *Natural Rubber Science and Technology*, ed. A.D. Roberts, pp. 1–34. Oxford University Press, Oxford.

Almeida, C.M.V.C. de, and C.F.G. Almeida. 1987. Coleta de cacau silvestre no Estado de Rondônia, Brasil. *Revista Theobroma* 17(2):65–92.

Almeida, C.M.V.C. de, J.P. Barriga, P.F.R. Machado, and B.G.D. Bartley. 1987. *Evolução do programa de conservação dos recursos geneticos de cacau na Amazônia brasileira*. Comissão Executiva do Plano da Lavoura Cacaueira (CEPLAC), Departamento Especial da Amazônia, Boletim Técnico 5, Belém, Brazil.

Almeida, L.C., and T. Anderhan. 1987. Recuperação de plantações de cacau com alta incidencia de vassoura-de-bruxa na Amazônia brasileira. In: *Proceedings of the 10th International Cocoa Research Conference, Santo Domingo, Dominican Republic, 17–23 May*, pp. 337–339.

Altieri, M.A., and L.C. Merrick. 1988. Agroecology and in situ conservation of native crop diversity in the Third World. In: *Biodiversity*, ed. E.O. Wilson and F.M. Peter, pp. 361–369. National Academy Press, Washington, D.C.

Alvim, P. 1977. The balance between conservation and utilization in the humid tropics with special reference to Amazonian Brazil. In: *Extinction Is Forever: The Status of Threatened and Endangered Plants in the Americas*, ed. G.T. Prance and T.S. Elias, pp. 347–352. New York Botanical Garden, Bronx.

——. 1989. Tecnologias apropriadas para a agricultura nos trópicos úmidos. *Agrotrópica* 1(1):1–22.

Alvim, P., and R. Alvim. 1980. Energy crops: a new challenge for tropical regions. *Renewable Energy Review Journal* 2(2):1–13.

Ameha, M. 1991. Significance of Ethiopian coffee genetic resources to coffee improvement. In: *Plant Genetic Resources of Ethiopia*, ed. J.M.M. Engels, J.G. Hawkes, and M. Worede, pp. 354–359. Cambridge University Press, Cambridge.

Amin, M.N., and V.S. Jaiswal. 1987. Rapid clonal propagation of guava through in vitro shoot proliferation on modal explants of mature trees. *Plant Cell, Tissue and Organ Culture* 9:235–243.

Ampofo, S.T., and K. Osei-Bonsu. 1987. Models for rehabilitating small scale cocoa farms in Ghana. In: *Proceedings of the 10th International Cocoa Research Conference, Santo Domingo, Dominican Republic, 17–23 May*, pp. 51–55.

Anderson, A.B. 1987. Forest management issues in the Brazilian Amazon. Unpublished report to the Ford Foundation.

——. 1990. Extraction and forest management by rural inhabitants in the Amazon estuary. In: *Alternatives to Deforestation: Steps toward Sustainable Use of the Amazon Rain Forest*, ed. A.B. Anderson, pp. 65–85, Columbia University Press, New York.

Anderson, D., and R. Fishwick. 1984. *Fuelwood Consumption and Deforestation in African Countries*. World Bank, Staff Working Paper 704, Washington, D.C.

Anderson, E. 1950. Variation in avocados at the Rodiles plantation. *Ceiba* 1:50–55.

Angel, P.M. 1978. *Incidencia de una explotación intensiva de la palma de chontaduro en la situación socio-económica de la población del litoral Pacífico vallecaucano: El chontaduro en Buenaventura.* Publicaciones del Jardín Botánico Valle, Boletín Divulgativo 3, Cali, Colombia.

Anonymous. 1924. L'agriculture au Congo Belge. *Bulletin Agricole du Congo Belge,* June, pp. 308–413.

Anonymous. 1974. Macadamia: a new crop for Costa Rica. *Turrialba* 2:5–6.

Anonymous. 1978. *Annual Report.* South Africa Department of Agricultural Technical Services, Pretoria.

Anonymous. 1988. The first macadamia. *Yearbook of the California Macadamia Society* 34:85–86.

Anonymous. 1989. *Marketing Non-Timber Tropical Forest Products: Prospects and Promise.* A workshop report of the Consultative Group on Biological Diversity in cooperation with Cultural Survival, Inc., Harvard University, Cambridge, Mass., 7 November, 1989.

Anthony, F., J. Berthaud, J. Guillaumet, and M. Lourd. 1987. Collecting wild *Coffea* species in Kenya and Tanzania. *Plant Genetic Resources Newsletter* 69:23–29.

Aragundi, J., C. Suarez, and G. Solorzano. 1987. Evidencia de resistencia a la moniliasis y escoba de bruja del cacao en el clon EET-233. In: *Proceedings of the 10th International Cocoa Research Conference, Santo Domingo, Dominican Republic, 17–23 May,* pp. 479–483.

Arasu, N.T. 1985. A decade of plant genetic resources activities in Malaysia. International Board for Plant Genetic Resources, Regional Committee for Southeast Asia. *Newsletter* 9(3):11–12.

Arasu, N.T., and N. Rajanaidu. 1975. Conservation and utilization of genetic resources in the oil palm. In: *South East Asian Plant Genetic Resources,* ed. J.T. Williams, C.H. Lamoureux, and N. Wulijarno-Soetjipto, pp. 182–186. IBPGR/BIOTROP/BPPP/LIPI, Bogor, Indonesia.

Arkcoll, D.B., and C.R. Clement. 1989. Potential new food crops from the Amazon. In: *New Crops for Food and Industry,* ed. G.E. Wickens, N. Haq, and P. Day, pp. 150–165, Chapman and Hall, London.

Arora, R.K. 1985. *Genetic Resources of Less Known Cultivated Food Plants.* National Bureau for Plant Genetic Resources, New Delhi.

Arora, R.K., and E.R. Nayar. 1984. *Wild Relatives of Crop Plants in India.* National Bureau for Plant Genetic Resources, New Delhi.

Ascenso, J.C. 1964. Outlines of the cacao selection and breeding programme in San Thome. *Economic Botany* 18:132–136.

———. 1966. Outlines of the oil palm breeding programme in Portuguese Guinea. *Euphytica* 15:268–277.

———. 1986a. Potential of the cashew crop—1. *Agriculture International* 38(11):324–327.

———. 1986b. Potential of the cashew crop—2. *Agriculture International* 38(12):368–371.

Ashton, P.S. 1988. A question of sustainable use. In: *People of the Rain Forest,* ed. J.S. Denslow and C. Padoch, pp. 185–196. University of California Press, Berkeley.

Aublet, F. 1775. *Histoire des plantes de la Guiane Française.* Pierre-François Didot, Paris.

Aubreville, A. 1962. Burseraceae. In: *Flore du Gabon,* vol. 3. Museum of Natural History, Paris.

Backer, C.A. 1910. Plantes exotiques naturalisées dans Java. *Annales du Jardin Botanique de Buitenzorg,* Suppl. 3(1):393–420.

Badillo, V.M. 1971. *Monografía de la familia Caricaceae*. Associación de Profesores, Maracay, Venezuela.

Baer, D.F. 1976. Systematics of the genus *Bixa* and geography of the cultivated annatto tree. Ph.D. dissertation, University of California, Los Angeles.

Baichoo, C.S. 1981. Tropical fruit tree crop production in Guyana. In: *Proceedings of the Caribbean Workshop on Traditional and Potential Fruit Tree Crop Development, St. Georges, Grenada, 9–14 November 1980*, pp. 217–237. Inter-American Institute for Cooperation on Agriculture (IICA), San José, Costa Rica.

Baker, H.G. 1970. *Plants and Civilization*. Wadsworth Publishing Co., Belmont, Calif.

Baker, J.H. 1980. The use of industrial enzymes in the conversion of starch. In: *Sago: The Equatorial Swamp as a Natural Resource*, ed. W.R. Stanton and M. Flach, pp. 215–221. Proceedings of the 2nd International Sago Symposium, Martinus Nijhoff, The Hague.

Baldwin, J.T. 1947a. *Hevea*: a first interpretation. *Journal of Heredity* 38:54–64.

———. 1947b. *Hevea rigidifolia*. *American Journal of Botany* 34(5):261–266.

Balée, W. 1986. Análise preliminar de inventário florestal e a etnobotânica Ka'apor (Maranhão). *Boletim do Museu Paraense Emilio Goeldi, Botânica* 2(2):141–167.

———. 1987. A etnobotânica quantitativa dos índios Tembé (Rio Gurupi, Pará). *Boletim do Museu Paraense Emilio Goeldi, Botânica* 3(1):29–50.

———. 1988. Indigenous adaptation to Amazonian palm forests. *Principes* 32(2): 47–54.

———. 1989a. The culture of Amazonian forests. In: *Resource Management in Amazonia: Indigenous and Folk Strategies*, ed. D.A. Posey and W. Balée, pp. 1–21. Advances in Economic Botany, vol. 7. New York Botanical Garden, Bronx.

———. 1989b. Nomenclatural patterns in Ka'apor ethnobotany. *Journal of Ethnobiology* 9(1):1–24.

Balée, W., and A. Gély. 1989. Managed forest succession in Amazonia: the Ka'apor case. In: *Resource Management in Amazonia: Indigenous and Folk Strategies*, ed. D.A. Posey and W. Balée, vol. 7, *Advances in Economic Botany*, pp. 129–158. New York Botanical Garden, Bronx.

Balée, W., and D. Moore. 1991. Similarity and variation in plant names in five Tupi-Guarani languages (eastern Amazonia). *Bulletin of the Florida Museum of Natural History*, Biological Sciences 55(4):209–262.

Balick, M.J. 1976. The palm heart as a new commercial crop from tropical America. *Principes* 20:24–28.

———. 1980. Economic botany of the Guahibo. I. Palmae. *Economic Botany* 33:361–376.

———. 1984a. Ethnobotany of palms in the Neotropics. *Advances in Economic Botany* 1:9–23.

———. 1984b. Palms, people, and progress. *Horizons* (Agency for International Development) 3(4):33–37.

Barcelos, E. 1986. *Características genético-ecológicas de populações naturais de Caiaué* (Elaeis oleifera, *H.B.K, Cortés*) na Amazônia brasileira. Master's Thesis, Instituto Nacional de Pesquisas da Amazônia/Universidade do Amazonas, Manaus, Brazil.

Barnes, R.D. 1988. Tropical forest genetics at the Oxford Forestry Institute. *Commonwealth Forestry Review* 67(3):231–241.

Barrau, J. 1958. *Subsistence Agriculture in Melanesia*. Bernice P. Bishop Museum, Bulletin 219, Honolulu, Hawaii.

———. 1959. The sago palms and other food plants of marsh dwellers in the South Pacific islands. *Economic Botany* 13:151–162.

———. 1961. *Subsistence Agriculture in Polynesia and Micronesia*. Bernice P. Bishop Museum, Bulletin 223, Honolulu, Hawaii.

Barrett, H.C., and A.M. Rhodes. 1976. A numerical taxonomic study of affinity re-lationships in cultivated citrus and its close relatives. *Systematic Botany* 1:130–136.

Barrett, S.W. 1980. Conservation in Amazonia. *Biological Conservation* 18: 209–235.

Barrow, J. 1831. *The Mutiny on the Bounty.* Blackie, London.

Barse, W.P. 1990. Preceramic occupations in the Orinoco River Valley. *Science* 250:1388–1390.

Bartley, B.G.D. 1963. Exploration for *Theobroma* in the Amazon Valley. *Genetica Agraria* 17:345–349.

——. 1981. Global concepts for genetic resources and breeding in cacao. In: *Proceedings of the 7th International Cocoa Research Conference, Douala, Cameroon,* pp. 519–525.

Bartley, P.G.D., P.F.R. Machado, D. Ahnert, J.P. Barriga, and C.M.V.C. de Almeida. 1987. Descrição de populações de cacau da Amazônia brasileira, 1: observações preliminares sobre populações de Alenquer, Pará. In: *Anais da 10a. Conferencia Internacional de Pesquisas em Cacau, Santo Domingo, 17–23 maio,* pp. 665–672.

Bates, H.W. 1864. *The Naturalist on the River Amazons.* John Murray, London.

Battacharya, D.K. 1987. Mango (*Mangifera indica*) kernel fat. In: *Non-Traditional Oilseeds and Oils in India,* ed. N.V. Bringi, pp. 73–96. Oxford and IBH Publishing Co., New Delhi.

Bavappa, K.V.A. 1980. Breeding and genetics of areca nut, *Areca catechu:* a review. *Journal of Plantation Crops* 8:13–23.

Beach, J.H. 1984. The reproductive biology of the peach or "pejibayé" palm (*Bactris gasipaes*) and a wild congener (*B. porschiana*) in the Atlantic lowlands of Costa Rica. *Principes* 28(3):107–109.

Beccari, O. 1986. *Wanderings in the Great Forests of Borneo.* Oxford University Press, Singapore.

Becker, S. 1958. The production of papain—an agricultural industry for tropical America. *Economic Botany* 12:62–79.

Beirnaert, A., and R. Vanderweyen. 1941. *Contribution à l'étude génétique et biometrique des variétés d'Elaeis guineensis.* INEAC, Serie Sci. 27, Brussels.

Bekey, R. 1986. Greenhouse thrips emerging as number one avocado pest. *Yearbook of the California Avocado Society* 70:99–102.

Bell, C.R., and B.J. Taylor. 1982. *Florida Wild Flowers and Roadside Plants.* Laurel Hill Press, Chapel Hill, N.C.

Bellachew, B. 1987. Coffee (*Coffea arabica* L.) genetic erosion and germplasm collection in Harerge region. *Germplasm Newsletter* (Plant Genetic Resource Centre, Ethiopia) 15:8–13.

Belt, T. 1888. *The Naturalist in Nicaragua.* Edward Bumpus, London.

Bennetzen, J.F., M.M. Qin, S. Ingels, and A.H. Ellingoe. 1988. Allele-specific and *Mutator*-associated instability at the *Rp1* disease-resistance locus of maize. *Nature* 332:369–370.

Bentley, B.L. 1983. *Bixa orellana* (Achiote, Annatto). In: *Costa Rican Natural History,* ed. Daniel H. Janzen, pp. 193–194. University of Chicago Press, Chicago.

Ben-Ya'acov, A. 1989. The first decade of clonal propagation of avocado rootstocks. *Proceedings of the Interamerican Society for Tropical Horticulture* 32:42–44.

Bergh, B.O. 1975. Avocados. In: *Advances in Fruit Breeding,* ed. J. Janick and J.N. Moore, pp. 541–567, Purdue University Press, West Lafayette, Penn.

——. 1976. Avocado, *Persea americana* (Lauraceae). In: *Evolution of Crop Plants,* ed. N.W. Simmonds, pp. 148–151. Longman, London.

------. 1985. Avocado varieties for California. *Yearbook of the California Avocado Society* 68:75–93.

Bergh, B.O., and N. Ellstrand. 1987. Taxonomy of the avocado. *Yearbook of the California Avocado Society* 70:135–145.

Bergh, B.O., R.W. Scora, and W.B. Storey. 1973. A comparison of leaf terpenes in *Persea* subgenus *Persea. Botanical Gazette* 134(2):130–134.

Berkes, F., D. Feeny, B.J. McCay, and J.M. Acheson. 1989. The benefits of the commons. *Nature* 340:91–93.

Berthaud, J. 1986. *Les ressources génétiques pour l'amélioration des caféiers africains diploides: évaluation de la richesse génétique des populations sylvestres et des mécanismes organisateurs, conséquences pour l'application.* Éditions de l'Orstom, Institut Français de Recherche Scientifique pour le Développement en Coopération, Collection Travaux et Documents 188, Paris.

Berthaud, J., and A. Charrier. 1988. Genetic Resources of *Coffea.* In: *Coffee,* ed. R. Clark and R. Macrae, vol. 4, *Agronomy,* pp. 1–42. Elsevier, London.

Berthaud, J., A. Charrier, J.L. Guillaumet, and M. Lourd. 1986. Les caféiers. In: *Gestion des ressources génétiques des plantes,* ed. J. Pernes, pp. 45–106. Vol. 1: Monographies, Agence de Coopération Culturelle et Technique, Paris.

Bhat, K.S. 1987. Growth and performance of cacao (*Theobroma cacao* L.) and arecanut (*Areca catechu* L.) under mixed cropping system. In: *Proceedings of the 10th International Cocoa Research Conference, Santo Domingo, Dominican Republic, 17–23 May,* pp. 15–19.

Bhattacharya, A.K., D.P. Ghosh, L.P. Sihare, G. Guha, K. Mukerjii, T. Bharati, A.P.D. Gupta, P.C.D. Gupta, R. Mookerjee, and S. Chandra. 1973. *Great Centres of Art: Calcutta.* Taraporevala, Bombay.

Blake, S.F. 1920. A preliminary revision of the North American and West Indian avocados (*Persea* spp.). *Journal of the Washington Academy of Sciences* 10(1):9–20.

Blombery, A., and T. Rodd. 1985. *Palms.* Angus and Robertson, London.

Boa, E.R. 1987. Fungal diseases of bamboo: a preliminary and provisional list. In: *Recent Research on Bamboos,* ed. A.N. Rao, G. Dhanarajan, and C.B. Sastry, pp. 271–279. CAF, Beijing/IDRC, Ottawa.

Boer, J.G. 1965. Palmae. In: *Flora of Suriname,* ed. J. Lanjouw, vol. 5(1), pp. 1–172. E.J. Brill, Leiden.

Bokdam, J., and A.F. Droogers. 1975. Contribution à l'étude ethnobotanique des Wagenia de Kisangani, Zaire. *Mededelingen Landbouwhogeschool Wagengingen* 75:1–74.

Bompard, J.M., and A. Kostermans. 1986. Local knowledge too late to save wild mangoes? *IUCN Bulletin* (International Union for Conservation of Nature and Natural Resources) 17(1–3):26.

Bonalume, R. 1989a. Rain forests: destruction area disputed. *Nature* 339:86.

------. 1989b. Amazonian forests: burning continues, slightly abated. *Nature* 339:569.

Bonanome, A., and S.M. Grundy. 1988. Effect of dietary stearic acid on plasma cholersterol and lipoprotein levels. *The New England Journal of Medicine* 318:1244–1248.

Bonaparte, E.E.N.A. 1981. Report of the meeting on the development of a coordinated international programme of research into witches' broom disease, Cartagena, Colombia. In: *Proceedings of the 8th International Cocoa Research Conference, Cartagena, Colombia,* p. 859.

Boom, B.M. 1986. The Chácobo Indians and their palms. *Advances in Economic Botany* 6:91–97.

------. 1989. Use of plant resources by the Chácobo. In: *Resource Management in Ama-*

zonia: Indigenous and Folk Strategies, ed. D.A. Posey and W. Balée, vol. 7, *Advances in Economic Botany*, pp. 78–96. New York Botanical Garden, Bronx.

Booth, W. 1989. Monitoring the fate of the forests from space. *Science* 243:1428–1429.

Boster, J. 1983. A comparison of the diversity of Jivaroan gardens of the tropical forest. *Human Ecology* 11(1):47–68.

Bowden, J., P.H. Gregory, and C.G. Johnson. 1971. Possible wind transport of coffee rust across the Atlantic ocean. *Nature* 229:500–501.

Brandt, S.A. 1984. New perspectives on the origins of food production in Ethiopia. In: *Hunters to Farmers: The Causes and Consequences of Food Production in Africa*, ed. J.D. Clark and S.A. Brandt, pp. 173–190. University of California Press, Berkeley.

Braudeau, J. 1974. The cocoa tree: agronomic aspects. In: *Phytophthora Disease of Cocoa*, ed. P.H. Gregory, pp. 1–12. Longman, London.

Bray, W. 1976. From predation to production: the nature of agricultural evolution in Mexico and Peru. In: *Problems in Economic and Social Archaeology*, ed. G.G. Sieveking, I.H. Longworth, and K.F. Wilson, pp. 73–95. Duckworth, London.

Brewbaker, J.L. 1979. Diseases of maize in the wet lowland tropics and the collapse of the classic Maya civilization. *Economic Botany* 33:101–118.

——. 1987a. *Leucaena*: a multipurpose tree genus for tropical agroforestry. In: *Agroforestry: A Decade of Development*, ed. H.A. Steppler and P.K. Nair, pp. 289–323. International Center for Research in Agroforestry, Nairobi.

——. 1987b. Species in the genus *Leucaena. Leucaena Research Reports* 7(2):6–20.

——. 1989. Leucana, can there be such a thing as a perfect tree? *Agroforestry Today* 1(14):4–7.

Brewbaker, J.L., and D.D. Gorrez. 1967. Genetics of self-incompatibility in the monocot genera *Ananas* (pineapple) and *Gasteria. American Journal of Botany* 54:611–616.

Brewbaker, J.L., and S. Kaye. 1981. Mimosine variations in species of the genus *Leucaena. Leucaena Research Reports* 2:66–68.

Brewbaker, J.L., and C.T. Sorensson. 1990. *Leucaena* Bentham: New tree crops from interspecific *Leucaena* hybrids. In: *Advances in New Crops, Proceedings of the 1st International Symposium on New Crops: Research, Development, Economics*, ed. J. Janick and J. Simon, pp. 283–289. Timber Press, Portland, Ore.

Brice, R.E., and K.P. Jones. 1988. The evolution of new uses for natural rubber. In: *Natural Rubber Science and Technology*, ed. A.D. Roberts, pp. 1–34. Oxford University Press, Oxford.

Briey, M.J. 1920. Le palmier à huile au Mayumbe. *Bulletin des Matières Grasses de l'Institut Colonial de Marseille* 5/6:227–249.

Brigham, W.T. 1887. *Guatemala: The Land of the Quetzal*. Charles Scribner's Sons, New York.

Bringi, N.V. 1987. Lesser known tree-borne oils seeds. In: *Non-Traditional Oilseeds and Oils in India*, ed. N.V. Bringi, pp. 216–248. Oxford and IBH Publishing Co., New Dehli.

Brinkgrieve, J.H. 1933. De kruidnagelcultuur in die Residentie Sumatras Westkust. *Landbouw* 8:645–600.

Brinkmann, W.F.F., J.A. Weinman, and M.N.G. Ribeiro. 1971. Air temperatures in Central Amazonia: 1. The daily record of air temperatures in a secondary forest near Manaus under cold front conditions (July 4th to July 13th, 1969). *Acta Amazonica* 1(2):51–56.

Britton, N.L. 1918. The flora of the American Virgin Islands. *Memoirs of the Brooklyn Botanic Garden* 1:19–118.

Britton, N.L., and P. Wilson. 1924. *Botany of Porto Rico and the Virgin Islands.* New York Academy of Sciences, vol. 5, no. 4.

Bromley, R. 1981. The colonization of humid tropical areas in Ecuador. *Singapore Journal of Tropical Geography* 2:15–25.

Brown, E., and H.H. Hunter. 1913. *Planting in Uganda: Coffee, Para Rubber, Cocoa.* Longmans, London.

Brown, L.R., and J.L. Jacobson. 1986. *Our Demographically Divided World.* Worldwatch Institute, Paper 74, Washington, D.C.

Brownlee, S. 1989. The best banana bred. *Atlantic Monthly* 264(3):22–28.

Brücher, H. 1989. *Useful Plants of Neotropical Origin and Their Wild Relatives.* Springer-Verlag, Berlin.

Brush, S. 1989. Rethinking crop genetic resource conservation. *Conservation Biology* 3(1):19–29.

———. 1991. Farmer conservation of New World crops: the case of Andean potatoes. *Diversity* 7(1 & 2):75–79.

Buchanan, K.M., and J.C. Pugh. 1955. *Land and People in Nigeria: The Human Geography of Nigeria and Its Environmental Background.* University of London Press, London.

Buchillet, D. 1988. Interpretação da doença e simbolismo ecológico entre os índios Desana. *Boletim do Museu Paraense Emílio Goeldi, Série Antropologia* 4(1): 27–42.

Buckley, D.P., D.M. O'Malley, V. Apsit, G.T. Prance, and K.S. Bawa. 1988. Genetics of Brazil nut (*Bertholletia excelsa* Humb. & Bonpl.: Lecythidaceae) I. Genetic variation in natural populations. *Theoretical and Applied Genetics* 76:923–928.

Buddenhagen, I.W. 1986. *Banana/Plantain Project Development by ACIAR.* Australian Centre for International Agricultural Research, Canberra.

Budowski, G. 1989. Developing the Chocó region of Colombia. In: *Fragile Lands of Latin America: Strategies for Sustainable Development,* ed. J.O. Browder, pp. 273–279. Westview Press, Boulder, Colo.

Bunyard, P. 1989. Guardians of the Amazon. *New Scientist* (16 December):38–41.

Burke, R., and D.T.M. Girvan. 1954. *The Farmer's Guide.* The University Press, Glasgow.

Burkill, I.H. 1935. *A Dictionary of the Economic Products of the Malay Peninsula,* vol. 1. Crown Agents for the Colonies, London.

Burley, J. 1986. Global needs and problems of collection, storage, and distribution of multipurpose tree germplasm. In: *Multipurpose Tree Germplasm,* ed. J. Burley and P. von Carlowitz, pp. 43–150. International Center for Research and Agroforestry/International Board for Plant Genetic Resources/German Technical Cooperation, Nairobi.

Busson, F. 1965. *Plantes alimentaires de l'Ouest Africain.* Ministère de la Coopération/Ministère d'État Chargé de la Recherche Scientifique et Technique/Ministère des Armées, Paris.

Camacho, S., R. Pineda, J.G. Rondón, E. Barragan, J.C. Toro, R.A. Montoya, E. García, and K.A. Okada. 1991. Management of agricultural genetic resources in Colombia. *Diversity* 7(1 & 2):37–38.

Cameron, I. 1987. *Lost Paradise: The Exploration of the Pacific.* Salem House Publishers, Topsfield, Mass.

Cameron, J.W., and R.K. Soost. 1976. Citrus. In: *Evolution of Crop Plants,* ed. N.W. Simmonds, pp. 261–265. Longman, London.

Campbell, C.W. 1984. Tropical fruits and nuts. In: *CRC Handbook of Tropical Food Crops,* ed. F.W. Martin, pp. 235–274. CRC Press, Boca Raton, Fla.

———. 1986. Tropical fruit crops in Florida—a rapidly changing situation. *Proceedings of the Florida Horticultural Society* 99:217–219.

Campbell, C.W., and S.E. Malo. 1973. Performance of sapodilla cultivars and seedling selections in Florida. *Proceedings of the Tropical Region, American Society of Horticultural Science* 17:220–226.

Carvalho, A. 1958. Recent advances in our knowledge of coffee trees. *Coffee and Tea Industries and the Flavor Field* 81:30–36.

Carvalho, A., and L.C. Monaco. 1968. Relaciones geneticas de especies seleccionadas de Coffea. *Café* 9(4):3–19.

Carvalho, A., L.C. Fazuoli, and F.A. Levy. 1988. Características de algumas introduções de *Coffea arabica* da Etiópia. In: *Anais do Encontro sobre Recursos Genéticos*, ed. S.M.C. Araujo and J.A. Osuna, p. 189. Faculdade de Ciências Agrárias e Veterinárias, UNESP, Jaboticabal, São Paulo/Centro Nacional de Recursos Genéticos e Biotecnologia, Brasília.

Carvalho, A., A.B. Eskes, J. Castillo, M.S. Sreenivasan, J.H. Echeverri, C.E. Fernandez, and L.C. Fazuoli. 1989. Breeding programs. In: *Coffee Rust: Epidemiology, Resistance, and Management*, ed. A.C. Kushalappa and A.B. Eskes, pp. 293–335. CRC Press, Boca Raton, Fla.

Carvalho, J.O.P., J.N.M. Silva, J.C.A. Lopes, V.M.J. Valcarcel, and N.R. de Graaf. 1986. Redução da densidade de uma floresta tropical úmida densa devido a exploração mecanizada. In: *Anais do Primeiro Simpósio do Tropico Umido*, Belém, Pará, 12–17 November 1984, vol. 2, pp. 269–281. Empresa Brasileira de Pesquisa Agropecuária, Brasília.

Castro Soares, L. 1956. *Excursion Guidebook No. 8: Amazonia*, trans. R.P. Momsen. International Geographical Union, Brazilian National Committee, Rio de Janeiro.

Cavalcante, P.B. 1976. *Frutas comestíveis da Amazônia*. Museu Paraense Emilio Goeldi, Belém, Brazil.

———. 1988. *Frutas comestíveis da Amazônia*. Museu Paraense Emilio Goeldi, Belém, Brazil.

Cavalcante, P.B., and D. Johnson. 1977. Edible palm fruits of the Brazilian Amazon. *Principes* 21(3):91–102.

Chadha, K.L. 1989. Current situation and future prospects of minor fruits in the Asia-Pacific region. Report presented at the Regional Expert Consultation on Fruits held 13–16 June at the Regional Office for Asia and the Pacific, Food and Agriculture Organization, Bangkok, Thailand.

Chadha, K.L., and O.P. Pareek. 1989. Genetic resources of fruit crops: achievements and gaps. *Indian Journal of Plant Genetic Resources* 1 (1 & 2):43–48.

Chandler, W.H. 1958. *Evergreen Orchards*. Lea and Febiger, Philadelphia.

Chapot, H. 1962. Avocado culture in Turkey. *Yearbook of the California Avocado Association* 51:93–96.

Charrier, A. 1978. La structure génétique des caféiers spontanés de la region Malagache: *Mascarocoffea*, leurs relations avec les caféiers d'origine africaine (*Eucoffea*). *Café, Cacao, Thé* 20:1–100.

———. 1980. The genetic resources of the genus *Coffea* in Africa. In: *Crop Genetic Resources in Africa, Proceedings of a Workshop Jointly Organized by the Association for the Advancement of Agricultural Sciences in Africa and the International Institute of Tropical Agriculture, IITA, Ibadan, Nigeria, 4–6 January, 1978*, pp. 57–70. International Institute of Tropical Agriculture, Ibadan, Nigeria.

Charrier, A., and J. Berthaud. 1985. Botanical classification of coffee. In: *Coffee: Botany, Biochemistry and Production of Beans and Beverage*, ed. M. Clifford and K.C. Willson, pp. 13–47. Croom Helm, London.

Chee, K.H., and R.L. Wastie. 1980. The status and future prospects of rubber diseases in tropical America. *Review of Plant Pathology* 59:541–548.

Cheesman, E.E. 1932. The economic botany of cacao: a critical survey of the literature to the end of 1930. *Supplement to Tropical Agriculture* 9:1–14.

——. 1944. Notes on the nomenclature, classification and possible relationships of cacao populations. *Tropical Agriculture* 21(8):144–159.

Chevalier, A. 1912. Énumération des plantes cultivées par les indigènes en Afrique tropicale et des espèces naturalisées dans le même pays et ayant probablement été cultivées a une époque plus ou moins reculée. *Bulletin de la Société d'Acclimatation* 59:65–79.

Chin, H.F., and H.S. Yong. 1985. *Malaysian Fruits in Colour*. Tropical Press, Kuala Lumpur.

Choke, H.C. 1973. Malaysia: locally originated tropical fruits. In: *Survey of Crop Genetic Resources in Their Centers of Origin*, ed. O.H. Frankel, pp. 160–166. IBP/FAO, Rome.

Clark, J.D. 1976. Prehistoric populations and pressures favouring plant domestication. In: *Origins of African Plant Domestication*, ed. J.R. Harlan, J. de Wet, and A.B.L. Stemler, pp. 67–105. Mouton, The Hague.

Clayton, W.D. 1979. *Gramineae. Part I of Flora of Tropical East Africa*, ed. E. Milne-Redhead and R.M. Polhill. Balkema, Rotterdam.

Cleary, D. 1990. *Anatomy of the Amazon Gold Rush*. University of Iowa Press, Iowa City.

Clement, C.R. 1986. El pejibaye: resultados y necesidades de investigación. In: *Useful Palms of Tropical America*, Newsletter, no. 2, pp. 2–4. EMBRAPA/CENARGEN, Brasília, Brazil.

——. 1988. Domestication of the pejibaye palm (*Bactris gasipaes*): past and present. *Advances in Economic Botany* 6:155–174.

——. 1989. The potential use of the pejibaye palm in agroforestry systems. *Agroforestry Systems* 7:201–212.

——. 1991. Amazonian fruits: neglected, threatened and potentially rich resources require urgent attention. *Diversity* 7(1 & 2):56–59.

Clement, C.R., C.H. Müller, and W.B. Chavez. 1982. Recursos genéticos de espécies frutíferas nativas da Amazônia brasileira. *Acta Amazonica* 12:677–695.

Clement, D., and C. Lanaud. 1987. Les cacaoyers de Guyane, prospections, originalité. In: *Proceedings of the 10th International Cocoa Research Conference, Santo Domingo, Dominican Republic, 17–23 May*, pp. 673–677.

Clemente, W. 1987. Biotechnology in the Third World: small is big. *Development: Seeds of Change* 4:25–26.

CMR. 1988. Vanilla breakthrough may herald biotech revolution in flavour area. *Chemical Marketing Reporter*, 9 May.

Cobley, L.S. 1956. *An Introduction to the Botany of Tropical Crops*. Longmans, Green and Company, London.

Coe, M.D., and R.A. Diehl. 1980. *In the Land of the Olmec. Vol. 2: The People of the River*. University of Texas Press, Austin.

Coenan, J., and J. Barrau. 1961. The breadfruit tree in Micronesia. *South Pacific Bulletin*, October:37–39.

Coffey, M. 1984. An integrated approach to the control of avocado root rot. *Yearbook of the California Avocado Society* 68:61–68.

——. 1987. Phytophthora root rot of avocado: an integrated approach to control in California. *Plant Disease* 71:1046–1052.

Coffey, M., F. Guillemet, G. Schieber, and G. Zentmyer. 1988. *Persea schiedeana* and Martin Grande. *Yearbook of the California Avocado Society* 72:107–120.

Cohen, M.N. 1977. *The Food Crisis in Pre-History*. Yale University Press, New Haven.

Coit, J.E. 1928. Sun-blotch of the avocado. *Yearbook of the California Avocado Association*, pp. 27–29.

Collins, J.L. 1960. *The Pineapple: Botany, Cultivation and Utilization.* Leonard Hill, London.

Conover, R.A., R.E. Litz, and S.E. Malo. 1986. 'Cariflora'—a papaya ringspot virus-tolerant papaya for South Florida and the Caribbean. *Hortscience* 21:1072.

Cope, F.W. 1976. Cacao, *Theobroma cacao* (Sterculiaceae). In: *Evolution of Crop Plants*, ed. N.W. Simmonds, pp. 285–289. Longman, London.

Coradin, L., and E. Lleras. 1986. Coleta de germoplasma de macaúba: situação atual. *Newsletter: Useful Plants of Tropical America* (FAO/CENARGEN) 2:5–6.

Cordero, L. 1950. *Enumeración botánica de las principales plantas, así útiles como nocivas, indígenas o aclimatadas, que se dan en las provincias del Azuay y del Cañar de la Republica del Ecuador.* Afrodisio Aguado, Madrid.

Corley, R.H.V. 1989. Assessment of new crops for plantations. In: *New Crops for Food and Industry*, ed. G.E. Wickens, N. Haq, and P. Day, pp. 53–65. Chapman and Hall, London.

Corner, E.J.H. 1951. *Wayside trees of Malaya.* Government Printing Office, Singapore.

——. 1964. *The Life of Plants.* Weidenfeld and Nicolson, London.

Coronel, R.E. 1983. *Promising Fruits of the Philippines.* College of Agriculture, University of the Philippines, Los Baños.

Corrales, F., and J. Mora-Urpí. 1990. Sobre el proto-pejibaye en Costa Rica. *Boletín Informativo* (Universidad de Costa Rica), *Serie Técnica Pejibaye (Guilielma)* 2(2):1–11.

Correa de Mello, J., and R. Spruce. 1869. Notes on the Papayaceae. *Journal of the Linnean Society, Botany* 10:1–18.

Correll, D.S. 1953. Vanilla—its botany, its history, cultivation, and economic importance. *Economic Botany* 7:291–358.

Cortés, S. 1919. *Flora de Colombia.* Libreria de el Mensajero, Bogotá.

Coudreau, H. 1883. *Les richesses de la Guyane Française.* Imprimerie du Gouvernement, Cayenne.

——. 1899. *Voyage entre Tocantins et Xingú: 3 avril 1898–3 novembre 1898.* A. Lahure, Paris.

Cowen, D.V. 1984. *Flowering Trees and Shrubs in India.* Thacker & Company, Bombay.

Crawford, D.L. 1937. *Hawaii's Crop Parade.* Advertiser Publishing Company, Honolulu.

Croat, T.B. 1978. *Flora of Barro Colorado Island.* Stanford University Press, Stanford.

Crofton, R.H. 1936. *A Pageant of the Spice Islands.* John Bale, Sons and Danielson, London.

Crosby, A.W. 1986. *Ecological Imperialism: The Biological Expansion of Europe, 900–1900.* Cambridge University Press, Cambridge.

Cruls, G. 1939. Impressões de uma visita a companhia Ford Industrial do Brasil. *Revista Brasileira de Geografia* 1(4):3–25.

Cuatrecasas, J. 1964. Cacao and its allies: a taxonomic revision of the genus *Theobroma. Contributions of the United States National Herbarium* 35:379–614.

Cut, A. 1977. *Cengkeh*—Eugenia caryophyllus. Banda, Acheh, Indonesia.

Dale, J.L. 1988. The status of disease indexing and the international distribution of banana germplasm. In: *Conservation and Movement of Vegetatively Propagated Germplasm: In Vitro Culture and Disease Aspects*, pp. 43–46. International Board for Plant Genetic Resources, Rome.

Dalziel, J. 1948. *The Useful Plants of West Tropical Africa*. The Crown Agents for the Colonies, London.

D'Arcy, W.G. 1977. Endangered landscapes in Panama and Central America. In: *Extinction Is Forever*: The status of Threatened and Endangered Plants in the Americas, ed. G.T. Prance and T.S. Elias, pp. 89–114. New York Botanical Garden, New York.

Das, P.K. 1982. The economics of cocoa mixed cropping with coconuts in India. In: *Proceedings of Plantations Crops Symposium (PLACROSYM V)*, vol. 5, pp. 397–408, Oxford and IBH Publishing Company, New Delhi.

——. 1986. Status of production and trade of cashew in India. *Agricultural Situation in India* 40(6):765–770.

Daugherty, H.E. 1972. The impact of man on the zoogeography of El Salvador. *Biological Conservation* 4:273–278.

——. 1973. *Conservación ambiental en El Salvador: recomendaciones para un programa de acción nacional*. Artes Gráficas Publicitarias, San Salvador, El Salvador.

Davidson, B. 1969. *A History of East and Central Africa to the Late Nineteenth Century*. Anchor Books, Garden City, N.Y.

Davis, T.A. 1986. Uses of semi-wild palms in Indonesia and elsewhere in South and Southeast Asia. *Advances in Economic Botany* 6:98–118.

Dean, W. 1987. *Brazil and the Struggle for Rubber*. Cambridge University Press, Cambridge.

De Beer, J.H., and M.J. McDermott. 1989. *The Economic Value of Non-Timber Forest Products in Southeast Asia*. Netherlands Committee for the IUCN, Amsterdam.

De Blank, S. 1952. A reconnaissance of the American oil palm *Elaeis melanococca* (Gaertner em. Bailey) = *Corozo oleifera* (Giseke), *Alfonsia oleifera* (H.B.K.). *Tropical Agriculture* 29:90–101.

De Candolle, A. 1855. *Géographie botanique raisonnée: ou exposition des faits principaux et des lois concernant la distribution géographique des plantes de l'époque actuelle*. V. Masson, Paris.

——. 1902. *Origin of Cultivated Plants*. Appleton, New York.

Decazy, B., and N. Coulibaly. 1981. Comportement de cultivars de cacaoyers a l'égard de quelques insectes depredateurs: possibilité d'une selection précoce des cacaoyers tolerant. In: *8ème Conférence Internationale sur la Recherche Cacaoyère, Cartagena, Colombia*, pp. 685–688.

Declert, C. 1984. Note technique: crites de l'état sanitaire d'une vanillerie. In: *INRA Report*, pp. 433–443. Institut National de la Recherche Agronomiques, Paris.

De Langhe, E. 1987. Towards an international strategy for genetic improvement in the genus *Musa*. In: *Banana and Plantain Breeding Strategies*, ed. G.J. Persley and E.A. de Langhe, pp. 19–23. ACIAR Proceedings 21, Australian Centre for International Agricultural Research, Canberra.

Delassus, M. 1962. *Rapport sur les maladies du vanillier Madagascar*. Institut de Recherches Agronomiques Tropicales et des Cultures Vivrières (IRAT), Paris.

Demeny, P. 1986. The world demographic situation. In: *World Population and U.S. Policy: The Choices Ahead*, ed. J. Menken, pp. 27–66. W.W. Norton, New York.

De Wilde, W.J.J. 1989. Sumatra. In: *Floristic Inventory of Tropical Countries*, ed. D.G. Campbell and H.D. Hammond, pp. 103–107. New York Botanical Garden, Bronx, New York.

Dias, C.V. 1959. Aspectos geográficos do comércio da castanha no médio Tocantins. *Revista Brasileira de Geografia* 21(4):77–91.

Dijkman, M.J. 1950. *Leucaena*—a promising soil-erosion control plant. *Economic Botany* 4:337–349.

Dillon, M.O. 1980. *Myroxylon*. In: *Flora of Panama*, ed. J.D. Dwyer and collaborators, pp. 735–737. Annals of the Missouri Botanical Gardens, vol. 67, St. Louis.

Diniz, T.D.A.S., and T.X. Bastos. 1974. Contribuição ao conhecimento do clima típica da castanha do Brasil. *Boletim Técnico do Instituto de Pesquisa Agropecuária do Norte* 64:1–83.

Dodson, C.H. 1967. Relationships between pollinators and orchid flowers. *Atas do Simpósio sôbre a Biota Amazônica* 5:1–72.

Donadio, L.C. 1984. The Brazilian avocado industry. *Yearbook of the California Avocado Society* 68:133–140.

———. 1987. Present status of Brazilian avocado industry. *Yearbook of the South African Avocado Growers' Association* 10:82–85.

Douglas, J.S., and R.A. Hart. 1985. *Forest Farming: Towards a Solution to Problems of World Hunger and Conservation.* Intermediate Technology Publications, London.

Dove, M.R. 1985. *Swidden Agriculture in Indonesia: The Subsistence Strategies of the Kalimantan Kantu.* Mouton, Berlin.

Doyle, J. 1985. *Altered Harvest: Agriculture, Genetics, and the Fate of the World's Food Supply.* Viking, New York.

Drummond, H.E. 1984a. *Freeze Loss Estimates.* Food and Resource Economics Department Staff Paper Series 265, University of Florida, Gainesville.

———. 1984b. *Estimated Citrus Tree Loss Caused by the December 1983 Freeze.* Food and Resource Economics Department Staff Paper Series 266, University of Florida, Gainesville.

———. 1985. *Estimated Value of Florida Citrus Fruit Loss Caused by the January, 1985 Freeze.* Food and Resource Economics Department Staff Paper Series 274, University of Florida, Gainesville.

Ducke, A. 1925. Plantes nouvelles ou peu connues de la région amazonienne. *Archivos do Jardim Botanico* (Rio de Janeiro) 4:1–211.

———. 1939. *Revision of the Genus* Hevea, *Mainly the Brazilian Species.* Serviço Florestal, Ministério da Agricultura, Rio de Janeiro.

———. 1946. *Plantas de cultura precolombiana na Amazônia brasileira: notas sobre as especies ou formas espontâneas que supostamente lhes teriam dada origem.* Instituto Agronômico do Norte, Boletim Técnico 8, Belém, Brazil.

Duke, J.A. 1981. *Handbook of Legumes of World Economic Importance.* Plenum, New York.

———. 1986. Oil palm in your future? In: *Useful Palms of Tropical America,* pp. 6–9, Empresa Brasileira de Pesquisa Agropecuária, Centro Nacional de Recuroses Genéticos e Biotecnologia, Brasília.

———. 1989. *Handbook of Nuts.* CRC Press, Boca Raton, Fla.

Dupaigne, P. 1979. Masticatoires et fruits tropicaux. *Fruits* 34:353–358.

Dutt, A.K., and U. Jamwal. 1987. Effect of coppicing at different heights on wood production in *Leucaena. Leucaena Research Reports* 8:27–28.

Earle, F.S., and W. Popenoe, 1915. Plant breeding in Cuba. *Journal of Heredity* 6(12):558–568.

Eckholm, E. 1975. *The Other Energy Crisis: Firewood.* Worldwatch Institute, Paper 1, Washington, D.C.

———. 1979. *Planting for the Future: Forestry for Human Needs.* Worldwatch Institute, Paper 26, Washington, D.C.

Eden, M.J. 1990. *Ecology and Land Management in Amazonia.* Belhaven Press, London.

Eden-Green, S.J., R. Balfas, T. Sutarjo, N. Susniahti, and N. Hasnam. 1985. Preliminary results on the transmission of xylem-limited bacteria causing Sumatra disease of cloves in Indonesia by tube-building cercopoids, *Hindola* spp. (Homoptera: Machaerotidae). Paper presented at the Sixth International Conference on Plant Pathogenic Bacteria, Beltsville, Maryland, U.S.A.

Eden-Green, S.J., R. Balfas, and J.A. Malius. 1986. Transmission of xylem-limited bacteria causing Sumatra disease of cloves in Indonesia by tube-building cercopoids, *Hindola* spp. (Homoptera: Machaerotidae). Paper presented at the Second International Workshop on Leafhoppers and Planthoppers of Economic Importance, Provo, Utah, U.S.A.

Eicher, C.K. 1989. *Sustainable Institutions for African Agricultural Development.* International Service for National Agricultural Research, Working Paper 19, The Hague.

Eiserle, R.J., and R.J. Barreto. 1969. Developing a substitute for national cinnamon. *Food Products Development* 3:88–90.

Ellstrand, N.C., and J.M. Lee. 1987. Cultivar identification of cherimoya (*Annona cherimola* Mill.) using isozyme markers. *Scientia Horticultur* 32:25–31.

Ellstrand, N.C., J.M. Lee, B.O. Bergh, M.D. Coffey, and G.A. Zentmyer. 1986. Isozymes confirm hybrid parentage for 'G755' selections. *Yearbook of the California Avocado Society* 70:199–203.

EMBRAPA. 1984. *Brazilian Agriculture and Agricultural Research.* Empresa Brasileira de Pesquisa Agropecuária, Brasília.

———. 1986. *Dendê: uma nova opção agrícola.* Empresa Brasileira de Pesquisa Agropecuária, Brasília.

Eskes, A.B. 1989. Resistance. In: *Coffee Rust: Epidemiology, Resistance, and Management,* ed. A.C. Kushalappa and A.B. Eskes, pp. 171–291. CRC Press, Boca Raton, Fla.

Evans, L.T. 1980. The natural history of crop yield. *American Scientist* 68:388–397.

Fairchild, D. 1939. *The World Was My Garden: Travels of a Plant Explorer.* Charles Scribner's Sons, New York.

Falesi, I.C. 1987. *Urucuzeiro: Recomendações básicas para seu cultivo.* Empresa Brasileira de Pesquisa Agropecuária, Belém, Brazil.

FAO. 1970. Report on the second consultation on forest tree breeding. *Unasylva* 24:11–32.

———. 1974 to date. *Forestry Occasional Papers: Forest Genetic Resources Information Series.* Food and Agriculture Organization, Rome.

———. 1981. *Map of the Fuelwood Situation in the Developing Countries.* Food and Agriculture Organization, Rome.

———. 1986. *Food and Fruit-Bearing Forest Species: 3. Examples from Latin America.* Forestry Paper 44/3, Food and Agriculture Organization, Rome.

———. 1988. Table of annual requirements until the year 2000: TFAP for Latin America and the Caribbean. *Unasylva* 159:11.

FAO/UNEP. 1982. *Tropical Forest Resources.* Food and Agriculture Organization, Rome/United Nations Environment Program, Nairobi.

Faulkner, O.T., and J.R. Mackie. 1933. *West African Agriculture.* Cambridge University Press, Cambridge.

Faulkner, R. 1976. Timber trees. In: *Evolution of Crop Plants,* ed. N.W. Simmonds, pp. 298–301. Longman, London.

Fawcett, W. 1891. *Economic Plants: An Index to Economic Products of the Vegetable Kingdom in Jamaica.* Government Printing Establishment, Kingston, Jamaica.

———. 1920. *Flora of Jamaica,* vol. 4, *Dicotyledon Families Leguminosae to Callitrichaceae.* British Museum, London.

Fawcett, W., and A.B. Rendle. 1926. *Flora of Jamaica, Containing Descriptions of the Flowering Plants Known from the Island,* vol. 5: *Dicotyledons, Families Buxaceae to Umbelliferae.* British Museum, London.

Fearnside, P.M. 1983. Land-use trends in the Brazilian Amazon region as factors in accelerating deforestation. *Environmental Conservation* 10(2):141–148.

——. 1986. *Human Carrying Capacity of the Brazilian Rainforest.* Columbia University Press, New York.

Ferdon, E.N. 1981. *Early Tahiti as the Explorers Saw It, 1767–1797.* University of Arizona Press, Tucson.

Ferwerda, F.P. 1976. Coffee, *Coffea* spp. (Rubiaceae). In: *Evolution of Crop Plants,* ed. N.W. Simmonds, pp. 257–260. Longman, London.

Feuillet, C. 1989. Diversity and distribution of the Guianan Passifloraceae. In: *Tropical Forests: Botanical Dynamics, Speciation and Diversity,* ed. L.B. Holm-Nielsen, I.C. Nielsen, and H. Baslev, pp. 311–318. Academic Press, London.

Fitch, M.M.M., and R.M. Manshardt. 1990. Somatic embryogenesis and plant regeneration from immature zygotic embryos of papaya (*Carica papaya*). *Plant Cell Reports* 9:320–324.

Fitch, M.M.M., R.M. Manshardt, D. Gonsalves, J.L. Slightom, and J.C. Sanford. 1990. Stable transformation of papaya via microprojectile bombardment. *Plant Cell Reports* 9:189–194.

Floch, M. 1983. *The Sago Palm.* Food and Agriculture Organization, Rome.

Fontaine, R.G. 1986. Management of humid tropical forests. *Unasylva* 154:16–21.

Ford, C. 1981. Work programme of CARDI on fruit tree crops. In: *Proceedings of the Caribbean Workshop on Traditional and Potential Fruit Tree Crop Development, St. Georges, Grenada, 9–14 September 1980,* pp. 133–136. Inter-American Institute for Cooperation on Agriculture, San José, Costa Rica.

Fosberg, F.R. 1960. Vegetation of Micronesia, I. General description of the vegetation of the Marianas Islands, and a detailed consideration of the vegetation of Guam. *Bulletin of the American Museum of Natural History* 119:1–75.

Foster, R.B. 1985. The seasonal rhythm of fruitfall on Barro Colorado Island. In: *The Ecology of a Tropical Forest: Seasonal Rhythms and Long-Term Changes,* ed. E.G. Leigh, Jr., A.S. Rand, and D.M. Windsor, pp. 151–164. Smithsonian Institution Press, Washington, D.C.

Fowler, G., and P. Mooney. 1990. *Shattering: Food, Politics, and the Loss of Genetic Diversity.* University of Arizona Press, Tucson.

Friedman, A. 1936. Avocado acclimatization in Palestine. *Yearbook of the California Avocado Association,* pp. 93–94.

Frikel, P. 1978. Areas de aboricultura pré-agrícola na Amazônia: notas preliminares. *Revista de Antropologia* 21(1):45–52.

Frolich, E.F., C.A. Schroeder, and G.A. Zentmyer. 1958. Graft compatibility in the genus *Persea. Yearbook of the California Avocado Society* 42:102–105.

Fuller, K.S. 1989. Debt-for-nature swaps. *Environmental Science and Technology* 23:1450–1451.

Galey, J. 1979. Industrialist in the wilderness: Henry Ford's Amazon venture. *Journal of Interamerican Studies and World Affairs* 21(2):261–289.

Gallegos, R. 1983. *Algunos aspectos del aguacate y su producción en Michoacán.* Universidad Autonóma Chapingo, Chapingo, Mexico.

Galloway, B.T. 1938. Plant immigrants make good in Florida's warm climate; wilderness now wonderland. In: *Native Plant Life and Plant Immigrants of Florida.* Supplementary Bulletin (New Series No. 16), Department of Agriculture, Tallahassee, Fla.

Galvão, E. 1979. The encounter of tribal and national societies in the Brazilian Amazon. In: *Brazil, Anthropological Perspectives: Essays in Honor of Charles Wagley,* ed. M.L. Margolis and W.E. Carter, pp. 25–38. Columbia University Press, New York.

Gandhi, M., and Y. Singh. 1989. *Brahma's Hair: On the Mythology of Indian Plants.* Rupa, Calcutta.

García, J., and G. Montes. 1988. *Coffee Boom, Government Expenditure, and Agri-*

cultural Prices: The Colombian Experience. International Food Policy Research Institute, Research Report 68, Washington, D.C.

Gascon, J.P., J.M. Noiret, and J. Meunier. 1989. Oil palm. In: *Oil Crops of the World: Their Breeding and Utilization*, ed. G. Röbbelen, R.K. Downey, and A. Ashri, pp. 475–493. McGraw-Hill, New York.

Gentry, A.H. 1972. Endangered plant species and habitats of Ecuador and Amazonian Peru. In: *Extinction Is Forever*, ed. G.T. Prance and T.S. Elias, pp. 136–149. New York Botanical Garden, Bronx.

———. 1986. An overview of neotropical phytogeographic patterns with an emphasis on Amazonia: In: *Anais do Primeiro Simpósio do Trópico Umido, Belém, Pará, 12 a 17 de novembro de 1984*, vol. 2, pp. 19–35. EMBRAPA, Brasília.

———. 1989. Speciation in tropical forests. In: *Tropical Forests: Botanical Dynamics, Speciation and Diversity*, ed. L.B. Holm-Nielsen, I.C. Nielsen, and H. Balslev, pp. 113–134. Academic Press, London.

Genú, P.J. de C., A.C. de Q. Pinto, F.R. Ferreira, T. Ogata, and R.G. Pedrazzi. 1987. Banco ativo de germoplasma de abacate. In: *Relatório técnico anual do Centro de Pesquisas Agropecuária dos Cerrados 1982–1985*, pp. 329–334. Centro de Pesquisa Agropecuária dos Cerrados, Empresa Brasileira de Pesquisa Agropecuária, Planaltina, D.F., Brazil.

Ghesquière, M., E. Barcelos, M.M. Santos, and P. Amblard. 1987. Polymorphisme enzymatique chez *Elaeis oleifera* H.B.K. (*E. melanococca*): analyse des populations du Bassin amazonien. *Oléagineux* 42(4):143–154.

Giacometti, D.C. 1987. Papaya breeding. *Acta Horticultura* 196:53–60.

Giacomo, R. 1982. Étude de la biologie florale du safoutier (*Dacryodes edulis*) au Gabon. Mimeo.

Gilbert, N.E., K.S. Dodds, and S. Subramaniam. 1973. Progress of breeding investigations with *Hevea brasiliensis*. V. Analysis of data from earlier crosses. *Journal of the Rubber Research Institute of Malaya* 23(5):365–380.

Gimlette, J.D. 1939. *A Dictionary of Malayan Medicine*. Oxford University Press, London.

Glendinning, D.R. 1967. Technical aspects of the breeding programme at the Cocoa Research Institute, Tafo, Ghana. 1. Breeding methods. *Euphytica* 16:76–82.

Gmitter, F.G., and X. Hu. 1990. The possible role of Yunnan, China, in the origin of contemporary *Citrus* species (Rutaceae). *Economic Botany* 44:267–277.

Godfrey-Sam-Aggrey, W. 1969. Avocado production in Ghana. *World Crops* 21:271–272.

Golley, F.B. 1983. Ecodevelopment. In: *Tropical Rain Forest Ecosystems: Structure and Function*, ed. F.B. Golley, pp. 335–344. Ecosystems of the World 14A, Elsevier, Amsterdam.

Golson, J. 1977. No room at the top: agricultural intensification in the New Guinea highlands. In: *Sunda and Sahul: Prehistoric Studies in Southeast Asia, Melanesia and Australia*, ed. J. Allen, J. Golson, and R. Jones, pp. 601–638. Academic Press, London.

Gomes e Sousa, A.F. 1929. Notas sôbre a flora da Guiné Portuguesa. *Boletim da Agência Geral das Colónias* 5(44):99–139.

———. 1930. *Subsidios para o conhecimento da flora da Guiné Portuguesa*. Memórias da Sociedade Broteriana, vol. 1, Instituto Botânica da Universidade de Coimbra, Coimbra, Portugal.

Gómez-Pompa, A. 1985. Tropical deforestation and Maya silviculture. *Tulane Studies in Zoology and Botany* 26:19–37.

——. 1987. On Mayan silviculture. *Mexican Studies* 3(1):1–17.

Gómez-Pompa, A., J. Salvador, and V. Sosa. 1987. The "Pet Kot": a man-made forest of the Maya. *Interciencia* 12(1):10–15.

Goodspeed, T.H. n.d. *Plant Hunters in the Andes*. Robert Hale, London.

Goulding, M. 1980. *The Fishes and the Forest: Explorations in Amazonian Natural History*. University of California Press, Berkeley.

Gowen, S. 1988. Bananas. *Biologist* 35:187–192.

Gradwohl, J., and R. Greenberg. 1988. *Saving the Tropical Forests*. Island Press, Washington, D.C.

Grattan, C.H. 1963. *The Southwest Pacific to 1900, a Modern History: Australia, New Zealand, the Islands, Antarctica*. University of Michigan Press, Ann Arbor.

Green, C.L. 1989. Opportunities and requirements for the development of new essential oil, spice and plant extractive industries. In: *New Crops for Food and Industry*, ed. G.E. Wickens, N. Haq, and P. Day, pp. 76–83. Chapman and Hall, London.

Green, G.A. 1927. Notes on the avocado in New Zealand. *Yearbook of the California Avocado Association*, pp. 39–40.

Green, G.M., and R.W. Sussman. 1990. Deforestation history of the eastern rain forests of Madagascar from satellite images. *Science* 248:212–215.

Greenhall, A.M. 1965. Sapucaia nut dispersal by greater spear-nosed bats in Trinidad. *Caribbean Journal of Science* 5(3–4):167–171.

Gregersen, H.M., and S.E. McGaughey. 1985. *Improving Policies and Financing Mechanisms for Forestry Development*. Inter-American Development Bank, Washington, D.C.

Grilli, E.R., B.B. Agostini, and M.J. Hooft-Welvaars. 1980. *The World Rubber Economy: Structure, Changes, and Prospects*. World Bank Staff Occasional Paper 30, Johns Hopkins University Press, Baltimore.

Grisebach, A.H.R. 1864. *Flora of the British West Indian Islands*. Lovell Reeve & Co., London.

Grosser, J.W., and F.G. Gmitter, Jr. 1990a. Wide-hybridization of *Citrus* via protoplast fusion: progress, strategies, and limitations. *Horticultural Biotechnology* 11 (Plant Biology):31–41.

——. 1990b. Protoplast fusion and citrus improvement. *Plant Breeding Reviews* 8:339–374.

——. 1990c. Somatic hybridization of *Citrus* with wild relatives for germplasm enhancement and cultivar development. *Hortscience* 25(2):147–151.

Grout, B.W., W.K. Shelton, and H.W. Pritchard. 1983. Orthodox behaviour of oil palm seed and crop preservation of the excised embryo for genetic conservation. *Annals of Botany* 52:381–384.

Guerra, A.T. 1955. *Estudo geográfico do território do Acre*. Instituto Brasileiro de Geografia e Estatística, Rio de Janeiro.

Guerra, I.A. 1957. Tipos de clima da região norte. In: *Enciclopédia dos municípios Brasileiros*, vol. 18. p. 18, Instituto Brasileiro de Geografia e Estatística, Rio de Janeiro.

Gulick, P., and D.H. Van Sloten. 1984. *Directory of Germplasm Collections, 6.1: Tropical and Subtropical Fruits and Tree Nuts*. International Board for Plant Genetic Resources, Rome.

Gupta, J.H., and A.S. Yadav. 1985. Screening of mango germplasm for their reaction against powdery mildew of mango. *Progressive Horticulture* 17(1):64–66.

Gupta, S.M. 1971. *Plant Myths and Traditions in India*. E.J. Brill, Leiden.

Hadiwijaya, T. 1979. *Cengkeh, data dan petunjuk ke arah swasembada*. Gunung Agung, Jakarta.

Hale, A. 1911. Balsam of Peru, a Central American contribution to the pharmacopaeia. *Bulletin of the Pan American Union*, May:880–891.

Hall, C. 1985. *Costa Rica: A Geographical Interpretation in Historical Perspective.* Westview Press, Boulder, Colo.

Hambali, G.G., M. Yatazawa, and A.T. Sunarto. 1989. Wild *Durio* germplasm for improving fruit quality and performance of *Durio zibethinus*. In: *Proceedings of the First PROSEA (Plant Resources of South-East Asia) International Symposium, May 22–25, 1989, Jakarta, Indonesia*, ed. J.S. Siemonsma and N. Wulijarni-Soetjipto, p. 261. PUDOC, Wageningen, Netherlands.

Hamilton, R.A. 1987. *Ten Tropical Fruits of Potential Value for Crop Diversification in Hawaii*. Hawaii Institute of Tropical Agriculture and Human Resources, Research Extension Series 85, Honolulu.

Hamilton, R.A., P.J. Ito, and C.L. Chia. 1983. *Macadamia: Hawaii's Dessert Nut*. Cooperative Extension Service, College of Tropical Agriculture and Human Resources Circular 485, University of Hawaii, Honolulu.

Hanbury, D. 1863. On the manufacture of balsam in Peru. *Pharmaceutical Journal* 5.2.5:241–248.

Hanke, D.E. 1990. Seeding the bamboo revolution. *Nature* 344:291–292.

Hanson, H. 1986. *Fifty Years around the Third World: Adventures and Reflections of an Overseas American*. Fraser, Burlington, Vt.

Hardon, J.J. 1969. Interspecific hybrids in the genus *Elaeis* II. Vegetative growth and yield of F1 hybrids *E. guineensis* × *E. oleifera*. *Euphytica* 18:380–388.

——. 1974. Oil palm (*Elaeis guineensis*). In: *Handbook of Plant Introduction in Tropical Crops*, ed. J. León, pp. 75–89. Agricultural Studies 93, Food and Agriculture Organization, Rome.

——. 1976. Oil palm. In: *Evolution of Crop Plants*, ed. N.W. Simmonds, pp. 225–229. Longman, London.

Hardon, J.J., R.H. Corley, and R.H.V. Lee. 1987. Breeding and selecting the oil palm. In: *Improving Vegetatively Propagated Crops*, ed. A.J. Abbott and R.K. Atkin, pp. 63–81. Academic Press, London.

Hardon, J.J., V. Rao, and N. Rajanaidu. 1985. A review of oil palm breeding. In: *Progress in Plant Breeding*, ed. G.E. Russell, pp. 139–163. Butterworths, Boston.

Hardy, F. 1960. *Cacao Manual*. Inter-American Institute of Agricultural Sciences, Turrialba, Costa Rica.

Harlan, J.R. 1975. *Crops and Man*. American Society of Agronomists and Crop Science Society of America, Madison, Wis.

——. 1976. Genetic resources in wild relatives of crops. *Crop Science* 16(3):329–333.

Harlan, J.R., and J.M.J. de Wet. 1971. Toward a rational classification of cultivated plants. *Taxon* 20(4):509–517.

Harlan, J.R., J.M.J. de Wet, and A.B.L. Stemler. 1976. Plant domestication and indigenous African agriculture. In: *Origins of African Plant Domestication*, ed. J.R. Harlan, J.M.J. de Wet, and A.B.L. Stemler, pp. 3–19. Mouton, The Hague.

Harris, D.R. 1976. Traditional systems of plant food production and the origins of agriculture in West Africa. In: *Origins of African Plant Domestication*, ed. J.W. Harlan, J.M.J. de Wet, and A.B.L. Stemler, pp. 311–356. Mouton, The Hague.

——. 1989. An evolutionary continuum of people-plant interaction. In: *Foraging and Farming: The Evolution of Plant Exploitation*, ed. D.R. Harris and G.C. Hillman, pp. 11–26. Unwin Hyman, London.

Harris, D.R., and G.C. Hillman, ed. 1989. *Foraging and Farming: The Evolution of Plant Exploitation*. Unwin Hyman, London.

Harrison Church, R.J. 1960. *West Africa: A Study of the Environment and of Man's Use of It*. Longmans, London.

Hartley, C.W.S. 1988. *The Oil Palm*. Longman/J. Wiley, New York.

Hartshorn, G.S., and L.J. Poveda. 1983. Checklist of trees. In: *Costa Rican Natural History*, ed. D.H. Janzen, pp. 158–183. University of Chicago Press, Chicago.

Hasanbahri, S. 1983. *Studi pertumbuhan dan penubahan morphologis setek batang bekerapa jenis bambu di wanagama I*. Fakultas Kehutanan, University of Jogjakarta, Indonesia.

Hashim, I. 1983. Biology and economic importance of South American leaf blight of *Hevea* rubber. In: *Exotic Plant Quarantine Pests and Procedures for Introduction of Plant Materials*, ed. K.G. Singh, pp. 27–34. ASEAN Plant Quarantine Centre and Training Institute, Serdang, Selangor, Malaysia.

Hatje, G. 1989. World importance of oil crops and their products. In: *Oil Crops of the World: Their Breeding and Utilization*, ed. G. Röbbelen, R.K. Downey, and A. Ashri, pp. 1–21. McGraw-Hill, New York.

Hawkes, J.G. 1983. *The Diversity of Crop Plants*. Harvard University Press, Cambridge.

Hawkins, R.E. 1986. *Encyclopedia of Indian Natural History*. Oxford University Press, Delhi, India.

Hecht, S.B. 1985. Environment, development and politics: capital accumulation and the livestock sector in eastern Amazonia. *World Development* 13(6):663–684.

Hecht, S.B., and A. Cockburn. 1989. *The Fate of the Forest: Developers, Destroyers and Defenders of the Amazon*. Verso, London.

Hedrick, U.P. 1919. *Sturtevant's Notes of Edible Plants*. Lyon & Co., Albany, N.Y.

Heinz, D.J. 1983. Use of agricultural land in Hawaii in relation to crop improvement. In: *Proceedings: First Fertilizer and Ornamentals Workshop, Kailua-Kona, Hawaii, January 12–14, 1983*, pp. 159–177. College of Tropical Agriculture and Human Resources, Research Extension Series 37, University of Hawaii, Honolulu.

Heiser, C.B., Jr. 1990. *Seed to Civilization: The Story of Food*. Harvard University Press, Cambridge.

Hemming, J. 1987. *Amazon Frontier: The Defeat of the Brazilian Indians*. Harvard University Press, Cambridge.

Hemsley, W.B. 1868. On the vegetable productions of Abyssinia. *Journal of Travel and Natural History* 1:309–318.

Hermann, B., J.C. Celhay, M. Guerin, J.M. Maclet, and J. Rentier. 1989. *Fleurs et plantes de Tahiti*. Les Editions du Pacifique, Singapore.

Hernandez, F. 1651. *Rerum medicarum novae Hispanae thesaurus*. Rome, Italy.

Hobson, L. 1972. Some macadamia observations: a visit to California. In: *Macadamia and Pecan Nut Symposium*, ed. G.F. Buchanan, pp. 7–10. Suppl. to *Hortus* 18.

Hodgson, R.W. 1927. Observations on avocado production made from a trip to Florida and Central America in February and March of 1927. *Yearbook of the California Avocado Association*, pp. 89–98.

Holdridge, L.R., W.C. Grenke, W.H. Hatheway, T. Laing, and J.A. Tosi, Jr. 1971. *Forest Environments in Tropical Life Zones: A Pilot Study*. Pergamon Press, Oxford.

Holliday, P. 1970. *South American Leaf Blight* (Microcyclus ulei) *of* Hevea brasiliensis. Commonwealth Mycological Institute, Phytopathological Paper 12, Kew, Surrey, U.K.

Holm, L., J.V. Pancho, J.P. Herberger, and D.L. Plucknett. 1979. *A Geographical Atlas of World Weeds*. J. Wiley, New York.

Holmes, E.B., and K. Newcombe. 1980. Potential and proposed development of sago (*Metroxylon* spp.) as a source of power alcohol in Papua New Guinea. In: *Sago: The Equatorial Swamp as a Natural Resource*, ed. W.R. Stanton and M. Flach, pp. 164–174. Proceedings of the 2nd International Sago Symposium, Martinus Nijhoff, The Hague.

Homma, A.K.O. 1989a. *Perspectivas da economia extrativista vegetal na amazônia.* Empresa Brasileira de Pesquisa Agropecuária, Belém, Pará, Brazil.

——. 1989b. A extração de recursos naturais renováveis; o caso do extrativismo vegetal na Amazônia. Ph.D. dissertation, Universidade Federal de Viçosa, Viçosa, Minas Gerais.

Hooker, W. 1989. *Hooker's Finest Fruits: A Selection of Paintings of Fruits by William Hooker (1799–1832).* Herbert Press, London.

Hope, G., J. Golson, and J. Allen. 1983. Palaeoecology and prehistory in New Guinea. *Journal of Human Evolution* 12:37–60.

Howarth, D. 1983. *Tahiti: A Paradise Lost.* Harvill Press, London.

Hsiung, W.Y. 1987. Bamboo in China: new prospects for an ancient resource. *Unasylva* 39:42–49.

Hsiung, W.Y., and Y.Y. Ren. 1983. Bamboos in development of Chinese civilization. *Bamboo Research* 2:88–100.

Hubbell, S.P., and R.B. Foster. 1986. Commonness and rarity in a neotropical forest: implications for tropical tree conservation. In: *Conservation Biology: The Science of Scarcity and Diversity,* ed. M.E. Soulé, pp. 205–231. Sinauer Associates, Sunderland, Mass.

Huber, J. 1898. Lista das plantas colligidas na Ilha de Marajó no anno de 1896. *Boletim do Museu Goeldi de Historia Natural e Ethnographia* 2:288–321.

——. 1906. Sur l'indigénat du *Theobroma cacao* dans les alluvions du Purus et sur quelques autres espèces du genre *Theobroma. Bulletin de l'Herbier Boissier* 6:272–274.

Huen-pu, W., C. Sing-chi, and W. Si-yu. 1989. China. In: *Floristic Inventory of Tropical Countries,* ed. D.G. Campbell and H.D. Hammond, pp. 35–43. New York Botanical Garden, Bronx.

Hunter, J.R. 1990. Status of cacao (*Theobroma cacao,* Sterculiaceae) in the Western Hemisphere. *Economic Botany* 44:425–439.

Hutton, E.M. 1984. Breeding and selecting leucaenas for acid tropical soils. *Pesquisa Agropecuária Brasileira* 19:262–274.

——. 1990. Field selection of acid soil–tolerant leucaena from *L. leucocephala* × *L. diversifolia* crosses in a tropical oxisol. *Tropical Agriculture* 67(1):2–8.

IAC. 1987. *Chão fecundo: 100 anos de história do Instituto Agronômico de Campinas.* Instituto Agronômico de Campinas, Campinas, Brazil.

IBDF. 1982. *Plano do sistema de unidades de conservação do Brasil: II etapa.* Instituto Brasileiro de Desenvolvimento Florestal, Brasília.

IBGE. 1981. *Sinopse estatística do Brasil.* Fundação Instituto Brasileiro de Geografia e Estatística, Rio de Janeiro.

——. 1986. *Anuário estatístico do Brasil.* Fundação Instituto Brasileiro de Geografia e Estatística, Rio de Janeiro.

IBPGR. 1981. *Genetic Resources of Cacao.* International Board for Plant Genetic Resources, Rome.

——. 1982. *Genetic Resources of Citrus: Report of a Working Group.* International Board for Plant Genetic Resources, Rome.

——. 1984. *Genetic Resources of Hevea.* International Board for Plant Genetic Resources, Rome.

——. 1985. *Ecogeographical Surveying and In Situ Conservation of Crop Relatives.* International Board for Plant Genetic Resources, Rome.

——. 1986. *Genetic Resources of Tropical and Sub-Tropical Fruits and Nuts (excluding Musa).* International Board for Plant Genetic Resources, Rome.

——. 1987. *Descriptors for Papaya.* International Board for Plant Genetic Resources, Rome.

———. 1988. *Descriptors for Citrus*. International Board for Plant Genetic Resources, Rome.

———. 1989. *Descriptors for Mango*. International Board for Plant Genetic Resources, Rome.

ICAR. 1988. *Potentialities of Oilpalm Cultivation in India*. Indian Council of Agricultural Research, New Delhi.

ICRAF. 1988. *The Potential of Agroforestry*. International Council for Research in Agroforestry, Nairobi.

IDRC. 1983. *Leucaena Research in the Asian Pacific Region*. International Development Research Centre, Singapore.

IFAR. 1991. *Research Needs for Bamboo and Rattan to the Year 2000*. International Fund for Agricultural Research, Arlington, Va.

IIHR. 1980. *Annual Report*. Indian Institute of Horticultural Research, Bangalore, India.

Iltis, H.H. 1988. Serendipity in the exploration of biodiversity: what good are weedy tomatoes? In: *Biodiversity*, ed. E.O. Wilson and F.M. Frances, pp. 98–105. National Academy Press, Washington, D.C.

Imle, E.P. 1966. Plant material distribution and quarantine measures for cocoa. *FAO Plant Protection Bulletin* 14(6):134–140.

———. 1978. *Hevea* rubber—past and future. *Economic Botany* 32:264–277.

Ingram, G.B., and J.T. Williams. 1984. In situ conservation of wild relatives of crops. In: *Crop Genetic Resources: Conservation and Evaluation*, ed. J.H.W. Holden and J.T. Williams, pp. 163–179. Allen and Unwin, London.

IPB. 1973. *Situasi cengkeh di propinsi aceh, bengkulu, lampung, Sulawesi utara dan Maluku*. Unpublished report, Facultas Pertanian, Institute Pertanian Bogor, Indonesia.

Irvine, D. 1989. Succession management and resource distribution in an Amazonian rain forest. In: *Resource Management in Amazonia: Indigenous and Folk Strategies*, ed. D.A. Posey and W. Balée, vol. 7, *Advances in Economic Botany*, pp. 223–237. New York Botanical Garden, Bronx.

Irvine, F.R. 1930. *Plants of the Gold Coast*. Oxford University Press, London.

———. 1961. *Wood Plants of Ghana with Special Reference to Their Uses*. Oxford University Press, London.

Isacco, E., A.L. Dallapicola, B.N. Goswamy, K. Goswamy, K.J. Khandalavala, W. Pink, and K. Vatsyayan. 1982. *Krishna: The Divine Lover, Myth and Legend through Indian Art*. B.I. Publications, Bombay.

ITC (International Trade Center). 1961. *Woody Plants of Ghana with Special Reference to Their Uses*. Oxford University Press, London.

———. 1986. *Essential Oils and Oleoresins: A Study of Selected Producers and Major Markets*. International Trade Center (UNCTAD/GATT), Geneva.

ITTO (International Tropical Timber Organization). 1990. *ITTO Guidelines for the Sustainable Management of Natural Tropical Forests*. Doc. PCF(VI)/16, Yokohama, Japan.

IUCN (International Union for the Conservation of Nature and Natural Resources). 1980. *World Conservation Strategy: Living Resource Conservation for Sustainable Development*. Gland, Switzerland.

Jacob, H.E. 1935. *The Saga of Coffee: The Biography of an Economic Product*. Allen and Unwin, London.

Jacobs, M. 1988. *The Tropical Rain Forest: A First Encounter*. Springer-Verlag, Berlin.

Jacobs, M.R. 1937. *Field Studies on the Gum Veins of the Eucalyptus*. Commonwealth Forestry Bureau, Bulletin 20, London.

Jain, H.K. 1988. *Role of Research in Transforming Traditional Agriculture: An Emerg-*

ing Perspective. International Service for National Agricultural Research (ISNAR), Reprint Series 4, The Hague.

Janick, J. 1989. Horticulture in Morocco: North Africa's California. *Hortscience* 24:18–22.

Janzen, D.H. 1983. *Crescentia alata* (jícaro, guacal, gourd tree). In: *Costa Rican Natural History,* ed. D.H. Janzen, pp. 222–224. University of Chicago Press, Chicago.

Jarret, F.M. 1959. Studies in *Artocarpus* and allied genera. III: A revision of subgenus *Artocarpus. Journal of Arnold Arboretum* 40: 113–155.

Jarret, R.L., and R. Fernandez. 1984. Shoot-tip vanilla culture for storage and exchange. *Plant Genetic Resources Newsletter* 57:25–27.

Jarret, R.L., R. Fernandez, and S. Salazar. 1986. In vitro conservation at CATIE. *Plant Genetic Resources Newsletter* (International Board for Plant Genetic Resources) 68:6–10.

Jarvie, J., and S. Koerniati. 1987. Clove germplasm collection in Maluku. *Regional Committee for Southeast Asia Newsletter* (International Board for Plant Genetic Resources) 11(1):6–8.

Jenkins, M.D. 1987. *Madagascar: An Environmental Profile.* International Union for Conservation of Nature and Natural Resources, Conservation Monitoring Centre, Cambridge, U.K.

Jimenez, J.M., J.J. Galindo, C. Ramirez, and G.A. Enriquez. 1987. Evaluación del combate biológico y químico de la moniliasis (*Moniliophthora roreri*) del cacao en Costa Rica. In: *Proceedings of the 10th International Cocoa Research Conference, Santo Domingo, Dominican Republic, 17–23 May,* pp. 453–456.

Johannessen, C.L. 1963. *Savannas of Interior Honduras.* University of California Press, Ibero-Americana 46, Berkeley.

——. 1966a. Pejibayes in commercial production. *Turrialba* 16(2):181–187.

——. 1966b. The domestication process in trees reproduced by seed: the pejibaye palm in Costa Rica. *Geographical Review* 56(4):363–376.

——. 1966c. Pejibaye palm: yields, prices and labor costs. *Economic Botany* 20(3): 302–315.

——. 1967. Pejibaye palm: physical and chemical analysis of the fruit. *Economic Botany* 21(4):371–378.

——. 1970. The dispersal of *Musa* in Central America: The domestication process in action. *Annals of the Association of American Geographers* 60:689–699.

Johnson, D.V. 1973a. The botany, origin, and spread of the cashew, *Anacardium occidentale* L. *Journal of Plantation Crops* 1(1–2):1–7.

——. 1973b. Geography and ecology of native cashew (*Anacardium occidentale*) in Northeast Brazil. *Revista Brasileira de Biologia* 33(4):485–494.

——. 1977. Distribution of sago-making in the Old World. In: *Proceedings of the 1st International Sago Symposium,* ed. K. San, pp. 65–75. Kermajuan Kanji Sdn Bhd., Kuala Lumpur.

——. 1987. Native palms for Brazilian development: three major utilization regions as examples. *Vida Silvestre Neotropical* 1(2):43–49.

Johnson, D.V., and P.K.R. Nair. 1985. Perennial crop-based agroforestry systems in Northeast Brazil. *Agroforestry Systems* 2:281–292.

Johnston, H. 1902. *The Uganda Protectorate,* vol. 1. Hutchinson, London.

Jones, D.T. 1985. Survey and collection of wild and cultivated *Citrus* of Sabah. Report Project No. MAL 69, World Wildlife Fund, Malaysia (mimeo).

Jones, L.H. 1987. Clonal propagation of plantation crops. In: *Improving Vegetatively Propagated Crops,* ed. A.J. Abbott and R.K. Atkin, pp. 385–405, Academic Press, London.

Jones, P.A. 1956. Notes on the varieties of *Coffea arabica* in Kenya. *Coffee Board of Kenya Monthly Bulletin*, November, pp. 1–4.

Jones, W.W., and J.K. Beaumont. 1937. Carbohydrate accumulation in relation to vegetative propagation of litchi. *Science* 86:313.

Joshi, P.C., I. Christin, C. Delvoie, B. Grdjic, G. Hosmalin, and S.Y. Rahim. 1980. *Madagascar: Evolution récente et perspectives économiques.* World Bank, Washington, D.C.

Juma, C. 1989. *The Gene Hunters.* Zed Books, London/Princeton University Press, Princeton.

Kahn, F., and A. Castro. 1985. The palm community in a forest of central Amazonia, Brazil. *Biotropica* 17(3):210–216.

Kajale, M.D. 1989. Mesolithic exploitation of wild plants in Sri Lanka: archaeobotanical study at the cave site of Beli-Lena. In: *Foraging and Farming: The Evolution of Plant Exploitation*, ed. D.R. Harris and G.C. Hillman, pp. 269–281. Unwin Hyman, London.

Kasim, R., Prayitno, Muchlas, and E.A. Haddad. 1980. Pencegahan cacar daun pada tanaman cengkeh secara kimiawi. *Pemberitaan, Lembaga Penelitian Tanaman Industri, Bogor, Indonesia* 38:1–8.

Katz, A.E. 1943. Natural and synthetic flavouring materials. *Spice Mill* 66:12, 34–35.

Keay, R.W.J., and A. Aubréville. 1959. *Carte de la végétation de l'Afrique au sud du Tropique du Cancer.* Oxford University Press, Oxford.

Keiding, H., and R.H. Kemp. 1977. Exploration, collection and investigation of gene resources: tropical pines and teak. Paper presented at the 3rd World Consultation on Forest Tree Breeding, Canberra.

Kemp, R.H. 1975. Central American pines—a case study. In: *Report on a Pilot Study on the Methodology of Conservation of Forest Gene Resources*, pp. 57–64, Food and Agriculture Organization, Rome.

Kennedy, A.J. 1985. The International Cocoa Genebank, Trinidad: its role in future cocoa research and development. *Cocoa Growers' Bulletin* 36:5–10.

Kennedy, J.D. 1936. *Forest Flora of Southern Nigeria.* Government Printer, Lagos.

Kerr, W.S. 1970. The avocado industry in southern California, a study of location, perception, and prospect. Ph.D. dissertation, University of Oklahoma, Norman.

Kerry, W.E., and C.R. Clement. 1980. Práticas agrícolas com consequências genéticas que possibilitaram aos Indios da Amazônia uma melhor adaptação as condições da região. *Acta Amazonica* 10:251–261.

Kervégant, D. 1937. Les plantes utiles et ornementales de la Martinique. *Bulletin Agricole de la Martinique* 5(1):1–94.

Kimber, C. 1966. Dooryard gardens of Martinique. *Yearbook of the Association of Pacific Coast Geographers* 28:97–118.

——. 1988. *Martinique Revisited: The Changing Plant Geographies of a West Indian Island.* Texas A&M University Press, College Station.

Kitamura, P.C., and C.H. Müller. 1984. *Castanhais nativos de Marabá-PA: fatores de depredação e bases para a sua preservação.* Empresa Brasileira de Pesquisa Agropecuária (EMBRAPA), Belém, Pará, Brazil.

Knapp, A.W. 1923. *The Cocoa and Chocolate Industry, the Tree, the Bean, the Beverage.* Sir Isaac Pitman and Sons, London.

Knight, E.E. 1917. Methods of avocado growing in the tropics applicable to California. *Annual Report of the California Avocado Association for the Year 1916*, pp. 95–98.

Knight, R.J. 1988a. Miscellaneous tropical fruits grown and marketed in Florida. *Proceedings of the Interamerican Society for Tropical Horticulture* 32:34–41.

————. 1988b. Achieving effective expansion of fruit production in the tropics and nearby regions. *Acta Horticulturae* 211:17–20.

Knight, R.J., M. Lamberts, and J.S. Bunch. 1984. World and local importance of some tropical fruit crops grown in Florida. *Proceedings of the Florida State Horticultural Society* 97:351–354.

Ko, W.H., and R.K. Kunimoto. 1986. A rapid method for screening macadamia seedlings for resistance to *Kretzschmaria clavus*. *Annals of the Phytopathological Society of Japan* 52:336–337.

Ko, W.H., R.K. Kunimoto, and I. Maeda. 1977. Root decay caused by *Kretzschmaria clavus*: its relation to macadamia decline. *Phytopathology* 67:18–21.

Ko, W.H., J. Tomita, and R.L. Short. 1986. Two natural hosts of *Kretzschmaria clavus* in Hawaiian forests. *Plant Pathology* 35(2):254–255.

Kochhar, S.L., and B.M. Singh. 1989. Plant resources for A.D. 2001. In: *Plants and Society*, ed. M.S. Swaminathan and S.L. Kochhar, pp. 556–617. Macmillan, London.

Konowicz, H., and J. Janick. 1984. In vitro propagation of *Vanilla planifolia*. *Hortscience* 19:58–59.

Kopp, L.E. 1966. A taxonomic revision of the genus *Persea* in the Western Hemisphere (Perseae-Lauraceae). *Memoirs of the New York Botanical Garden* 14:1–120.

Kostermans, A.J.G. 1958. The genus *Durio*. *Reinwardtia* 4:387–460.

————. 1973. A forgotten Ceylonese cinnamon-tree (*Cinnamomum cappare-coronde* Bl.). *Ceylon Journal of Science (Biological Sciences)* 10:119–121.

————. 1980. A note on two species of *Cinnamomum* (Lauraceae) described in *Hortus Indicus Malabaricus*. A.A. Balkenna, Rotterdam.

Kotzé, J.M., and J.M. Darvas. 1983. Integrated control of avocado root rot. *Yearbook of the California Avocado Society* 67:83–86.

Kowalska, M.T., and D. Puett. 1990. Potential biomedical applications for tropical fruit products. *Tropical Fruit World* 1(4):126.

Krome, I.B. 1956. Florida mangos. *Ceiba* 4(6):337–339.

Krug, C.A., and R.A. De Poerck. 1968. *World Coffee Survey*. Food and Agriculture Organization, Agricultural Studies 76, Rome.

Krüssmann, G. 1985. *Manual of Cultivated Conifers*. Timber Press, Portland, Ore.

Kushalappa, A.C. 1989. Introduction. In: *Coffee Rust: Epidemiology, Resistance, and Management*, ed. A.C. Kushalappa and A.B. Eskes, pp. 1–11. CRC Press, Boca Raton, Fla.

Lahane, B.N., L.L. Relwani, D.K. Raina, and H.L. Gadekar. 1987. Initial evaluation of *L. leucocephala* cultivars for fodder production. *Leucaena Research Papers* 8:29–30.

Laker, H.A., and T.N. Sreenivasan. 1987. The resistance of some cocoa clones to *Crinipellis perniciosa* in Trinidad. In: *Proceedings of the 10th International Cocoa Research Conference, Santo Domingo, Dominican Republic, 17–23 May*, pp. 637–641.

Lakshminarayana, S., and M.A. Moreno-Rivera. 1979. Promising Mexican guava selection rich in vitamin C. *Proceedings of the Florida State Horticultural Society* 92:300–303.

Lam, H.J. 1932. Burseraceae of the Malay archipelago and peninsula. *Bulletin du Jardin Botanique de Buitenzorg* 111:281–561.

Lambourne, J. 1934. The avocado pear. *Malayan Agriculture Journal* 22:131–140.

————. 1937. The rambutan and its propagation. *Malayan Agriculture Journal* 25:11–17.

Lande, R. 1988. Genetics and demography in biological conservation. *Science* 241:1455–1460.

Landrum, L.R. 1991. The guava. *National Geographic Research* 7:116.

Lang, J. 1988. *Inside Development in Latin America: A Report from the Dominican Republic*. University of North Carolina Press, Chapel Hill.

Langenheim, J.H. 1973. Leguminous resin-producing trees in Africa and South America. In: *Tropical Forest Ecosystems in Africa and South America: A Comparative Review*, ed. B.J. Meggers, E.S. Ayensu, and W.D. Duckworth, pp. 89–104. Smithsonian Institution Press, Washington, D.C.

——. 1990. Plant resins. *American Scientist* 78:16–24.

Lanly, J.P. 1982. *Tropical Forest Resources*. Food and Agriculture Organization, Forestry Paper 30, Rome.

Laraia, R., and R. Da Matta. 1978. Indios e castanheiros: a empresa extrativa e os índios no médio Tocantins. Paz e Terra, Rio de Janeiro.

Laryea, A.A. 1981. Technology transfer to cocoa farmers in West Africa. In: *Proceedings of the 8th International Cocoa Research Conference, Cartagena, Colombia*, pp. 583–591.

Lathrap, D.W. 1977. Our father the cayman, our mother the gourd; Spinden revisited, or a unitary model for the emergence of agriculture in the New World. In: *Origins of Agriculture*, ed. C.A. Reed, pp. 713–751, Mouton, The Hague.

Lavabre, E.M. 1981. Les ennemis du cacaoyer en Afrique. In: *7ème Conférence Internationale sur la Recherche Cacaoyère, Douala, Cameroon*, pp. 423–427.

Lavelle, P. 1987. Biological processes and productivity of soils in the humid tropics. In: *The Geophysiology of Amazonia: Vegetation and Climate Interactions*, ed. R.E. Dickinson, pp. 175–214. J. Wiley, New York.

Lazaroff, L. 1989. Strategy for development of a new crop. In: *New Crops for Food and Industry*, ed. G.E. Wickens, N. Haq, and P. Day, pp. 108–119. Chapman and Hall, London.

Leader-Williams, N., and S.D. Albon. 1988. Allocation of resources for conservation. *Nature* 336:533–535.

Le Cointe, P. 1947. *Amazônia brasileira III: árvores e plantas úteis*. Companhia Editora Nacional, São Paulo.

Ledec, G., and R. Goodland. 1988. *Wildlands: Their Protection and Management in Economic Development*. World Bank, Washington, D.C.

Ledig, F.T. 1988. The conservation of diversity in forest trees: why and how should genes be conserved? *Bioscience* 38:471–478.

Lee, D.W. 1980. Durians. *Horticulture* 58(7):46–51.

Legg, J.T. 1981. The British cocoa swollen-shoot project in Ghana. In: *Proceedings of the 7th International Cocoa Research Conference, Douala, Cameroon*, pp. 399–405.

Lele, U. 1988. *Agricultural Growth, Domestic Policies, the External Environment, and Assistance to Africa: Lessons of a Quarter Century*. World Bank, MADIA (Managing Agricultural Development in Africa) Discussion Paper 1, Washington, D.C.

Lele, U., and L.R. Meyers. 1988. *Growth and Structural Change in East Africa: Domestic Policies, Agricultural Performance, and World Bank Assistance, 1963–1986, Parts I and II*. World Bank, MADIA (Managing Agricultural Development in Africa) Discussion Paper 3, Washington, D.C.

León, J. 1960. Taxonomy of cacao and related genera. In: *Cacao Manual*, ed. F. Hardy, IAAS, Turrialba, Costa Rica.

——. 1987. *Botánica de los cultivos tropicales*. Instituto Interamericano de Cooperación para la Agricultura (IICA), San José, Costa Rica.

Leung, A.V. 1980. *Encyclopedia of Common Natural Ingredients Used in Food, Drugs, and Cosmetics*. Wiley, New York.

Levingstone, R., and R. Zamora. 1983. Medicine trees of the tropics. *Unasylva* 35(140):7–10.

Lévi-Strauss, C. 1963a. The tribes of the upper Xingu River. In: *Handbook of South American Indians*, ed. J.H. Steward, vol. 3, pp. 321–348. Cooper Square Publishers, New York.

———. 1963b. The Tupí-Cawahíb. In: *Handbook of South American Indians*, ed. J.H. Steward, vol. 3, pp. 299–305. Cooper Square Publishers, New York.

Lim, L. 1989. The reproductive biology of rambutan, *Nephelium lappaceum* L. (Sapindaceae). *Gardens Bulletin* (Singapore) 57:181–192.

Linares, E., and R.A. Bye. 1987. A study of four medicinal plant complexes of Mexico and adjacent U.S.A. *Journal of Ethnopharmacology* 19:153–184.

Litz, R.E. 1984. In vitro somatic embryogenesis from nucellar callus of monoembryonic mango. *Hortscience* 19:715–717.

———. 1986. Papaya (*Carica papaya* L.). In: *Biotechnology in Agriculture and Forestry*, ed. Y.P.S. Bajaj, vol. 1, pp. 220–232. Spinger-Verlag, Berlin.

———. 1987. Application of tissue culture to tropical fruits. *Plant Biology* 3:407–418.

Litz, R.E., and C.W. Campbell, P.K. Soderholm, and K. Norstog. 1983. Tropical and subtropical plant germplasm resources in South Florida. *Plant Molecular Biology Reporter* 1(4):44–51.

Litz, R.E., and R.A. Conover. 1979. Development of systems for obtaining parasexual *Carica* hybrids. *Proceedings of the Florida State Horticultural Society* 92:281–283.

———. 1982. In vitro somatic embryogenesis and plant regeneration from *Carica papaya* L. ovular callus. *Plant Science Letters* 26:153–158.

Litz, R.E., G.A. Moore, and C. Srinivasan. 1985. In vitro systems for propagation and improvement of tropical fruit and palms. *Horticultural Reviews* 7:157–200.

Lleras, E., and L. Coradin. 1986. Native neotropical oil palms: state of the art and perspectives for Latin America. *Advances in Economic Botany* 6:201–213.

Lockwood, G. 1985. Selection and breeding. In: *Cocoa Production: Present Constraints and Priorities for Research*, ed. R.A. Lass and G.A.R. Wood, pp. 29–37, World Bank Technical Paper 39, Washington, D.C.

Lohmann, L., and M. Colchester. 1990. Paved with good intentions: TFAP's road to oblivion. *The Ecologist* 20(3):91–98.

Lopes, J.C.A., J.O.P. Carvalho, J.N.M. Silva, and H.B. Costa. 1986. Sociabilidade entre 18 espécies comerciais ocorrentes na Floresta Nacional do Tapajós. In: *Anais do Primeiro Simpósio do Trópico Umido, Belém, Pará, 12 a 17 de novembro de 1984*, vol. 2, *Flora e floresta*, pp. 263–268. Empresa Brasileira de Pesquisa Agropecuária (EMBRAPA), Centro de Pesquisa Agropecuária do Trópico Umido (CPATU), Belém, Pará, Brazil.

Lopez, L.E. 1981. Necessidades de colectar poblaciones de cacao y de su conservación en un banco internacional de germoplasma. In: *8a Conferencia Internacional de Investigación en Cacao, Cartagena, Colombia*, pp. 663–666.

Lopez, O., J. Mulato, and J.I. Lopez. n.d. Recolección de germoplasma de cacao (*Theobroma cacao* L.) en Mexico. Unpublished report.

Loureiro, A.A., M.F. Silva, and J.C. Alencar. 1979. *Essências madeireiras da Amazônia*, vol. 1. Instituto Nacional de Pesquisas' da Amazônia, Manaus, Brazil.

Lovejoy, T.E. 1984. Aid debtor nations' ecology. *New York Times*, 4 October.

Lovejoy, T.E., R.O. Bierregaard, Jr., A.B. Rylands, J.R. Malcolm, C.E. Quintela, L.H. Harpers, K.S. Brown, Jr., A.H. Powell, G.V.N. Powell, H.O.R. Schubart, and M.B. Hays. 1986. Edge and other effects of isolation on Amazon forest fragments. In: *Conservation Biology: The Science of Scarcity and Diversity*, ed. M.E. Soulé, pp. 257–285. Sinauer Associates, Sunderland, Mass.

LPTI. 1974. Survey determinasi dan observasi tipe cengkeh. Unpublished report, Lembaga Penelitian Tanaman Industri, Bogor, and Institute Pertanian Bogor, Indonesia.

Lukmonohadi, B., Diyoyo, and E. Waluyo. 1983. Pembibitan tanaman cengkeh dengan cara stek di kebun Branggah Banaran. *Buletin Cengkeh dan Tembakau, Yayasan Cengkeh, Indonesia* 4:4–19.

Lynch, S.J., and M.J. Mustard. 1956. *Mangos in Florida.* Department of Agriculture, State of Florida/University of Miami, Tallahassee.

Mabbett, T. 1988. Gene genius finds a home in Trinidad. *Coffee and Cocoa International* 6:45.

McClure, F.A. 1966. *The Bamboos: A Fresh Perspective.* Harvard University Press, Cambridge.

——. 1973. *Genera of Bamboos Native to the New World (Gramineae, Bambusoideae).* Smithsonian Institution, Washington, D.C.

McConachie, I. 1980. The macadamia story. *California Macadamia Society Yearbook* 36:41–75.

McCune, S. 1949. Sequence of plantation agriculture in Ceylon. *Economic Geography* 25:226–235.

McCurrach, J.C. 1960. *Palms of the World.* Harper and Rowe, New York.

MacDicken, K.G. 1990. Agroforestry management in the humid tropics. In: *Agroforestry: Classification and Management*, ed. K.G. MacDicken and N.T. Vergara, pp. 98–149. Wiley, New York.

McGeary, F.M., and B.E.J. Wheeler. 1988. Growth rates of, and mycelial interactions between, isolates of *Crinipellis perniciosa* from cocoa. *Plant Pathology* 37:489–498.

Maciel, U.N., and P.L.B. Lisboa. 1989. Estudo florístico de 1 hectare de mata de terra firme no km 15 da rodovia Presidente Médici-Costa Marques (RO-429), Rondônia. *Boletim do Museu Paraense Emílio Goeldi, Botânica* 5(1):25–37.

Maclaren, W.A. 1924. *Rubber, Tea and Cacao with Special Sections on Coffee, Spices and Tobacco.* Van Nostrand, New York.

Macmillan, H.F. 1935. *Tropical Planting and Gardening with Special Reference to Ceylon.* Macmillan, London.

McNeely, J.A., K.R. Miller, and J.W. Thorsell. 1987. Objectives, selection, and management of protected areas in tropical forest habitats. In: *Primate Conservation in the Tropical Rain Forest*, ed. C.W. Marsh and R.A. Mittermeier, pp. 181–204. Alan R. Liss, New York.

McNeely, J.A. K.R. Miller, W.V. Reid, and R.A. Mittermeier. 1990. *Conserving the World's Biological Diversity.* IUCN, Switzerland; WRI/CI/WWF-US/World Bank, Washington, D.C.

MacNeish, R.S. 1967. A summary of the subsistence. In: *Prehistory of the Tehuacan Valley*, ed. D.S. Beyers, vol. 1, pp. 290–309. University of Texas Press, Austin.

McVaugh, R. 1956. Tropical American Myrtaceae. *Fieldiana (Botany)* 29:145–228.

Magalhães, B. 1980. *O café na história, no folclore e nas belas-artes.* Companhia Editora Nacional, São Paulo.

Magalhães, J.C.A., J.B.R. Sampaio, and J.E. Silva. 1987. Estabelecimento da cafeicultura na região dos Cerrados. In: *Relatório técnico anual do Centro de Pesquisas Agropecuária dos Cerrados 1982–1985*, pp. 345–349. Centro de Pesquisa Agropecuária dos Cerrados, Empresa Brasileira de Pesquisa Agropecuária, Planaltina, D.F., Brazil.

Mahar, D.J. 1989. *Government Policies and Deforestation in Brazil's Amazon Region.* World Bank, Washington, D.C.

Mahdeem, H. 1990. Zill's *Annona* project. *Tropical Fruit World* 1(4):109.

Mahta, U., I.V.R. Rao, and M. Ram. 1982. Somatic embryogenesis in bamboo. In: *Proceedings of the International Congress on Plant Tissue and Cell Culture*, ed. A. Fujiwara, pp. 109–110. Matuzen, Tokyo.

Maistre, J. 1964. *Les plantes à épices.* Techniques Agricoles et Productions Tropicales, Maisonneuve et Larose, Paris.

Manokaran, N. 1990. *The State of the Rattan and Bamboo Trade.* Rattan Information

Centre, Occasional Paper 7, Forest Research Institute, Kepong-Kuala Lumpur, Malaysia.

Mansfeld, R. 1959. *Vorlufige Verzeichnis landwirtschaftlich oder gartner kultivierte Pflanzenarten*. Kulturpflanze Suppl. 2, Berlin.

Manshardt, R.M., and T.F. Wenslaff. 1989. Interspecific hybridization of papaya with other *Carica* species. *Journal of the American Society for Horticultural Science* 114(4):689–694.

Marshall, R.C. 1934. *Trees of Trinidad and Tobago*. Government Printing Office, Port of Spain, Trinidad.

Martin, F.W., ed. 1984. *CRC Handbook of Tropical Food Crops*. CRC Press, Boca Raton, Fl.

Martin, F.W., C.W. Campbell, and R. Ruberté. 1987. *Perennial Edible Fruits of the Tropics*. Agricultural Research Service, U.S. Department of Agriculture, Handbook 642, Washington, D.C.

Master, H. 1981. Village forestry inventory in Bangladesh. Food and Agriculture Organization/United Nations Development Program Report, Rome.

Mathews, V.M. 1979. Multiple plantlets in lateral bud and leaf explants in in vitro culture of pineapple. *Scientia Horticulturae*, 11:319–328.

Matthews, E. 1983. Global vegetation and land use. *Journal of Climate and Applied Meteorology* 22:474–483.

Mayo, N. 1938. *Native Plant Life and Plant Immigrants of Florida*. Department of Agriculture, Supplementary Bulletin (New Series 16), Tallahassee, Fl.

Meadley, J. 1989. The commercial implications of new crops. In: *New Crops for Food and Industry*, ed. G.E. Wickens, N. Haq, and P. Day, pp. 23–28. Chapman and Hall, London.

Medina, J.T. 1988. *The Discovery of the Amazon*. Dover, New York.

Mee, M. 1988. *In Search of Flowers of the Amazon Forests*. Nonesuch Expeditions, Woodbridge, Suffolk, England.

Meggers, B.J., and C. Evans. 1957. *Archaeological Investigations at the Mouth of the Amazon*. Bureau of American Ethnology, Bulletin 167, Smithsonian Institution, Washington, D.C.

Mendenzez, T., and K. Shepherd. 1975. Breeding new bananas. *World Crops* 27:104–112.

Mendes, A.C.B., N.C.A. Ribeiro, J.J.S. Garcia, and O. Trevizan. 1987. Danos de Conotrachelus? humeropictus (Coleoptera, Curculionidae): nova praga do cacaueiro na Amazônia brasileira. In: *Proceedings of the 10th International Cocoa Research Conference, Santo Domingo, Dominican Republic, 17–23 May*, pp. 535–539.

Menz, K.M., and E.M. Fleming. 1989. *Economic Prospects for Vanilla in the South Pacific*. Australian Council for International Agricultural Research, Technical Report 11, Canberra.

Métraux, A. 1963a. Tribes of the Middle and Upper Amazon river. In: *Handbook of South American Indians*, ed. J.H. Steward, vol. 3, pp. 687–712. Cooper Square Publishers, New York.

———. 1963b. Tribes of eastern Bolivia and Madeira. In: *Handbook of South American Indians*, ed. J.H. Steward, vol. 3, pp. 381–454. Cooper Square Publishers, New York.

Meunier, J. 1969. Étude des populations naturelles d'*Elaeis guineensis* en Côte d'Ivoire. *Oléagineux* 24:195–201.

———. 1975. Le "palmier à huile" américain *Elaeis melanococca*. *Oléagineux* 30:51–61.

Meyer, F.G. 1967. Recent introductions of wild arabica coffee germ plasm from Ethiopia for updating coffee research. In: *Proceedings of the International Symposium on*

Plant Introduction, Escuela Agrícola Panamericana, Tegucigalpa, Honduras, November 30–December 2, 1966, pp. 119–127. Escuela Agrícola Panamericana, Tegucigalpa, Honduras.

Milad, Y. 1936. Avocado culture in Egypt. *Yearbook of the California Avocado Association*, pp. 76–77.

Miles, D.H., J.M.R. Del Medeiros, V. Chittawong, C. Swithenbank, Z. Lidert, J.A. Weeks, J.L. Atwood, and P.A. Hedin. 1990. 3'-formyl-2',4',6'-trihydroxy-5'-methyldihyrochalcone, a prospective new agrochemical from *Psidium acutangulum*. *Journal of Natural Products* 53(6):1548–1551.

Miller, C. 1990. Natural history, economic botany, and germplasm conservation of the Brazil nut tree *(Bertholletia excelsa*, Humb & Bonpl.). M.S. thesis, University of Florida, Gainesville.

Miller, W.W. 1955. The macadamia nut: a new commercial crop for California. *Yearbook of the California Avocado Society* 39:143–145.

Milton, K. 1984. Protein and carbohydrate resources of the Maku Indians of northwestern Amazonia. *American Anthropologist* 86(1):7–27.

Minc, L.D., and J. Vandermeer. 1990. The origin and spread of agriculture. In: *Agroecology*, ed. C.R. Carroll, J.H. Vandermeer, and P. Rosset, pp. 65–111. McGraw Hill, New York.

Mirov, N.T. 1967. *The Genus* Pinus. Ronald Press, New York.

Mitchell, J.D. 1990. The poisonous Anacardiaceae genera of the world. *Advances in Economic Botany* 8:103–129.

Mitchell, J.D., and S.A. Mori. 1987. *The Cashew and Its Relatives (Anacardium: Anacardiaceae)*. Memoirs of the New York Botanical Garden, vol. 42, 76 pp.

Mohd, N.G. 1984. Progress report on *Hevea* germplasm of 1981. *Planters' Bulletin of the Rubber Research Institute of Malaysia* 178:26–28.

Mongi, H.O., and P.A. Huxley, ed. 1979. *Soils Research in Agroforestry*. International Council for Research in Agroforestry, Nairobi.

Montenegro, H.W.S. 1960. *A cultura do abacateiro*. Editora Melhoramentos, São Paulo.

Moore, M.P. 1978. Political culture and agricultural policy: the case of cinnamon in Sri Lanka. *Agricultural Administration* 5:121–129.

Moorehead, A. 1967. *The Fatal Impact: An Account of the Invasion of the South Pacific 1767–1840*. Dell, New York.

———. 1971. *The White Nile*. Hamish Hamilton, London.

Moran, E.F. 1981. *Developing the Amazon*. Indiana University Press, Bloomington.

Mora-Urpí, J. 1982. Polinización en *Bactris gasipaes* H.B.K. (Palmae): nota adicional. *Revista de Biología Tropical* 30(2):174–176.

———. 1989. El palmito de pejibaye: un cultivo Costarriquense. *Boletín Informativo, Serie Técnica* (Universidad de Costa Rica) 1(1):3–12.

Mora-Urpí, J., and E.M. Solis. 1980. Polinización en *Bactris gasipaes* H.B.K. (Palmae). *Revista de Biología Tropical* 28:153–174.

Mora-Urpí, J., E. Vargas, C.A. López, M. Villaplana, G. Allón, and C. Blanco. 1984. *The Pejibaye Palm (Bactris gasipaes H.B.K.)*. Food and Agriculture Organization, Rome.

Mori, S.A. 1989. Eastern, extra-Amazonian Brazil. In: *Floristic Inventory of Tropical Countries*, ed. D.G. Campbell and H.D. Hammond, pp. 427–454. New York Botanical Garden, New York.

———. 1990. Diversificação e conservação das Lecythidaceae neotropicais. *Acta Botanica Bras.* 4(1):45–68.

Mori, S.A., and G.T. Prance. 1987. Species diversity, phenology, plant-animal interactions, and their correlation with climate, as illustrated by the Brazil nut family (Lecythidaceae). In: *The Geophysiology of Amazonia: Vegetation and Climate Interactions*, ed. R.E. Dickinson, pp. 69–89, J. Wiley, New York.

——. 1990. Taxonomy, ecology, and economic botany of the Brazil nut (*Bertholletia excelsa* Humb. and Bonpl.: Lecythidaceae). *Advances in Economic Botany* 8:130–150.

Mori, S.A., G.T. Prance, and C.H. de Zeeuw. 1990. Lecythidaceae, Part II. Flora Neotropica, Monograph 21 (II), New York Botanical Garden, Bronx.

Morren, C. 1839. On the production of vanilla in Europe. *Annals and Magazine of Natural History* 13:1–9.

Mors, W.B., and C.T. Rizzini. 1966. *Useful Plants of Brazil*. Holden-Day, San Francisco.

Morschel, J.R. 1971. *Introduction to Plant Quarantine*. Australian Government Publishing Service, Canberra.

Morton, J.F. 1961. The cashew's brighter future. *Economic Botany* 15:57–58.

——. 1987. *Fruits of Warm Climates*. Creative Resource Systems, Winterville, North Carolina.

Muchaili, J.H. 1989. Management problems in the introduction of new crops: the Zambian experience. In: *New Crops for Food and Industry*, ed. G.E. Wickens, N. Haq, and P. Day, pp. 29–35. Chapman and Hall, London.

Mukherjee, S.K. 1972. Origin of mango (*Mangifera indica*). *Economic Botany* 26:260–268.

——. 1985. *Systematic and Ecogeographic Studies of Crop Genepools: 1.* Mangifera L. International Board for Plant Genetic Resources, Rome.

Mukherjee, S.K., R.N. Singh, P.K. Majumder, and D.K. Sharma. 1968. Present position regarding breeding of mango in India. *Euphytica* 17:462–467.

Mukherji, S. 1949. A monograph on the genus *Mangifera* L. *Lloydia* 12:73–136.

Müller, C.H. 1981. *Castanha-do-Brasil; estudos agronômicos*. Centro de Pesquisa Agropecuária do Trópico Umido, Documentos 1, Belém, Brazil.

Müller, F. 1895. *Select Extra-Tropical Plants, Readily Eligible for Industrial Culture or Naturalisation*. R.S. Barain, Melbourne.

Muller, R.A. 1981. La lutte intégrée contre les maladies et les ennemis du cacaoyer. In: *8ème Conférence Internationale sur la Recherche Cacaoyère, Cartagena, Colombia*, pp. 285–288.

Mumford, J.D. 1985. The cocoa pod borer (cocoa moth). In: *Cocoa Production: Present Constraints and Priorities for Research*, ed. R.A. Lass and G.A.R. Wood, pp. 75–77, World Bank, Technical Paper 39, Washington, D.C.

Murillo, M. 1990. Utilización de la harina de pejibaye en la alimentación de las aves. *Boletín Informativo* (Universidad de Costa Rica), *Serie Técnica Pejibaye (Guilielma)* 2(1):4–6.

Murray, G.F. 1987a. The domestication of wood in Haiti: a case study in applied evolution. In: *Anthropological Praxis: Translating Knowledge into Action*, ed. R.M. Wulff and S.J. Fiske, pp. 223–240. Westview Press, Boulder, Colo.

——. 1987b. Land tenure and agroforestry in Haiti: a case study in anthropological design. In: *Land, Trees, and Tenure: Proceedings of an International Workshop on Tenure Issues in Agroforestry*. International Council for Research in Agroforestry (ICRAF)/ Land Tenure Center, Nairobi.

Myers, J.G. 1930. Notes on wild cacao in Surinam and in British Guiana. *Bulletin of Miscellaneous Information* (Royal Botanic Gardens, Kew) 1:1–10.

Myers, N. 1980a. The present status and future prospects of tropical moist forests. *Environmental Conservation* 7(2):101–114.

———. 1980b. *Conversion of Tropical Forests*. National Academy of Sciences, Washington, D.C.

———. 1983. *A Wealth of Wild Species: Storehouse for Human Welfare*. Westview Press, Boulder, Colo.

———. 1985a. *The Primary Source: Tropical Forests and Our Future*. W.W. Norton, New York.

———. 1985b. Tropical deforestation and species extinctions. *Futures* 17:451–463.

———. 1986. Forestland farming in western Amazonia: stable and sustainable. *Forest Ecology and Management* 15:81–93.

———. 1988a. Tropical deforestation and climatic change. *Environmental Conservation* 15(4):293–298.

———. 1988b. The future of forests. In: *The Fragile Environment*, ed. L. Friday and R.A. Laskey, pp. 22–40. Cambridge University Press, Cambridge.

———. 1988c. Mass extinction—profound problem, splendid opportunity. *Oryx* 22: 205–215.

Nadgauda, R.S., V.A. Parasharami, and A.F. Mascarenhas. 1990. Precocious flowering and seeding behaviour in tissue-cultured bamboos. *Nature* 344:335–336.

Nagao, M., W. Ko, H. Hirae, A. Hara, and W. Nishijima. 1990. Macadamia quick decline update. *Hawaii Mac Facts* 3(2):5–6.

Nair, P.K. 1980. *Agroforestry Species*. International Council for Research in Agroforestry, Nairobi.

———. 1990. *The Prospects for Agroforestry in the Tropics*. World Bank, Technical Paper 131, Washington, D.C.

———. 1991. State of the art of agroforestry systems. *Forest Ecology and Management* 45:4–29.

Namkoong, G. 1986. Genetics and the forests of the future. *Unasylva* 152:2–18.

Napompeth, B., K.G. MacDicken, M. McFadden, and I.N. Oka. 1987. *A Regional Research Plan for* Leucaena *Psyllid Control*. F/FRED *Leucaena* Psyllid Research Workshop, Manila, Philippines, June.

NAS. 1975. *Underexploited Tropical Plants with Promising Economic Value*. National Academy of Sciences, Washington, D.C.

———. 1980. *Firewood Crops: Shrub and Tree Species for Energy Production*. National Academy of Sciences, Washington, D.C.

Nayar, N.M. 1983. Conservation of plantation crops genetic resources. In: *Conservation of Tropical Plant Resources*, ed. S.K. Jain and K.L. Mehra, pp. 246–253, Botanical Survey of India, Howrah.

Neal, M.C. 1965. *In Gardens of Hawaii*. Bernice B. Bishop Museum Press, Honolulu, Hawaii.

Needham, J. 1986. *Science and Civilization in China*, vol. 6, part 1. *Biology and Biological Technology; Botany*. Cambridge University Press, Cambridge.

Nelson, B.W., M.L. Absy, E.M. Barbosa, and G.T. Prance. 1985. Observations on flower visitors to *Bertholletia excelsa* H.B.K. and *Couratari tenuicarpa* A.C. Sm. (Lecythidaceae). *Acta Amazonica* 15(½), suplemento: 225–234.

Neumann, R.P., and G.E. Machlis. 1989. Land-use and threats to parks in the neotropics. *Environmental Conservation* 16:13–18.

Ngatchou, J.N., and J. Kengu. 1989. Review of the African plum tree (*Dacryodes edulis*). In: *New Crops for Food and Industry*, ed. G.E. Wickens, N. Haq, and P. Day, pp. 265–271. Chapman and Hall, London.

Nimuendajú, C. 1963a. Tribes of the lower and middle Xingu River. In: *Handbook of South American Indians*, ed. J.H. Steward, vol. 3, pp. 213–243. Cooper Square Publishers, New York.

——. 1963b. The Maué and Arapium. In: *Handbook of South American Indians*, ed. J.H. Steward, vol. 3, pp. 245–254. Cooper Square Publishers, New York.

——. 1963c. The Cayabí, Tapanyuna, and Apiaca. In: *Handbook of South American Indians*, ed. J.H. Steward, vol. 3, pp. 307–320. Cooper Square Publishers, New York.

Nimuendajú, C., and A. Métraux. 1963. The Amanayé. In: *Handbook of South American Indians*, ed. J.H. Steward, vol. 3, pp. 199–202. Cooper Square Publishers, New York.

Nodari, R.O., and M.P. Guerra. 1986. O palmiteiro no sul do Brasil: situação e perspectivas. In: *Useful Palms of Tropical America*, pp. 9–10. Newsletter 2, EMBRAPA/CENARGEN, Brasília, Brazil.

Northwood, P.J. 1964. Some observations in flowering and fruit setting in cashew, *Anacardium occidentale*. *Tropical Agriculture* (Trinidad) 43:35–42.

NRC. 1984. *Leucaena: Promising Forage and Tree Crop for the Tropics*. National Research Council, Washington, D.C.

——. 1989. *Lost Crops of the Incas*. National Research Council, National Academy Press, Washington, D.C.

——. 1991. *Managing Global Genetic Resources*. National Research Council, National Academy Press, Washington, D.C.

NRCC. 1988. *Annual Report 1988*. National Research Centre for Cashew, Puttur, Karnataka, India.

——. n.d. *Tea Mosquito: A Major Pest of Cashew*. Extension Folder 2, National Research Centre for Cashew, Puttur, Karnataka, India.

Nutman, F.J., and F.M. Roberts. 1971. The clove industry and diseases of the clove tree. *Pest Articles and News Summaries* 17:147–165.

Nyabyenda, P. 1988. Priorités de recherches a l'ISAR et possibilités de coopération avec l'UNR. In: *Rapport et recommendations du Séminaire de Concertation sur la recherche a la faculté d'agronomie (UNR)*, pp. 87–108. Université Nationale du Rwanda, Butare, Rwanda.

Oakes, A.J. 1968. *Leucaena leucocephala*: description, culture, utilization. *Advancing Frontiers of Plant Science* 10:1–114.

O'Brian, P. 1987. *Joseph Banks: A Life*. Collins Hanill, London.

Ochse, J.J., and R.C.B. van den Brink, 1931. *Vruchten en Vruchtenteelt in Nederlandsch-Oost-Indië*. Kloff, Batavia (Jakarta), Indonesia.

Ochse, J.J., M.J. Soule, M.J. Dijkman, and C. Wehlburg. 1961. *Tropical and Subtropical Agriculture*, vol. 1. MacMillan, New York.

Ogbe, F.D. 1990. Symposium on conserving African plant diversity hosted by IITA in Nigeria. *Diversity* 6(3/4):7–8.

O'Hare, J. 1986. Coffee and the Holy Ghost Fathers. *Kenya Past and Present* 18:13–15.

Ohler, J.G. 1969. Macadamia nuts: review article. *Tropical Abstracts* 24(12):781–791.

Ojima, M., F.A.C. Dall'Orto, and O. Nigitano. 1976. *Germinação de sementes de nogueira macadamia*. Boletim Técnico Instituto Agronômico Bras. No. 33.

Oldfield, M.L., and J.B. Alcorn. 1987. Conservation of traditional agroecosystems: can age-old farming practices conserve crop genetic resources? *Bioscience* 37:199–208.

Oliveira, J.C., P.E. Carnier, and G.M. Assis. 1988. Preservação de germoplasma de maracujazeiros. In: *Anais do encontro sôbre recursos genéticos*, ed. S.M.C. Araujo and J.A. Osuna, p. 200. Faculdade de Ciências Agrárias e Veterinárias, UNESP, Jaboticabal, São Paulo/Centro Nacional de Recursos Genéticos e Biotecnologia, Brasília.

Ollennu, L.A., and G.K. Owusu. 1987. Studies of the reinfection of replanted cocoa by cocoa swollen shoot virus in Ghana. In: *Proceedings of the 10th International Cocoa*

Research Conference, Santo Domingo, Dominican Republic, 17–23 May, pp. 515–520.

Ooi, S.C., E.B. da Silva, A.A. Müller, and J.C. Nascimento. 1981. Oil palm genetic resources: native *E. oleifera* populations in Brazil offer promising sources. *Pesquisa Agropecuária Brasileira* 16(3):385–395.

Oppenheimer, J.R. 1985. *Cebus capucinus*: home range, population dynamics, and interspecific relationships. In: *The Ecology of a Tropical Forest: Seasonal Rhythms and Long-Term Changes*, ed. E.G. Leigh, A.S. Rand, and D.M. Windsor, pp. 253–272. Smithsonian Institution Press, Washington, D.C.

Oram, P.A. 1988. *International Agricultural Research Needs in Sub-Saharan Africa: Current Problems and Future Imperatives—Issues and Options for the CGIAR.* Consultative Group on International Agricultural Research, World Bank, Washington, D.C.

Oshima, J. 1982. The culture of moso bamboo in Japan, part. 1. *Journal of the American Bamboo Society* 3:2–28.

Oster, G., and S. Oster. 1985. The great breadfruit scheme: a beautiful tree still bears the stigma of its past. *Natural History* 94(3):35–41.

Owen, D.F. 1973. *Man in Tropical Africa: The Environmental Predicament.* Oxford University Press, New York.

Palca, J. 1988. Trees to benefit from Third World debt. *Nature* 332:389.

Pandey, S.N. 1986. Mango cultivars: nomenclature and registration. *Acta Horticulturae* 182:259–263.

Pannetier, C., and C. Lenaud. 1976. Diverse aspects de l'utilization possible de culture in vitro pour la multiplication de "Cayenne lisse." *Fruits* 31:739–750.

Parham, B.E.V. 1972. *Plants of Samoa.* New Zealand Department of Scientific Research, Information Series 85, Wellington.

Parker, R.N. 1986. *Forty Common Indian Trees and How to Know Them.* Logos Press, New Delhi.

Paroda, R.S., and B. Mal. 1989. New plant sources for food and industry in India. In: *New Crops for Food and Industry*, ed. G.E. Wickens, N. Haq, and P. Day, pp. 135–149. Chapman and Hall, London.

Parry, J.W. 1969. *Spices*, 2 vols. Food Trade Press, London.

Parsons, J.J. 1952. The settlement of the Sinú Valley of Colombia. *Geographical Review* 42:67–86.

———. 1955. The Miskito pine savanna of Nicaragua and Honduras. *Annals of the Association of American Geographers* 45:36–63.

———. 1991. Giant American bamboo in the vernacular architecture of Colombia and Ecuador. *Geographical Review* 81(2):131–152.

Pascal, J.P. 1988. *Wet Evergreen Forests of the Western Ghats of India: Ecology, Structure, Floristic Composition and Succession.* Travaux Section Science of Technologie, Institut Français de Pondicherry, India.

Patiño, V.M. 1963a. *Plantas cultivadas y animales domésticos en América equinoccial*, vol. 2. Imprenta Departamental, Cali, Colombia.

———. 1963b. *Plantas cultivadas y animales domésticos en América equinoccial*, vol. 1. Imprenta Departamental, Cali, Colombia.

Pearce, F. 1990. Brazil, where the ice cream comes from. *New Scientist* 127(1724):45–48.

Peña, J.E., H. Nadel, and V. Torres. 1990. Pests of *Annona* species. *Tropical Fruit World* 1(4):121–122.

Pérez, C.A.A., and M.G. Salán. 1986. *Informe final del proyecto de recolección de algunos cultivos nativos de Guatemala.* Facultad de Agronomía, Universidad de San

Carlos de Guatemala/Instituto de Ciencia y Tecnología Agrícolas (ICTA), Guatemala City, Guatemala.

Pérez-Arbeláez, E. 1956. *Plantas útiles de Colombia.* Camacho Roldán, Bogotá.

Persad, C. 1987. Studies on the control of black pod disease of cocoa in Trinidad. In: *Proceedings of the 10th International Cocoa Research Conference, Santo Domingo, Dominican Republic, 17–23 May,* pp. 437–440.

Pesce, C. 1985. *Oil Palms and Other Oilseeds of the Amazon*; tran. and ed. D.V. Johnson. Reference Publications, Algonac, Mich.

Pétard, P. 1986. *Quelques plantes utiles de Polynésie française et Raau Tahiti.* Éditions Haere Po No Tahiti, Papeete, Tahiti.

Peters, C.M., A.H. Gentry, and R.O. Mendelsohn. 1989. Valuation of an Amazonian rainforest. *Nature* 339:655–656.

Peters, W.J., and L.F. Neuenschwander. 1988. *Slash and Burn Farming in the Third World Forest.* University of Idaho Press, Moscow.

Phillips, W., and J.J. Galindo. 1987. Evaluación de la resistencia de cultivares de cacao a *Moniliophthora roreri.* In: *Proceedings of the 10th International Cocoa Research Conference, Santo Domingo, Dominican Republic, 17–23 May,* pp. 685–689.

Piang, G. 1936. The avocado in the Philippines. *Yearbook of the California Avocado Association,* pp. 103–116.

Pickersgill, B. 1976. Pineapple. In: *Evolution of Crop Plants,* ed. N.W. Simmonds, pp. 14–18. Longman, London.

Pimentel, D., W. Dazhong, S. Eigenbrode, H. Lang, D. Emerson, and M. Karasik. 1986. Deforestation: interdependency of fuelwood and agriculture. *Oikos* 46:404–412.

Pimm, S.L., and M.E. Gilpin. 1989. Theoretical issues in conservation biology. In: *Perspectives in Ecological Theory,* ed. J. Roughgarden, R.M. May, and S.A. Levin, pp. 287–305. Princeton University Press, Princeton.

Pittier, H. 1926. *Manual de las plantas usuales de Venezuela.* Litografía del Comercio, Caracas.

———. 1935. Degeneration of cacao through natural hybridization. *Journal of Heredity* 26:385–390.

Platt, B.S. 1962. *Tables of Representative Values of Foods Commonly Used in Tropical Countries.* Medical Research Council, Special Report Series 302, London.

Plotkin, M.J. 1988. The outlook for new agricultural and industrial products from the tropics. In: *Biodiversity,* ed. E.O. Wilson and F.M. Peter, pp. 106–116. National Academy Press, Washington, D.C.

Plotkin, M.J., and M.J. Balick. 1984. Medicinal uses of South American palms. *Journal of Ethnopharmacology* 10:157–179.

Plucknett, D.L. 1979. Cassia: a tropical essential oil crop. In: *New Agricultural Crops,* ed. G.A. Ritchie, pp. 149–186. American Association for the Advancement of Science, Selected Symposium 38, Westview Press, Boulder, Colo.

Plucknett, D.L., and N.J.H. Smith. 1982. Agricultural research and Third World food production. *Science* 217:215–220.

———. 1984. Networking in international agricultural research. *Science* 225:989–993.

———. 1986. Historical perspectives on multiple cropping. In: *Multiple Cropping,* ed. C.A. Francis, pp. 20–39. Macmillan, New York.

———. 1987. Food security and genetic diversity. Paper presented at Symposium on Genetic Diversity and Food Security: A Global Perspective. Annual Meetings of the American Association for the Advancement of Science (AAAS), Chicago, February.

———. 1988. *Plant Quarantine and the International Transfer of Germplasm.* Consultative Group on International Agricultural Research, Study Paper 25, Washington, D.C.

——. 1989. Quarantine and the exchange of crop genetic resources. *Bioscience* 39:16–23.

Plucknett, D.L., N.J.H. Smith, and S.O. Ozgediz. 1990. *Networking in International Agricultural Research*. Cornell University Press, Ithaca.

Plucknett, D.L., N.J.H. Smith, J.T. Williams, and N.M. Anishetty. 1987. *Gene Banks and the World's Food*. Princeton University Press, Princeton.

Pool, P.A. 1986. *Research Report 1982–1986*. Republic of Indonesia: U.K. Overseas Development Administration Clove Diseases Project, Puslitbangtri, Bogor, Indonesia.

——. 1988. Variation, reproductive biology and yielding behaviour in cloves (*Syzygium aromaticum*). Ph.D. dissertation, University of Reading, Reading, U.K.

Pool, P.A., S.J. Eden-Green, and M.T. Muhammad. 1986. Variation in clove (*Syzygium aromaticum*) germplasm in the Moluccan Islands. *Euphytica* 35:149–159.

Pope, G.T. 1929. *The Macadamia Nut in Hawaii*. Bulletin 59 of the Hawaii Agricultural Experiment Station, University of Hawaii, Honolulu.

Popenoe, F.O. 1915. Varieties of the avocado. *Report of the First Semi-Annual Meeting of the California Avocado Association*, pp. 44–69.

Popenoe, W. 1913. The cherimoya in California. *Pomona College Journal of Economic Botany* 2:277–300.

——. 1914. *Brazilian Expedition 1913–1914*. Office of Foreign Seed and Plant Introduction, U.S. Department of Agriculture, Washington, D.C.

——. 1918. Exploring Guatemala for desirable new avocados. *Annual Report of the California Avocado Association for the Year 1917*, pp. 104–188.

——. 1920. *Manual of Tropical and Subtropical Fruits*. Macmillan, New York.

——. 1921. Letter from Ecuador. *Annual Report of the California Avocado Association for the Years 1920–1921*, pp. 76–81.

——. 1922. Avocados of the Chota Valley, Ecuador. *Annual Report of the California Avocado Association for the Years 1921–1922*, pp. 35–39.

——. 1926. The parent Fuerte tree at Atlixco, Mexico. *Annual Report of the California Avocado Association for the Years 1925–1926*, pp. 24–34.

——. 1927. Wild avocados. *Yearbook of the California Avocado Association*, pp. 51–54.

——. 1928. Names of Guatemalan avocados. *Yearbook of the California Avocado Association*, pp. 41–43.

——. 1934. Early history of the avocado. *Yearbook of the California Avocado Association*, pp. 106–110.

——. 1935. Origin of the cultivated races of avocados. *Yearbook of the California Avocado Association*, pp. 184–194.

——. 1939. Looking forward. *Yearbook of the California Avocado Society*, pp. 31–43.

——. 1941. The avocado—a horticultural problem. *Tropical Agriculture* 18(1):3–7.

——. 1952. The chinini, coyo, chucte or yas. *Ceiba* 1:310–311.

——. 1956. Tribute to Florida. *Ceiba* 4(6):309–315.

Popenoe, W., and O. Jimenez. 1921. The pejibaye, a neglected food plant of tropical America. *Journal of Heredity* 12:154–166.

Portères, R. 1962. Berceaux agricoles primaires sur le continent africain. *Journal of African History* 3:195–210.

Posey, D. 1983. Indigenous knowledge and development: an ideological bridge to the future. *Ciência e Cultura* 35(7):877–894.

——. 1984. A preliminary report on diversified management of tropical forest by the Kayapó Indians of the Brazilian Amazon. *Advances in Economic Botany* 1:112–126.

——. 1985. Indigenous management of tropical forest ecosystems: the case of the Kayapó Indians of the Brazilian Amazon. *Agroforestry Systems* 3:139–158.

———. 1988. Kayapó Indian natural-resource management. In: *People of the Tropical Rain Forest*, ed. J.S. Denslow and C. Padoch, pp. 89–90. University of California Press, Berkeley.

Postel, S., and L. Heise. 1988. Reforesting the earth. In: *State of the World 1988*, ed. L.R. Brown, W.U. Chandler, A. Durning, C. Flavin, L. Heise, J. Jacobson, S. Postel, C.P. Shea, L. Starke, and E.C. Wolf, pp. 83–100. W.W. Norton, New York.

Pound, B., and L. Martinez. 1983. *Leucaena: Its Cultivation and Use*. Overseas Development Administration, London.

Pound, F.J. 1938. *Cacao and Witchbroom Disease* (Marasmius perniciosus) *of South America with Notes on Other Species of* Theobroma. Yuille's Printerie, Port-of-Spain, Trinidad and Tobago.

Powell, J.M. 1976. Ethnobotany. In: *New Guinea Vegetation*, ed. K. Pailmans, pp. 23–105. Elsevier, Amsterdam.

Prance, G.T. 1984. The pejibaye, *Guilielma gasipaes* (HBK), Bailen, and the papaya, *Carica papaya* L. In: *Pre-Columbian Plant Migration*, ed. D. Stone, pp. 85–104. Papers of the Peabody Museum of Archeology and Ethnology, vol. 76, Harvard University, Cambridge.

———. 1989. Economic prospects from tropical rainforest ethnobotany. In: *Fragile Lands of Latin America: Strategies for Sustainable Development*, ed. J.O. Browder, pp. 61–74. Westview Press, Boulder, Colo.

———. 1990. Consensus for conservation. *Nature* 345:384.

Prescott-Allen, C., and R. Prescott-Allen. 1986. *The First Resource: Wild Species in the North American Economy*. Yale University Press, New Haven.

———. 1990. How many plants feed the world? *Conservation Biology* 4:365–374.

Preuss, P. 1901. *Expedition nach Central und Sud-Amerika*. Kolonial-Wirtschaftlichen Kommitee, Berlin.

Price, R.O. 1917. The avocado in Mexico. *Annual Report of the California Avocado Association for the Year 1916*, pp. 71–73.

Prior, C. 1984. Approaches to the control of diseases of cocoa in Papua New Guinea. *Journal of Plant Protection in the Tropics* 1(1):39–46.

———. 1985. Vascular-streak dieback disease. In: *Cocoa Production: Present Constraints and Priorities for Research*, ed. R.A. Lass and G.A.R. Woods, pp. 51–55. World Bank, Technical Paper 39, Washington, D.C.

Pulgar, J. 1987. *Geografía del Perú*. Peisa, Lima.

Purseglove, J.W. 1974. *Tropical Crops: Dicotyledons*. J. Wiley, New York.

———. 1975. *Tropical Crops: Monocotyledons*. J. Wiley, New York.

———. 1976. The origins and migration of crops in tropical Africa. In: *Origins of African Plant Domestication*, ed. J.W. Harlan, J.M.J. de Wet, and A.B.L. Stemler, pp. 291–309. Mouton, The Hague.

Purseglove, J.W., E.G. Brown, C.L. Green, and S.R.J. Robbins. 1981a. *Spices*, vol. 1. Longman, Harlow, U.K.

———. 1981b. *Spices*, vol. 2. Longman, London.

Quick, G.R. 1989. Oilseeds as energy crops. In: *Oil Crops of the World: Their Breeding and Utilization*, ed. G. Röbbelen, R.K. Downey, and A. Ashri, pp. 118–131. McGraw-Hill, New York.

Quilter, J., B. Ojeda, D.M. Pearsall, D.H. Sandweiss, J.G. Jones, and E. Wing. 1991. Subsistence economy of El Paraíso, an early Peruvian site. *Science* 251:277–283.

Radell, D.R., and J.J. Parsons. 1971. Realejo: a forgotten colonial port and shipbuilding center in Nicaragua. *Hispanic American Historical Review* 51:295–312.

Raeburn, P. 1990. Seeds of despair. *Issues in Science and Technology* 4(2):71–76.

RAFI (Rural Advancement Fund International). 1987. Vanilla and biotechnology. *Development: Seeds of Change*. 1987(4):35–36.

Ragone, D. 1987. Collecting breadfruit in the Central Pacific. *Bulletin of the Pacific Tropical Botanical Garden* 17(2):37–41.

——. 1988. *Breadfruit Varieties in the Pacific Atolls.* United Nations Development Program (UNDP), Integrated Atoll Development Project, Noumea.

——. 1989. Status of the breadfruit collections at Kahanu Gardens. *Bulletin of the National Tropical Botanical Garden* 19(2):46–52.

Rahman, A.M. 1987. Bamboo blight in the village groves of Bangladesh. In: *Recent Research on Bamboos*, ed. A.N. Rao, G. Dhanarajan, and C.B. Sastry, pp. 266–270. CAF, Beijing/IDRC, Ottawa.

Raimbault, M. 1980. Bioconversion of starch into protein. In: *Sago: The Equatorial Swamp as a Natural Resource*, ed. W.R. Stanton and M. Flach. Proceedings of the 2nd International Sago Symposium, pp. 222–229. Martinus Nijhoff, The Hague.

Rajanaidu, N. 1983. *Elaeis oleifera* collection in South and Central America. *Plant Genetic Resources Newsletter* (International Board for Plant Genetic Resources) 56:42–51.

——. 1987. Collecting oil palm (*Elaeis guineensis*) in Tanzania and Madagascar. *Plant Genetic Resources Newsletter* (International Board for Plant Genetic Resources) 72:38–40.

——. 1990. Major developments in oil palm (*Elaeis guineensis*) breeding. *Mitteilungen Institut für aus dem Allgemeine Botanik Hamburg* 23a:39–52.

Rajanaidu, N., S.C. Ooi, and M.J. Lawrence. 1982. Genotype-environment interaction and conservation of oil palm genetic resources. In: *Crop Improvement Research*, ed. T.C. Yap, K.M. Graham, and J. Sukaimi, pp. 2–18. SABRAO, Kuala Lumpur, Malaysia.

Ram, M., P.K. Majumder, and R.N. Singh. 1985. Papaya germplasm collection in India. *Newsletter of the Regional Committee for Southeast Asia* (International Board for Plant Genetic Resources) 9(1):6–7.

Rangel, J.F. 1982. *CEPLAC/CACAU Ano 25.* Instituto Interamericano de Cooperação para a Agricultura (IICA), Brasília.

Rao, I.V.R., and I.U. Rao. 1990. Tissue culture approaches to the mass-propagation and genetic improvement of bamboos. In: *Bamboo: Current Research*, ed. I.V. Rao, R. Gnanaharan, and C.B. Sastry, pp. 151–158. Kerala Forest Research Institute, India/IDRC, Ottawa.

Rao, I.V.R, A.M. Yusoff, A.N. Rao, and C.B. Sastry. 1989. *Propagation of Bamboo and Rattan through Tissue Culture.* IDRC Bamboo and Rattan Research Network, Singapore.

Ratnam, R. 1961. Introduction of criollo cacao into Madras State. *South Indian Horticulture* 9(4):24–29.

Rauwerdink, J.B. 1986. An essay on *Metroxylon*, the sago palm. *Principes* 30(4):165–180.

Record, S.J., and R.W. Hess, 1986. *Timbers of the New World.* Ayer Company, Salem, N.H.

Redford, K.H. 1992. The empty forest. *Bioscience* 42:412–422.

Reed, D. 1989. Debt-for-nature swaps: opportunities and controversies. Typescript, World Wildlife Fund, Washington, D.C.

Repetto, R. 1988. *The Forest for the Trees? Government Policies and the Misuse of Forest Resources.* World Resources Institute, Washington, D.C.

Ribeiro, M.N.G. 1976. Aspectos climatológicos de Manaus. *Acta Amazonica* 6(2):229–233.

Richards, P.W. 1977. Tropical forests and woodlands: an overview. *Agro-Ecosystems* 3:225–238.

Ridley, H.N. 1912. *Spices.* Macmillan, London.

Rindos, D. 1984. *The Origins of Agriculture: An Evolutionary Perspective.* Academic Press, Orlando, Fla.

Rivera, S. 1986. Informe generale sobre el proyecto de recolección de cacao criollo en Guatemala. International Board for Plant Genetic Resources, Rome, AGPG: IBPGR/86/156.

Rizvi, S.A.A. 1987. *The Wonder That Was India;* vol. 2. Sidgwick and Jackson, London.

Roberts, L. 1988a. Hard choices on biodiversity. *Science* 241:1759–1761.

——. 1988b. Beef and chocolate: a partial reprieve. *Science* 240:1149.

Rodrigues, R.A. 1986. Inventário florístico em áreas do projeto Albras-Alunorte, Bacarena, Pa. In: *Anais do Primeiro Simpósio do Trópico Umido, Belém, Pará, 12 a 17 denovembro de 1984,* vol. 2, *Flora e floresta,* pp. 153-166. Empresa Brasileira de Pesquisa Agropecuária (EMBRAPA), Centro de Pesquisa Agropecuária do Trópico Umido (CPATU), Belém, Pará, Brazil.

Rodriguez, F. 1982. *El aguacate.* AGT Editor, Mexico City.

Roig, J.T., and J.I. Mesa. 1962. *Diccionario botánico de nombres vulgares Cubanos.* Havana, Cuba.

Rosengarten, F. 1969. *The Book of Spices.* Livingstone Publishing Company, Wynnewood, Penn.

——. 1984. *The Book of Edible Nuts.* Walker and Co., New York.

Rowe, P.R. 1987. Banana breeding in Honduras. In: *Banana and Plantain Breeding Strategies,* ed. G.J. Persley and E. De Langhe, pp. 74–77. ACIAR Proceedings 21, Australian Centre for International Agricultural Research, Canberra.

Rowley, H.K., and K. Nakamura. 1988. Adverse weather reduces 1987–88 harvest. *Yearbook of the California Macadamia Society* 34:72–81.

Ruddle, K., D.V. Johnson, P.K. Townsend, and J.D. Rees. 1978. *Palm Sago: A Tropical Palm from Marginal Lands.* University of Hawaii Press, Honolulu.

Rudgard, S.A. 1986. Witches' broom disease of cocoa in Rondonia, Brazil: pod losses. *Tropical Pest Management* 32:24–26.

Ruehle, G.D. 1959. *Growing Guavas in Florida.* Agricultural Extension Service, Bulletin 170, Gainesville, Fla.

Ruehle, G.D., and R.B. Ledin. 1956. *Mango Growing in Florida.* Agricultural Experiment Stations, Bulletin 574, University of Florida, Gainesville.

Rumphius, G.R. 1741. *Herbarium Amboinense,* vol. 2. Joannes Burmannus, Amsterdam.

Russel-Smith, A. 1986. Mission report from the Workshop on the Biological and Genetic Control of the Leucaena Psyllid. *PANESA Newsletter* (Pasture Network for Eastern and Southern Africa) 3:12–15.

Rylands, A.B. 1990. Uma mapa que protege a Amazônia. *Ciência Hoje* 11(65):6–7.

Sader, S.A., and A.T. Joyce. 1988. Deforestation rates and trends in Costa Rica. *Biotropica* 20:11–19.

Saenz, B., and F. Soleibe. 1987. Substitución de café por cacao en la zona marginal baja cafetera de Colombia. In: *Proceedings of the International Cocoa Research Conference, Santo Domingo, Dominican Republic, 17–23 May,* pp. 21–25.

Sahertian, S.A. 1980. *Klasifikasi dan hubungan antar tipe-tipe cengkeh berdasarkan identitas morfologi dan analisa jarak taxonomi.* Tesis Sarjana Pertanian, Fakultas Pertanian/Kehutanan, Universitas Pattimura, Ambon, Indonesia.

Samson, J.A. 1986. *Tropical Fruits.* Longman, New York.

Sanchez, P., K. Jaffe, C. Giron, and C. Marin. 1987. Proyecto colección de cacao silvestre en el Amazonas venezolano. International Board for Plant Genetic Resources, Informe Nr 1–PR 3/11, IBPGR-Cocoa, Rome.

Sanchez, P., K. Jaffe, R. Gonzalez, I. Garcia, M.C. Muller, and J.J. Tortolero. 1988. Proyecto colección de cacao silvestre en el Amazonas venezolano. International Board for Plant Genetic Resources, Informe Nr 3–PR 3/11, IBPGR-Cocoa, Rome.

Santa-Anna Nery, F.J. 1885. *Le pays des amazones, l'El-Dorado, les terres a caoutchouc.* L. Frinzine, Paris.

Sastrapradja, D., and J.P. Mogea. 1977. Present uses and future development of *Metroxylon sagu* in Indonesia. In: *Proceedings of the 1st International Sago Symposium,* ed. K. Tan, pp. 112–117. Kermajuan Kanji Sdn Bhd., Kuala Lumpur.

Sastrapradja, S. 1975. Tropical fruit germplasm in South East Asia. In: *South East Asian Plant Genetic Resources,* ed. J.T. Williams, C.H. Lamoureux, and N. Wulijarno-Soetjipto, pp. 33–46. IBPGR/BIOTROP/LIPI, Bogor, Indonesia.

Sastrapradja, S., S.H.A. Lubis, E. Djajasukma, H. Soetarno, and I. Lubis. 1981. *Vegetables.* International Board for Plant Genetic Resources, Rome.

Sastrapradja, S., U. Sutisna, G. Panggabean, J.P. Mogea, S. Sukardjo, and A.T. Sunarto. 1980. *Fruits.* International Board for Plant Genetic Resources, Rome.

Sauer, C.O. 1966. *The Early Spanish Main.* University of California Press, Berkeley.

———. 1969. *Agricultural Origins and Dispersals: The Domestication of Animals and Foodstuffs.* M.I.T. Press, Cambridge.

Sauer, J.D. 1988. *Plant Migration: The Dynamics of Geographic Patterning in Seed Plant Species.* University of California Press, Berkeley.

Schieber, E., M.D. Coffey, F.B. Guillemet, and G.A. Zentmyer. 1984. Collecting *Persea schiedeana* in the Baja and Alta Verapaz, Guatemala. *Yearbook of the California Avocado Society* 68:103–107.

Schieber, E., and G.A. Zentmyer. 1974. Collecting "Matul-oj" types of *Persea* in Guatemala. *Yearbook of the California Avocado Society* 57:101–105.

———. 1976. Exploring for *Persea* in Matugalpa, Nicaragua. *Yearbook of the California Avocado Society* 59:118–120.

———. 1978a. Exploring for *Persea* in Latin America. *Yearbook of the California Avocado Society* 62:60–65.

———. 1978b. Hunting for *Persea steyermarkii* in the mountains of Guatemala. *Yearbook of the California Avocado Society* 62:67–71.

———. 1979. Exploring for *Persea* in northern Guatemala. *Yearbook of the California Avocado Society* 63:41–45.

———. 1980. Exploring for *Persea* in Orizaba, Mexico. *Yearbook of the California Avocado Society* 64:79–84.

———. 1981a. Exploring for "aguacate de mico" in Central America. *Yearbook of the California Avocado Society* 65:49–55.

———. 1981b. Exploring for *Persea* on Volcano Quetzaltepeque, Guatemala. *Yearbook of the California Avocado Society* 65:57–63.

———. 1983. *Persea* exploration in Middle America: an interview and discussion. *Yearbook of the California Avocado Society* 67:93–99.

Schieber, E., G.A. Zentmyer, and M.D. Coffey. 1983. Variability in Mexican avocados (Matuloj) in Guatemala. *Yearbook of the California Avocado Society* 67:87–91.

Schmid, M. 1989. Vietnam, Kampuchea and Laos. In: *Floristic Inventory of Tropical Countries,* ed. D.G. Campbell and H.D. Hammond, pp. 83–90. New York Botanical Garden, New York.

Schmidt, J. 1942. O clima da Amazônia. *Revista Brasileira de Geografia* 4(3):465–500.

Schneider, S.H. 1989. The greenhouse effect: science and policy. *Science* 243:771–780.

Schroeder, C.A. 1935. Effects of sun-blotch on the anatomy of the avocado stem. *Yearbook of the California Avocado Association,* pp. 125–129.

———. 1967. Avocado introduction in southern California. In: *Proceedings of the Inter-*

national Symposium on Plant Introduction, Tegucigalpa, Honduras, November 30–December 2, 1966, pp. 61–69. Escuela Agrícola Panamericana, Tegucigalpa, Honduras.

——. 1977. No yas—a threat to the avocado. *Yearbook of the California Avocado Society* 61:37–42.

Schroeder, C.A., and M.R. Schroeder. 1982. Some historical aspects of the avocado in Australia. *Yearbook of the California Avocado Society* 66:119–124.

Schultes, R.E. 1945. Estudio preliminar del género *Hevea* en Colombia. *Revista de la Academia Colombiana de Ciencias Exactas Físicas y Naturales* 6:331–338.

——. 1956. The Amazon Indian and evolution in *Hevea* and related genera. *Journal of the Arnold Arboretum* 37:123–148.

——. 1970. The history of taxonomic studies in *Hevea*. *Botanical Review* 36:197–276.

——. 1977a. The odyssey of the cultivated rubber tree. *Endeavour*, n.s. 1(3/4):133–138.

——. 1977b. Wild *Hevea*: an untapped source of germ plasm. *Journal of the Rubber Research Institute of Sri Lanka* 54:227–257.

——. 1984a. Amazonian cultigens and their northward and westward migration in pre-Columbian times. In: *Pre-Columbian Plant Migration*, ed. D. Stone, pp. 19–37. Papers of the Peabody Museum of Archeology and Ethnology, vol. 76. Harvard University, Cambridge.

——. 1984b. The tree that changed the world in one century. *Arnoldia* 44 (2):3–16.

——. 1987a. Studies in the genus *Hevea*. VIII. Notes on infraspecific variants of *Hevea brasiliensis* (Euphorbiaceae). *Economic Botany* 41(2):125–147.

——. 1987b. Little-known cultivated plants of the Colombian Amazonia. *Economic Botany* 41(3):446–450.

——. 1988. *Where the Gods Reign: Plants and Peoples of the Colombian Amazon.* Synergetic Press, Oracle, Ariz.

——. 1990. *A Brief Taxonomic View of the Genus* Hevea. Malaysian Rubber Research and Development Board, Monograph 14, Kuala Lumpur, Malaysia.

Schultes, R.E., and R.F. Raffauf. 1990. *The Healing Forest: Medicinal and Toxic Plants of the Northwest Amazonia.* Dioscorides Press, Portland, Ore.

Schwartzman, S. 1989. Extractive reserves: the rubber tappers' strategy for sustainable use of the Amazon rainforest. In: *Fragile Lands of Latin America: Strategies for Sustainable Development*, ed. J.O. Browder, pp. 150–163. Westview Press, Boulder, Colo.

Sedgley, M., and D.M. Alexander. 1983. Avocado breeding research in Australia. *Yearbook of the California Avocado Society* 67:129–135.

SEDUE. 1987. *Información básica sobre las áreas naturales protegidas de México.* Secretaria de Desarrollo Urbano y Ecología, Mexico City.

Seemann, B. 1856. *Popular History of the Palms and Their Allies.* Lovell Reeve, London.

Segura, R.R. 1986. Banana export trade of the Philippines. *PCARRD Monitor* (Philippine Council for Agriculture and Resources Research and Development) 14(8):8–10.

Seibert, B. 1989. Indigenous fruit trees of Kalimantan in traditional culture. In: *Proceedings of the First PROSEA (Plant Resources of South-East Asia) International Symposium, May 22–25, 1989, Jakarta, Indonesia*, ed. J.S. Siemonsma and N. Wulijarni-Soetjipto, pp. 299–300. PUDOC, Wageningen, Netherlands.

Seibert, R.J. 1948. The uses of *Hevea* for food in relation to its domestication. *Annals of the Missouri Botanical Garden* 35:117–121.

Senanayake, U.M., R.A. Edwards, and T.H. Lee. 1976. Cinnamon. *Food Technology in Australia*, September, pp. 333–338.

Serrão, E.A. 1986. Pastagem em área de floresta no trópico úmido brasileiro: conhecimentos atuais. In: *Anais do Primeiro Simpósio do Trópico Umido, Belém, Pará, 12 a 17 de novembro de 1984*, vol. 5, *Pastagem e produção animal*, pp. 147–174. Empressa Brasileira de Pasquisa Agropecuária (EMBRAPA), Centro de Pesquisa Agropecuária do Trópico Umido (CPATU), Belém, Pará, Brazil.

Shachar, Z. 1982. The avocado in Israel. *Yearbook of the California Avocado Society* 66:103–108.

Shaffer, M. 1987. Minimum viable populations: coping with uncertainty. In: *Viable Populations for Conservation*, ed. M.E. Soulé, pp. 69–86. Cambridge University Press, Cambridge.

Shamel, A.D. 1936. The parent Fuerte avocado tree. *Yearbook of the California Avocado Association*, pp. 86–92.

Sharma, D.K. 1987. Mango breeding. *Acta Horticulturae* 196:161–167.

Sharma, Y.M.L. 1985. Project report for forest working and captive plantations. Nagaland Pulp and Paper Company.

——. 1987. Inventory and resources of bamboos. In: *Recent Research on Bamboos*, ed. A.N. Rao, G. Dhanarajan, and C.B. Sastry, pp. 4–17. Chinese Academy of Forestry, Beijing/IDRC, Ottawa.

Sheffield, F.M.L. 1950. The clove trees of the Seychelles. *East African Agricultural Journal* 16:3–8.

Shepherd, J. 1988. The ever-whirling wheel of change . . . or, time turned topsy-turvy. *Yearbook of the California Avocado Society* 72:159–163.

Shepherd, K. 1968. Banana breeding in the West Indies. *Pest Articles and News Summaries* 14(4):370–379.

Shepherd, K., J.L.L. Dantas, and E.J. Alves. 1987. Banana breeding in Brazil. In: *Banana and Plantain Breeding Strategies*, ed. G.J. Persely and E. de Langhe, pp. 78–83. ACIAR Proceedings 21, Australian Centre for International Agricultural Research, Canberra.

Silva, D. 1973. *A castanha do Pará como fator inicial de desenvolvimento de Marabá: perspectivas atuais*. Instituto de Geografia, Geografia Econômica 12, Universidade de São Paulo, São Paulo.

Silva, M.F. da. 1976. Insetos que visitam o cupuaçu, *Theobroma grandiflorum* (Wild. ex Spreng.) Schum. (Sterculiaceae), e índice de ataque nas folhas. *Acta Amazonica* 6(1):49–54.

Silva, M.F.F., P.L. Lisboa, and R.C. Lisboa. 1977. *Nomes vulgares de plantas amazônicas*. Instituto Nacional de Pesquisas da Amazônia, Manaus.

Silva, M.F.F., and N.A. Rosa. 1986. Estudos botânicos na área do Projeto Ferro Carajás, Serra Norte, Pará. II. Regeneração de "castanheira" em mata primária na bacia do Itacaiúnas. In: *Anais do Primeiro Simpósio do Trópico Umido, Belém, Pará, 12 a 17 de novembro de 1984*, vol. 2, *Flora e floresta*, pp. 167–170. Empresa Brasileira de Pesquisa Agropecuária (EMBRAPA), Centro de Pesquisa Agropecuária do Trópico Umido (CPATU), Belém, Pará, Brazil.

Sim, E.S. 1990. Sago research in Sarawok. *Sago Study* 1(1):10.

Simberloff, D. 1986. Are we on the verge of a mass extinction in tropical rain forests? In: *Dynamics of Extinction*, ed. D.K. Elliott, pp. 165–180. J. Wiley, New York.

Simmonds, N.W. 1962. *The Evolution of Bananas*. Longman, London.

——. 1966. *Bananas*. Longman, London.

——. 1976. Bananas, *Musa* (Musaceae). In: *Evolution of Crop Plants*, ed. N.W. Simmonds, pp. 211–215. Longman, London.

Singh, B. 1981. *Establishment of First Gene Sanctuary in India for Citrus in Garo Hills*. Concept Publishing, New Delhi.

Singh, J.P., and S.K. Sharma. 1981. Screening and chemical basis of resistance in guava varieties to anthracnose (*Glomerella cingulata*). *Harayana Journal of Horticultural Science* 10:155–157.

Singh, L.B. 1976. Mango, *Mangifera indica* (Anacardiaceae). In: *Evolution of Crop Plants*, ed. N.W. Simmonds, pp. 7–9, Longman, London.

S'Jacob, L.G. 1931. Proven over kunstmatige kruis en zelfbestuiving bij *Hevea brasiliensis*. *Archives of Rubber Cultivation* 15:261–288.

Skutch, A.F. 1980. *A Naturalist on a Tropical Farm*. University of California Press, Berkeley.

———. 1983. *Birds of Tropical America*. University of Texas Press, Austin.

Smith, A.E. 1986. *International Trade in Clove, Nutmeg, Mace, Cinnamon, Cassia and Their Derivatives*. Tropical Development and Research Institute, London.

Smith, N.J.H. 1976. Utilization of game along Brazil's Transamazon Highway. *Acta Amazonica* 6(4):455–466.

———. 1980. Anthrosols and human carrying capacity in Amazonia. *Annals of the Association of American Geographers* 70(4):553–566.

———. 1981a. *Man, Fishes, and the Amazon*. Columbia University Press, New York.

———. 1981b. *Wood: An Ancient Fuel with a New Future*. Paper 42, Worldwatch Institute, Washington, D.C.

———. 1981c. Fuel forests: a spreading energy resource in developing countries. *Interciencia* 6(5):336–343.

———. 1982. *Rainforest Corridors: The Transamazon Colonization Scheme*. University of California Press, Berkeley.

———. 1986. *Botanic Gardens and Germplasm Conservation*. University of Hawaii Press, Honolulu.

———. In press. Historical dimensions to sustainable agriculture and silviculture in Amazonia. In: *Amazonia: A Dynamic Habitat, Past, Present, and Future*, ed. A.C. Roosevelt and A.O. Haller, American Association for the Advancement of Science, Washington, D.C.

Smith, N.J.H., and R.E. Schultes. 1990. Deforestation and shrinking crop gene pools in Amazonia. *Environmental Conservation* 17(3):227–234.

Soderholm, P.K., and F. Vasquez. 1985. Cacao germplasm collection and distribution in USA. *Plant Genetic Resources Newsletter* (International Board for Plant Genetic Resources) 63:8–14.

Soegeng-Reksodihardjo, W. 1962. The species of *Durio* with edible fruits. *Economic Botany* 16:270–282.

Soejarno, R. 1980. Potency of sago as a food-energy source in Indonesia. In: *Sago: The Equatorial Swamp as a Natural Resource*, ed. W.R. Stanton and M. Flach, pp. 35–38. Martinus Nijhoff, The Hague.

Soost, R.K. 1987. Breeding citrus—genetics and nucellar embryony. In: *Improving Vegetatively Propagated Crops*, ed. A.J. Abbott and R.K. Atkin, pp. 83–110. Academic Press, London.

Sorensson, C.T. 1989a. Status and mechanisms of self-incompatibility in *Leucaena* species. *Plant Cell Incompatibility Newsletter* 21:77–85.

———. 1989b. Breeding strategies for *Leucaena* species hybrids. In: *Proceedings of the First International Conference on Leucaena*, July 10–13, University of West Indies, Port of Spain, Trinidad and Tobago.

Soria, J. 1966. Principales variedades de cacao cultivadas en América tropical. *Turrialba* 16:261–265.

———. 1970. Principal varieties of cocoa cultivated in tropical America. *Cocoa Growers' Bulletin* 15:12–21.

——. 1973. Primitive cultivars of cacao in America. In: *Survey of Crop Genetic Resources in Their Centres of Diversity*, ed. O.H. Frankel, pp. 119–125. FAO/IBP, Rome.

Sousa, G.S. 1971. *Tratado descritivo do Brasil em 1587*. Companhia Editora Nacional/ Editôra da Universidade de São Paulo, São Paulo.

Spears, J. 1983. Sustainable land use and strategy options for management and conservation of the moist tropical eco-systems. Paper presented at the International Symposium on Tropical Afforestation, University of Wageningen, Netherlands.

Spruce, R. 1908a. *Notes of a Botanist on the Amazon and Andes*; vol. 1. Macmillan, London.

Stahel, G. 1920. Een wild cacaobosch aan de Mamaboen-Kreek (Boven Coppename). *Die Indische Mercuur* (Amsterdam) 43 (39):681–682.

Standley, P.C. 1924. *Trees and Shrubs of Mexico*. Contributions from the National Herbarium, vol. 23, Smithsonian Institution, Washington, D.C.

——. 1927. *The Flora of Barro Colorado Island, Panama*. Smithsonian Miscellaneous Collections, vol. 78, no. 8, Smithsonian Institution, Washington, D.C.

——. 1928. *Flora of the Panama Canal Zone*. Contributions from the National Herbarium, vol. 27, Washington, D.C.

——. 1930a. A second list of the trees of Honduras. *Tropical Woods* 21:9–41.

——. 1930b. The woody plants of Siguatepeque, Honduras. *Journal of the Arnold Arboretum* 11:15–46.

——. 1931. Flora of the Lancetilla Valley, Honduras. *Field Museum of Natural History, Botany Series* 10:7–49.

Standley, P.C., and S. Calderón. 1927. *Lista preliminar de las plantas de El Salvador*. Tipografía La Unión, San Salvador, El Salvador.

Standley, P.C., and J.A. Steyermark. 1949. *Flora of Guatemala*. Chicago Natural History Museum, Fieldiana: Botany vol. 24, part 6.

Stearman, A.M. 1989. *Yuqui: Forest Nomads in a Changing World*. Holt, Rinehart and Winston, New York.

Stearn, W.T., and R.A. Roach. 1989. *Hooker's Finest Fruits: A Selection of Paintings of Fruits by William Hooker (1779–1832)*. Herbert Press, London.

Stein, N. 1978. *Coniferen im westlichen malayischen archipelago*. Biogeographica 11, W. Junk, The Hague.

Stern, K., and L. Roche. 1974. *Genetics of Forest Ecosystems*. Springer-Verlag, New York.

Steward, J.H. 1963. Culture areas of the tropical forest. In: *Handbook of South American Indians*, ed. J.H. Steward, vol. 3, pp. 883–899. Cooper Square Publishers, New York.

Stone, B. 1970. *Flora of Guam*, vol. 6. University of Guam, Micronesia.

Stone, D. 1949. *The Boruca of Costa Rica*. Papers of the Peabody Museum of Archaeology and Ethnology, no. 26, Harvard University, Cambridge.

Storey, W.B. 1954. The macadamia nut industry in Hawaii. *Yearbook of the California Avocado Society* 38:63–67.

——. 1959. Progress report on the macadamia. *Yearbook of the California Avocado Society* 43:67–71.

——. 1965. The *ternifolia* group of *Macadamia* species. *Pacific Science* 19:507–514.

——. 1979. Macadamia fallacies and facts. *Yearbook of the California Macadamia Society* 25:68–71.

Storey, W.B., B. Bergh, and G.A. Zentmyer. 1987. The origin, indigenous range, and dissemination of the avocado. *Yearbook of the California Avocado Society* 70:127–133.

References

Storey, W.B., and W.B. Saleeb. 1970. Interspecific hybridization in *Macadamia*. *Yearbook of the California Macadamia Society* 16:75–85.

Stover, R.H., and N.W. Simmonds. 1987. *Bananas*. Longman/John Wiley, New York.

Sun, M. 1988. Costa Rica's campaign for conservation. *Science* 239:1366–1369.

Sutton, S.Y. 1989. Nicaragua. In: *Floristic Inventory of Tropical Countries*, ed. D.G. Campbell and H.D. Hammond, pp. 299–304. New York Botanical Garden, Bronx.

Svarstad, H. 1987. Biotechnology: consequences for West African countries of cocoa smallholders. *Development: Seeds of Change* 4:28–32.

Swinbanks, D., and A. Anderson. 1989. Amazon forests: Japan and Brazil team up. *Nature* 338:103.

Swingle, W.T., and P.C. Reece. 1967. The botany of citrus and its wild relatives. In: *The Citrus Industry*, vol. 1, pp. 190–430. University of California Press, Berkeley.

Sykes, S.R. 1988. Overview of the family Rutaceae. In: *Citrus Breeding Workshop*, ed. R.R. Walker, pp. 93–100, Melbourne, Australia.

Sylvain, P.G. 1955. Some observations on *Coffea arabica* L. in Ethiopia. *Turrialba* 5(1–2):37–53.

Takahashi, M., and J.C. Ripperton. 1949. *Koa haole*, its establishment, culture and utilization as a forage crop. *Bulletin of the Hawaii Agricultural Experiment Station* 100:56.

Tan, H. 1987. Strategies in rubber breeding. In: *Improving Vegetatively Propagated Crops*, ed. A.J. Abbott and R.K. Atkin, pp. 27–62. Academic Press, London.

Tan, S.C. 1984. *World Rubber Structure and Stabilization*. World Bank Staff Commodity Working Paper 10, Washington, D.C.

Tangley, L. 1987. Fighting Central America's other war: rising above national political strife, CATIE leads a fight against the region's agricultural and environmental problems. *Bioscience* 37:772–777.

———. 1988. Beyond national parks: conservationists increasingly see a need for new and innovative approaches to wildlands protection. *Bioscience* 38:146–147.

Targioni-Tozzetti, A. 1855. Historical notes on the introduction of various plants into the agriculture and horticulture of Tuscany. *Journal of the Horticultural Society of London* 9:133–181.

Teixeira, L.O., and R.S. Secco. 1989. Contribuição ao conhecimento morfológico, fitogeográfico e ecológico de *Hevea camporum* Ducke (Euphorbiaceae). *Boletim do Museu Paraense Emílio Goeldi, Botânica* 5(1):69–79.

Terborgh, J. 1983. *Five New World Primates: A Study in Comparative Ecology*. Princeton University Press, Princeton.

———. 1986. Keystone plant resources in the tropical forest. In: *Conservation Biology: The Science of Scarcity and Diversity*, ed. M.E. Soulé, pp. 330–344. Sinauer Associates, Sunderland, Mass.

Tétényi, P. 1970. *Infraspecific Chemical Taxa of Medicinal Plants*. Chemical Publishing Co., New York.

Theakston, F.E. 1976. *Carica papaya*—pawpaw. In: *The Propagation of Tropical Fruit Trees*, ed. R.J. Gardner, S.A. Chandor, and staff of the Commonwealth Bureau of Horticulture and Plantation Crops, pp. 304–320. FAO/CAB (Commonwealth Agricultural Bureaux), Slough, U.K.

Thomas, A.S. 1942. The wild *arabica* coffee on the Boma Plateau, Anglo-Egyptian Sudan. *Empire Journal of Experimental Agriculture* 10:207–212.

Thomas, P. 1975. *Hindu Religion: Customs and Manners*. D.B. Taraporevala and Sons, Bombay.

Thomas, T. 1989. Food industry and agriculture. In: *New Crops for Food and Industry*, ed. G.E. Wickens, N. Haq, and P. Day, pp. 13–22. Chapman and Hall, London.

Thresh, J.M., G.K. Owusu, A. Boamah, and G. Lockwood. 1988. Ghanian cocoa varieties and swollen shoot virus. *Crop Protection* 7:219–231.

Thurston, H.D. 1973. Threatening plant diseases. *Annual Review of Phytopathology* 11:27–52.

Tidbury, G.E. 1949. *The Clove Tree*. Crosby, Lockwood and Sons, London.

———. 1976. *Durio zibethinus*—durian. In: *The Propagation of Tropical Fruit Trees*, ed. R.J. Gardner, S.A. Chandor, and the Staff of the Commonwealth Bureau of Horticulture and Plantation Crops. FAO/CAB (Commonwealth Agricultural Bureaux), Slough, U.K.

Torre, J.D. de la. 1984. *Guía para cultivar aguacate en El Bajío*. Secretaría de Agricultura y Recursos Hidráulicos, Folleto para Productores 10, Mexico City.

Toxopeus, H., and A.J. Kennedy. 1989. Effective use of cacao genetic resources as exemplified by the International Cocoa Genebank, Trinidad. In: *Utilization of Genetic Resources: Suitable Approaches, Agronomical Evaluation and Use*, pp. 59–81. Food and Agriculture Organization, Plant Production and Protection Paper 94, Rome.

Toy, L.R. 1931. *Avocados in Florida*. Department of Agriculture, Tallahassee, Fla.

Tracy, M.D. 1985. *The Pejibaye Fruit: Problems and Prospects for Its Development in Costa Rica*. M.S. thesis, University of Texas, Austin.

Tsay, H.S., and S.Y. Su. 1985. Anther culture of papaya (*Carica papaya* L.). *Plant Cell Reports* 4:28–30.

Tudge, C. 1988. *Food Crops for the Future*. Basil Blackwell, Oxford.

Ubidia, J.A. 1985. The use of bamboo in Ecuador: past, present and future. *Journal of the American Bamboo Society* 6:64–103.

Ueno, I., and T. Akihama. 1988. Potential of wild relatives of *citrus*. In: *Crop Genetic Resources of East Asia*, ed. S. Suzuki, pp. 157–159. International Board for Plant Genetic Resources, Rome.

Uhl, N.W., and J. Dransfield. 1987. *Genera Palmarum: A Classification of Palms Based on the Work of Harold E. Moore, Jr.* Allen Press, Lawrence, Kans.

Ullasa, B.A., H.S. Sohi, R.D. Rawal, and M.D. Subramanyan. 1983. Foliar disease of papaya: reactions of *Carica* genotypes. *Indian Journal of Plant Pathology* 1(1):1–4.

UNCTAD/GATT. 1982. *Spices—A Survey of the World Market;* vol. 1, *Selected Markets in Europe*. UNCTAD (United Nations Conference on Trade and Development)/ GATT (General Agreement on Tariffs and Trade), International Trade Center, Geneva, Switzerland.

UNESCO. 1985. *Action Plan for Biosphere Reserves*. United Nations Educational, Scientific and Cultural Organization, Man and the Biosphere Program, Paris.

Urata, U. 1954. *Pollination Requirements of Macadamia*. Technical Bulletin of the Hawaii Agricultural Experiment Station, No. 22, University of Hawaii, Honolulu.

Urquhart, D.H. 1961. *Cocoa*. J. Wiley, New York.

USDA. 1961. *Charcoal: Production, Marketing and Use*. Forest Products Lab Report 2213, U.S. Department of Agriculture, Washington, D.C.

Vakili, N.G. 1967. The experimental formation of polyploidy and its effect in the genus *Musa*. *American Journal of Botany* 54:24–36.

Valmayor, R.V. n.d. *The Mango: Its Botany and Production*. College of Agriculture, University of the Philippines, Laguna.

Valmayor, R.V., and O.C. Pascua. 1985. Southeast Asian banana and plantain cultivar names and synonyms—a preliminary list. *Regional Committee for Southeast Asia* (International Board for Plant Genetic Resources) 9(1):9–11.

Van den Berg, M.E., and M.H.L. Silva. 1986. Plantas medicinais do Amazonas. In: *Anais do Primeiro Simpósio do Trópico Umido*, 12 a 17 de novembro de 1984, vol. 2, *Flora et floresta*, pp. 127–133. Empresa Brasileira de Pesquisa Agropecuária (EM-

BRAPA), Centro de Pesquisa Agropecuária do Trópico Umido (CPATU), Belém, Pará, Brazil.

Vandermeer, J. 1983. Pejibaye palm (pejibaye). In: *Costa Rican Natural History,* ed. D.H. Janzen, pp. 98–101, University of Chicago Press, Chicago.

Van Hall, C.J.J. 1914. *Cocoa.* Macmillan, London.

Van Roosmalen, M.G.M. 1985. *Fruits of the Guianan Flora.* Institute of Systematic Botany, Utrecht University/Silvicultural Department, Agricultural University, Wageningen.

Van Sloten, D.H., and P. Gulick. 1984. *Directory of Germplasm Collections 6.1. Tropical and Subtropical Fruits and Nuts.* International Board for Plant Genetic Resources, Rome.

Vargas, E. 1989a. Enfermedades del tallo de palmito. *Boletín Informativo* (Universidad de Costa Rica), *Serie Técnica Pejibaye (Guilielma)* 1(1):12–13.

———. 1989b. Enfermedades del follage. *Boletín Informativo* (Universidad de Costa Rica), *Serie Técnica Pejibaye (Guilielma)* 1(2):11.

Vasquez, R., and A.H. Gentry. 1989. Use and misuse of forest-harvested fruits in the Iquitos area. *Conservation Biology* 3(4):350–361.

Vega, R., and S. Kitto. 1988. Tissue culture of babaco, *Carica rentugana* (Heilborn) Badillo. *Horticultural Science* 23:675.

Vello, F., and H.M. Rocha. 1967. *II expedição à Amazônia brasileira.* Centro de Pesquisas do Cacau (CEPLAC), Itabuna, Brazil, Communicação Técnica No. 4.

Vello, F., and L.F. Silva. 1968. *Relatório de viagem à região amazônica.* Comissão Executiva do Plano da Lavoura Cacaueira (CEPLAC), Comunicação Técnica 22, Itabuna, Brazil.

Venturieri, G.A., J.H.I. Martel, and G.M.E. Machado. 1987. Enxertia do cupuaçuzeiro (*Theobroma grandiflorum,* Wild ex Spreng., Schum.), com uso de gemas e garfos com e sem toalete. *Acta Amazonica* 17:27–40.

Vermeer, D.E. 1979. The tradition of experimentation in swidden cultivation among the Tiv of Nigeria. In: *Applied Geography Conferences,* ed. J.W. Frazier and B.J. Epstein, pp. 244–256. State University of New York, Buffalo.

Vickers, W.T., and T. Plowman. 1984. *Useful Plants of the Siona and Secoya Indians.* Field Museum of Natural History, Fieldiana Botany n.s. 15, Chicago.

Vietmeyer, N.D. 1986. Lesser-known plants of potential use in agriculture. *Science* 232:1379–1384.

Vitousek, P.M. 1988. Diversity and biological invasions of oceanic islands. In: *Biodiversity,* ed. E.O. Wilson and F.M. Peter, pp. 181–189. National Academy Press, Washington, D.C.

Vles, R.O., and J.J. Gottenbos. 1989. Nutritional characteristics and food uses of vegetable oils. In: *Oil Crops of the World: Their Breeding and Utilization,* ed. G. Röbbelen, R.K. Downey, and A. Ashri, pp. 63–86. McGraw-Hill, New York.

Vuylsteke, D.R. 1989. *Shoot-Tip Culture for the Propagation, Conservation and Exchange of* Musa *Germplasm.* International Board for Plant Genetic Resources, Rome.

Wakasa, K. 1979. Variation in the plant differentiated from the tissue culture of pineapple. *Japanese Journal of Breeding* 29:13–22.

Waldman, S. 1990. The great cherry caper: how, after years, red dye no. 3 was banned. *Newsweek,* 5 February, p. 48.

Walker, E.P. 1975. *Mammals of the World,* vol. 2. Johns Hopkins University Press, Baltimore.

Wallace, A.R. 1853. *Palm Trees of the Amazon and Their Uses.* John Van Voorst, London.

———. [1869] 1986. *The Malay Archipelago: The Land of the Orang-Utan and the Bird of Paradise.* Reprint. Oxford University Press, Singapore.

Wallis, E.S., I.M. Wood, and D.E. Blyth. 1989. New crops: a suggested framework for their selection, evaluation, and commercial development. In: *New Crops for Food and Industry*, ed. G.E. Wickens, N. Haq, and P. Day, pp. 36–52. Chapman and Hall, London.

Walsh, J. 1987. Bolivia swaps debt for conservation. *Science* 237:596–597.

Watson, E. 1985. *Cultivos tropicales adaptados a la selva alta peruana, particularmente al Alto Huallaga*. Banco Agrario del Perú, Lima.

Watt, G. 1908. *The Commercial Products of India*. John Murray, London.

WCED. 1987. *Our Common Future*. The World Commission on Environment and Development, Oxford University Press, Oxford.

Webber, H.J. 1918. Cold resistance of the avocado. In: *Annual Report of the California Avocado Association for the Year 1917*, pp. 49–51. Riverside, Calif.

Weber, G.F. 1973. *Bacterial and Fungal Diseases of Plants in the Tropics*. University of Florida Press, Gainesville.

Weinstein, B. 1983. *The Amazon Rubber Boom 1850–1920*. Stanford University Press, Stanford.

Westoby, J. 1989. *Introduction to World Forestry*. Basil Blackwell, Oxford.

Wheeler, B.E.J., and R. Mepsted. 1988. Pathogenic variability amongst isolates of *Crinipellis perniciosa* from cocoa (*Theobroma cacao*). *Plant Pathology* 37:475–488.

Wheeler, L.C. 1977. Sementes de *Hevea* para alimentação humana. *Acta Amazonica* 7(1):139–143.

Whitehead, C. 1959. The rambutan, a description of the characteristics and potential of the more important varieties. *Malayan Agriculture Journal* 42:53–75.

Whitmore, T.C. 1973. *Palms of Malaysia*. Oxford University Press, Kuala Lumpur.

——. 1985. *Tropical Rain Forests of the Far East*. Clarendon Press, Oxford.

——. 1990. *An Introduction to Tropical Rain Forests*. Clarendon Press, Oxford.

Whitten, A.J., M. Mustafa, and G.S. Henderson. 1987. *The Ecology of Sulawesi*. Gadjah Mada University Press, Yogyakarta, Indonesia.

Whittle, T. 1988. *The Plant Hunters*. PAJ Publications, New York.

Wiedner, D.L. 1962. *A History of Africa South of the Sahara*. Vintage Books, New York.

Wilkes, G. 1989. Germplasm preservation: objectives and needs. In: *Biotic Diversity and Germplasm Preservation: Global Imperatives*, ed. L. Knutson and A.K. Stoner, pp. 13–41. Kluwer, Dordrecht.

Williams, J.A. 1978. Cocoa genetic resources in Nigeria. In: *Crop Genetic Resources in Africa*, pp. 80–89. Association for the Advancement of Agricultural Sciences in Africa/International Institute of Tropical Agriculture, Ibadan, Nigeria.

Williams, J.T. 1981. Cacao genetic resources in Latin America. *Plant Genetic Resources Newsletter* 45:20–22.

——. 1987. Banana and plantain germplasm conservation and movement and needs for research. In: *Banana and Plantain Breeding Strategies*, ed. G.J. Persely and E. de Langhe, pp. 177–181. ACIAR Proceedings 21, Australian Centre for International Agricultural Research, Canberra.

——. 1988. In situ conservation of plant genetic resources—some global perspectives. *Indian Journal of Plant Genetic Resources* 1(1 & 2):23–31.

——. 1991. The time has come to clarify and implement strategies for plant conservation. *Diversity* 7(4):37–39.

Williams, J.T., and A.B. Damania. 1981. *Directory of Germplasm Collections 5. Industrial Crops: I. Cacao, Coconut, Pepper, Sugarcane and Tea*. International Board for Plant Genetic Resources, Rome.

Williams, L.O. 1950. Two new Perseas from Central America. *Ceiba* 1:55–58.

——. 1977. The avocados, a synopsis of the genus *Persea*, subg. *Persea*. *Economic Botany* 31:315–320.

Williams, R.O., and R.O. Williams, Jr. 1951. *The Useful and Ornamental Plants in Trinidad and Tobago.* Guardian Commercial Printery, Port of Spain, Trinidad.

Willis, E.O. 1976. Effects of a cold wave on an Amazonian avifauna in the upper Paraguay drainage, western Mato Grosso, and suggestions on Oscine-Suboscine relationships. *Acta Amazonica* 6(3):379–394.

Wilson, E.O. 1988. The current state of biological diversity. In: *Biodiversity*, ed. E.O. Wilson and F.M. Peter, pp. 3–18. National Academy Press, Washington, D.C.

Wilson, G.F. 1987. Status of bananas and plantains in West Africa. In: *Banana and Plantain Breeding Strategies*, ed. G.J. Persely and E. de Langhe, pp. 29–35. ACIAR Proceedings 21, Australian Centre for International Agricultural Research, Canberra.

Wilson, T.B. 1954. The marketing of durian fruit from Parit District, Perak. *Malaysian Agriculture Journal* 37:211–217.

Wisler, G.C., F.W. Zettler, and L. Mu. 1987a. Viruses infecting vanilla in French Polynesia. *American Orchid Society Bulletin* 56:381–387.

——. 1987b. Virus infections of *Vanilla* and other orchids in French Polynesia. *Plant Disease* 71:1125–1129.

Wisniewski, A., and C.F.M. Melo. 1986. *Borrachas naturais brasileiras. VI. Borrachas do genero Hevea.* Centro de Pesquisa Agropecuária do Trópico Umido (CPATU), Empresa Brasileira de Pesquisa Agropecuária (EMBRAPA), Belém, Brazil.

Wit, F. 1976. Clove. In: *Evolution of Crop Plants*, ed. N.W. Simmonds, pp. 216–218. Longman, London.

Withers, L.A. 1987. In vitro methods for collecting germplasm in the field. *Plant Genetic Resources Newsletter* (International Board for Plant Genetic Newsletter) 69:2–6.

——. 1989. *In vitro* conservation and germplasm utilisation. In: *The Use of Plant Genetic Resources*, ed. A.H.D. Brown, O.H. Frankel, D.R. Marshall, and J.T. Williams, pp. 309–334. Cambridge University Press, Cambridge.

Wolfe, H.S. 1937. Fifty years of tropical fruit culture. *Proceedings of the Florida State Horticultural Society* 50:72–78.

Wolfe, H.S., L.R. Toy, and A.L. Stahl. 1934. *Avocado Production in Florida.* Agricultural Experiment Station, Bulletin 272, University of Florida, Gainesville.

Wolstenholme, B.N. 1987. Some aspects of avocado research world-wide. *Yearbook of the South African Avocado Growers' Association* 10:8–11.

Wondimu, M. 1987. Variation in physiologic groups of *Coffea arabica* L. for resistance to rust (*Hemileia vastatrix* B & Br.) and their pattern of distribution in Ethiopia. *Germplasm Newsletter* (Plant Genetic Resource Centre, Ethiopia) 15:13–19.

Wood, G.A.R., and R.A. Lass. 1987. *Cocoa.* Longman/J. Wiley, New York.

Wood, P.J., J. Burley, and A. Grainger. 1982. *Technologies and Technology Systems for Reforestation of Degraded Tropical Lands.* Office of Technology Assessment, U.S. Congress, Washington, D.C.

Woodson, R.E., Jr., and R.W. Schery. 1948. Flora of Panama (Lauraceae). *Annals of the Missouri Botanical Garden* 35(1):1–106.

Worede, M. 1991. An Ethiopian perspective on conservation and utilization of plant genetic resources. In: *Plant Genetic Resources of Ethiopia*, ed. J.M.M. Engels, J.G. Hawkes, and M. Worede, pp. 3–19. Cambridge University Press, Cambridge.

World Bank. 1978. *Forestry: Sector Policy Paper.* World Bank, Washington, D.C.

——. 1980. *Madagascar: évolution récente et perspectives économiques.* World Bank, Washington, D.C.

——. 1982. *Natural Rubber: Sector Policy Paper.* World Bank, Washington, D.C.

——. 1983. *Outlook for Primary Commodities.* World Bank Staff Commodity Working Paper 9, Washington, D.C.

——. 1984. *World Development Report.* World Bank, Washington, D.C.

——. 1987. *The World Bank Atlas 1987.* World Bank, Washington, D.C.

——. 1988. *The Jengka Triangle Projects in Malaysia: Impact Evaluation Report.* Operations Evaluation Department, World Bank, Washington, D.C.

WRI. 1985a. *Tropical Forests: A Call for Action. Part I, the Plan.* World Resources Institute, Washington, D.C.

——. 1985b. *Tropical Forests: A Call for Action. Part II, Case Studies.* World Resources Institute, Washington, D.C.

——. 1985c. *Tropical Forests: A Call for Action. Part III, Country Investment Profiles.* World Resources Institute, Washington, D.C.

Wright, M. 1984. A note on insects associated with cocoa in Ecuador. *Tropical Pest Management* 30:29–31.

Wrigley, G. 1988. *Coffee.* Longman/J. Wiley, New York.

Wu, B., and N. Ma. 1987. Bamboo research in China. In: *Recent Research on Bamboos,* ed. A.N. Rao, G. Dhanarajan, and C.B. Sastry, pp. 18–23. CAF, Beijing/IDRC, Ottawa.

Wycherley, P.R. 1969. Breeding of *Hevea. Journal of the Rubber Research Institute of Malaya* 21:38–55.

——. 1976. Rubber, *Hevea brasiliensis* (Euphorbiaceae). In: *Evolution of Crop Plants,* ed. N.W. Simmonds, pp. 77–81. Longman, London.

Xianpu, W., and F. Yenfeng. 1987. Establishment of rare plant centre of Sichuan, China. *Botanic Gardens Conservation* 1(1):14.

Yen, D.E. 1973. The origins of oceanic agriculture. *Archaeology and Physical Anthropology in Oceania* 8:68–85.

Yudodibroto, H. 1987. Bamboo research in Indonesia. In: *Recent Research on Bamboos,* ed. A.N. Rao, G. Dhanarajan, and C.B. Sastry, pp. 33–44. CAF, Beijing/IDRC, Ottawa.

Yuncker, T.F. 1959. *Plants of Tonga.* Bernice P. Bishop Museum, Bulletin 220, Honolulu, Hawaii.

Zakri, A.H., M.N. Ghani, and A. Halim. 1987. Characterization, evaluation and utilization of industrial crops germplasm in the SABRAO region. In: *The Breeding of Horticultural Crops,* ed. W. Chang, R.T. Opeña, and J. Bay-Pertersen, pp. 255–261. Food and Fertilizer Technology Center for the Asian and Pacific Region, Taipei, Taiwan, China.

Zapata, A. 1971. Pejibaye palm from the Pacific coast of Colombia: a detailed chemical analysis. *Economic Botany* 21:371–378.

Zarate, P.S. 1984. Taxonomic revision of the genus *Leucaena* Benth. from Mexico. *Bulletin of the International Group for the Study of Mimosoideae* 12:24–34.

Zeiger, D.X., and G.A. Zentmeyer. 1967. Epidemic decline of breadfruit in the Pacific Islands. *Food and Agriculture Organization Plant Protection Bulletin* 15:22–29.

Zentmyer, G.A. 1954. 1953 collections in Central America and Mexico for resistance to avocado root rot. *Yearbook of the California Avocado Society* 38:45–48.

——. 1957. The search for resistant rootstock in Latin America. *Yearbook of the California Avocado Society* 41:101–106.

——. 1984. Avocado diseases. *Tropical Pest Management* 30:388–400.

——. 1985. Origin and distribution of *Phytophthora cinnamomi. Yearbook of the California Avocado Society* 69:89–94.

——. 1987. Taxonomic relationships and distribution of species of *Phytophthora* causing black pod of cacao. In: *Proceedings of the 10th International Cocoa Research Conference, Santo Domingo, Dominican Republic, 17–23 May,* pp. 391–395.

Zentmyer, G.A., and E. Schieber. 1982. *Persea* explorations in Honduras. *Yearbook of the California Avocado Society* 66:93–102.

——. 1987. The search for resistance to Phytophthora root rot in Latin America. *Yearbook of the South African Avocado Growers' Association* 10:109–110.

Zettler, F.W., N. Ko, G.C. Wisler, M.S. Elliott, and S. Wong. 1990. Viruses of orchids and their control. *Plant Disease* 74:621–626.

Zeven, A.C. 1972. The partial and complete domestication of the oil palm (*Elaeis guineensis*). *Economic Botany* 26:274–279.

Zhang, G., and F. Chen. 1985. Studies on bamboo hybridization. In: *Recent Research on Bamboos*, ed. A.N. Rao, G. Dhanarajan, and C.B. Sastry, pp. 179–184. Chinese Academy of Forestry, Beijing/IDRC, Ottawa.

Zobel, B.J., G. Van Wyk, and P. Stahl. 1987. *Growing Exotic Forests*. Wiley, New York.

Zohary, D. 1982. *Plants of the Bible*. Cambridge University Press, Cambridge.

Index

Spelled out names of organizations are given in Appendix 2.